T0261446

Vauxhall/Opel Mokka
Owners Workshop Manual

John S. Mead

Models covered *(6413 - 400)*

Mokka mini-SUV with two- and four-wheel-drive

Petrol: 1.4 litre (1364cc) & 1.6 litre (1598cc)
Diesel: 1.6 litre (1598cc) & 1.7 litre (1686cc)

Does NOT cover 'Mokka X' range introduced October 2016 or dual fuel (LPG) models

© Haynes Publishing 2018

ABCDE
FGHIJ
KLMNO
PQRST

A book in the **Haynes Owners Workshop Manual Series**

ISBN **978 1 78521 413 4**

British Library Cataloguing in Publication Data
A catalogue record for this book is available from the British Library.

Printed in Malaysia

Haynes Publishing
Sparkford, Yeovil, Somerset BA22 7JJ, England

Haynes North America, Inc
859 Lawrence Drive, Newbury Park, California 91320, USA

Printed using NORBRITE BOOK 48.8gsm (CODE: 40N6533) from NORPAC; procurement system certified under Sustainable Forestry Initiative standard. Paper produced is certified to the SFI Certified Fiber Sourcing Standard (CERT - 0094271)

Contents

LIVING WITH YOUR VAUXHALL/OPEL MOKKA

Roadside Repairs

Weekly checks

Lubricants and fluids

Tyre pressures

MAINTENANCE

Routine Maintenance and Servicing

Contents

The Vauxhall/Opel Mokka 5-door hatchback, sports utility vehicle was introduced in the UK in November 2012. The Mokka range is available with two sizes of petrol engines and two sizes of diesel engines. Covered in this manual are the 1.4 litre (1364 cc) and 1.6 litre (1598 cc) petrol engines, both being of double overhead camshaft (DOHC) 16-valve design. Both engines feature multi-point fuel injection and are equipped with an extensive range of emissions control systems. In addition, the 1.4 litre engine is a turbocharged unit. The diesel engine was initially a 1.7 litre (1686 cc), double overhead camshaft (DOHC) 16-valve design, but this was replaced by a 1.6 litre (1598 cc) DOHC 16-valve unit in 2015.

All models covered in this manual have front-wheel drive, or four-wheel drive, with fully independent front suspension, and semi-independent rear suspension with a torsion beam and trailing arms.

A five- or six-speed manual transmission is fitted as standard across the range, with a six-speed automatic transmission optionally available on certain models.

A wide range of standard and optional equipment is available within the Mokka range including air conditioning, central locking, electric windows, an electric sunroof, an anti-lock braking system, electronic stability program and supplementary restraint system.

For the home mechanic, the Mokka is a straightforward vehicle to maintain, and most of the items requiring frequent attention are easily accessible.

Your Vauxhall/Opel Mokka Manual

The aim of this manual is to help you get the best value from your vehicle. It can do so in several ways. It can help you decide what work must be done (even should you choose to get it done by a garage), provide information on routine maintenance and servicing, and give a logical course of action and diagnosis when random faults occur. However, it is hoped that you will use the manual by tackling the work yourself. On simpler jobs, it may even be quicker than booking the car into a garage and going there twice, to leave and collect it. Perhaps most important, a lot of money can be saved by avoiding the costs a garage must charge to cover its labour and overheads.

The manual has drawings and descriptions to show the function of the various components, so that their layout can be understood. Then the tasks are described and photographed in a clear step-by-step sequence.

References to the 'left' or 'right' are in the sense of a person in the driver's seat, facing forward.

Acknowledgements

Thanks are due to Draper Tools Limited, who provided some of the workshop tools, and to all those people at Sparkford who helped in the production of this Manual.

We take great pride in the accuracy of information given in this manual, but vehicle manufacturers make alterations and design changes during the production run of a particular vehicle of which they do not inform us. No liability can be accepted by the authors or publishers for loss, damage or injury caused by any errors in, or omissions from, the information given.

Working on your car can be dangerous. This page shows just some of the potential risks and hazards, with the aim of creating a safety-conscious attitude.

General hazards

Scalding

• Don't remove the radiator or expansion tank cap while the engine is hot.
• Engine oil, transmission fluid or power steering fluid may also be dangerously hot if the engine has recently been running.

Burning

• Beware of burns from the exhaust system and from any part of the engine. Brake discs and drums can also be extremely hot immediately after use.

Crushing

• When working under or near a raised vehicle, always supplement the jack with axle stands, or use drive-on ramps.
Never venture under a car which is only supported by a jack.
• Take care if loosening or tightening high-torque nuts when the vehicle is on stands. Initial loosening and final tightening should be done with the wheels on the ground.

Fire

• Fuel is highly flammable; fuel vapour is explosive.
• Don't let fuel spill onto a hot engine.
• Do not smoke or allow naked lights (including pilot lights) anywhere near a vehicle being worked on. Also beware of creating sparks (electrically or by use of tools).
• Fuel vapour is heavier than air, so don't work on the fuel system with the vehicle over an inspection pit.
• Another cause of fire is an electrical overload or short-circuit. Take care when repairing or modifying the vehicle wiring.
• Keep a fire extinguisher handy, of a type suitable for use on fuel and electrical fires.

Electric shock

• Ignition HT and Xenon headlight voltages can be dangerous, especially to people with heart problems or a pacemaker. Don't work on or near these systems with the engine running or the ignition switched on.

• Mains voltage is also dangerous. Make sure that any mains-operated equipment is correctly earthed. Mains power points should be protected by a residual current device (RCD) circuit breaker.

Fume or gas intoxication

• Exhaust fumes are poisonous; they can contain carbon monoxide, which is rapidly fatal if inhaled. Never run the engine in a confined space such as a garage with the doors shut.
• Fuel vapour is also poisonous, as are the vapours from some cleaning solvents and paint thinners.

Poisonous or irritant substances

• Avoid skin contact with battery acid and with any fuel, fluid or lubricant, especially antifreeze, brake hydraulic fluid and Diesel fuel. Don't syphon them by mouth. If such a substance is swallowed or gets into the eyes, seek medical advice.
• Prolonged contact with used engine oil can cause skin cancer. Wear gloves or use a barrier cream if necessary. Change out of oil-soaked clothes and do not keep oily rags in your pocket.
• Air conditioning refrigerant forms a poisonous gas if exposed to a naked flame (including a cigarette). It can also cause skin burns on contact.

Asbestos

• Asbestos dust can cause cancer if inhaled or swallowed. Asbestos may be found in gaskets and in brake and clutch linings. When dealing with such components it is safest to assume that they contain asbestos.

Special hazards

Hydrofluoric acid

• This extremely corrosive acid is formed when certain types of synthetic rubber, found in some O-rings, oil seals, fuel hoses etc, are exposed to temperatures above 4000C. The rubber changes into a charred or sticky substance containing the acid. *Once formed, the acid remains dangerous for years. If it gets onto the skin, it may be necessary to amputate the limb concerned.*
• When dealing with a vehicle which has suffered a fire, or with components salvaged from such a vehicle, wear protective gloves and discard them after use.

The battery

• Batteries contain sulphuric acid, which attacks clothing, eyes and skin. Take care when topping-up or carrying the battery.
• The hydrogen gas given off by the battery is highly explosive. Never cause a spark or allow a naked light nearby. Be careful when connecting and disconnecting battery chargers or jump leads.

Air bags

• Air bags can cause injury if they go off accidentally. Take care when removing the steering wheel and trim panels. Special storage instructions may apply.

Diesel injection equipment

• Diesel injection pumps supply fuel at very high pressure. Take care when working on the fuel injectors and fuel pipes.

⚠ *Warning: Never expose the hands, face or any other part of the body to injector spray; the fuel can penetrate the skin with potentially fatal results.*

Remember...

DO

• Do use eye protection when using power tools, and when working under the vehicle.

• Do wear gloves or use barrier cream to protect your hands when necessary.

• Do get someone to check periodically that all is well when working alone on the vehicle.

• Do keep loose clothing and long hair well out of the way of moving mechanical parts.

• Do remove rings, wristwatch etc, before working on the vehicle – especially the electrical system.

• Do ensure that any lifting or jacking equipment has a safe working load rating adequate for the job.

DON'T

• Don't attempt to lift a heavy component which may be beyond your capability – get assistance.

• Don't rush to finish a job, or take unverified short cuts.

• Don't use ill-fitting tools which may slip and cause injury.

• Don't leave tools or parts lying around where someone can trip over them. Mop up oil and fuel spills at once.

• Don't allow children or pets to play in or near a vehicle being worked on.

The following pages are intended to help in dealing with common roadside emergencies and breakdowns. You will find more detailed fault finding information at the back of the manual, and repair information in the main chapters.

If your car won't start and the starter motor doesn't turn

☐ Open the bonnet and make sure that the battery terminals are clean and tight.
☐ Switch on the headlights and try to start the engine. If the headlights go very dim when you're trying to start, the battery is probably flat. Get out of trouble by jump starting (see next page) using a friend's car.

If your car won't start even though the starter motor turns as normal

☐ Is there fuel in the tank?
☐ Is there moisture on electrical components under the bonnet? Switch off the ignition, then wipe off any obvious dampness with a dry cloth. Spray a water-repellent aerosol product (WD-40 or equivalent) on ignition and fuel system electrical connectors like those shown in the photos.

1 On petrol engines, check that the wiring to the ignition module is connected firmly.

2 Check that the airflow meter wiring is connected securely.

3 Check the security and condition of the battery connections.

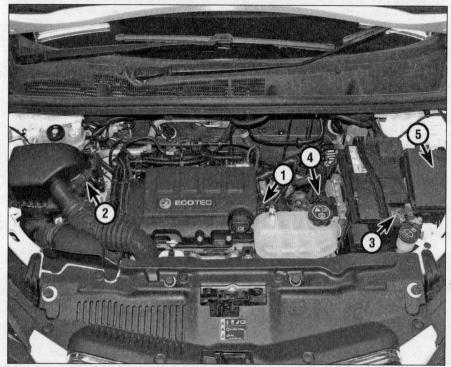

Check that electrical connections are secure (with the ignition switched off) and spray with water dispersant if you suspect a problem due to damp.

4 Check all multiplugs and wiring connectors for security.

5 Check that all fuses are still in good condition and none have blown.

Jump starting

When jump-starting a car using a booster battery, observe the following precautions:

✔ Before connecting the booster battery, make sure that the ignition is switched off.

Caution: Remove the key in case the central locking engages when the jump leads are connected

✔ Ensure that all electrical equipment (lights, heater, wipers, etc) is switched off.

✔ Take note of any special precautions printed on the battery case.

✔ Make sure that the booster battery is the same voltage as the discharged one in the vehicle.

✔ If the battery is being jump-started from the battery in another vehicle, the two vehicles MUST NOT TOUCH each other.

✔ Make sure that the transmission is in neutral (or PARK, in the case of automatic transmission).

Jump starting will get you out of trouble, but you must correct whatever made the battery go flat in the first place. There are three possibilities:

1 The battery has been drained by repeated attempts to start, or by leaving the lights on.

2 The charging system is not working properly (alternator drivebelt slack or broken, alternator wiring fault or alternator itself faulty).

3 The battery itself is at fault (electrolyte low, or battery worn out).

1 Connect one end of the red jump lead to the positive (+) terminal of the flat battery

2 Connect the other end of the red lead to the positive (+) terminal of the booster battery.

3 Connect one end of the black jump lead to the negative (-) terminal of the booster battery

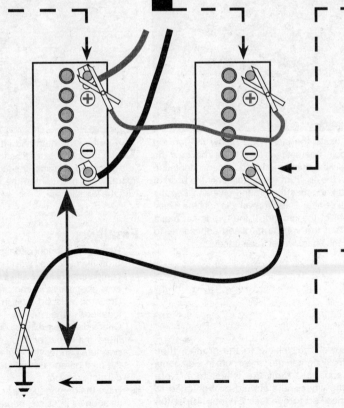

4 Connect the other end of the black jump lead to a bolt or bracket on the engine block, well away from the battery, on the vehicle to be started.

5 Make sure that the jump leads will not come into contact with the fan, drive-belts or other moving parts of the engine.

6 Start the engine using the booster battery and run it at idle speed. Switch on the lights, rear window demister and heater blower motor, then disconnect the jump leads in the reverse order of connection. Turn off the lights etc.

Wheel changing

Note: *Most Mokka models are equipped with a puncture repair kit and do not have a spare wheel and jack. If your vehicle has a puncture repair kit, refer to the information contained on the next page.*

 Warning: Do not change a wheel in a situation where you risk being hit by other traffic. On busy roads, try to stop in a lay-by or a gateway. Be wary of passing traffic while changing the wheel – it is easy to become distracted by the job in hand.

Preparation

- ☐ When a puncture occurs, stop as soon as it is safe to do so.
- ☐ Park on firm level ground, if possible, and well out of the way of other traffic.
- ☐ Use hazard warning lights if necessary.

- ☐ If you have one, use a warning triangle to alert other drivers of your presence.
- ☐ Apply the handbrake and engage first or reverse gear (or Park on models with automatic transmission).

- ☐ Chock the wheel diagonally opposite the one being removed – a couple of large stones will do for this.
- ☐ If the ground is soft, use a flat piece of wood to spread the load under the jack.

Changing the wheel

1 Lift out the floor covering and take out the tool carrier stored in the spare wheel.

2 Unscrew the spare wheel clamp nut and take out the spare wheel.

3 Using a screwdriver inserted in the slots of the wheel nut trim caps, prise off the caps.

4 Slacken each wheel nut by half a turn.

5 Locate the jack head below the jacking point nearest the wheel to be changed; the jacking point is indicated by an indentation in the sill. Turn the handle until the base of the jack touches the ground ensuring that the jack is vertical. Raise the vehicle until the wheel is clear of the ground. If the tyre is flat make sure that the vehicle is raised sufficiently to allow the spare wheel to be fitted.

6 Remove the nuts and lift the wheel from the vehicle. Place it beneath the sill as a precaution against the jack failing. Fit the spare wheel and tighten the nuts moderately with the wheelbrace.

Finally . . .

- ☐ Remove the wheel chocks.
- ☐ Stow the punctured wheel in the luggage compartment and secure it to the lashing eyes using the strap contained in the tool kit.
- ☐ Stow the jack and tools in the correct locations in the car.
- ☐ Check the tyre pressure on the wheel just fitted. If it is low, or if you don't have a pressure gauge with you, drive slowly to the next garage and inflate the tyre to the correct pressure.
- ☐ Have the damaged tyre or wheel repaired as soon as possible, or another puncture will leave you stranded.

7 Lower the vehicle to the ground, then finally tighten the wheel nuts in a diagonal sequence. Refit the wheel trim. Note that the wheel nuts should be tightened to the specified torque (see Chapter 10) at the earliest opportunity.

Using the puncture repair kit

⚠️ *Warning: Do not attempt to repair a punctured tyre in a situation where you risk being hit by other traffic. On busy roads, try to stop in a lay-by or a gateway. Be wary of passing traffic while using the kit – it is easy to become distracted by the job in hand.*

Preparation

☐ When a puncture occurs, stop as soon as it is safe to do so.

☐ Park on firm level ground, if possible, and well out of the way of other traffic.

☐ Use hazard warning lights if necessary.

☐ If you have one, use a warning triangle to alert other drivers of your presence.

☐ Apply the handbrake and engage first or reverse gear (or Park on models with automatic transmission).

Repairing the puncture

1 Open the cover on the right-hand side of the luggage compartment and take out the puncture repair kit tray. Remove the sealant bottle and air compressor from the tray.

2 Remove the air hose and electrical cable from the underside of the compressor.

3 Screw the air hose onto the sealant bottle connection.

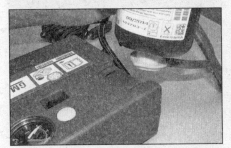

4 Fit the sealant bottle in the retainer on the compressor, then position the compressor near the punctured tyre.

5 Unscrew the dust cap from the punctured tyre and screw the sealant bottle air hose onto the tyre valve.

6 Ensure that the switch on the compressor is set to '0', then plug the compressor electrical cable into the accessory socket or cigarette lighter socket.

7 Switch on the ignition, then set the switch to 'I' to start the compressor. To avoid discharging the battery when the compressor is running, it is advisable to start the engine. The pump will initially pump the sealant into the tyre which will take approximately 30 seconds, and then start to inflate the tyre. During the initial 30 second period, the pressure gauge on the pump will indicate up to 6 bar (87 psi) and then drop. The correct tyre pressure (see end of 'Weekly checks') should be obtained within 10 minutes. The compressor can then be switched off by returning the switch to the 'O' position.

8 If it is necessary to release the pressure in the tyre, press the yellow button next to the pressure gauge on the compressor.

Important notes

☐ If the correct tyre pressure is not obtained within 10 minutes, it is likely that the tyre is too badly damaged to be repaired with the kit.

☐ The maximum speed sticker attached to the air compressor should be placed in the driver's field of view. Do not exceed the permitted maximum speed until an undamaged wheel and tyre have been fitted.

☐ On completion, disconnect the air hose and continue driving immediately so that the sealant is evenly distributed around the inside of the tyre.

☐ After driving approximately 6 miles (but no more than 10 minutes) stop and check the tyre pressure by connecting the air hose to the tyre valve. As long as the pressure indicated on the gauge is more than 1.3 bar (19 psi) it may be adjusted to the correct value using the air compressor. If the pressure has fallen below 1.3 bar (19 psi) the repair has not been successful and the car should not be driven. It will therefore be necessary to seek roadside assistance.

Identifying leaks

Puddles on the garage floor or drive, or obvious wetness under the bonnet or underneath the car, suggest a leak that needs investigating. It can sometimes be difficult to decide where the leak is coming from, especially if an engine undershield is fitted. Leaking oil or fluid can also be blown rearwards by the passage of air under the car, giving a false impression of where the problem lies.

 Warning: Most automotive oils and fluids are poisonous. Wash them off skin, and change out of contaminated clothing, without delay.

 The smell of a fluid leaking from the car may provide a clue to what's leaking. Some fluids are distinctively coloured. It may help to remove the engine undershield, clean the car carefully and to park it over some clean paper overnight as an aid to locating the source of the leak.
Remember that some leaks may only occur while the engine is running.

Sump oil

Engine oil may leak from the drain plug...

Oil from filter

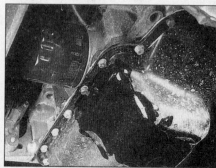

...or from the base of the oil filter.

Gearbox oil

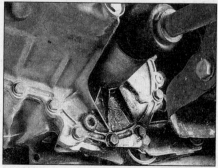

Gearbox oil can leak from the seals at the inboard ends of the driveshafts.

Antifreeze

Leaking antifreeze often leaves a crystalline deposit like this.

Brake fluid

A leak occurring at a wheel is almost certainly brake fluid.

Towing

Caution: On vehicles equipped with four-wheel drive, special precautions apply. If in doubt, do not tow, or transmission damage may result.
When all else fails, you may find yourself having to get a tow home – or of course you may be helping somebody else. Long-distance recovery should only be done by a garage or breakdown service. For shorter distances, DIY towing using another car is easy enough, but observe the following points:
☐ Use a proper tow-rope – they are not expensive. The vehicle being towed must display an ON TOW sign in its rear window.

☐ Always turn the ignition key to the 'on' position when the vehicle is being towed, so that the steering lock is released, and the direction indicator and brake lights work.
☐ The towing eye is located in the right-hand side of the luggage compartment, or under the floor panel in the luggage compartment on vehicles with a spare wheel. To fit the eye, unclip the access cover from the front or rear bumper and screw the eye firmly into position.
☐ Before being towed, release the handbrake and select neutral on the transmission. On models with automatic transmission, special precautions apply. If in doubt, do not tow, or transmission damage may result.

☐ Note that greater-than-usual pedal pressure will be required to operate the brakes, since the vacuum servo unit is only operational with the engine running.
☐ The driver of the car being towed must keep the tow-rope taut at all times to avoid snatching.
☐ Make sure that both drivers know the route before setting off.
☐ Only drive at moderate speeds and keep the distance towed to a minimum. Drive smoothly and allow plenty of time for slowing down at junctions.

Introduction

There are some very simple checks which need only take a few minutes to carry out, but which could save you a lot of inconvenience and expense.

These *Weekly checks* require no great skill or special tools, and the small amount of time they take to perform could prove to be very well spent, for example:

☐ Keeping an eye on tyre condition and pressures, will not only help to stop them wearing out prematurely, but could also save your life.

☐ Many breakdowns are caused by electrical problems. Battery-related faults are particularly common, and a quick check on a regular basis will often prevent the majority of these.

☐ If your car develops a brake fluid leak, the first time you might know about it is when your brakes don't work properly. Checking the level regularly will give advance warning of this kind of problem.

☐ If the oil or coolant levels run low, the cost of repairing any engine damage will be far greater than fixing the leak, for example.

Underbonnet check points

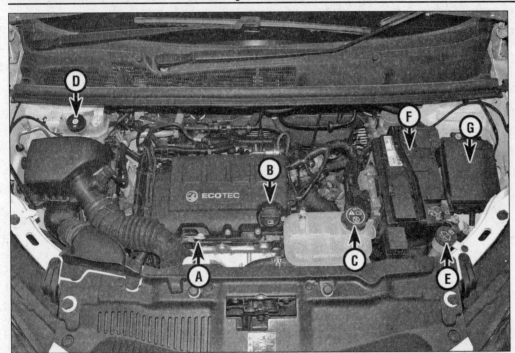

◀ 1.4 litre petrol engine models

A *Engine oil level dipstick*

B *Engine oil filler cap*

C *Coolant expansion tank*

D *Brake (and clutch) fluid reservoir*

E *Screen washer fluid reservoir*

F *Battery*

G *Fuse/relay box*

◀ 1.6 litre petrol engine models

A *Engine oil level dipstick*

B *Engine oil filler cap*

C *Coolant expansion tank*

D *Brake (and clutch) fluid reservoir*

E *Screen washer fluid reservoir*

F *Battery*

G *Fuse/relay box*

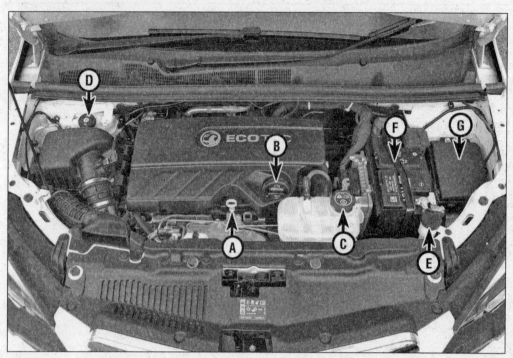

◄ 1.6 litre diesel engine models

A Engine oil level dipstick
B Engine oil filler cap
C Coolant expansion tank
D Brake (and clutch) fluid reservoir
E Screen washer fluid reservoir
F Battery
G Fuse/relay box

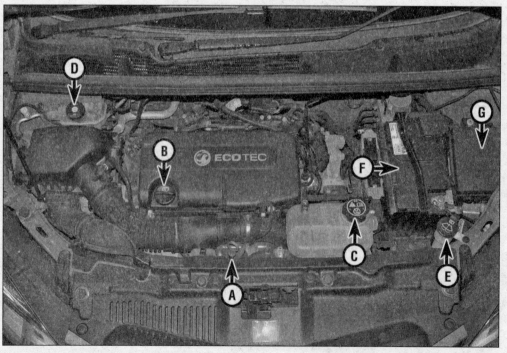

◄ 1.7 litre diesel engine models

A Engine oil level dipstick
B Engine oil filler cap
C Coolant expansion tank
D Brake (and clutch) fluid reservoir
E Screen washer fluid reservoir
F Battery
G Fuse/relay box

Engine oil level

Before you start
✔ Make sure that the car is on level ground.
✔ Check the oil level before the car is driven, or at least 5 minutes after the engine has been switched off.

HAYNES HiNT *If the oil is checked immediately after driving the vehicle, some of the oil will remain in the upper engine components, resulting in an inaccurate reading on the dipstick.*

The correct oil
Modern engines place great demands on their oil. It is very important that the correct oil for your car is used (see *Lubricants and fluids*).

Car care
● If you have to add oil frequently, you should check whether you have any oil leaks. Place some clean paper under the car overnight, and check for stains in the morning. If there are no leaks, then the engine may be burning oil.

● Always maintain the level between the upper and lower dipstick marks (see photo 3). If the level is too low, severe engine damage may occur. Oil seal failure may result if the engine is overfilled by adding too much oil.

1 The dipstick is located at the front of the engine (see *Underbonnet check points*), and is brightly coloured for easy identification. Withdraw the dipstick.

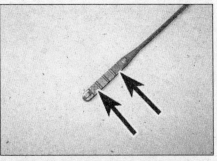

3 Note the oil level on the end of the dipstick, which should be between the upper (MAX) mark and lower (MIN) mark. Approximately 1.0 litre of oil will raise the level from the lower mark to the upper mark.

2 Using a clean rag or paper towel remove all oil from the dipstick. Insert the clean dipstick into the tube as far as it will go, then withdraw it again.

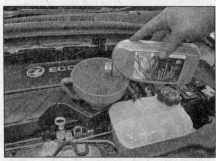

4 Oil is added through the filler cap. Unscrew the cap and top-up the level; a funnel may help to reduce spillage. Add the oil slowly, checking the level on the dipstick often. Don't overfill (see *Car care*).

Coolant level

⚠️ *Warning: Do not attempt to remove the expansion tank pressure cap when the engine is hot, as there is a very great risk of scalding. Do not leave open containers of coolant about, as it is poisonous.*

Car care
● Adding coolant should not be necessary on a regular basis. If frequent topping-up is required, it is likely there is a leak. Check the radiator, all hoses and joint faces for signs of staining or wetness, and rectify as necessary.

● It is important that antifreeze is used in the cooling system all year round, not just during the winter months. Don't top-up with water alone, as the antifreeze will become too diluted.

1 The coolant level varies with the temperature of the engine. When the engine is cold, the coolant level should be slightly above the minimum level mark on the side of the tank (arrowed). When the engine is hot, the level will rise.

2 If topping-up is necessary, remove the pressure cap (see *Warning*) from the expansion tank, which is located at the front of the engine compartment.

3 Add a mixture of water and antifreeze to the expansion tank until the coolant level is up to the minimum mark. Once the level is correct, securely refit the cap.

Tyre condition and pressure

It is very important that tyres are in good condition, and at the correct pressure - having a tyre failure at any speed is highly dangerous. Tyre wear is influenced by driving style - harsh braking and acceleration, or fast cornering, will all produce more rapid tyre wear. As a general rule, the front tyres wear out faster than the rears. Interchanging the tyres from front to rear ("rotating" the tyres) may result in more even wear. However, if this is completely effective, you may have the expense of replacing all four tyres at once!

Remove any nails or stones embedded in the tread before they penetrate the tyre to cause deflation. If removal of a nail does reveal that the tyre has been punctured, refit the nail so that its point of penetration is marked. Then immediately change the wheel, and have the tyre repaired by a tyre dealer.

Regularly check the tyres for damage in the form of cuts or bulges, especially in the sidewalls. Periodically remove the wheels, and clean any dirt or mud from the inside and outside surfaces. Examine the wheel rims for signs of rusting, corrosion or other damage. Light alloy wheels are easily damaged by "kerbing" whilst parking; steel wheels may also become dented or buckled. A new wheel is very often the only way to overcome severe damage.

New tyres should be balanced when they are fitted, but it may become necessary to re-balance them as they wear, or if the balance weights fitted to the wheel rim should fall off. Unbalanced tyres will wear more quickly, as will the steering and suspension components. Wheel imbalance is normally signified by vibration, particularly at a certain speed (typically around 50 mph). If this vibration is felt only through the steering, then it is likely that just the front wheels need balancing. If, however, the vibration is felt through the whole car, the rear wheels could be out of balance. Wheel balancing should be carried out by a tyre dealer or garage.

1 *Tread Depth - visual check*
The original tyres have tread wear safety bands (B), which will appear when the tread depth reaches approximately 1.6 mm. The band positions are indicated by a triangular mark on the tyre sidewall (A).

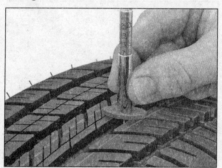

2 *Tread Depth - manual check*
Alternatively, tread wear can be monitored with a simple, inexpensive device known as a tread depth indicator gauge.

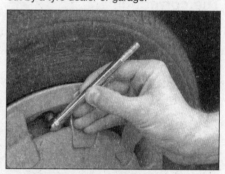

3 *Tyre Pressure Check*
Check the tyre pressures regularly with the tyres cold. Do not adjust the tyre pressures immediately after the vehicle has been used, or an inaccurate setting will result.

Tyre tread wear patterns

Shoulder Wear

Underinflation (wear on both sides)
Under-inflation will cause overheating of the tyre, because the tyre will flex too much, and the tread will not sit correctly on the road surface. This will cause a loss of grip and excessive wear, not to mention the danger of sudden tyre failure due to heat build-up.
Check and adjust pressures
Incorrect wheel camber (wear on one side)
Repair or renew suspension parts
Hard cornering
Reduce speed!

Centre Wear

Overinflation
Over-inflation will cause rapid wear of the centre part of the tyre tread, coupled with reduced grip, harsher ride, and the danger of shock damage occurring in the tyre casing.
Check and adjust pressures

If you sometimes have to inflate your car's tyres to the higher pressures specified for maximum load or sustained high speed, don't forget to reduce the pressures to normal afterwards.

Uneven Wear

Front tyres may wear unevenly as a result of wheel misalignment. Most tyre dealers and garages can check and adjust the wheel alignment (or "tracking") for a modest charge.
Incorrect camber or castor
Repair or renew suspension parts
Malfunctioning suspension
Repair or renew suspension parts
Unbalanced wheel
Balance tyres
Incorrect toe setting
Adjust front wheel alignment
Note: *The feathered edge of the tread which typifies toe wear is best checked by feel.*

Brake and clutch fluid level

 Warning:
• **Brake fluid can harm your eyes and damage painted surfaces, so use extreme caution when handling and pouring it.**
• **Do not use fluid that has been standing open for some time, as it absorbs moisture from the air, which can cause a dangerous loss of braking effectiveness.**

Safety first!

● If the reservoir requires repeated topping-up this is an indication of a fluid leak somewhere in the system, which should be investigated immediately.

● If a leak is suspected, the car should not be driven until the braking system has been checked. Never take any risks where brakes are concerned.

 • **Make sure that your car is on level ground.**

• **The fluid level in the reservoir will drop slightly as the brake pads and shoes wear down, but the fluid level must never be allowed to drop below the MIN mark.**

1 The upper (MAX) and lower (MIN) fluid level markings are on the side of the reservoir, which is located at the rear of the engine compartment. The fluid level must be kept between the marks at all times.

2 If topping-up is necessary, first wipe clean the area around the filler cap to prevent dirt entering the hydraulic system. Unscrew the reservoir cap.

3 Carefully add fluid, avoiding spilling it on the surrounding paintwork. Use only the specified type of hydraulic fluid. After filling to the correct level, refit the cap and tighten it securely. Wipe off any spilt fluid.

Screen washer fluid level

● Screenwash additives not only keep the windscreen clean during bad weather, they also prevent the washer system freezing in cold weather – which is when you are likely to need it most. Don't top-up using plain water, as the screenwash will become diluted, and will freeze in cold weather.

 Warning: On no account use engine coolant antifreeze in the screen washer system – this may damage the paintwork.

1 The washer fluid reservoir is located in the left-hand front corner of the engine compartment. If topping-up is necessary, open the filler cap.

2 When topping-up, add water and a screenwash additive in the quantities recommended on the bottle.

Wiper blades

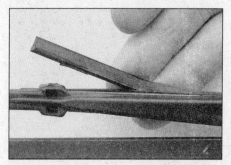

1 Check the condition of the wiper blades: if they are cracked or show signs of deterioration, or if the glass swept area is smeared, renew them. For maximum clarity of vision, wiper blades should be renewed annually.

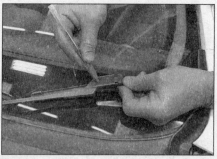

2 To remove a windscreen wiper blade, lift the wiper from the screen and press the retaining tab with a ballpoint pen or similar...

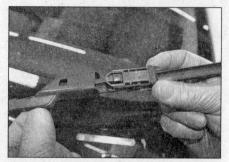

3 ...then slide the blade from the end of the wiper arm, taking care not to allow the wiper arm to spring back and damage the windscreen. Fit the new blade, pushing it onto the arm until the retaining tab engages.

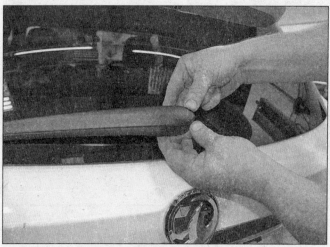

4 To remove the rear wiper blade, pull the blade cover off the spindle end of the wiper arm...

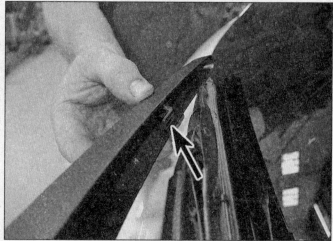

5 ...then disengage the retaining tab and remove the cover.

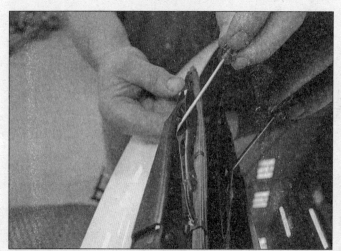

6 Using a small screwdriver or similar, press the retaining tab...

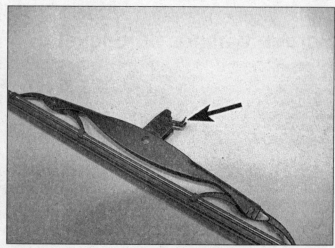

7 ...(shown here with the blade removed) and withdraw the blade from the arm. Fit the new blade ensuring that the retaining tab engages, then refit the blade cover.

Battery

Caution: Before carrying out any work on the vehicle battery, read the precautions given in 'Safety first!' at the start of this manual. If the battery is to be disconnected, refer to Chapter 5A, Section 4 before proceeding.

✔ Make sure that the battery tray is in good condition, and that the clamp is tight. Corrosion on the tray, retaining clamp and the battery itself can be removed with a solution of water and baking soda. Thoroughly rinse all cleaned areas with water. Any metal parts damaged by corrosion should be covered with a zinc-based primer, and then painted.

✔ Periodically (approximately every three months), check the charge condition of the battery as described in Chapter 5A, Section 3.

✔ If the battery is flat, and you need to jump start your vehicle, see *Roadside Repairs*.

 Battery corrosion can be kept to a minimum by applying a layer of petroleum jelly to the clamps and terminals after they are reconnected.

1 Lift the plastic cover to gain access to the battery positive terminal, which is located on the left-hand side of the engine compartment. The exterior of the battery should be inspected periodically for damage such as a cracked case or cover.

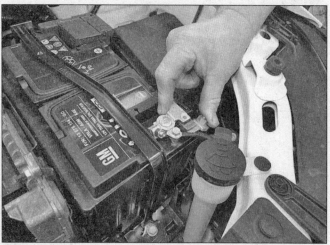

2 Check the battery lead clamps for tightness to ensure good electrical connections, and check the leads for signs of damage.

3 If corrosion (white, fluffy deposits) is evident, remove the cables from the battery terminals, clean them with a small wire brush, then refit them. Automotive stores sell a tool for cleaning the battery post...

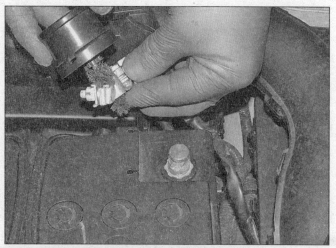

4 ...as well as the battery cable clamps.

Electrical systems

✔ Check all external lights and the horn. Refer to Chapter 12 for details if any of the circuits are found to be inoperative.

✔ Visually check all accessible wiring connectors, harnesses and retaining clips for security, and for signs of chafing or damage.

HAYNES HiNT *If you need to check your brake lights and indicators unaided, back up to a wall or garage door and operate the lights. The reflected light should show if they are working properly.*

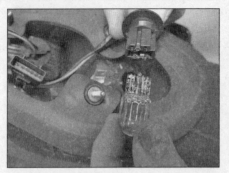

1 If a single indicator light, brake light, sidelight or headlight has failed, it is likely that a bulb has blown, and will need to be renewed. Refer to Chapter 12, Section 5 for details. If both brake lights have failed, it is possible that the switch has failed (see Chapter 9, Section 15).

2 If more than one indicator or headlight has failed, it is likely that either a fuse has blown or that there is a fault in the circuit (see Chapter 12, Section 2). The main fuses are located in the fuse/relay box on the left-hand side of the engine compartment. Release the tabs and lift off the cover for access.

3 Additional fuses and relays are located under the facia behind the glovebox...

4 ...and behind a cover in the left-hand side of the luggage compartment. Refer to the wiring diagrams at the end of Chapter 12 for further details of the fuses and circuits protected.

5 To renew a blown fuse, remove it, where applicable, using the plastic tool provided. Fit a new fuse of the same rating, available from car accessory shops. It is important that you find the reason that the fuse blew (see electrical fault finding in Chapter 12, Section 2).

Lubricants and fluids

Engine (petrol and diesel):

All except 1.4 litre petrol engines	Multigrade engine oil, viscosity SAE 5W/30 to General Motors dexos 2™ specification
1.4 litre petrol engines.	Multigrade engine oil, viscosity SAE 5W/30 to General Motors dexos 1™ Generation 2 specification
Manual transmission	Vauxhall/Opel gear oil (09 120 541)
Automatic transmission.	Vauxhall/Opel automatic transmission fluid (91 17 946)
Transfer gearbox (4WD models)	Vauxhall/Opel transmission oil (Ident TR08)
Final drive differential (4WD models)	Vauxhall/Opel transmission oil
Cooling system.	Vauxhall/Opel Dexcool long-life coolant
Brake/clutch fluid reservoir.	Hydraulic fluid to DOT 4

Tyre pressures

Details of the tyre pressures applicable to your vehicle are given on a sticker attached to the inside of the fuel filler flap or to the front door pillar on the left-hand side.

Chapter 1 Part A
Routine maintenance and servicing – petrol models

Contents

Degrees of difficulty

| Easy, suitable for novice with little experience | | Fairly easy, suitable for beginner with some experience | | Fairly difficult, suitable for competent DIY mechanic | | Difficult, suitable for experienced DIY mechanic | | Very difficult, suitable for expert DIY or professional | |

Servicing specifications

Lubricants and fluids............................... Refer to *Lubricants and fluids* on page 0•18

Capacities

Engine oil (with filter):
    1.4 litre engines .... 4.0 litres
    1.6 litre engines .... 4.5 litres
Difference between MAX and MIN dipstick marks................ 1.0 litre
Cooling system:
    1.4 litre engines .... 7.3 litres
    1.6 litre engines .... 5.6 litres
Manual transmission:
    5-speed .... 1.8 litres
    6-speed (at oil change)..................................... 1.76 litres
Automatic transmission (at fluid change)...................... 4.0 to 6.0 litres
Transfer gearbox (4WD models)........................... 0.35 litres
Final drive differential (4WD models) .... 0.5 litres
Fuel tank .... 52 litres

Cooling system

Antifreeze mixture:
    50% antifreeze .... Protection down to -40°C

Ignition system

Spark plugs:
  Type:
    1.4 litre engines... NGK IFR7X-7G
    1.6 litre engines... Champion RC 10MCC R6
    Electrode gap .... 0.85 to 0.95 mm

Brakes

Friction material minimum thickness .... 2.0 mm

Torque wrench settings

	Nm	lbf ft
Oil filter housing cap to filter housing.........................	25	18
Roadwheel nuts	140	103
Spark plugs	25	18
Sump drain plug...	14	10

1 Maintenance schedule

1 The maintenance intervals in this manual are provided with the assumption that you, not the dealer, will be carrying out the work. These are the minimum maintenance intervals based on the standard service schedule recommended by the manufacturer for vehicles driven daily. If you wish to keep your vehicle in peak condition at all times, you may wish to perform some of these procedures more often. We encourage frequent maintenance, because it enhances the efficiency, performance and resale value of your vehicle.

2 If the vehicle is driven in dusty areas, used to tow a trailer, or driven frequently at slow speeds (idling in traffic) or on short journeys, more frequent maintenance intervals are recommended.

3 When the vehicle is new, it should be serviced by a dealer service department (or other workshop recognised by the vehicle manufacturer as providing the same standard of service) in order to preserve the warranty. The vehicle manufacturer may reject warranty claims if you are unable to prove that servicing has been carried out as and when specified, using only original equipment parts or parts certified to be of equivalent quality.

Every 250 miles or weekly
☐ Refer to *Weekly checks*.

Every 10 000 miles or 6 months – whichever comes first
☐ Renew the engine oil and filter (Section 5)*

Note: *Vauxhall/Opel recommend that the engine oil and filter are changed every 20 000 miles or 12 months if the vehicle is being operated under the standard service schedule. However, oil and filter changes are good for the engine and we recommend that the oil and filter are renewed more frequently, especially if the vehicle is used on a lot of short journeys.*

Every 20 000 miles or 12 months – whichever comes first
☐ Check all underbonnet and underbody components, pipes and hoses for leaks (Section 6)
☐ Check the condition of the brake pads (renew if necessary), the calipers and discs (Section 7)
☐ Check the condition of all brake fluid pipes and hoses (Section 8)
☐ Check the condition of the front suspension and steering components, particularly the rubber gaiters and seals (Section 9)
☐ Check the condition of the driveshaft joint gaiters, and the driveshaft joints (Section 10)
☐ Check the condition of the exhaust system components (Section 11)
☐ Check the condition of the rear suspension components (Section 12)
☐ Check the bodywork and underbody for damage and corrosion, and check the condition of the underbody corrosion protection (Section 13)
☐ Check the tightness of the roadwheel nuts (Section 14)
☐ Lubricate all door, bonnet and tailgate hinges and locks (Section 15)
☐ Check the operation of the horn, all lights, and the wipers and washers (Section 16)
☐ Carry out a road test (Section 17)
☐ Reset the service interval indicator (Section 18)

Every 40 000 miles or 2 years – whichever comes first
☐ Renew the pollen filter (Section 19)
☐ Check the auxiliary drivebelt(s) and tensioner (Section 20)
☐ Check the operation of the handbrake and adjust if necessary (Section 21).
☐ Check the headlight beam alignment (Section 22)

Every 2 years, regardless of mileage
☐ Renew the battery for the remote control handset (Section 23)
☐ Renew the brake and clutch fluid (Section 24)
☐ Renew the coolant (Section 25)*

Note: *Vehicles using Vauxhall/Opel silicate-free coolant do not need the coolant renewed on a regular basis.*

Every 40 000 miles or 4 years – whichever comes first
☐ Renew the air cleaner filter element (Section 26)
☐ Renew the spark plugs (Section 27)
☐ Renew the timing belt, tensioner and idler pulleys – 1.6 litre engines (Section 28)*

Note: *The manufacturer's specified timing belt renewal interval is 100 000 miles or 6 years. However, it is strongly recommended that the interval used is 40 000 miles on vehicles which are subjected to intensive use, ie, mainly short journeys or a lot of stop-start driving. The actual belt renewal interval is therefore very much up to the individual owner, but bear in mind that severe engine damage will result if the belt breaks.*

Every 80 000 miles or 6 years – whichever comes first
☐ Renew the auxiliary drivebelt and tensioner (Section 29)

Every 100 000 miles or 10 years – whichever comes first
☐ Check, and if necessary adjust, the valve clearances – 1.6 litre engines (Section 30)

2 Component locations

Underbonnet view of a 1.4 litre engine

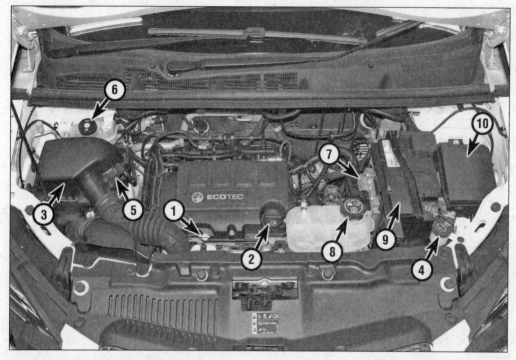

1 Engine oil level dipstick
2 Engine oil filler cap
3 Air cleaner assembly
4 Screen washer fluid reservoir
5 Mass airflow sensor
6 Brake (and clutch) fluid reservoir
7 Engine management ECU
8 Coolant expansion tank
9 Battery
10 Fuse/relay box

Underbonnet view of a 1.6 litre engine

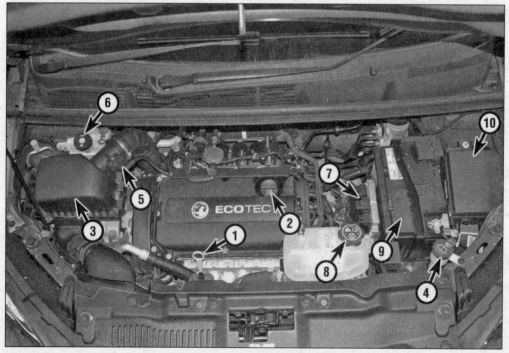

1 Engine oil level dipstick
2 Engine oil filler cap
3 Air cleaner assembly
4 Screen washer fluid reservoir
5 Mass airflow sensor
6 Brake (and clutch) fluid reservoir
7 Engine management ECU
8 Coolant expansion tank
9 Battery
10 Fuse/relay box

Front underbody view (1.4 litre 4WD model)

1 Exhaust front pipe
2 Catalytic converter
3 Steering track rods
4 Front suspension lower arms
5 Front brake calipers
6 Transfer gearbox
7 Engine mounting rear torque link
8 Right-hand driveshaft
9 Manual transmission
10 Engine oil drain plug
11 Air conditioning compressor
12 Front subframe

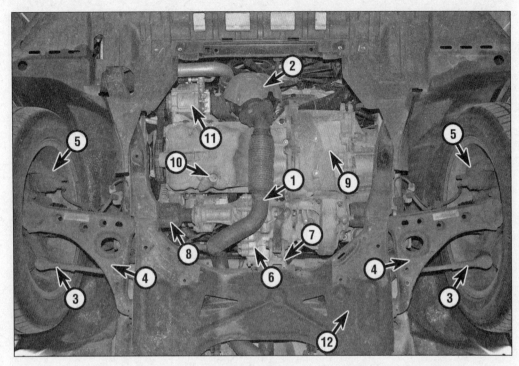

Front underbody view (1.6 litre FWD model)

1 Exhaust front pipe
2 Catalytic converter
3 Steering track rods
4 Front suspension lower arms
5 Front brake calipers
6 Engine mounting rear torque link
7 Right-hand driveshaft
8 Manual transmission
9 Engine oil drain plug
10 Air conditioning compressor
11 Front subframe

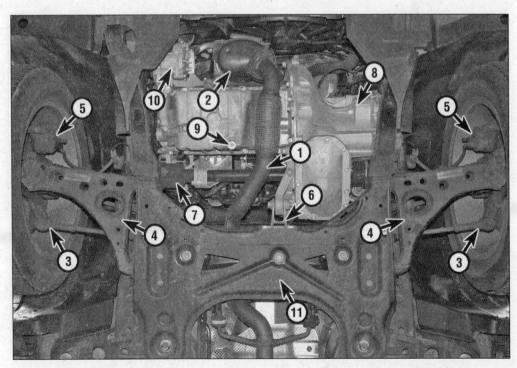

Rear underbody view (1.4 litre 4WD model)

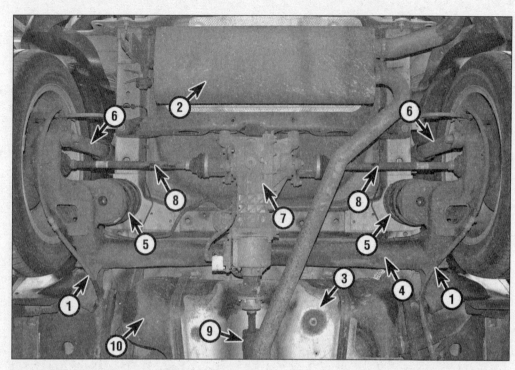

1 Handbrake cables
2 Exhaust rear silencer
3 Exhaust heat shield
4 Rear axle
5 Rear coil springs
6 Shock absorbers
7 Final drive
8 Rear driveshafts
9 Propeller shaft
10 Fuel tank

Rear underbody view (1.6 litre FWD model)

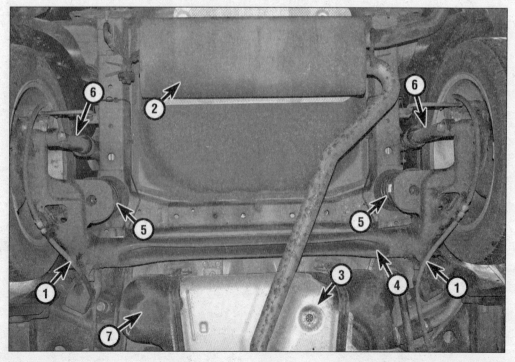

1 Handbrake cables
2 Exhaust rear silencer
3 Exhaust heat shield
4 Rear axle
5 Rear coil springs
6 Shock absorbers
7 Fuel tank

Maintenance procedures

3   General Information

1 This Chapter is designed to help the home mechanic maintain his/her vehicle for safety, economy, long life and peak performance.

2 The Chapter contains a master maintenance schedule, followed by Sections dealing specifically with each task in the schedule. Visual checks, adjustments, component renewal and other helpful items are included. Refer to the accompanying illustrations of the engine compartment and the underside of the vehicle for the locations of the various components.

3 Servicing your vehicle in accordance with the mileage/time maintenance schedule and the following Sections will provide a planned maintenance programme, which should result in a long and reliable service life. This is a comprehensive plan, so maintaining some items but not others at the specified service intervals, will not produce the same results.

4 As you service your vehicle, you will discover that many of the procedures can – and should – be grouped together, because of the particular procedure being performed, or because of the proximity of two otherwise-unrelated components to one another. For example, if the vehicle is raised for any reason, the exhaust can be inspected at the same time as the suspension and steering components.

5 The first step in this maintenance programme is to prepare yourself before the actual work begins. Read through all the Sections relevant to the work to be carried out, then make a list and gather all the parts and tools required. If a problem is encountered, seek advice from a parts specialist, or a dealer service department.

4   Regular maintenance

1 If, from the time the vehicle is new, the routine maintenance schedule is followed closely, and frequent checks are made of fluid levels and high-wear items, as suggested throughout this manual, the engine will be kept in relatively good running condition, and the need for additional work will be minimised.

2 It is possible that there will be times when the engine is running poorly due to the lack of regular maintenance. This is even more likely if a used vehicle, which has not received regular and frequent maintenance checks, is purchased. In such cases, additional work may need to be carried out, outside of the regular maintenance intervals.

3 If engine wear is suspected, a compression test will provide valuable information regarding the overall performance of the main internal components. Such a test can be used as a basis to decide on the extent of the work to be carried out. If, for example, a compression test indicates serious internal engine wear, conventional maintenance as described in this Chapter will not greatly improve performance of the engine, and may prove a waste of time and money, unless extensive overhaul work is carried out first.

4 The following series of operations are those most often required to improve the performance of a generally poor-running engine:

Primary operations

a) Clean, inspect and test the battery (refer to 'Weekly checks').

b) Check all the engine-related fluids (refer to 'Weekly checks').

c) Check the condition and tension of the auxiliary drivebelt (Section 20).

d) Renew the spark plugs (Section 27).

e) Check the condition of the air cleaner element, and renew if necessary (Section 26).

f) Check the condition of all hoses, and check for fluid leaks (Section 6).

5 If the above operations do not prove fully effective, carry out the following secondary operations:

Secondary operations

6 All items listed under *Primary operations*, plus the following:

a) Check the charging system (refer to Chapter 5A, Section 5).

b) Check the ignition system (refer to Chapter 5B, Section 2).

c) Check the fuel, exhaust and emission control systems (refer to Chapter 4A or Chapter 4C).

Every 10 000 miles or 6 months

5   Engine oil and filter renewal

> **HAYNES HiNT** *Frequent oil and filter changes are the most important preventative maintenance procedures which can be undertaken by the DIY owner. As engine oil ages, it becomes diluted and contaminated, which leads to premature engine wear.*

1 Before starting this procedure, gather together all the necessary tools and materials. Also make sure that you have plenty of clean rags and newspapers handy, to mop up any spills. Ideally, the engine oil should be warm, as it will drain more easily, and more built-up sludge will be removed with it. Take care not to touch the exhaust or any other hot parts of the engine when working under the vehicle. To avoid any possibility of scalding, and to protect yourself from possible skin irritants and other harmful contaminants in used engine oils, it is advisable to wear gloves when carrying out this work.

2 Access to the underside of the vehicle will be greatly improved if it can be raised on a lift, driven onto ramps, or jacked up and supported on axle stands (see *Jacking and vehicle support*). Whichever method is chosen, make sure that the vehicle remains level, or if it is at an angle, that the drain plug is at the lowest point. The drain plug is located at the rear of the sump.

3 Remove the oil filler cap from the camshaft cover (twist it through a quarter-turn anti-clockwise and withdraw it).

4 Using a spanner, or preferably a suitable socket and bar, slacken the drain plug about half a turn. Position the draining container under the drain plug, then remove the plug completely **(see Haynes Hint)**.

5 Allow some time for the oil to drain, noting that it may be necessary to reposition the container as the oil flow slows to a trickle.

As the drain plug releases from the threads, move it away quickly so that the stream of oil running out of the sump goes into the drain pan and not up your sleeve.

5.7a Unscrew and remove the oil filter housing cap, then separate the element from the cap...

5.7b...and remove the O-ring seal

5.9a Locate a new O-ring seal in the groove on the filter housing cap...

5.9b...locate the new element in the cap...

5.9c...and lubricate the threads of the filter housing cap with clean engine oil

5.10 Renew the drain plug O-ring if it shows signs of damage which may prevent an oil-tight seal

6 Position another container under the oil filter. The filter is located at the front left-hand side of the cylinder block and is accessible from above. For improved access, undo the two bolts and move the coolant expansion tank to one side.

7 Unscrew the oil filter housing cap and withdraw it, together with the element, then separate the element, and remove the O-ring seal from the cap **(see illustrations)**.

8 Use a clean rag to remove all oil, dirt and sludge from the oil filter housing.

9 Locate a new O-ring seal in the groove on the filter housing cap, then locate the new element in the cap. Lubricate the threads of the filter housing cap with clean engine oil, then insert the filter and cap into the housing **(see illustrations)**. Screw on the cap and tighten it to the specified torque.

10 After all the oil has drained, wipe the drain plug and the sealing O-ring with a clean rag. Examine the condition of the O-ring, and renew it if it shows signs of damage which may prevent an oil-tight seal **(see illustration)**. Clean the area around the drain plug opening, and refit the plug complete with the O-ring. Tighten the plug to the specified torque.

11 Remove the old oil and all tools from under the vehicle then lower the vehicle to the ground.

12 Fill the engine through the filler hole in the camshaft cover, using the correct grade and type of oil (refer to *Weekly checks* for details of topping-up). Pour in half the specified quantity of oil first, then wait a few minutes for the oil to drain into the sump. Continue to add oil, a small quantity at a time, until the level is up to the lower mark on the dipstick.

Adding approximately a further 1.0 litre will bring the level up to the upper mark on the dipstick.

13 Start the engine and run it until it reaches normal operating temperature. While the engine is warming up, check for leaks around the oil filter and the sump drain plug.

14 Stop the engine, and wait at least five minutes for the oil to settle in the sump once more. With the new oil circulated and the filter now completely full, recheck the level on the dipstick, and add more oil as necessary.

15 Dispose of the used engine oil and filter safely, with reference to *General repair procedures*. Do not discard the old filter with domestic household waste. The facility for waste oil disposal provided by many local council recycling centres generally has a filter receptacle alongside.

Every 20 000 miles or 12 months

6 Hose and fluid leak check

1 Visually inspect the engine joint faces, gaskets and seals for any signs of water or oil leaks. Pay particular attention to the areas around the camshaft cover, cylinder head, oil filter and sump joint faces. Similarly, check the transmission and the air conditioning compressor for oil leakage. Bear in mind

that, over a period of time, some very slight seepage from these areas is to be expected; what you are really looking for is any indication of a serious leak. Should a leak be found, renew the offending gasket or oil seal by referring to the appropriate Chapters in this manual.

2 Also check the security and condition of all the engine-related pipes and hoses. Ensure that all cable ties or securing clips are in place, and in good condition. Clips which are broken or missing can lead to chafing of

the hoses pipes or wiring, which could cause more serious problems in the future.

3 Carefully check the radiator hoses and heater hoses along their entire length. Renew any hose which is cracked, swollen or deteriorated. Cracks will show up better if the hose is squeezed. Pay close attention to the hose clips that secure the hoses to the cooling system components. Hose clips can pinch and puncture hoses, resulting in cooling system leaks. If wire-type hose clips are used, it may be a good idea to replace them with screw-type clips.

4 Inspect all the cooling system components (hoses, joint faces, etc) for leaks. Where any problems of this nature are found on system components, renew the component or gasket with reference to Chapter 3.

5 Where applicable, inspect the automatic transmission fluid cooler hoses for leaks or deterioration.

6 With the vehicle raised, inspect the petrol tank and filler neck for punctures, cracks and other damage. The connection between the filler neck and tank is especially critical. Sometimes, a rubber filler neck or connecting hose will leak due to loose retaining clamps or deteriorated rubber.

7 Carefully check all rubber hoses and metal fuel lines leading away from the petrol tank. Check for loose connections, deteriorated hoses, crimped lines and other damage. Pay particular attention to the vent pipes and hoses, which often loop up around the filler neck and can become blocked or crimped. Follow the lines to the front of the vehicle, carefully inspecting them all the way. Renew damaged sections as necessary. Similarly, whilst the vehicle is raised, take the opportunity to inspect all underbody brake fluid pipes and hoses.

8 From within the engine compartment, check the security of all fuel hose attachments and pipe unions, and inspect the fuel hoses and vacuum hoses for kinks, chafing and deterioration.

7   Brake pad and disc check

1 Firmly apply the handbrake, then jack up the front and rear of the vehicle and support it securely on axle stands (see *Jacking and vehicle support*). Remove the roadwheels.

2 For a quick check, the pad thickness can be carried out via the inspection hole on the caliper **(see Haynes Hint)**. Using a steel rule, measure the thickness of the pad friction material. This must not be less than that indicated in the Specifications.

3 The view through the caliper inspection hole gives a rough indication of the state of the brake pads. For a comprehensive check, the brake pads should be removed and cleaned.

The operation of the caliper can then also be checked, and the condition of the brake disc itself can be fully examined on both sides. Chapter 9, Section 6 contains a detailed description of how the brake disc should be checked for wear and/or damage.

4 If any pad's friction material is worn to the specified thickness or less, all four pads must be renewed as a set. Refer to Chapter 9, Section 4, or Chapter 9, Section 5, for details.

5 On completion, refit the roadwheels and lower the vehicle to the ground.

8   Brake fluid pipe and hose check

1 The brake hydraulic system includes a number of metal pipes, which run from the master cylinder to the hydraulic modulator of the anti-lock braking system (ABS) and then to the front and rear brake assemblies. Flexible hoses are fitted between the pipes and the front and rear brake assemblies, to allow for steering and suspension movement.

2 When checking the system, first look for signs of leakage at the pipe or hose unions, then examine the flexible hoses for signs of cracking, chafing or deterioration of the rubber. Bend the hoses sharply between the fingers (but do not actually bend them double, or the casing may be damaged) and check that this does not reveal previously-hidden cracks, cuts or splits. Check that the pipes and hoses are securely fastened in their clips.

3 Carefully working along the length of the metal pipes, look for dents, kinks, damage of any sort, or corrosion. Light corrosion can be polished off, but if the depth of pitting is significant, the pipe must be renewed.

9   Front suspension and steering check

1 Apply the handbrake, then raise the front of the vehicle and securely support it on axle stands (see *Jacking and vehicle support*).

2 Inspect the balljoint dust covers and the steering gear gaiters for splits, chafing or deterioration **(see illustrations)**.

For a quick check, the thickness of friction material remaining on the brake pads can be measured through the aperture in the caliper body.

3 Any wear of these components will cause loss of lubricant, and may allow water to enter the components, resulting in rapid deterioration of the balljoints or steering gear.

4 Grasp each roadwheel at the 12 o'clock and 6 o'clock positions, and try to rock it **(see illustration)**. Very slight free play may be felt, but if the movement is appreciable, further investigation is necessary to determine the source. Continue rocking the wheel while an assistant depresses the footbrake. If the movement is now eliminated or significantly reduced, it is likely that the hub bearings are at fault. If the free play is still evident with the footbrake depressed, then there is wear in the suspension joints or mountings.

5 Now grasp each wheel at the 9 o'clock and 3 o'clock positions, and try to rock it as before. Any movement felt now may again be caused by wear in the hub bearings or the steering track rod end balljoints. If the track rod end balljoint is worn, the visual movement will be obvious.

6 Using a large screwdriver or flat bar, check for wear in the suspension mounting bushes by levering between the relevant suspension component and its attachment point. Some movement is to be expected, as the mountings are made of rubber, but excessive wear should be obvious. Also check the condition of any visible rubber bushes, looking for splits, cracks or contamination of the rubber.

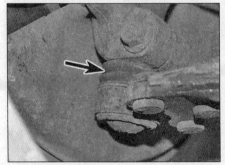

9.2a Inspect the balljoint dust covers...

9.2b ...and the steering gear gaiters for splits, chafing or deterioration

9.4 Check for wear in the hub bearings by grasping the wheel and trying to rock it

7 Check for any signs of fluid leakage around the suspension struts, or from the rubber gaiters around the piston rods **(see illustration)**. Should any fluid be noticed, the suspension strut is defective internally, and should be renewed. Note that suspension struts should always be renewed in pairs on the same axle.

8 With the vehicle standing on its wheels, have an assistant turn the steering wheel back-and-forth about an eighth of a turn each way. There should be very little, if any, lost movement between the steering wheel and roadwheels. If this is not the case, closely observe the joints and mountings previously described. In addition, check the steering column universal joints for wear, and also check the rack-and-pinion steering gear itself.

9 The efficiency of each suspension strut may be checked by bouncing the vehicle at each front corner. Generally speaking, the body will return to its normal position and stop after being depressed. If it rises and returns on a rebound, the suspension strut is probably suspect. Also examine the suspension strut upper mountings for any signs of wear.

10 Driveshaft check

1 Firmly apply the handbrake, then jack up the front of the car and support it securely on axle stands (see *Jacking and vehicle support*).

2 Turn the steering onto full lock then slowly rotate the roadwheel. Inspect the condition of the outer constant velocity (CV) joint rubber gaiters while squeezing the gaiters to open out the folds **(see illustration)**. Check for signs of cracking, splits or deterioration of the rubber which may allow the grease to escape and lead to water and grit entry into the joint. Also check the security and condition of the retaining clips. Repeat these checks on the inner CV joints. If any damage or deterioration is found, the gaiters should be renewed as described in Chapter 8, Section 4.

3 At the same time, check the general condition

9.7 Check for signs of fluid leakage around the suspension struts, or from the rubber gaiters around the piston rods

of the CV joints themselves by first holding the driveshaft and attempting to rotate the wheel. Repeat this check by holding the inner joint and attempting to rotate the driveshaft. Any appreciable movement indicates wear in the joints, wear in the driveshaft splines or loose driveshaft retaining nut.

4 On 4wd models, chock the front wheels, then jack up the rear of the car and support it securely on axle stands (see *Jacking and vehicle support*).

5 Using the same general procedures described previously for the front driveshafts, thoroughly inspect the rear driveshafts and the propeller shaft. Refer to the information contained in Chapter 8 for additional information.

11 Exhaust system check

1 With the engine cold (at least an hour after the vehicle has been driven), check the complete exhaust system from the engine to the end of the tailpipe. The exhaust system is most easily checked with the vehicle raised on a hoist, or suitably-supported on axle stands, so that the exhaust components are readily visible and accessible (see *Jacking and vehicle support*).

2 Check the exhaust pipes and connections for evidence of leaks, severe corrosion and damage. Make sure that all brackets and mountings are in good condition, and that all relevant nuts and bolts are tight. Leakage at any of the joints or in other parts of the system will usually show up as a black sooty stain in the vicinity of the leak.

3 Rattles and other noises can often be traced to the exhaust system, especially the brackets and mountings **(see illustration)**. Try to move the pipes and silencers. If the components are able to come into contact with the body or suspension parts, secure the system with new mountings. Otherwise separate the joints (if possible) and twist the pipes as necessary to provide additional clearance.

12 Rear suspension check

1 Chock the front wheels, then jack up the rear of the vehicle and support securely on axle stands (see *Jacking and vehicle support*).

2 Inspect the rear suspension components for any signs of obvious wear or damage. Pay particular attention to the rubber mounting bushes, and renew if necessary.

3 Grasp each roadwheel at the 12 o'clock and 6 o'clock positions **(see illustration 9.4)**, and try to rock it. Any excess movement indicates wear in the hub bearings. Wear may also be accompanied by a rumbling sound when the wheel is spun, or a noticeable roughness if the wheel is turned slowly. The hub bearing can be renewed as described in Chapter 10, Section 9.

4 Check for any signs of fluid leakage around the shock absorber bodies. Should any fluid be noticed, the shock absorber is defective internally, and should be renewed.

Note: *Shock absorbers should always be renewed in pairs on the same axle.*

5 With the vehicle standing on its wheels, the efficiency of each shock absorber may be checked by bouncing the vehicle at each rear

10.2 Check the condition of the driveshaft gaiters

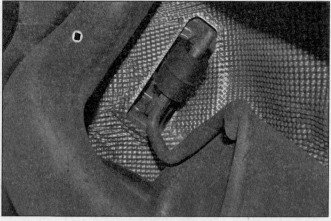

11.3 Exhaust mountings

corner. Generally speaking, the body will return to its normal position and stop after being depressed. If it rises and returns on a rebound, the shock absorber is probably suspect.

13 Bodywork and underbody condition check

Note: *This work should be carried out by a Vauxhall/Opel dealer in order to validate the vehicle warranty. The work includes a thorough inspection of the vehicle paintwork and underbody for damage and corrosion.*

Bodywork damage/ corrosion check

1 Once the car has been washed, and all tar spots and other surface blemishes have been cleaned off, carefully check all paintwork, looking closely for chips or scratches. Pay particular attention to vulnerable areas such as the front body panels, and around the wheel arches. Any damage to the paintwork must be rectified as soon as possible, to comply with the terms of the manufacturer's anti-corrosion warranties; check with a Vauxhall/Opel dealer for details.
2 If a chip or light scratch is found which is recent and still free from rust, it can be touched-up using the appropriate touch-up stick which can be obtained from Vauxhall/ Opel dealers. Any more serious damage, or rusted stone chips, can be repaired as described in Chapter 11, but if damage or corrosion is so severe that a panel must be renewed, seek professional advice as soon as possible.
3 Always check that the door and ventilation opening drain holes and pipes are completely clear, so that water can drain out.

Corrosion protection check

4 The wax-based underbody protective coating should be inspected annually, preferably just prior to Winter, when the underbody should be washed down as thoroughly as possible without disturbing the protective coating. Any damage to the coating should be repaired using a suitable wax-based sealer. If any of the body panels are disturbed for repair or renewal, do not forget to re-apply the coating. Wax should be injected into door cavities, sills and box sections, to maintain the level of protection provided by the vehicle manufacturer – seek the advice of a Vauxhall/ Opel dealer.

14 Roadwheel nut tightness check

1 Where applicable, remove the wheel trims from the wheels.
2 Using a torque wrench on each wheel nut in turn, ensure that the nuts are tightened to the specified torque.

3 Where applicable, refit the wheel trims on completion, making sure they are fitted correctly.

15 Hinge and lock lubrication

1 Work around the vehicle and lubricate the hinges of the bonnet, doors and tailgate with a light machine oil.
2 Lightly lubricate the bonnet release mechanism and exposed section of inner cable with a smear of grease.
3 Check the security and operation of all hinges, latches and locks, adjusting them where required. Check the operation of the central locking system.
4 Check the condition and operation of the tailgate struts, renewing them both if either is leaking or no longer able to support the tailgate securely when raised.

16 Electrical systems check

1 Check the operation of all the electrical equipment, ie, lights, direction indicators, horn, etc. Refer to the appropriate Sections of Chapter 12 for details if any of the circuits are found to be inoperative.
2 Note that the stop-light switch is described in Chapter 9, Section 15.
3 Check all accessible wiring connectors, harnesses and retaining clips for security, and for signs of chafing or damage. Rectify any faults found.

17 Road test

Instruments/ electrical equipment

1 Check the operation of all instruments, warning lights and electrical equipment.
2 Make sure that all instruments read correctly, and switch on all electrical equipment in turn, to check that it functions properly.

Steering and suspension

3 Check for any abnormalities in the steering, suspension, handling or road 'feel'.
4 Drive the vehicle, and check that there are no unusual vibrations or noises.
5 Check that the steering feels positive, with no excessive 'sloppiness', or roughness, and check for any suspension noises when cornering and driving over bumps.

Drivetrain

6 Check the performance of the engine, clutch, transmission and driveshafts.

7 Listen for any unusual noises from the engine, clutch and transmission.
8 Make sure that the engine runs smoothly when idling, and that there is no hesitation when accelerating.
9 Check that the clutch action is smooth and progressive, that the drive is taken up smoothly, and that the pedal travel is not excessive. Also listen for any noises when the clutch pedal is depressed.
10 Check that all gears can be engaged smoothly without noise, and that the gear lever action is smooth and not abnormally vague or 'notchy'.
11 On automatic transmission models, make sure that all gearchanges occur smoothly, without snatching, and without an increase in engine speed between changes. Check that all of the gear positions can be selected with the vehicle at rest. If any problems are found, they should be referred to a Vauxhall/Opel dealer.
12 Listen for a metallic clicking sound from the front of the vehicle, as the vehicle is driven slowly in a circle with the steering on full-lock. Carry out this check in both directions. If a clicking noise is heard, this indicates wear in a driveshaft joint (see Chapter 8, Section 5).

Braking system

13 Make sure that the vehicle does not pull to one side when braking, and that the wheels do not lock when braking hard.
14 Check that there is no vibration through the steering when braking.
15 Check that the handbrake operates correctly, without excessive movement of the lever, and that it holds the vehicle stationary on a slope.
16 Test the operation of the brake servo unit as follows. Depress the footbrake four or five times to exhaust the vacuum, then start the engine. As the engine starts, there should be a noticeable 'give' in the brake pedal as vacuum builds-up. Allow the engine to run for at least two minutes, and then switch it off. If the brake pedal is now depressed again, it should be possible to detect a hiss from the servo as the pedal is depressed. After about four or five applications, no further hissing should be heard, and the pedal should feel considerably harder.

18 Service interval indicator reset

1 Switch on the ignition, but don't start the engine.
2 Press the 'Menu' button on the direction indicator stalk until 'Vehicle Information Menu' is displayed in the driver information centre.
3 Turn the adjuster wheel on the direction indicator stalk to select 'Remaining Oil Life'.
4 Press the 'SET/CLR' button on the stalk to reset the system.
5 On completion, switch off the ignition.

19.2 Press the sides of the glovebox door and lower it past the stops

19.3 Release the retaining catches and remove the pollen filter housing cover

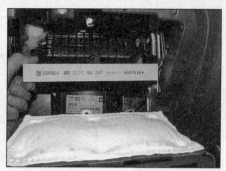

19.4 Remove the filter from the housing

Every 40 000 miles or 2 years

19 Pollen filter renewal

1 Open the glovebox fully, then detach the damper from the retaining peg by depressing the tabs and sliding the damper off.
2 Press together the sides of the glovebox door and lower the door past the stops **(see illustration)**.
3 If fitted, undo the pollen filter housing cover retaining screw, then release the retaining catches and lift off the filter housing cover **(see illustration)**.
4 Slide the pollen filter out of the housing **(see illustration)**.
5 Fit the new filter using a reversal of the removal procedure. Make sure that the filter is fitted the correct way up as indicated by the arrows on the edge of the filter.

20 Auxiliary drivebelt check

1 A single drivebelt is fitted at the right-hand end of the engine. The drivebelt drives the alternator, air conditioning compressor and coolant pump from the crankshaft pulley, and is tensioned by an automatic tensioner.

2 Due to their function and material make-up, drivebelts are prone to failure after a long period of time and should therefore be inspected regularly.
3 Apply the handbrake, then jack up the front of the vehicle and support it on axle stands (see *Jacking and vehicle support*). Remove the right-hand front roadwheel, then remove the wheel arch liner as described in Chapter 11, Section 21 for access to the right-hand side of the engine.
4 With the engine stopped, inspect the full length of the drivebelt for cracks and separation of the belt plies. It will be necessary to turn the engine (using a spanner or socket and bar on the crankshaft pulley bolt) so that the belt can be inspected thoroughly. Twist the belt between the pulleys so that both sides can be viewed. Also check for fraying, and glazing which gives the belt a shiny appearance. Check the pulleys for nicks, cracks, distortion and corrosion. If the belt show signs of wear or damage, it should be renewed as a precaution against breakage in service as described in Section 29.

21 Handbrake operation and adjustment check

1 With the vehicle on a slight slope, apply the handbrake lever by up to 4 clicks of the

ratchet, and check that it holds the vehicle stationary, then release the lever and check that there is no resistance to movement of the vehicle.
2 If necessary, adjust the handbrake as follows.
3 Chock the front wheels then jack up the rear of the car and securely support it on axle stands (see *Jacking and vehicle support*).
4 Fully depress, then release the brake pedal at least five times. Similarly, fully apply, then release the handbrake at least five times.
5 Referring to the information contained in Chapter 11, Section 26, remove the cup holder and support base from the centre console for access to the handbrake cable adjuster nut **(see illustrations)**.
6 Move the handbrake lever to the fully released position, then turn the cable adjuster nut anti-clockwise to remove all tension from the cables.
7 With the handbrake lever set on the third notch of the ratchet mechanism, rotate the adjuster nut clockwise until a reasonable amount of force is required to turn each wheel. **Note:** *The force required should be equal for each wheel.*
8 Now pull the handbrake lever up to the fourth notch of the ratchet mechanism and check that both rear wheels are locked. Once this is so, fully release the handbrake lever and check that the wheels rotate freely. Check the

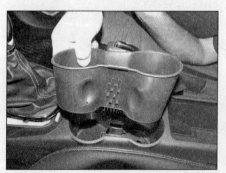

21.5a Remove the cup holder...

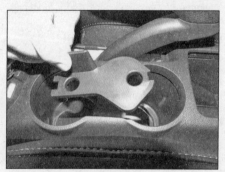

21.5b ...and support base from the centre console...

21.5c ...for access to the handbrake cable adjuster nut

adjustment by applying the handbrake fully whilst counting the clicks emitted from the handbrake ratchet and, if necessary, re-adjust.
9 On completion of adjustment, refit the centre console components (see Chapter 11, Section 26) then lower the vehicle to the ground.

Every 2 years

23 Remote control battery renewal

1 Using a screwdriver, prise the battery cover from the ignition key fob **(see illustrations)**.
2 Note how the circular battery is fitted, then carefully remove it from the contacts.
3 Fit the new battery (type CR 20 32) and refit the cover making sure that it clips fully onto the base.
4 After changing the battery, lock and unlock the driver's door with the key in the lock. The remote control unit will be synchronised when the ignition is switched on.

24 Hydraulic fluid renewal

> ⚠️ **Warning: Hydraulic fluid can harm your eyes and damage painted surfaces, so use extreme caution when handling and pouring it. Do not use fluid that has been standing open for some time, as it absorbs moisture from the air. Excess moisture can cause a dangerous loss of braking effectiveness.**

Note: *The brake and clutch hydraulic systems share a common reservoir.*
1 The procedure is similar to that for the bleeding of the hydraulic system as described in Chapter 9, Section 2 (brake) and Chapter 6, Section 2 (clutch).
2 Working as described in Chapter 9, Section 2, open the first bleed screw in the sequence, and pump the brake pedal gently until nearly all the old fluid has been emptied from the master cylinder reservoir. Top-up to the MAX level with new fluid, and continue pumping until only the new fluid remains in the reservoir, and new fluid can be seen emerging from the bleed screw. Tighten the screw, and top the reservoir level up to the MAX level line.

> **HAYNES HiNT** *Old hydraulic fluid is invariably much darker in colour than the new, making it easy to distinguish the two.*

3 Work through all the remaining bleed screws in the sequence until new fluid can be seen at all of them. Be careful to keep the master cylinder reservoir topped-up to above the MIN level at all times, or air may enter the

system and greatly increase the length of the task.
4 Bleed the fluid from the clutch hydraulic system as described in Chapter 6, Section 2.
5 When the operation is complete, check that all bleed screws are securely tightened, and that their dust caps are refitted. Wash off all traces of spilt fluid, and recheck the master cylinder reservoir fluid level.
6 Check the operation of the brakes and clutch before taking the car on the road.

25 Coolant renewal

Note: *Vauxhall/Opel do not specify renewal intervals for the antifreeze mixture, as the mixture used to fill the system when the vehicle is new is designed to last the lifetime of the vehicle. However, it is strongly recommended that the coolant is renewed at the intervals specified in the 'Maintenance schedule', as a precaution against possible engine corrosion problems. This is particularly advisable if the coolant has been renewed using an antifreeze other than that specified by Vauxhall/Opel. With many antifreeze types, the corrosion inhibitors become progressively less effective with age. It is up to the individual owner whether or not to follow this advice.*

Cooling system draining

> ⚠️ **Warning: Wait until the engine is cold before starting this procedure. Do not allow antifreeze to come in contact with your skin, or with the painted surfaces of the vehicle. Rinse off spills immediately with plenty of water. Never leave antifreeze lying around in an open container, or in a puddle in the driveway or on the garage floor. Children**

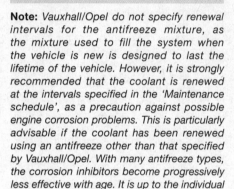

23.1a Using a screwdriver, release the battery cover from the ignition key fob...

22 Headlight beam alignment check

1 Accurate adjustment of the headlight beam

and pets are attracted by its sweet smell, but antifreeze can be fatal if ingested.
1 To drain the cooling system, first cover the expansion tank cap with a wad of rag, and slowly turn the cap anti-clockwise to relieve the pressure in the cooling system (a hissing sound will normally be heard). Wait until any pressure remaining in the system is released, then continue to turn the cap until it can be removed.
2 Chock the rear wheels, then jack up the front of the vehicle and support securely on axle stands (see *Jacking and vehicle support*).
3 Remove the engine undertray as described in Chapter 11, Section 21.
4 The coolant drain plug is located at the bottom of the radiator right-hand end tank. Position a container beneath the radiator then unscrew the drain plug and allow the coolant to drain.
5 When the flow of coolant stops, refit and tighten the drain plug.
6 As no cylinder block drain plug is fitted, it is not possible to drain all of the coolant. Due consideration must be made for this when refilling the system, in order to maintain the correct concentration of antifreeze.
7 If the coolant has been drained for a reason other than renewal, then provided it is clean and less than two years old, it can be re-used.

Cooling system flushing

8 If coolant renewal has been neglected, or if the antifreeze mixture has become diluted, then in time, the cooling system may gradually lose efficiency, as the coolant passages become restricted due to rust, scale deposits, and other sediment. The cooling system efficiency can be restored by flushing the system clean.
9 The radiator should be flushed independently of the engine, to avoid unnecessary contamination.

is only possible using optical beam-setting equipment, and this work should therefore be carried out by a Vauxhall/Opel dealer or service station with the necessary facilities. Refer to Chapter 12, Section 8 for further information.

23.1b...then remove the battery cover

25.19 Radiator bleed screw location (shown with radiator removed)

Radiator flushing

10 Disconnect the top and bottom hoses and any other relevant hoses from the radiator, with reference to Chapter 3, Section 2.

11 Insert a garden hose into the radiator top inlet. Direct a flow of clean water through the radiator, and continue flushing until clean water emerges from the radiator bottom outlet.

12 If after a reasonable period, the water still does not run clear, the radiator can be flushed with a good proprietary cleaning agent. It is important that the manufacturer's instructions are followed carefully. If the contamination is particularly bad, remove the radiator, insert the hose in the radiator bottom outlet, and reverse-flush the radiator.

Engine flushing

13 To flush the engine, the thermostat must be removed, because it will be shut, and would otherwise prevent the flow of water around the engine. The thermostat can be removed as described in Chapter 3, Section 4. Take care not to introduce dirt or debris into the system if this approach is used.

14 With the bottom hose disconnected from the radiator, insert a garden hose into the thermostat opening. Direct a clean flow of water through the engine, and continue flushing until clean water emerges from the radiator bottom hose.

15 On completion of flushing, refit the thermostat with reference to Chapter 3, Section 4 and reconnect the hoses.

Cooling system filling

16 Before attempting to fill the cooling system, make sure that all hoses and clips are in good condition, and that the clips are tight. Note that an antifreeze mixture must be used all year round, to prevent corrosion of the engine components.

17 If not already done, refit the engine undertray, then lower the car to the ground.

18 Remove the expansion tank filler cap.

19 Open the bleed screw at the upper left-hand side of the radiator (see illustration).

20 Fill the system by slowly pouring the coolant into the expansion tank until it is up to the filler neck.

21 As soon as coolant begins to run from the radiator bleed screw, close the screw.

22 Top up the expansion tank until the coolant level is up to the minimum level mark on the side of the tank.

23 Start the engine and run it at idling speed. Add further coolant to the expansion tank until the level stabilizes at the minimum mark.

24 Refit and tighten the expansion tank filler cap.

25 Increase the engine speed to 2500 rpm for 30 to 40 seconds, then switch the engine off.

26 Top up the expansion tank until the coolant level is up to the maximum level mark on the side of the tank.

27 Start the engine again and continue running it at idling speed and allow it to warm-up. When the cooling fan cuts-in, briefly run the engine again at 2000 to 2500 rpm, then allow it to return to idle. Repeat this three times.

28 Stop the engine, and allow it to cool, then re-check the coolant level with reference to Weekly Checks. Top-up the level if necessary and refit the expansion tank filler cap.

Antifreeze mixture

29 Always use a Vauxhall/Opel approved antifreeze which is suitable for use in mixed-metal cooling systems. The quantity of antifreeze and level of protection is given in the Specifications.

Note: Vauxhall/Opel recommend the use of Dexcool long-life coolant.

30 Before adding antifreeze, the cooling system should be completely drained, preferably flushed, and all hoses checked for condition and security.

31 After filling with antifreeze, a label should be attached to the expansion tank, stating the type and concentration of antifreeze used, and the date installed. Any subsequent topping-up should be made with the same type and concentration of antifreeze.

32 Do not use engine antifreeze in the windscreen/tailgate washer system, as it will cause damage to the vehicle paintwork. A screenwash additive should be added to the washer system in the quantities stated on the bottle.

Every 40 000 miles or 4 years

26 Air cleaner element renewal

1 The air cleaner is located at the right-hand side of the engine compartment.

2 Undo the two screws securing the air cleaner cover to the air cleaner housing (see illustration).

3 Lift the cover up and disengage the locating lugs on the left-hand side (see illustration). Move the cover to one side.

4 Lift the filter element out of the air cleaner housing (see illustration).

5 Wipe out the air cleaner housing and the cover.

6 Fit the new filter element, noting that the rubber locating flange should be uppermost. Engage the cover locating lugs, then secure the cover with the two screws.

26.2 Undo the screws securing the air cleaner cover to the air cleaner housing

26.3 Lift up the air cleaner cover and disengage the locating lugs

26.4 Lift out the filter element

27.4a Unscrew the spark plugs...

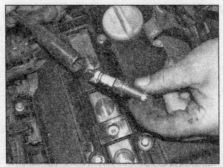

27.4b...and remove them

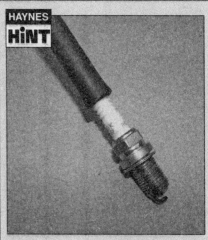

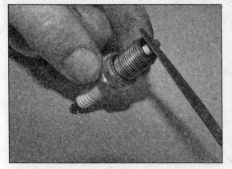

27.9a Check the electrode gap using a feeler gauge...

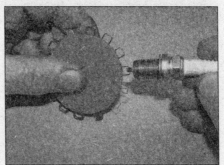

27.9b...or a wire gauge

It is very often difficult to insert spark plugs into their holes without crossthreading them. To avoid this possibility, fit a short length of rubber hose over the end of the spark plug. The flexible hose acts as a universal joint to help align the plug with the plug hole. Should the plug begin to cross-thread, the hose will slip on the spark plug, preventing thread damage to the aluminium cylinder head.

27 Spark plug renewal

1 The correct functioning of the spark plugs is vital for the correct running and efficiency of the engine. It is essential that the plugs fitted are appropriate for the engine; a suitable type is specified at the beginning of this Chapter. If the correct type is used and the engine is in good condition, the spark plugs should not need attention between scheduled replacement intervals. Spark plug cleaning is rarely necessary, and should not be attempted unless specialised equipment is available, as damage can easily be caused to the firing ends.
2 Remove the ignition module from the spark plugs as described in Chapter 5B, Section 3.
3 It is advisable to remove the dirt from the spark plug recesses using a clean brush, vacuum cleaner or compressed air before removing the plugs, to prevent dirt dropping into the cylinders.
4 Unscrew the spark plugs from the cylinder head using a spark plug spanner, suitable box spanner or a deep socket and extension bar **(see illustrations)**. Keep the socket aligned with the spark plug – if it is forcibly moved to

one side, the ceramic insulator may be broken off.
5 Examination of the spark plugs will give a good indication of the condition of the engine. As each plug is removed, examine it as follows. If the insulator nose of the spark plug is clean and white, with no deposits, this is indicative of a weak mixture or too hot a plug (a hot plug transfers heat away from the electrode slowly, a cold plug transfers heat away quickly).
6 If the tip and insulator nose are covered with hard black-looking deposits, then this is indicative that the mixture is too rich. Should the plug be black and oily, then it is likely that the engine is fairly worn, as well as the mixture being too rich.
7 If the insulator nose is covered with light tan to greyish-brown deposits, then the mixture is correct and it is likely that the engine is in good condition.
8 The spark plug electrode gap is of considerable importance. If the gap is too large or too small, the size of the spark and its efficiency will be seriously impaired and it will not perform correctly under all engine speed and load conditions. For the best results, the spark plug gap should be set in accordance with the Specifications at the beginning of this Chapter.
9 To set the gap, measure it with a feeler

blade or spark plug gap gauge and then carefully bend the outer plug electrode until the correct gap is achieved. The centre electrode should never be bent, as this may crack the insulator and cause plug failure, if nothing worse. If using feeler blades, the gap is correct when the appropriate-size blade is a firm sliding fit **(see illustrations)**.
10 Special spark plug electrode gap adjusting tools are available from most motor accessory shops, or from some spark plug manufacturers.
11 Before fitting the spark plugs, check that the connector sleeves on the top of the plug are tight, and that the plug exterior surfaces and threads are clean.
12 Screw in the spark plugs by hand where possible, then tighten them to the specified torque. Take extra care to enter the plug threads correctly, as the cylinder head is of light alloy construction **(see Haynes Hint)**.
13 On completion, refit the ignition module as described in Chapter 5B, Section 3.

28 Timing belt, tensioner and idler pulley renewal – 1.6 litre engines

1 Refer to the procedures contained in Chapter 2B, Section 8.

29.5a Turn the tensioner clockwise against the spring tension...

29.5b...and insert a suitable locking pin/bolt through the special hole provided – 1.4 litre engines

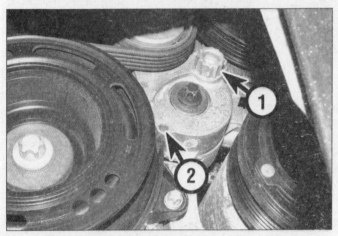

29.5c Auxiliary drivebelt tensioner – 1.6 litre engines

1 Raised projection on tensioner arm
2 Locking pin/bolt hole

29.7 Auxiliary drivebelt routing

Every 80 000 miles or 6 years

29 Auxiliary drivebelt renewal

Note: *The manufacturers recommend that the tensioner pulley is checked and if necessary renewed at the same time as the drivebelt.*

1 If not already done, apply the handbrake, then jack up the front of the vehicle and support it on axle stands (see *Jacking and vehicle support*). Remove the right-hand front roadwheel, then remove the wheel arch liner as described in Chapter 11, Section 21 for access to the right-hand side of the engine.
2 For additional working clearance, remove the air cleaner assembly as described in Chapter 4A, Section 2.
3 If the drivebelt is to be re-used, mark it to indicate its normal running direction.
4 On 1.4 litre engines, support the engine and remove the right-hand engine mounting as described in Chapter 2A, Section 16.
Note: *On these engines, the engine mounting locates within the auxiliary drivebelt.*
5 Note the routing of the drivebelt, then, using a Torx key or spanner (as applicable) on the pulley centre bolt, or the raised projection on the tensioner arm, turn the tensioner clockwise (1.4 litre engines) or anti-clockwise (1.6 litre engines) against the spring tension. Hold the tensioner in this position by inserting a suitable locking pin/bolt through the special hole provided **(see illustrations)**.
6 Slip the drivebelt off the pulleys and remove it from the engine.
7 Locate the drivebelt onto the pulleys in the correct routing **(see illustration)**. If the drivebelt is being re-used, make sure it is fitted the correct way around.
8 Turn back the tensioner and remove the locking pin/bolt then release it, making sure that the drivebelt ribs locate correctly on each of the pulley grooves.
9 Refit the air cleaner housing (if removed), then refit the wheel arch liner and roadwheel, then lower the vehicle to the ground. Tighten the roadwheel nuts to the specified torque.

Every 100 000 miles or 10 years

30 Valve clearance check and adjustment – 1.6 litre engines

1 Refer to the procedures contained in Chapter 2B, Section 11.

Chapter 1 Part B
Routine maintenance and servicing – diesel models

Contents

Degrees of difficulty

| **Easy,** suitable for novice with little experience | | **Fairly easy,** suitable for beginner with some experience | | **Fairly difficult,** suitable for competent DIY mechanic | | **Difficult,** suitable for experienced DIY mechanic | | **Very difficult,** suitable for expert DIY or professional | |

Servicing specifications

Lubricants and fluids . . . .   Refer to *Lubricants and fluids* on page 0•18

Capacities

Engine oil (with filter):
    1.6 litre engines . . . .   5.0 litres
    1.7 litre engines . . . .   5.4 litres
Difference between MIN and MAX dipstick marks . . . .   1.0 litre
Cooling system:
    1.6 litre engines . . . .   6.2 litres
    1.7 litre engines . . . .   7.1 litres
Manual transmission (at oil change) . . . .   1.76 litres
Automatic transmission (at fluid change) . . . .   4.0 to 6.0 litres
Transfer gearbox (4WD models) . . . .   0.35 litres
Final drive differential (4WD models) . . . .   0.5 litres
Fuel tank . . . .   52 litres

Cooling system

Antifreeze mixture:
    50% antifreeze . . . .   Protection down to -40°C

Brakes

Friction material minimum thickness . . . .   2.0 mm

Torque wrench settings

	Nm	lbf ft
Engine oil filter housing cap	25	18
Engine oil filter housing cap drain plug (1.6 litre engines)	10	7
Fuel filter base	25	18
Roadwheel nuts	140	103
Sump drain plug:		
1.6 litre engines	25	18
1.7 litre engines	20	15

1 Maintenance schedule

1 The maintenance intervals in this manual are provided with the assumption that you, not the dealer, will be carrying out the work. These are the minimum maintenance intervals based on the standard service schedule recommended by the manufacturer for vehicles driven daily. If you wish to keep your vehicle in peak condition at all times, you may wish to perform some of these procedures more often. We encourage frequent maintenance, because it enhances the efficiency, performance and resale value of your vehicle.

2 If the vehicle is driven in dusty areas, used to tow a trailer, or driven frequently at slow speeds (idling in traffic) or on short journeys, more frequent maintenance intervals are recommended.

3 When the vehicle is new, it should be serviced by a dealer service department (or other workshop recognised by the vehicle manufacturer as providing the same standard of service) in order to preserve the warranty. The vehicle manufacturer may reject warranty claims if you are unable to prove that servicing has been carried out as and when specified, using only original equipment parts or parts certified to be of equivalent quality.

Every 250 miles or weekly

☐ Refer to *Weekly checks*

Every 10 000 miles or 6 months – whichever comes first

☐ Renew the engine oil and filter (Section 5)

Note: *Vauxhall/Opel recommend that the engine oil and filter are changed every 20 000 miles or 12 months if the vehicle is being operated under the standard service schedule. However, oil and filter changes are good for the engine and we recommend that the oil and filter are renewed more frequently, especially if the vehicle is used on a lot of short journeys.*

Every 20 000 miles or 12 months – whichever comes first

☐ Check all underbonnet and underbody components, pipes and hoses for leaks (Section 6)
☐ Drain the water from the fuel filter (Section 7)
☐ Check the condition of the brake pads (renew if necessary), the calipers and discs (Section 8)
☐ Check the condition of all brake fluid pipes and hoses (Section 9)
☐ Check the condition of the front suspension and steering components, particularly the rubber gaiters and seals (Section 10)
☐ Check the condition of the driveshaft joint gaiters, and the driveshaft joints (Section 11)
☐ Check the condition of the exhaust system components (Section 12)
☐ Check the condition of the rear suspension components (Section 13)
☐ Check the bodywork and underbody for damage and corrosion, and check the condition of the underbody corrosion protection (Section 14)
☐ Check the tightness of the roadwheel nuts (Section 15)
☐ Lubricate all door, bonnet, boot lid and tailgate hinges and locks (Section 16)
☐ Check the operation of the horn, all lights, and the wipers and washers (Section 17)
☐ Carry out a road test (Section 18)
☐ Reset the service interval indicator (Section 19)

Every 40 000 miles or 2 years – whichever comes first

☐ Renew the pollen filter (Section 20)
☐ Renew the fuel filter (Section 21)
☐ Check the auxiliary drivebelt and tensioner (Section 22)
☐ Check the operation of the handbrake and adjust if necessary (Section 23).
☐ Check the headlight beam alignment (Section 24)

Every 2 years, regardless of mileage

☐ Renew the battery for the remote control handset (Section 25)
☐ Renew the brake and clutch fluid (Section 26)
☐ Renew the coolant (Section 27)

Note: *Vehicles using Vauxhall/Opel silicate-free coolant do not need the coolant renewed on a regular basis.*

Every 40 000 miles or 4 years – whichever comes first

☐ Renew the air cleaner filter element (Section 28)
☐ Renew the timing belt, tensioner and idler pulleys – 1.7 litre engines (Section 29)

Note: *The manufacturer's specified timing belt renewal interval is 100 000 miles or 6 years. However, it is strongly recommended that the interval used is 40 000 miles on vehicles which are subjected to intensive use, ie, mainly short journeys or a lot of stop-start driving. The actual belt renewal interval is therefore very much up to the individual owner, but bear in mind that severe engine damage will result if the belt breaks.*

Every 80 000 miles or 6 years – whichever comes first

☐ Renew the auxiliary drivebelt and tensioner (Section 30)

Every 100 000 miles or 10 years – whichever comes first

☐ Check, and if necessary adjust, the valve clearances – 1.7 litre engines (Section 31)

2 Component locations

Underbonnet view of a 1.6 litre engine

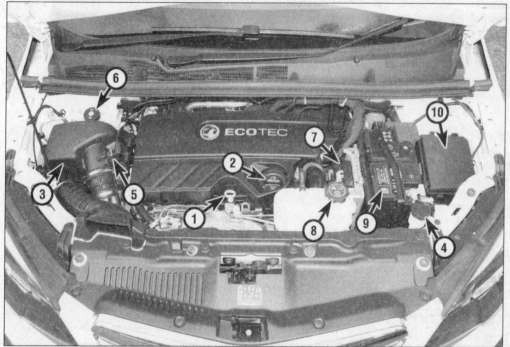

1   Engine oil level dipstick
2   Engine oil filler cap
3   Air cleaner assembly
4   Screen washer fluid
    reservoir
5   Mass airflow sensor
6   Brake (and clutch) fluid
    reservoir
7   Engine management
    ECU
8   Coolant expansion tank
9   Battery
10 Fuse/relay box

Underbonnet view of a 1.7 litre engine

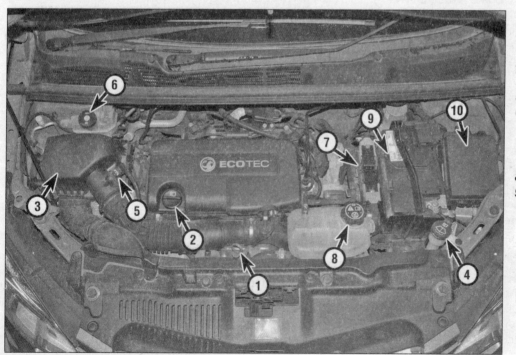

1   Engine oil level dipstick
2   Engine oil filler cap
3   Air cleaner assembly
4   Screen washer fluid
    reservoir
5   Mass airflow sensor
6   Brake (and clutch) fluid
    reservoir
7   Engine management
    ECU
8   Coolant expansion tank
9   Battery
10 Fuse/relay box

Front underbody view (1.6 litre FWD model)

1 Exhaust front pipe
2 Catalytic converter/
   diesel particulate filter
3 Steering track rods
4 Front suspension lower
   arms
5 Front brake calipers
6 Oil level sensor
7 Engine mounting rear
   torque link
8 Right-hand driveshaft
9 Manual transmission
10 Engine oil drain plug
11 Oil filter cap
12 Air conditioning
   compressor
13 Front subframe

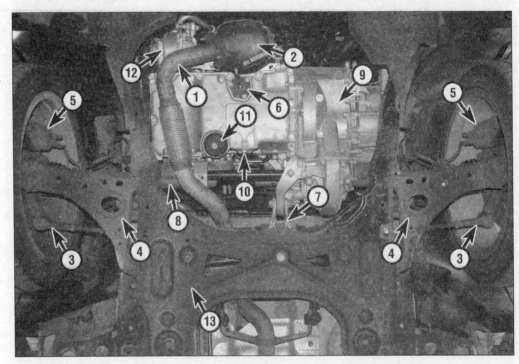

Front underbody view (1.7 litre FWD model)

1 Exhaust front pipe
2 Catalytic converter
3 Steering track rods
4 Front suspension lower
   arms
5 Front brake calipers
6 Engine mounting rear
   torque link
7 Right-hand driveshaft
8 Manual transmission
9 Engine oil drain plug
10 Alternator
11 Front subframe

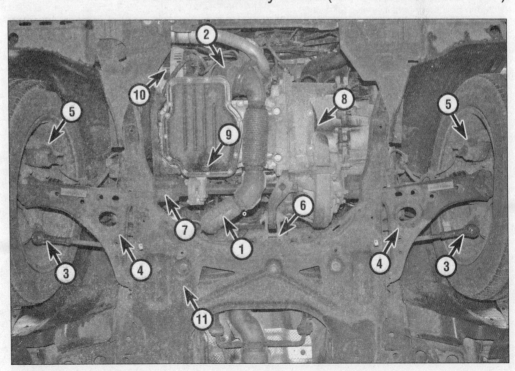

Rear underbody view (FWD model)

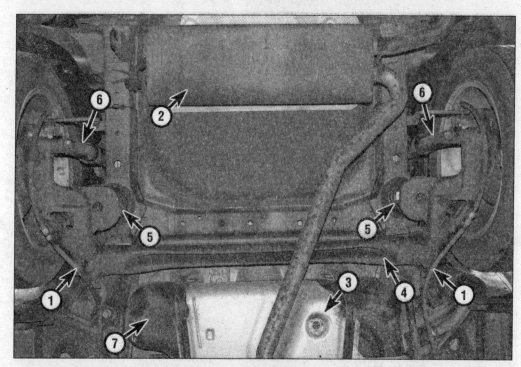

1 Handbrake cables
2 Exhaust rear silencer
3 Exhaust heat shield
4 Rear axle
5 Rear coil springs
6 Shock absorbers
7 Fuel tank

Maintenance procedures

3 General Information

1 This Chapter is designed to help the home mechanic maintain his/her vehicle for safety, economy, long life and peak performance.
2 The Chapter contains a master maintenance schedule, followed by Sections dealing specifically with each task in the schedule. Visual checks, adjustments, component renewal and other helpful items are included. Refer to the accompanying illustrations of the engine compartment and the underside of the vehicle for the locations of the various components.
3 Servicing your vehicle in accordance with the mileage/time maintenance schedule and the following Sections will provide a planned maintenance programme, which should result in a long and reliable service life. This is a comprehensive plan, so maintaining some items but not others at the specified service intervals, will not produce the same results.
4 As you service your vehicle, you will discover that many of the procedures can – and should – be grouped together, because of the particular procedure being performed, or because of the proximity of two otherwise-unrelated components to one another. For example, if the vehicle is raised for any reason, the exhaust can be inspected at the same time as the suspension and steering components.
5 The first step in this maintenance

programme is to prepare yourself before the actual work begins. Read through all the Sections relevant to the work to be carried out, then make a list and gather all the parts and tools required. If a problem is encountered, seek advice from a parts specialist, or a dealer service department.

4 Regular maintenance

1 If, from the time the vehicle is new, the routine maintenance schedule is followed closely, and frequent checks are made of fluid levels and high-wear items, as suggested throughout this manual, the engine will be kept in relatively good running condition, and the need for additional work will be minimised.
2 It is possible that there will be times when the engine is running poorly due to the lack of regular maintenance. This is even more likely if a used vehicle, which has not received regular and frequent maintenance checks, is purchased. In such cases, additional work may need to be carried out, outside of the regular maintenance intervals.
3 If engine wear is suspected, a compression test will provide valuable information regarding the overall performance of the main internal components. Such a test can be used as a basis to decide on the extent of the work to be carried out. If, for example, a compression test indicates serious internal engine wear,

conventional maintenance as described in this Chapter will not greatly improve the performance of the engine, and may prove a waste of time and money, unless extensive overhaul work is carried out first.
4 The following series of operations are those most often required to improve the performance of a generally poor-running engine:

Primary operations

a) Clean, inspect and test the battery (refer to 'Weekly checks').
b) Check all the engine-related fluids (refer to 'Weekly checks').
c) Check the condition and tension of the auxiliary drivebelt (Section 22).
d) Check the condition of the air filter, and renew if necessary (Section 28).
e) Renew the fuel filter (Section 21).
f) Check the condition of all hoses, and check for fluid leaks (Section 6).

5 If the above operations do not prove fully effective, carry out the following secondary operations:

Secondary operations

6 All items listed under *Primary operations*, plus the following:
a) Check the charging system (refer to Chapter 5A, Section 5).
b) Check the pre/post-heating system (refer to Chapter 5A, Section 16).
c) Check the fuel, exhaust and emission control systems (refer to Chapter 4B or 4C).

Every 10 000 miles

5   Engine oil and filter renewal

 Frequent oil and filter changes are the most important preventative maintenance procedures which can be undertaken by the DIY owner. As engine oil ages, it becomes diluted and contaminated, which leads to premature engine wear.

1 Before starting this procedure, gather together all the necessary tools and materials. Also make sure that you have plenty of clean rags and newspapers handy, to mop up any spills. Ideally, the engine oil should be warm, as it will drain more easily, and more built-up sludge will be removed with it. Take care not to touch the exhaust or any other hot parts of the engine when working under the vehicle. To avoid any possibility of scalding, and to protect yourself from possible skin irritants and other harmful contaminants in used engine oils, it is advisable to wear gloves when carrying out this work.

2 Access to the underside of the vehicle will be greatly improved if it can be raised on a lift, driven onto ramps, or jacked up and supported on axle stands (see *Jacking and vehicle support*). Whichever method is chosen, make sure that the vehicle remains level, or if it is at an angle, that the drain plug is at the lowest point.

1.6 litre engines

3 Remove the oil filler cap from the camshaft cover, then position a draining container under the sump drain plug.

4 Using a spanner, or preferably a suitable socket and bar, slacken the drain plug about half a turn, then remove the plug completely **(see illustrations** and **Haynes Hint)**. Note that the sealing washer is integral with the drain plug and a new plug must be fitted.

 As the drain plug releases from the threads, move it away quickly so that the stream of oil running out of the sump goes into the drain pan and not up your sleeve.

5 Allow some time for the oil to drain, noting that it may be necessary to reposition the container as the oil flow slows to a trickle.

6 After all the oil has drained, clean the area around the drain plug opening, and screw in the new drain plug. Tighten the plug to the specified torque, using a torque wrench.

7 Reposition the container under the oil filter which is located in the sump.

8 Using a suitable Torx bit, slacken the filter drain plug located in the centre of the oil filter cap, then unscrew the plug completely **(see illustrations)**. Allow the oil in the filter housing to drain.

9 Using a large socket, unscrew and remove the oil filter housing cap, withdraw the filter element, and remove the O-ring seal from the cap **(see illustrations)**.

10 Use a clean rag to remove all oil, dirt and sludge from the oil filter housing and cap.

11 Locate a new O-ring seal in the groove on the filter housing cap, then lubricate the seal and the cap threads with clean engine oil **(see illustration)**.

5.4a Slacken the drain plug using a spanner or socket and bar...

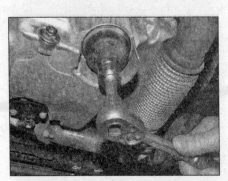

5.4b ...then remove the plug completely

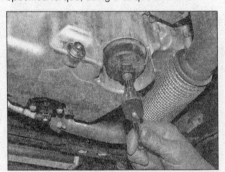

5.8a Slacken the drain plug in the oil filter cap...

5.8b ...then remove the drain plug and allow the oil in the filter housing to drain

5.9a Unscrew the oil filter housing cap...

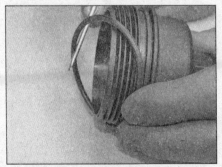

5.9b ...and remove the O-ring seal from the cap

5.11 Lubricate the new O-ring seal and filter housing cap threads with clean engine oil

5.12a Fit the new filter to the housing...

5.12b ...then screw on the cap and tighten it to the specified torque

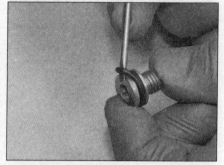

5.13 Renew the oil filter cap drain plug O-ring

5.15 Unscrew the oil filter housing cap and withdraw it, together with the element

12 Fit the new filter to the housing, then screw on the cap and tighten it to the specified torque (see illustrations).

13 Renew the O-ring seal on the oil filter cap drain plug, then refit the drain plug and tighten it to the specified torque (see illustration).

5.20 Renew the drain plug O-ring if it shows signs of damage which may prevent an oil-tight seal

1.7 litre engines

14 Position a draining container beneath the oil filter housing located at the rear of the engine.

15 Using a large socket, unscrew the oil filter

housing cap and withdraw it, together with the element (see illustration).

16 Separate the element, and remove the O-ring seal from the cap, and the two smaller O-ring seals from the base of the cap centre tube.

17 Remove the oil filler cap from the camshaft cover, then reposition the draining container under the sump drain plug.

18 Using a spanner, or preferably a suitable socket and bar, slacken the drain plug about half a turn, then remove the plug completely.

19 Allow some time for the oil to drain, noting that it may be necessary to reposition the container as the oil flow slows to a trickle.

20 After all the oil has drained, wipe the drain plug and the sealing washer/O-ring with a clean rag. Examine the condition of the sealing washer/O-ring, and renew it if it shows signs of damage which may prevent an oil-tight seal (see illustration). Clean the area around the drain plug opening, and refit the plug complete with the washer/O-ring. Tighten the plug to the specified torque, using a torque wrench.

21 Use a clean rag to remove all oil, dirt and sludge from the oil filter housing and cap.

22 Locate a new O-ring seal in the groove on the filter housing cap, and two new smaller O-ring seals to the cap centre tube. Lubricate the O-ring seals and the filter housing cap threads with clean engine oil (see illustrations).

23 Fit the new filter to the filter housing cap, then screw on the cap and tighten it to the specified torque (see illustrations).

5.22a Locate a new O-ring seal in the groove on the filter housing cap...

5.22b...and new smaller O-ring seals in the cap centre tube upper groove...

5.22c...and lower groove

5.23a Fit the new filter to the filter housing cap...

5.23b...and push it fully into position

All engines

24 Remove the old oil and all tools from under the vehicle, then lower the vehicle to the ground.

25 Fill the engine through the camshaft cover, using the correct grade and type of oil (refer to *Lubricants and fluids*). Pour in half the specified quantity of oil first, then wait a few minutes for the oil to drain into the sump. Continue to add oil, a small quantity at a time, until the level is up to the lower mark on the dipstick. Adding approximately a further 1.0 litre will bring the level up to the upper mark on the dipstick.

26 Start the engine and run it until it reaches normal operating temperature. While the engine is warming up, check for leaks around the oil filter and the sump drain plug.

27 Stop the engine, and wait at least five minutes for the oil to settle in the sump once more. With the new oil circulated and the filter now completely full, recheck the level on the dipstick, and add more oil as necessary.

28 Dispose of the used engine oil and filter safely, with reference to *General repair procedures*. Do not discard the old filter with domestic household waste. The facility for waste oil disposal provided by many local council recycling centres generally has a filter receptacle alongside.

Every 20 000 miles or 12 months

6   Hose and fluid leak check

1 Visually inspect the engine joint faces, gaskets and seals for any signs of water or oil leaks. Pay particular attention to the areas around the camshaft cover, cylinder head, oil filter and sump joint faces. Similarly, check the transmission and the air conditioning compressor for oil leakage. Bear in mind that, over a period of time, some very slight seepage from these areas is to be expected; what you are really looking for is any indication of a serious leak. Should a leak be found, renew the offending gasket or oil seal by referring to the appropriate Chapters in this manual.

2 Also check the security and condition of all the engine-related pipes and hoses. Ensure that all cable ties or securing clips are in place, and in good condition. Clips which are broken or missing can lead to chafing of the hoses pipes or wiring, which could cause more serious problems in the future.

3 Carefully check the radiator hoses and heater hoses along their entire length. Renew any hose which is cracked, swollen or deteriorated. Cracks will show up better if the hose is squeezed. Pay close attention to the hose clips that secure the hoses to the cooling system components. Hose clips can pinch and puncture hoses, resulting in cooling system leaks. If wire-type hose clips are used, it may be a good idea to replace them with screw-type clips.

4 Inspect all the cooling system components (hoses, joint faces, etc) for leaks. Where any problems of this nature are found on system components, renew the component or gasket with reference to Chapter 3.

5 Where applicable, inspect the automatic transmission fluid cooler hoses for leaks or deterioration.

6 With the vehicle raised, inspect the fuel tank and filler neck for punctures, cracks and other damage. The connection between the filler neck and tank is especially critical. Sometimes, a rubber filler neck or connecting hose will leak due to loose retaining clamps or deteriorated rubber.

7 Carefully check all rubber hoses and metal fuel lines leading away from the fuel tank. Check for loose connections, deteriorated hoses, crimped lines and other damage. Pay particular attention to the vent pipes and hoses, which often loop up around the filler neck and can become blocked or crimped. Follow the lines to the front of the vehicle, carefully inspecting them all the way. Renew damaged sections as necessary. Similarly, whilst the vehicle is raised, take the opportunity to inspect all underbody brake fluid pipes and hoses.

8 From within the engine compartment, check the security of all fuel hose attachments and pipe unions, and inspect the fuel hoses and vacuum hoses for kinks, chafing and deterioration.

7   Fuel filter water draining

Caution: Before starting any work on the fuel filter, wipe clean the filter assembly and the area around it; it is essential that no dirt or other foreign matter is allowed into the system. Obtain a suitable container into which the filter can be drained and place rags or similar material under the filter assembly to catch any spillages.

1 The fuel filter is located under the car, attached to the left-hand side of the fuel tank.

2 Chock the front wheels, then jack up the rear of the vehicle, and support it securely on axle stands (see *Jacking and vehicle support*).

3 Using a screwdriver, unscrew and remove the filter drain screw at the base of the filter housing and allow the filter to drain until clean fuel, free of dirt or water, emerges **(see illustration)**.

4 Refit the drain screw, and tighten the screw securely. Lower the vehicle to the ground.

5 On completion, dispose of the drained fuel safely. Check all disturbed components to ensure that there are no leaks (of air or fuel) when the engine is restarted.

8   Brake pad and disc check

1 Firmly apply the handbrake, then jack up the front and rear of the vehicle and support it securely on axle stands (see *Jacking and vehicle support*). Remove the roadwheels.

2 For a quick check, the pad thickness can be carried out via the inspection hole on the caliper **(see Haynes Hint)**. Using a steel rule, measure the thickness of the pad friction material. This must not be less than that indicated in the Specifications.

3 The view through the caliper inspection hole gives a rough indication of the state of the brake pads. For a comprehensive check, the brake pads should be removed and cleaned. The operation of the caliper can then also be checked, and the condition of the brake disc itself can be fully examined on both sides. Chapter 9, Section 6 contains a detailed description of how the brake disc should be checked for wear and/or damage.

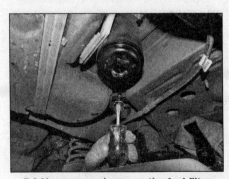

7.3 Unscrew and remove the fuel filter drain screw at the base of the filter housing

HAYNES HINT

For a quick check, the thickness of friction material remaining on the brake pads can be measured through the aperture in the caliper body.

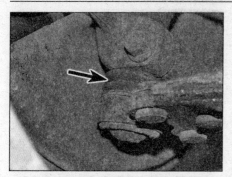

10.2a Inspect the balljoint dust covers...

10.2b...and the steering gear gaiters for splits, chafing or deterioration

10.4 Check for wear in the hub bearings by grasping the wheel and trying to rock it

10.7 Check for signs of fluid leakage around the suspension struts, or from the rubber gaiters around the piston rods

4 If any pad's friction material is worn to the specified thickness or less, all four pads must be renewed as a set. Refer to Chapter 9, Section 4, or Chapter 9, Section 5, for details.
5 On completion, refit the roadwheels and lower the vehicle to the ground.

9 Brake fluid pipe and hose check

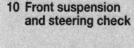

1 The brake hydraulic system includes a number of metal pipes, which run from the master cylinder to the hydraulic modulator of the anti-lock braking system (ABS) and then to the front and rear brake assemblies. Flexible hoses are fitted between the pipes and the front and rear brake assemblies, to allow for steering and suspension movement.
2 When checking the system, first look for signs of leakage at the pipe or hose unions, then examine the flexible hoses for signs of cracking, chafing or deterioration of the rubber. Bend the hoses sharply between the fingers (but do not actually bend them double, or the casing may be damaged) and check that this does not reveal previously-hidden cracks, cuts or splits. Check that the pipes and hoses are securely fastened in their clips.
3 Carefully working along the length of the metal pipes, look for dents, kinks, damage of any sort, or corrosion. Light corrosion can be polished off, but if the depth of pitting is significant, the pipe must be renewed.

10 Front suspension and steering check

1 Apply the handbrake, then raise the front of the vehicle and securely support it on axle stands (see *Jacking and vehicle support*).
2 Inspect the balljoint dust covers and the steering gear gaiters for splits, chafing or deterioration **(see illustrations)**.
3 Any wear of these components will cause loss of lubricant, and may allow water to enter the components, resulting in rapid deterioration of the balljoints or steering gear.
4 Grasp each roadwheel at the 12 o'clock and 6 o'clock positions, and try to rock it **(see illustration)**. Very slight free play may be felt, but if the movement is appreciable, further investigation is necessary to determine

11.2 Check the condition of the driveshaft gaiters

the source. Continue rocking the wheel while an assistant depresses the footbrake. If the movement is now eliminated or significantly reduced, it is likely that the hub bearings are at fault. If the free play is still evident with the footbrake depressed, then there is wear in the suspension joints or mountings.
5 Now grasp each wheel at the 9 o'clock and 3 o'clock positions, and try to rock it as before. Any movement felt now may again be caused by wear in the hub bearings or the steering track rod end balljoints. If the track rod end balljoint is worn, the visual movement will be obvious.
6 Using a large screwdriver or flat bar, check for wear in the suspension mounting bushes by levering between the relevant suspension component and its attachment point. Some movement is to be expected, as the mountings are made of rubber, but excessive wear should be obvious. Also check the condition of any visible rubber bushes, looking for splits, cracks or contamination of the rubber.
7 Check for any signs of fluid leakage around the suspension struts, or from the rubber gaiters around the piston rods **(see illustration)**. Should any fluid be noticed, the suspension strut is defective internally, and should be renewed. Note that suspension struts should always be renewed in pairs on the same axle.
8 With the vehicle standing on its wheels, have an assistant turn the steering wheel back-and-forth about an eighth of a turn each way. There should be very little, if any, lost movement between the steering wheel and roadwheels. If this is not the case, closely observe the joints and mountings previously described. In addition, check the steering column universal joints for wear, and also check the rack-and-pinion steering gear itself.
9 The efficiency of each suspension strut may be checked by bouncing the vehicle at each front corner. Generally speaking, the body will return to its normal position and stop after being depressed. If it rises and returns on a rebound, the suspension strut is probably suspect. Also examine the suspension strut upper mountings for any signs of wear.

11 Driveshaft check

1 Firmly apply the handbrake, then jack up the front of the car and support it securely on axle stands (see *Jacking and vehicle support*).
2 Turn the steering onto full lock then slowly rotate the roadwheel. Inspect the condition of the outer constant velocity (CV) joint rubber gaiters while squeezing the gaiters to open out the folds **(see illustration)**. Check for signs of cracking, splits or deterioration of the rubber which may allow the grease to escape and lead to water and grit entry into the joint. Also check the security and condition of the

retaining clips. Repeat these checks on the inner CV joints. If any damage or deterioration is found, the gaiters should be renewed as described in Chapter 8, Section 4.

3 At the same time, check the general condition of the CV joints themselves by first holding the driveshaft and attempting to rotate the wheel. Repeat this check by holding the inner joint and attempting to rotate the driveshaft. Any appreciable movement indicates wear in the joints, wear in the drive-shaft splines or loose driveshaft retaining nut.

4 On 4WD models, chock the front wheels, then jack up the rear of the car and support it securely on axle stands (see *Jacking and vehicle support*).

5 Using the same general procedures described previously for the front driveshafts, thoroughly inspect the rear driveshafts and the propeller shaft. Refer to the information contained in Chapter 8 for additional information.

12 Exhaust system check

1 With the engine cold (at least an hour after the vehicle has been driven), check the complete exhaust system from the engine to the end of the tailpipe. The exhaust system is most easily checked with the vehicle raised on a hoist, or suitably supported on axle stands, so that the exhaust components are readily visible and accessible (see *Jacking and vehicle support*).

2 Check the exhaust pipes and connections for evidence of leaks, severe corrosion and damage. Make sure that all brackets and mountings are in good condition, and that all relevant nuts and bolts are tight. Leakage at any of the joints or in other parts of the system will usually show up as a black sooty stain in the vicinity of the leak.

3 Rattles and other noises can often be traced to the exhaust system, especially the brackets and mountings **(see illustration)**. Try to move the pipes and silencers. If the components are able to come into contact with the body or suspension parts, secure the system with new mountings. Otherwise separate the joints (if possible) and twist the pipes as necessary to provide additional clearance.

13 Rear suspension check

1 Chock the front wheels, then jack up the rear of the vehicle and support securely on axle stands (see *Jacking and vehicle support*).

2 Inspect the rear suspension components for any signs of obvious wear or damage. Pay particular attention to the rubber mounting bushes, and renew if necessary.

3 Grasp each roadwheel at the 12 o'clock and 6 o'clock positions **(see illustration 10.4)**, and try to rock it. Any excess movement indicates wear in the hub bearings. Wear may also be accompanied by a rumbling sound when the wheel is spun, or a noticeable roughness if the wheel is turned slowly. The hub bearing can be renewed as described in Chapter 10, Section 9.

4 Check for any signs of fluid leakage around the shock absorber bodies. Should any fluid be noticed, the shock absorber is defective internally, and should be renewed.

Note: *Shock absorbers should always be renewed in pairs on the same axle.*

5 With the vehicle standing on its wheels, the efficiency of each shock absorber may be checked by bouncing the vehicle at each rear corner. Generally speaking, the body will return to its normal position and stop after being depressed. If it rises and returns on a rebound, the shock absorber is probably suspect.

14 Bodywork and underbody condition check

Note: *This work should be carried out by a Vauxhall/Opel dealer in order to validate the vehicle warranty. The work includes a thorough inspection of the vehicle paintwork and underbody for damage and corrosion.*

Bodywork damage/ corrosion check

1 Once the car has been washed, and all tar spots and other surface blemishes have been cleaned off, carefully check all paintwork, looking closely for chips or scratches. Pay particular attention to vulnerable areas such as the front body panels, and around the wheel arches. Any damage to the paintwork must be rectified as soon as possible, to comply with the terms of the manufacturer's anti-corrosion warranties; check with a Vauxhall/Opel dealer for details.

2 If a chip or light scratch is found which is recent and still free from rust, it can be touched-up using the appropriate touch-up stick which can be obtained from Vauxhall/Opel dealers. Any more serious damage, or rusted stone chips, can be repaired as described in Chapter 11, but if damage or corrosion is so severe that a panel must be renewed, seek professional advice as soon as possible.

3 Always check that the door and ventilation opening drain holes and pipes are completely clear, so that water can drain out.

Corrosion protection check

4 The wax-based underbody protective coating should be inspected annually, preferably just prior to Winter, when the underbody should be washed down as thoroughly as possible without disturbing the

12.3 Exhaust mountings

protective coating. Any damage to the coating should be repaired using a suitable wax-based sealer. If any of the body panels are disturbed for repair or renewal, do not forget to re-apply the coating. Wax should be injected into door cavities, sills and box sections, to maintain the level of protection provided by the vehicle manufacturer – seek the advice of a Vauxhall/Opel dealer.

15 Roadwheel nut tightness check

1 Where applicable, remove the wheel trims from the wheels.

2 Using a torque wrench on each wheel nut in turn, ensure that the nuts are tightened to the specified torque.

3 Where applicable, refit the wheel trims on completion, making sure they are fitted correctly.

16 Hinge and lock lubrication

1 Work around the vehicle and lubricate the hinges of the bonnet, doors, boot lid and tailgate with a light machine oil.

2 Lightly lubricate the bonnet release mechanism and exposed section of inner cable with a smear of grease.

3 Check the security and operation of all hinges, latches and locks, adjusting them where required. Check the operation of the central locking system.

4 Check the condition and operation of the tailgate struts, renewing them both if either is leaking or no longer able to support the tailgate securely when raised.

17 Electrical systems check

1 Check the operation of all the electrical equipment, ie, lights, direction indicators, horn, etc. Refer to Chapter 12, Section 2 for details if any of the circuits are found to be inoperative.

2 Note that the stop-light switch is described in Chapter 9, Section 15.

3 Check all accessible wiring connectors, harnesses and retaining clips for security, and for signs of chafing or damage. Rectify any faults found.

18 Road test

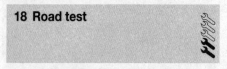

Instruments/electrical equipment

1 Check the operation of all instruments, warning lights and electrical equipment.

2 Make sure that all instruments read correctly, and switch on all electrical equipment in turn, to check that it functions properly.

Steering and suspension

3 Check for any abnormalities in the steering, suspension, handling or road 'feel'.

4 Drive the vehicle, and check that there are no unusual vibrations or noises.

5 Check that the steering feels positive, with no excessive 'sloppiness', or roughness, and check for any suspension noises when cornering and driving over bumps.

Drivetrain

6 Check the performance of the engine, clutch, transmission and driveshafts.

7 Listen for any unusual noises from the engine, clutch and transmission.

8 Make sure that the engine runs smoothly when idling, and that there is no hesitation when accelerating.

9 Check that, where applicable, the clutch action is smooth and progressive, that the drive is taken up smoothly, and that the pedal travel is not excessive. Also listen for any noises when the clutch pedal is depressed.

10 Check that all gears can be engaged smoothly without noise, and that the gear lever action is smooth and not abnormally vague or 'notchy'.

11 On automatic transmission models, make sure that all gearchanges occur smoothly, without snatching, and without an increase in engine speed between changes. Check that all of the gear positions can be selected with the vehicle at rest. If any problems are found, they should be referred to a Vauxhall/Opel dealer.

12 Listen for a metallic clicking sound from the front of the vehicle, as the vehicle is driven slowly in a circle with the steering on full-lock. Carry out this check in both directions. If a clicking noise is heard, this indicates wear in a driveshaft joint (see Chapter 8, Section 5).

Braking system

13 Make sure that the vehicle does not pull to one side when braking, and that the wheels do not lock when braking hard.

14 Check that there is no vibration through the steering when braking.

15 Check that the handbrake operates correctly, without excessive movement of the lever (manually operated handbrake), and that it holds the vehicle stationary on a slope.

16 Test the operation of the brake servo unit as follows. Depress the footbrake four or five times to exhaust the vacuum, then start the engine. As the engine starts, there should be a noticeable 'give' in the brake pedal as vacuum builds-up. Allow the engine to run for at least two minutes, and then switch it off. If the brake pedal is now depressed again, it should be possible to detect a hiss from the servo as the pedal is depressed. After about four or five applications, no further hissing should be heard, and the pedal should feel considerably harder.

19 Service interval indicator reset

1 Switch on the ignition, but don't start the engine.

2 Press the 'Menu' button on the direction indicator stalk until 'Vehicle Information Menu' is displayed in the driver information centre.

3 Turn the adjuster wheel on the direction indicator stalk to select 'Remaining Oil Life'.

4 Press the 'SET/CLR' button on the stalk to reset the system.

5 On completion, switch off the ignition.

Every 40 000 miles or 2 years

20 Pollen filter renewal

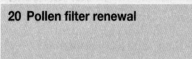

1 Open the glovebox fully, then detach the damper from the retaining peg by depressing the tabs and sliding the damper off.

2 Press together the sides of the glovebox door and lower the door past the stops **(see illustration)**.

3 If fitted, undo the pollen filter housing cover retaining screw, then release the retaining catches and lift off the filter housing cover **(see illustration)**.

4 Slide the pollen filter out of the housing **(see illustration)**.

5 Fit the new filter using a reversal of the removal procedure. Make sure that the filter is fitted the correct way up as indicated by the arrows on the edge of the filter.

21 Fuel filter renewal

Caution: Before starting any work on the fuel filter, wipe clean the filter assembly and the area around it; it is essential *that no dirt or other foreign matter is allowed into the system. Obtain a suitable container into which the filter can be drained and place rags or similar material under the filter assembly to catch any spillages.*

1 The fuel filter is located under the car, attached to the left-hand side of the fuel tank.

2 Unscrew the fuel tank filler cap to relieve any residual pressure in the fuel system.

3 Chock the front wheels, then jack up the rear of the car, and support it securely on axle stands (see *Jacking and vehicle support*).

4 Drain the water from the filter housing as described in Section 7.

20.2 Press the sides of the glovebox door and lower it past the stops

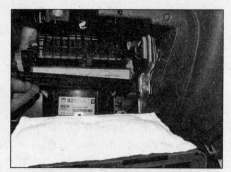

20.3 Release the retaining catches and remove the pollen filter housing cover

20.4 Remove the filter from the housing

21.5a Unscrew the fuel filter base...

21.5b ...remove it from the housing...

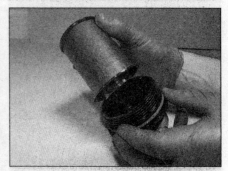

21.5c ...then remove the filter from the base

21.6 Remove the sealing ring from the filter base

21.8 Locate the new filter in the housing base

21.9 Refit the filter and base to the housing and tighten it to the specified torque

5 When the filter has drained as much as possible, unscrew and remove the filter base, then remove the old filter from the base **(see illustrations)**. Be prepared for considerable fuel spillage as the filter base is removed.

6 Remove the sealing ring from the filter base **(see illustration)**.

7 Thoroughly clean the filter base and filter housing.

8 Fit a new sealing ring to the filter base, then locate the new filter in the base **(see illustration)**.

9 Refit the filter and base and tighten the base to the specified torque **(see illustration)**. Refit the drain screw on the filter base and tighten it securely.

10 Lower the car to the ground, refit the fuel tank filler cap, then prime and bleed the fuel system as described in Chapter 4B, Section 5.

22 Auxiliary drivebelt check

1 Due to their function and material make-up, drivebelts are prone to failure after a long period of time and should therefore be inspected regularly.

2 Apply the handbrake, then jack up the front of the vehicle and support it on axle stands (see *Jacking and vehicle support*). Remove the right-hand front roadwheel.

3 Remove the right-hand front wheel arch liner as described in Chapter 11, Section 21.

4 With the engine stopped, inspect the full length of the drivebelt for cracks and separation of the belt plies, turning the engine over so that the belt can be inspected thoroughly. On 1.6 litre engines it will be necessary to use a suitable forked tool engaged with the holes on the crankshaft pulley to enable the engine to be turned. On 1.7 litre engines the engine can be turned using a spanner or socket and bar on the crankshaft pulley bolt. In all cases the engine must only be turned in the normal direction of rotation (clockwise as viewed from the right-hand side of the car). Twist the belt between the pulleys so that both sides can be viewed. Also check for fraying, and glazing

23.5a Remove the cup holder...

which gives the belt a shiny appearance. Check the pulleys for nicks, cracks, distortion and corrosion. If the belt shows signs of wear or damage, it should be renewed as a precaution against breakage in service as described in Section 30.

23 Handbrake operation and adjustment check

1 With the vehicle on a slight slope, apply the handbrake lever by up to 4 clicks of the ratchet, and check that it holds the vehicle stationary, then release the lever and check that there is no resistance to movement of the vehicle.

2 If necessary, adjust the handbrake as follows.

3 Chock the front wheels then jack up the rear of the car and securely support it on axle stands (see *Jacking and vehicle support*).

4 Fully depress, then release the brake pedal at least five times. Similarly, fully apply, then release the handbrake at least five times.

5 Referring to the information contained in Chapter 11, Section 26, remove the cup holder and support base from the centre console for access to the handbrake cable adjuster nut **(see illustrations)**.

6 Move the handbrake lever to the fully released position, then turn the cable adjuster

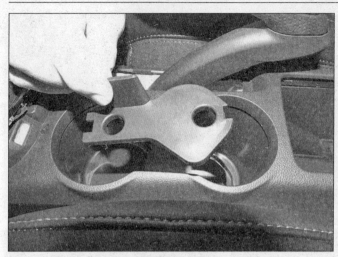

23.5b ...and support base from the centre console...

23.5c ...for access to the handbrake cable adjuster nut

nut anti-clockwise to remove all tension from the cables.

7 With the handbrake lever set on the third notch of the ratchet mechanism, rotate the adjuster nut clockwise until a reasonable amount of force is required to turn each wheel.

Note: *The force required should be equal for each wheel.*

8 Now pull the handbrake lever up to the fourth notch of the ratchet mechanism and check that both rear wheels are locked. Once this is so, fully release the handbrake lever and check that the wheels rotate freely. Check the adjustment by applying the handbrake fully whilst counting the clicks emitted from the handbrake ratchet and, if necessary, re-adjust.

9 On completion of adjustment, refit the centre console components (see Chapter 11, Section 26) then lower the vehicle to the ground.

24 Headlight beam alignment check

1 Accurate adjustment of the headlight beam is only possible using optical beam-setting equipment, and this work should therefore be carried out by a Vauxhall/Opel dealer or service station with the necessary facilities. Refer to Chapter 12, Section 8 for further information.

Every 2 years

25 Remote control battery renewal

1 Using a screwdriver, prise the battery cover from the ignition key fob **(see illustrations)**.

2 Note how the circular battery is fitted, then carefully remove it from the contacts.

3 Fit the new battery (type CR 20 32) and refit the cover making sure that it clips fully onto the base.

4 After changing the battery, lock and unlock the driver's door with the key in the lock. The remote control unit will be synchronised when the ignition is switched on.

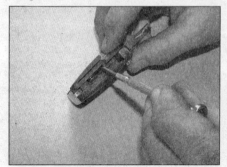

25.1a **Using a screwdriver, release the battery cover from the ignition key fob...**

26 Hydraulic fluid renewal

⚠ *Warning: Hydraulic fluid can harm your eyes and damage painted surfaces, so use extreme caution when handling and pouring it. Do not use fluid that has been standing open for some time, as it absorbs moisture from the air. Excess moisture can cause a dangerous loss of braking effectiveness.*

Note: *The brake and clutch hydraulic systems share a common reservoir.*

1 The procedure is similar to that for the

25.1b...then remove the battery cover

bleeding of the hydraulic system as described in Chapter 9, Section 2 (brake) and Chapter 6, Section 2 (clutch).

2 Working as described in Chapter 9, open the first bleed screw in the sequence, and pump the brake pedal gently until nearly all the old fluid has been emptied from the master cylinder reservoir. Top-up to the MAX level with new fluid, and continue pumping until only the new fluid remains in the reservoir, and new fluid can be seen emerging from the bleed screw. Tighten the screw, and top the reservoir level up to the MAX level line.

HAYNES HINT *Old hydraulic fluid is invariably much darker in colour than the new, making it easy to distinguish the two.*

3 Work through all the remaining bleed screws in the sequence until new fluid can be seen at all of them. Be careful to keep the master cylinder reservoir topped-up to above the MIN level at all times, or air may enter the system and greatly increase the length of the task.

4 Bleed the fluid from the clutch hydraulic system as described in Chapter 6, Section 2.

5 When the operation is complete, check that all bleed screws are securely tightened, and that their dust caps are refitted. Wash off all traces of spilt fluid, and recheck the master cylinder reservoir fluid level.

6 Check the operation of the brakes and clutch before taking the car on the road.

27 Coolant renewal

Note: *Vauxhall/Opel do not specify renewal intervals for the antifreeze mixture, as the mixture used to fill the system when the vehicle is new is designed to last the lifetime of the vehicle. However, it is strongly recommended that the coolant is renewed at the intervals specified in the 'Maintenance schedule', as a precaution against possible engine corrosion problems. This is particularly advisable if the coolant has been renewed using an antifreeze other than that specified by Vauxhall/Opel. With many antifreeze types, the corrosion inhibitors become progressively less effective with age. It is up to the individual owner whether or not to follow this advice.*

Cooling system draining

⚠️ **Warning: Wait until the engine is cold before starting this procedure. Do not allow antifreeze to come in contact with your skin, or with the painted surfaces of the vehicle. Rinse off spills immediately with plenty of water. Never leave antifreeze lying around in an open container, or in a puddle in the driveway or on the garage floor. Children and pets are attracted by its sweet smell, but antifreeze can be fatal if ingested.**

1 To drain the cooling system, first cover the expansion tank cap with a wad of rag, and slowly turn the cap anti-clockwise to relieve the pressure in the cooling system (a hissing sound will normally be heard). Wait until any pressure remaining in the system is released, then continue to turn the cap until it can be removed.

2 Chock the rear wheels, then jack up the front of the vehicle and support securely on axle stands (see *Jacking and vehicle support*).

3 Remove the engine undertray as described in Chapter 11, Section 21.

4 The coolant drain plug is located at the bottom of the radiator right-hand end tank. Position a container beneath the radiator then unscrew the drain plug and allow the coolant to drain.

5 When the flow of coolant stops, refit and tighten the drain plug.

6 As no cylinder block drain plug is fitted, it is not possible to drain all of the coolant. Due consideration must be made for this when refilling the system, in order to maintain the correct concentration of antifreeze.

7 If the coolant has been drained for a reason other than renewal, then provided it is

clean and less than two years old, it can be re-used.

Cooling system flushing

8 If coolant renewal has been neglected, or if the antifreeze mixture has become diluted, then in time, the cooling system may gradually lose efficiency, as the coolant passages become restricted due to rust, scale deposits, and other sediment. The cooling system efficiency can be restored by flushing the system clean.

9 The radiator should be flushed independently of the engine, to avoid unnecessary contamination.

Radiator flushing

10 Disconnect the top and bottom hoses and any other relevant hoses from the radiator, with reference to Chapter 3, Section 2.

11 Insert a garden hose into the radiator top inlet. Direct a flow of clean water through the radiator, and continue flushing until clean water emerges from the radiator bottom outlet.

12 If after a reasonable period, the water still does not run clear, the radiator can be flushed with a good proprietary cleaning agent. It is important that the manufacturer's instructions are followed carefully. If the contamination is particularly bad, remove the radiator, insert the hose in the radiator bottom outlet, and reverse-flush the radiator.

Engine flushing

13 To flush the engine, the thermostat must be removed, because it will be shut, and would otherwise prevent the flow of water around the engine. The thermostat can be removed as described in Chapter 3, Section 4. Take care not to introduce dirt or debris into the system if this approach is used.

14 With the bottom hose disconnected from the radiator, insert a garden hose into the thermostat opening. Direct a clean flow of water through the engine, and continue flushing until clean water emerges from the radiator bottom hose.

15 On completion of flushing, refit the thermostat with reference to Chapter 3, Section 4 and reconnect the hoses.

Cooling system filling

16 Before attempting to fill the cooling system, make sure that all hoses and clips are in good condition, and that the clips are tight. Note that an antifreeze mixture must be used all year round, to prevent corrosion of the engine components.

17 If not already done, refit the engine undertray, then lower the car to the ground.

18 Remove the expansion tank filler cap.

19 Open the bleed screw at the upper left-hand side of the radiator **(see illustration)**.

20 Fill the system by slowly pouring the coolant into the expansion tank until it is up to the filler neck.

27.19 Radiator bleed screw location (shown with radiator removed)

21 As soon as coolant begins to run from the radiator bleed screw, close the screw.

22 Top up the expansion tank until the coolant level is up to the minimum level mark on the side of the tank.

23 Start the engine and run it at idling speed. Add further coolant to the expansion tank until the level stabilizes at the minimum mark.

24 Refit and tighten the expansion tank filler cap.

25 Increase the engine speed to 2500 rpm for 30 to 40 seconds, then switch the engine off.

26 Top up the expansion tank until the coolant level is up to the maximum level mark on the side of the tank.

27 Start the engine again and continue running it at idling speed and allow it to warm-up. When the cooling fan cuts-in, briefly run the engine again at 2000 to 2500 rpm, then allow it to return to idle. Repeat this three times.

28 Stop the engine, and allow it to cool, then re-check the coolant level with reference to *Weekly checks*. Top-up the level if necessary and refit the expansion tank filler cap.

Antifreeze mixture

29 Always use a Vauxhall/Opel approved antifreeze which is suitable for use in mixed-metal cooling systems. The quantity of antifreeze and level of protection is given in the Specifications.

Note: *Vauxhall/Opel recommend the use of Dexcool long-life coolant.*

30 Before adding antifreeze, the cooling system should be completely drained, preferably flushed, and all hoses checked for condition and security.

31 After filling with antifreeze, a label should be attached to the expansion tank, stating the type and concentration of antifreeze used, and the date installed. Any subsequent topping-up should be made with the same type and concentration of antifreeze.

32 Do not use engine antifreeze in the windscreen/tailgate washer system, as it will cause damage to the vehicle paintwork. A screenwash additive should be added to the washer system in the quantities stated on the bottle.

28.2 Disconnect the mass airflow sensor wiring connector

28.3a Pull the inlet air duct upward to release it from the rubber grommets…

28.3b …and the air cleaner assembly

28.4 Undo the screws securing the air cleaner cover to the air cleaner housing

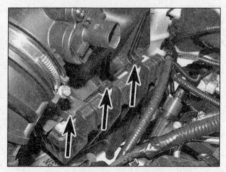

28.5 Lift up the air cleaner cover and disengage the locating lugs

28.6 Lift out the filter element

Every 40 000 miles or 4 years

28 Air cleaner element renewal

1 The air cleaner is located at the right-hand side of the engine compartment.
2 Disconnect the wiring connector at the mass airflow sensor **(see illustration)**.
3 On 1.6 litre engines, remove the inlet air duct by pulling it upward to release it from the rubber grommets and the air cleaner assembly **(see illustrations)**.

4 Undo the two screws securing the air cleaner cover to the air cleaner housing **(see illustration)**.
5 Lift the cover up and disengage the locating lugs on the left-hand side **(see illustration)**. Move the cover to one side.
6 Lift the filter element out of the air cleaner housing **(see illustration)**.
7 Wipe out the air cleaner housing and the cover.
8 Fit the new filter element, noting that the rubber locating flange should be uppermost. Engage the cover locating

lugs, then secure the cover with the two screws.
9 Refit the air inlet duct (where applicable) and reconnect the mass airflow sensor wiring connector.

29 Timing belt, tensioner and idler pulley renewal – 1.7 litre engines

1 Refer to the procedures contained in Chapter 2D, Section 8.

Every 80 000 miles or 6 years

30 Auxiliary drivebelt renewal

1 Apply the handbrake, then jack up the front of the vehicle and support it on axle stands (see *Jacking and vehicle support*). Remove the right-hand front roadwheel, then remove the right-hand front wheel arch liner as described in Chapter 11, Section 21.
2 For additional working clearance, remove the air cleaner assembly as described in Chapter 4B, Section 3.
3 If the drivebelt is to be re-used, mark it to indicate its normal running direction.

1.6 litre engines

4 Using a socket or spanner on the tensioner raised projection, turn the tensioner anti-clockwise against the spring tension. Hold the tensioner in this position by inserting a suitable locking pin/bolt through the special hole provided **(see illustration)**.

1.7 litre engines

5 Using a socket or spanner on the pulley centre bolt, turn the tensioner clockwise against the spring tension. Hold the tensioner in this position by inserting a suitable locking pin/bolt through the special hole provided.

30.4 On 1.6 litre engines, turn the tensioner anti-clockwise to relieve the drivebelt tension

All engines

6 Noting its routing, slip the drivebelt off the pulleys and remove it from the engine.

7 Locate the drivebelt onto the pulleys in the correct routing **(see illustration)**. If the drivebelt is being re-used, make sure it is fitted the correct way around.

8 Turn back the tensioner and remove the locking pin/bolt then release it, making sure that the drivebelt ribs locate correctly on each of the pulley grooves.

9 Refit the air cleaner assembly (if removed), then refit the wheel arch liner and roadwheel, then lower the vehicle to the ground.

30.7 Auxiliary drivebelt routing – 1.6 litre engines

Every 100 000 miles or 10 years

31 Valve clearance check and adjustment – 1.7 litre engines 🔧

1 Refer to the procedures contained in Chapter 2D, Section 10.

Chapter 2 Part A
1.4 litre petrol engine in-car repair procedures

Contents

Degrees of difficulty

Easy, suitable for novice with little experience	**Fairly easy,** suitable for beginner with some experience	**Fairly difficult,** suitable for competent DIY mechanic	**Difficult,** suitable for experienced DIY mechanic	**Very difficult,** suitable for expert DIY or professional

Specifications

General

Engine type.	Four-cylinder, in-line, water-cooled. Double overhead camshafts, chain-driven, acting on rocker arms and hydraulic valve lifters
Manufacturer's engine code *.	A14 NET (LUJ)
Bore	72.50 mm
Stroke	82.60 mm
Capacity	1364 cc
Firing order.	1-3-4-2 (No 1 cylinder at timing chain end of engine)
Direction of crankshaft rotation	Clockwise (viewed from timing chain end of engine)
Compression ratio	9.5: 1

* For details of engine code location, see 'Vehicle identification' in Reference

Compression pressures

Nominal pressure	12.0 to 14.0 bar
Maximum difference between any two cylinders.	1.0 bar

Lubrication system

Minimum oil pressure at 80ºC	1.5 bar at idle speed
Oil pump type.	Variable displacement vane-type, driven directly from crankshaft
Oil pump slide pin spring free length	61.0 mm
Oil pump clearances:	
Maximum vane rotor-to-housing axial clearance.	0.1 mm
Maximum vane-to-housing axial clearance	0.09 mm
Maximum vane ring-to-housing axial clearance.	0.4 mm
Maximum pump slide-to-housing axial clearance	0.08 mm
Maximum pump slide seal-to-housing axial clearance	0.09 mm
Maximum vane rotor-to-vane radial clearance.	0.05 mm
Maximum vane-to-pump slide radial clearance.	0.2 mm

Torque wrench settings

	Nm	lbf ft
Air conditioning compressor mounting bracket bolts	22	16
Air conditioning compressor-to-mounting bracket	22	16
Auxiliary drivebelt tensioner mounting bolts:		
M8 bolt	22	16
M10 bolt	55	41
Big-end bearing cap bolts*:		
Stage 1	25	18
Stage 2	Angle-tighten a further 45°	
Camshaft bearing cap bolts..............................	8	6
Camshaft cover bolts....................................	8	6
Camshaft sprocket bolt*:		
Stage 1	50	37
Stage 2	Angle-tighten a further 60°	
Camshaft VVT solenoid valve bolts	8	6
Coolant pump bolts	8	6
Coolant pump pulley bolts...............................	22	16
Crankshaft pulley bolt*:		
Stage 1	150	111
Stage 2	Angle-tighten a further 60°	
Cylinder block baseplate-to-cylinder block*:		
M6 bolts:		
Stage 1	10	7
Stage 2	Angle-tighten a further 60°	
Stage 3	Angle-tighten a further 15°	
M8 bolts:		
Stage 1	25	18
Stage 2	Angle-tighten a further 60°	
Stage 3	Angle-tighten a further 15°	
Cylinder block closure bolt (for TDC setting tool)	40	30
Cylinder head bolts*:		
Stage 1	35	26
Stage 2	Angle-tighten a further 180°	
Driveplate bolts*:		
Stage 1	35	26
Stage 2	Angle-tighten a further 30°	
Stage 3	Angle-tighten a further 15°	
Engine mountings:		
Left-hand mounting:		
Mounting-to-body bolts/nut	58	43
Mounting bracket-to-transmission bracket*:		
Stage 1	50	37
Stage 2	Angle-tighten a further 60°	
Rear mounting/torque link:		
Mounting-to-transmission bracket*:		
Stage 1	80	59
Stage 2	Angle-tighten a further 60°	
Mounting-to-subframe*:		
Stage 1	100	74
Stage 2	Angle-tighten a further 100°	
Transmission bracket bolts*:		
Stage 1	100	74
Stage 2	Angle-tighten a further 45°	
Right-hand mounting:		
Mounting bracket-to-timing cover*:		
Stage 1	60	44
Stage 2	Angle-tighten a further 60°	
Mounting-to-body...............................	62	46
Mounting-to-mounting bracket*:		
Stage 1	50	37
Stage 2	Angle-tighten a further 60°	
Engine-to-transmission bolts:		
M10 bolts	40	32
M12 bolts	60	44
Exhaust front pipe-to-catalytic converter nuts*	22	16
Flywheel bolts*:		
Stage 1	60	44
Stage 2	Angle-tighten a further 45°	
Stage 3	Angle-tighten a further 15°	

Torque wrench settings (continued)	Nm	lbf ft
Oil filter housing cap to filter housing	25	18
Oil filter housing to cylinder block	10	7
Oil pressure relief valve cap	50	37
Oil pump cover bolts	8	6
Roadwheel nuts	140	103
Sump-to-cylinder block baseplate/timing cover	10	7
Sump-to-transmission	40	30
Timing chain guide rail bolts.	8	6
Timing chain sliding rail-to-cylinder head	8	6
Timing chain tension rail pivot bolt.	20	15
Timing chain tensioner bolts	8	6
Timing chain tensioner closure bolt (for locking pin access)	50	37
Timing cover bolts:		
M6 bolts	8	6
M10 bolts	35	26

*Use new fasteners

1 General Information

How to use this Chapter

1 This Part of Chapter 2 describes the repair procedures which can reasonably be carried out on the engine while it remains in the vehicle. If the engine has been removed from the vehicle and is being dismantled as described in Chapter 2E, any preliminary dismantling procedures can be ignored.

2 Note that, while it may be possible physically to overhaul items such as the piston/connecting rod assemblies while the engine is in the vehicle, such tasks are not usually carried out as separate operations, and usually require the execution of several additional procedures (not to mention the cleaning of components and of oil ways); for this reason, all such tasks are classed as major overhaul procedures, and are described in Chapter 2E.

3 Chapter 2E describes the removal of the engine/transmission unit from the vehicle, and the full overhaul procedures which can then be carried out.

Engine description

4 The engine is of four-cylinder in-line type, with double overhead camshafts (DOHC). The engine is turbocharged and mounted transversely at the front of the vehicle.

5 The cylinder block is made of cast iron and the cylinder bores are an integral part of the block. An cast iron baseplate is bolted to the cylinder block and forms the lower half of the crankcase.

6 The crankshaft runs in five shell-type main bearings with crankshaft endfloat being controlled by thrust-washers which are an integral part of No 4 main bearing shells.

7 The connecting rods are attached to the crankshaft by horizontally-split shell-type big-end bearings. The pistons are attached to the connecting rods by gudgeon pins, which are a sliding fit in the connecting rod small-end eyes and retained by circlips. The aluminium-alloy pistons are fitted with three piston rings – two compression rings and an oil control ring.

8 The camshafts are driven from the crankshaft by a hydraulically tensioned timing chain. Each cylinder has four valves (two inlet and two exhaust), operated via rocker arms which are supported at their pivot ends by hydraulic self-adjusting valve lifters (tappets). One camshaft operates the inlet valves, and the other operates the exhaust valves.

9 The inlet and exhaust valves are each closed by a single valve spring, and operate in guides pressed into the cylinder head.

10 A variable valve timing (VVT) system is employed. The VVT system allows the inlet and exhaust camshaft timing to be varied under the control of the engine management system, to boost both low-speed torque and top-end power, as well as reducing exhaust emissions. The VVT camshaft adjuster is integral with each camshaft timing chain sprocket, and is supplied with two pressurised oil feeds through passages in the camshaft itself. Two electro-magnetic solenoid valves, one for each camshaft and operated by the engine management system, are fitted to the timing cover, and are used to control the supply of oil to each camshaft adjuster through the two oil feeds. Each adjuster contains two chambers – depending on which of the two oil feeds is enabled by the solenoid valve, the oil pressure will turn the camshaft clockwise (advance) or anti-clockwise (retard) to adjust the valve timing as required. If pressure is removed from both feeds, this induces a timing 'hold' condition. Thus the valve timing is infinitely variable within a given range.

11 A variable displacement vane-type oil pump is located in the timing cover attached to the cylinder block, and is driven directly from the crankshaft.

12 The coolant pump is located externally on the timing cover, and is driven by the auxiliary drivebelt.

Repair operations possible with the engine in the car

13 The following operations can be carried out without having to remove the engine from the vehicle.

a) Removal and refitting of the cylinder head.
b) Removal and refitting of the timing cover.
c) Removal and refitting of the timing chain, tensioner and sprockets.
d) Removal and refitting of the camshafts.
e) Removal and refitting of the sump.
f) Removal and refitting of the big-end bearings, connecting rods, and pistons*.
g) Removal and refitting of the oil pump.
h) Removal and refitting of the oil filter housing.
i) Renewal of the crankshaft oil seals.
j) Renewal of the engine mountings.
k) Removal and refitting of the flywheel.

*Although the operation marked with an asterisk can be carried out with the engine in the vehicle (after removal of the sump), it is preferable for the engine to be removed, in the interests of cleanliness and improved access. For this reason, the procedure is described in Chapter 2E.

2 Compression test – description and interpretation

1 When engine performance is down, or if misfiring occurs which cannot be attributed to the ignition or fuel systems, a compression test can provide diagnostic clues as to the engine's condition. If the test is performed regularly, it can give warning of trouble before any other symptoms become apparent.

2 The engine must be fully warmed-up to operating temperature, the oil level must be correct and the battery must be fully-charged.

3 Open the cover on the engine compartment fuse/relay box and remove the fuel pump relay. Refer to the wiring diagrams in Chapter 12 for information on fuse and relay locations. Now start the engine and allow it to run until it stalls.

3.5 Pull out the spring clip and disconnect the crankcase ventilation hose from the inlet manifold

3.6 Open the quick-release connector and disconnect the crankcase ventilation hose from the turbocharger air inlet

3.7 Release the retaining clip and disconnect the vacuum hose from the turbocharger

4 Remove the throttle housing as described in Chapter 4A, Section 8.

5 Remove the spark plugs as described in Chapter 1A, Section 27.

6 Fit a compression tester to the No 1 cylinder spark plug hole – the type of tester which screws into the spark plug thread is preferable.

7 Arrange for an assistant to crank the engine over for several seconds on the starter motor while you observe the compression gauge reading. The compression will build-up fairly quickly in a healthy engine. Low compression on the first stroke, followed by gradually-increasing pressure on successive strokes, indicates worn piston rings. A low compression on the first stroke which does not rise on successive strokes, indicates leaking valves or a blown head gasket (a cracked cylinder head could also be the cause). Deposits on the underside of the valve heads can also cause low compression. Record the highest gauge reading obtained, then repeat the procedure for the remaining cylinders.

8 Due to the variety of testers available, and the fluctuation in starter motor speed when cranking the engine, different readings are often obtained when carrying out the compression test. However, the most important factor is that the compression pressures are uniform in all cylinders, and that is what this test is mainly concerned with.

9 Add some engine oil (about three squirts from a plunger type oil can) to each cylinder

through the spark plug holes, and then repeat the test.

10 If the compression increases after the oil is added, the piston rings are probably worn. If the compression does not increase significantly, the leakage is occurring at the valves or the head gasket. Leakage past the valves may be caused by burned valve seats and/or faces, or warped, cracked or bent valves.

11 If two adjacent cylinders have equally low compressions, it is most likely that the head gasket has blown between them. The appearance of coolant in the combustion chambers or on the engine oil dipstick would verify this condition.

12 If one cylinder is about 20 percent lower than the other, and the engine has a slightly rough idle, a worn lobe on the camshaft could be the cause.

13 On completion of the checks, refit the spark plugs, throttle housing and fuel pump relay.

3 Camshaft cover – removal and refitting

Note: *A new camshaft cover rubber seal will be required for refitting, and a suitable silicone sealant will be required to seal the timing cover-to-cylinder head upper joint.*

Removal

1 Disconnect the battery negative lead as described in Chapter 5A Section 4.

2 Remove the ignition module from the centre of the camshaft cover as described in Chapter 5B, Section 3.

3 Slacken the retaining clips and remove the air cleaner outlet air duct from the air cleaner and turbocharger. Suitably cover the turbocharger air inlet to prevent the possible entry of foreign material.

4 Open the two crankcase ventilation hose retaining clips at the rear of the camshaft cover.

5 Pull out the retaining spring clip and disconnect the crankcase ventilation hose from the inlet manifold **(see illustration)**.

6 Open the quick-release connector and disconnect the other end of the crankcase ventilation hose from the turbocharger air inlet **(see illustration)**. Suitably cover the turbocharger inlet to prevent the possible entry of foreign material.

7 Release the retaining clips and disconnect the vacuum hose from the turbocharger and from the charge air bypass regulator solenoid on the underside of the inlet manifold **(see illustration)**.

8 Lift the crankcase ventilation hose, together with the vacuum hose out of the support clips and remove the two hoses from the engine.

9 Disconnect the wiring connectors at the inlet and exhaust camshaft VVT solenoid valves, inlet and exhaust camshaft position sensors and thermostat. Unclip the wiring harness trough from the camshaft cover and move it to one side **(see illustrations)**.

10 Remove the engine oil level dipstick from the camshaft cover.

3.9a Disconnect the wiring connectors at the inlet and exhaust camshaft VVT solenoid valves...

3.9b...inlet and exhaust camshaft position sensors...

3.9c...and thermostat...

11 Progressively slacken the fifteen camshaft cover retaining bolts until they are all fully unscrewed. Note that the bolts are captive and will remain in place in the cover as it is removed.

12 Lift the camshaft cover up and off the cylinder head **(see illustration)**.

Refitting

13 Remove the old seal from the camshaft cover, then examine the inside of the cover for a build-up of oil sludge or any other contamination, and if necessary clean the cover with paraffin, or a water-soluble solvent. Dry the cover thoroughly before refitting.

14 Fit a new rubber seal to the cover ensuring that it is correctly located in the camshaft cover groove **(see illustration)**.

15 Inspect the joint between the timing cover and cylinder head, and cut off any projecting timing cover gasket using a sharp knife.

16 Thoroughly clean the mating faces of the camshaft cover and cylinder head.

17 Apply a 2 mm diameter bead of silicone sealant to the joint between the timing cover and cylinder head on each side **(see illustration)**.

18 Locate the camshaft cover on the cylinder head and screw in the retaining bolts. Progressively and evenly tighten the retaining bolts securely.

19 Insert the engine oil level dipstick back into its location in the camshaft cover.

20 Reconnect the wiring connectors to the components listed in paragraph 9, then clip the wiring harness trough back into place in the camshaft cover.

21 Refit the crankcase ventilation hose and vacuum hose and secure with the retaining clips, then refit the air cleaner outlet air duct.

22 Refit the ignition module to the centre of the camshaft cover as described in Chapter 5B, Section 3.

23 Reconnect the battery negative terminal on completion.

---

4 Valve timing – checking and adjustment

Note: *Certain special tools will be required for this operation. Read through the entire procedure to familiarise yourself with the work involved, then either obtain the manufacturer's special tools, or suitable alternatives. New gaskets and sealing rings will also be required for all disturbed components.*

General

1 To accurately check, and if necessary adjust, the valve timing, various special tools are required to retain the camshafts, camshaft sprockets, camshaft sensor phase discs and crankshaft in the top dead centre (TDC) position, with No 1 piston on the compression stroke.

2 The Vauxhall/Opel tool numbers are as follows:

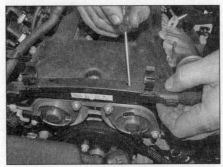

3.9d ...then unclip the wiring harness trough from the camshaft cover

3.14 Fit a new rubber seal to the camshaft cover

Crankshaft TDC positioning tool	EN-952
Camshaft setting tool	EN-953-A
Camshaft sprocket holding tool	EN-49977-200
Camshaft sensor phase disc setting tool	EN-49977-100

3 Although the Vauxhall/Opel tools are specifically designed for the purpose, suitable relatively inexpensive aftermarket alternatives are readily available **(see illustration)**. It will be necessary to use the Vauxhall/Opel tools or the aftermarket alternatives for any work on the timing chain, camshafts or cylinder head.

Checking

4 Disconnect the battery negative lead as described in Chapter 5A Section 4.

5 Remove the air cleaner assembly as described in Chapter 4A, Section 2.

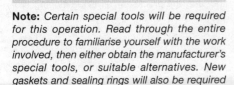

4.3 Typical valve timing checking and adjusting tool set

3.12 Lift the camshaft cover up and off the cylinder head

3.17 Apply a bead of silicone sealant to the joint between the timing cover and cylinder head on each side

6 Remove the camshaft cover as described in Section 3.

7 Firmly apply the handbrake, then jack up the front of the car and support it securely on axle stands (see *Jacking and vehicle support*). Remove the right-hand front roadwheel, then remove the wheel arch liner as described in Chapter 11, Section 21 for access to the crankshaft pulley.

8 Using a socket or spanner on the crankshaft pulley bolt, turn the crankshaft in the normal direction of rotation (clockwise as viewed from the right-hand side of the car) until the hole in the crankshaft pulley is located just before the cast lug on the timing cover **(see illustration)**.

9 Check that No 1 piston is on the compression stroke by observing the offset slots at the left-hand end of the camshafts.

4.8 Turn the crankshaft until the hole in the crankshaft pulley is located just before the cast lug on the timing cover

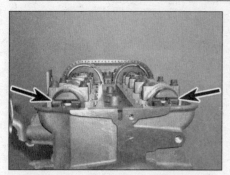

4.9 With No 1 piston on compression, the offset slots in the camshafts should be above the cylinder head mating face

4.10 Undo the closure bolt from the crankshaft TDC position setting hole

4.11 Insert the crankshaft TDC positioning tool into the position setting hole

4.12a Insert the camshaft setting tool into the slots in the ends of the camshafts...

4.12b...ensuring that the tool is inserted fully, up to its stop

The slots should be positioned above the camshaft cover mating face of the cylinder head **(see illustration)**. If they are not, No 1 piston is on the exhaust stroke and the crankshaft should be turned through a further

full turn, stopping once again when the hole in the crankshaft pulley is located just before the cast lug on the timing cover.

10 Undo the closure bolt from the crankshaft TDC position setting hole. The bolt is located

on the front facing side of the cylinder block baseplate, adjacent to the timing cover joint **(see illustration)**. Note that a new closure bolt sealing ring will be required for refitting.

11 Insert the crankshaft TDC positioning tool into the position setting hole and push it into contact with the crankshaft **(see illustration)**. Slowly turn the crankshaft in the normal direction of rotation until the tool engages with the TDC slot in the crankshaft, and moves fully in, up to its stop.

12 Insert the camshaft setting tool into the slots in the left-hand end of the camshafts. Ensure that the tool is inserted fully, up to its stop, to lock both camshafts **(see illustrations)**.

13 If it is not possible to insert the camshaft setting tool, then the valve timing must be adjusted as described in paragraphs 15 to 34 below.

14 If all is satisfactory so far, the position of the camshaft position sensor phase discs should be checked. Using the phase disc setting tool, check to see if the tool will engage with the phase disc on each camshaft, and also seat squarely on the timing cover surface. If it does, proceed to paragraph 31. If adjustment is required, proceed as follows.

Adjustment

Note: *New camshaft sprocket retaining bolts will be required for this operation.*

15 Remove the camshaft sensor phase disc setting tool, but leave the camshaft setting tool and the crankshaft TDC positioning tool in place.

16 Undo the four bolts securing the VVT solenoid valves to the timing cover. Carefully rotate the inlet camshaft solenoid valve anti-clockwise slightly while at the same time pulling it squarely out of the timing cover. Repeat this procedure to remove the exhaust camshaft solenoid valve, but this time, rotate it clockwise during removal **(see illustrations)**. Remove the sealing ring from the rear of each solenoid valve, noting that new sealing rings will be required for refitting.

17 Slacken the sprocket retaining bolts for both camshafts, using a spanner to counterhold each camshaft as the bolts are slackened **(see illustrations)**.

18 Unscrew and remove both camshaft

4.16a Undo the VVT solenoid valve retaining bolts...

4.16b...and remove the solenoid valves

4.17a Slacken the camshaft sprocket retaining bolts...

4.17b...while using a spanner to counterhold each camshaft

sprocket retaining bolts together with the camshaft position sensor phase discs.

19 Fit the phase discs and the new bolts to both camshaft sprockets and tighten them finger tight only at this stage. Check that the phase discs can still be turned.

20 Again, using the spanner on the camshaft flats, turn the camshafts slightly, as necessary, until the camshaft setting tool can be reinserted into the slots in the ends of the camshafts.

21 Undo the two bolts and remove the timing chain sliding rail from the top of the cylinder head **(see illustration)**.

22 Slacken the bolt at the rear of the camshaft sprocket holding tool, place the tool in position on the cylinder head, then fit and tighten the two retaining bolts. Slide the toothed portion of the tool to engage with the inlet camshaft sprocket teeth. Hold the toothed portion fully into engagement with the sprocket teeth and tighten the bolt **(see illustrations)**.

23 Position the camshaft sensor phase discs as shown to allow the phase disc setting tool to locate over them. Fit the phase disc setting tool and secure it to the timing cover with the three bolts **(see illustrations)**. Ensure that the tool correctly engages with the phase discs and seats squarely on the timing cover surface.

24 Tighten the camshaft sprocket retaining bolts to the specified torque, then through the specified angle in the two stages given in the Specifications **(see illustrations)**. The inlet camshaft sprocket bolt should be tightened first. Counterhold the camshafts using the spanner on the camshaft flats as the sprocket bolts are tightened.

25 Remove the camshaft sensor phase disc setting tool and the camshaft sprocket holding tool.

26 Refit the timing chain sliding rail to the top of the cylinder head and secure with the two bolts, tightened to the specified torque.

27 Remove the camshaft setting tool and the crankshaft TDC positioning tool.

28 Turn the crankshaft through two complete revolutions, stopping just before the hole in the crankshaft pulley aligns with the cast lug on the timing cover **(see illustration 4.8)**.

4.21 Undo the two bolts and remove the timing chain sliding rail

4.22b Slide the toothed portion of the tool to engage with the inlet camshaft sprocket teeth

29 Slowly turn the crankshaft further until the TDC position setting tool can once again be inserted to lock the crankshaft.

4.23a Correct positioning of the camshaft sensor phase discs...

4.22a Slacken the bolt at the rear of the camshaft sprocket holding tool and place the tool in position

4.22c Hold the toothed portion fully into engagement with the sprocket teeth and tighten the bolt

30 It should now be possible to re-insert the camshaft setting tool into the slots in the camshafts, and to fit the camshaft sensor

4.23b...to allow the phase disc setting tool to locate over them

4.23c Fit the phase disc setting tool and secure it to the timing cover with the three bolts

4.24a Tighten the camshaft sprocket retaining bolts to the specified torque...

4.24b...then through the specified angle

4.32 Correct positioning of the camshaft VVT solenoid valves

5.2 Turn the tensioner clockwise and insert a suitable locking pin/bolt through the special hole provided

phase disc setting tool over the discs. If this is not possible repeat the entire adjustment procedure.

31 If all is satisfactory, remove all the setting/ positioning tools and refit the cylinder block closure bolt using a new sealing ring. Tighten the closure bolt to the specified torque.

32 Locate a new sealing ring on each camshaft VVT solenoid valve. Lubricate the sealing rings with clean engine oil and carefully insert them into their locations in the timing cover. Position the solenoid valves as shown, then insert the four retaining bolts and tighten them securely **(see illustration)**.

33 Refit the wheel arch liner and the roadwheel. Tighten the roadwheel nuts to the specified torque, then lower the vehicle to the ground.

34 Refit the camshaft cover as described in Section 3.

35 Refit the air cleaner assembly as described in Chapter 4A, Section 2, then reconnect the battery negative terminal.

To make a sprocket holding tool, obtain two lengths of steel strip about 6 mm thick by about 30 mm wide or similar, one 600 mm long, the other 200 mm long (all dimensions are approximate). Bolt the two strips together to form a forked end, leaving the bolt slack so that the shorter strip can pivot freely. At the other end of each 'prong' of the fork, drill a suitable hole and fit a nut and bolt to allow the tool to engage with the spokes in the sprocket.

5 Crankshaft pulley – removal and refitting

Removal

1 Firmly apply the handbrake, then jack up the front of the car and support it securely on axle stands (see *Jacking and vehicle support*). Remove the right-hand roadwheel, then remove the wheel arch liner as described in Chapter 11, Section 21.

2 Using a socket or spanner on the pulley centre bolt, turn the auxiliary drivebelt tensioner clockwise against the spring tension. Hold the tensioner in this position by inserting a suitable locking pin/bolt through the special hole provided **(see illustration)**.

3 With the drivebelt tension released, slip the belt off the crankshaft pulley. Note that it is not necessary to completely remove the drivebelt, as this would entail removal of the right-hand engine mounting.

4 Using a socket or spanner on the crankshaft pulley bolt, turn the crankshaft in the normal direction of rotation (clockwise as viewed from the right-hand side of the car) until the hole in the crankshaft pulley is aligned with the cast lug on the timing cover **(see illustration 4.8)**.

5 It will now be necessary to hold the crankshaft pulley to enable the retaining bolt to be removed. Vauxhall/Opel special tools EN-49979 and EN-956-1 are available for this purpose, however, a home-made tool can easily be fabricated **(see Tool Tip)**.

5.6a Unscrew the crankshaft pulley retaining bolt...

6 Using the holding tool to prevent rotation of the crankshaft, slacken the pulley retaining bolt. Remove the holding tool, unscrew the retaining bolt and remove the pulley **(see illustrations)**. Note that a new pulley retaining bolt will be required for refitting.

Refitting

7 Align the hole in the crankshaft pulley with the cast lug on the timing cover and locate the pulley on the crankshaft. Note that the pulley flange must engage with the hexagon of the oil pump vane rotor and the two flats of the crankshaft. It may be necessary to carefully reposition the oil pump vane rotor slightly using a screwdriver, until the pulley flange will fully engage. When the pulley is fully engaged, the distance from the pulley inner rim to the cast lug on the timing cover should be no more than 5.5 mm.

8 Fit the new pulley retaining bolt and tighten it to the specified torque, then through the specified angle while preventing crankshaft rotation using the method employed on removal.

9 Refit the auxiliary drivebelt over the crankshaft pulley and ensure that it is correctly seated in the other pulleys.

10 Turn back the tensioner and remove the locking pin/bolt then release it, making sure that the drivebelt ribs locate correctly on each of the pulley grooves.

11 Refit the wheel arch liner lower cover and the roadwheel and tighten the wheel nuts to the specified torque.

12 Lower the car to the ground.

6 Timing cover and chain – removal and refitting

Note: *The special tools described in Section 4 will also be required for this operation. Read through the entire procedure and also the procedures contained in Section 4 to familiarise yourself with the work involved, then either obtain the manufacturer's special tools or suitable alternatives. New gaskets and sealing rings will also be required for all disturbed components, together with new camshaft sprocket retaining bolts, crankshaft pulley retaining bolt and right-hand engine*

5.6b ...and remove the pulley

6.6 Air conditioning compressor mounting bracket retaining bolts

6.13 Undo the upper and lower mounting bolts and remove the auxiliary drivebelt tensioner

6.17 Undo the fifteen retaining bolts and remove the timing cover from the engine

mounting bracket retaining bolts. A tube of silicone sealant will be needed to seal the joint between the cylinder block and cylinder head.

Removal

1 Carry out the operations described in Section 4, paragraphs 1 to 12 to set the engine at TDC for No 1 piston on the compression stroke.

2 Drain the cooling system as described in Chapter 1A, Section 25. Tighten the drain plug after draining the system.

3 Position a container beneath the engine sump, then unscrew the drain plug and drain the engine oil. Clean, refit and tighten the drain plug on completion.

4 Using a socket or spanner on the pulley centre bolt, turn the auxiliary drivebelt tensioner clockwise against the spring tension. Hold the tensioner in this position by inserting a suitable locking pin/bolt through the special hole provided **(see illustration 5.2)**. Note the routing of the drivebelt, then remove it from the pulleys – mark the drivebelt for fitted direction. **Note:** *The drivebelt cannot be removed completely until the right-hand engine mounting has been removed.*

5 Disconnect the wiring from the air conditioning compressor and release the wiring harness from the cable clips. Unbolt the compressor from the cylinder block bracket and support the compressor clear of the engine. Do not disconnect the refrigerant lines from the compressor.

6 Undo the three retaining bolts and remove the air conditioning compressor mounting bracket from the cylinder block and timing cover **(see illustration)**.

7 Remove the crankshaft pulley as described in Section 5.

8 Remove the alternator as described in Chapter 5A, Section 8.

9 Remove the sump as described in Section 11.

10 The engine must now be supported while the right-hand engine mounting is removed. To do this, use a hoist attached to the top of the engine, or make up a wooden frame to locate on the crankcase and use a trolley jack.

11 Remove the right-hand engine mounting as described in Section 16, then remove the auxiliary drivebelt.

12 Undo the three bolts and remove the right-hand engine mounting bracket from the timing cover. Note that new bolts will be required for refitting.

13 Undo the upper and lower mounting bolts and remove the auxiliary drivebelt tensioner from the timing cover **(see illustration)**.

14 Remove the coolant pump as described in Chapter 3, Section 8.

15 Undo the four bolts securing the VVT solenoid valves to the timing cover. Carefully rotate the inlet camshaft solenoid valve anti-clockwise slightly while at the same time pulling it squarely out of the timing

cover. Repeat this procedure to remove the exhaust camshaft solenoid valve, but this time, rotate it clockwise during removal **(see illustrations 4.16a and 4.16b)**. Remove the sealing ring from the rear of each solenoid valve, noting that new sealing rings will be required for refitting.

16 Slacken the sprocket retaining bolts for both camshafts, using a spanner to counterhold each camshaft as the bolts are slackened **(see illustrations 4.17a and 4.17b)**.

17 Undo the fifteen timing cover retaining bolts and remove the timing cover from the engine **(see illustration)**. The cover will be initially tight as it is located on dowels and secured by sealant. If necessary, gently tap it off using a soft-faced mallet.

18 Fully push back the timing chain tensioner plunger and secure it in the released position by inserting a 2 mm diameter drill or roll pin in the hole on the tensioner body **(see illustration)**.

19 Undo the two bolts and remove the timing chain sliding rail from the top of the cylinder head **(see illustrations)**.

20 Undo the two bolts and remove the timing chain (front) guide rail from the cylinder block **(see illustrations)**.

21 Undo the lower pivot bolt and remove the timing chain (rear) tension rail from the cylinder block **(see illustrations)**.

22 Slide the drive sprocket and timing chain

6.18 Secure the timing chain tensioner plunger in the released position by inserting a drill or roll pin in the tensioner body hole

6.19a Undo the two bolts...

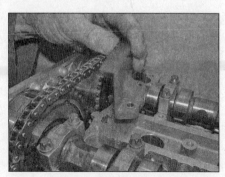

6.19b ...and remove the timing chain sliding rail

6.20a Undo the two bolts...

6.20b...and remove the timing chain (front) guide rail

6.21a Undo the lower pivot bolt...

6.21b...and remove the timing chain (rear) tension rail from the cylinder block

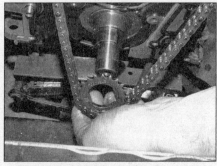

6.22a Slide the drive sprocket and timing chain off the crankshaft...

6.22b...then lift the chain off the camshaft sprockets

off the crankshaft, then lift the chain off the camshaft sprockets **(see illustrations)**.

23 Remove the composite gasket from the cylinder block baseplate, cylinder block, and cylinder head, using a plastic spatula if necessary to release the sealant. Note that if the cylinder head has ever been previously removed, the gasket will be a two-piece type, with a split at the cylinder head-to cylinder block joint. A new gasket will be required for refitting.

24 Thoroughly clean the timing cover and remove all traces of gasket and sealant from all the mating surfaces. Similarly clean the cylinder block baseplate, cylinder block and cylinder head mating surfaces. Ensure that all traces of old sealant are removed, particularly from the area of the cylinder head-to-cylinder block joint.

25 Inspect the timing chain, sprockets, sliding rail, guide rail and tension rail for any sign of wear or deformation, and renew any suspect components as necessary. Renew the crankshaft timing chain end oil seal in the timing cover as a matter of course using the procedures described in Section 13.

26 It is advisable to check the condition of the timing chain tensioner at this stage, as described in Section 7.

27 Obtain all new gaskets and components as necessary ready for refitting.

Refitting

28 Apply a 2 mm diameter bead of silicone sealant to the joint between the cylinder block and cylinder head on each side. The bead should be long enough to fill the joint in the area covered by the timing cover gasket.

Similarly, apply silicone sealant to the joint between the cylinder block and baseplate.

29 Ensure that the locating dowels are in position, then locate the new timing cover gasket in place on the cylinder head, cylinder block and baseplate **(see illustration)**.

30 Engage the timing chain with the camshaft sprockets, then place the drive sprocket in the chain with the markings facing outward. Slide the drive sprocket onto the crankshaft **(see illustration)**.

31 Place the timing chain tension rail in position, refit the lower pivot bolt and tighten the bolt to the specified torque **(see illustration)**.

32 Attach the timing chain guide rail to the cylinder block and secure with the two bolts tightened securely **(see illustration)**.

33 Remove the roll pin used to secure the

6.29 Locate the new timing cover gasket in place on the cylinder head, cylinder block and baseplate

6.30 Engage the timing chain with the camshaft sprockets, then place the drive sprocket in the chain

6.31 Place the timing chain tension rail in position, refit the lower pivot bolt and tighten it to the specified torque

timing chain tensioner plunger in the released position **(see illustration)**.

34 Locate the timing cover in position and refit all the retaining bolts, finger tight only at this stage. Note that the two M10 bolts are fitted at the lower right-hand side of the timing cover and the longer M6 bolt is fitted adjacent to the crankshaft pulley location **(see illustration)**. Now tighten all the bolts progressively to the specified torque.

35 Align the hole in the crankshaft pulley with the cast lug on the timing cover and locate the pulley on the crankshaft. Note that the pulley flange must engage with the hexagon of the oil pump rotor and the two flats of the crankshaft. It may be necessary to carefully reposition the oil pump rotor slightly using a screwdriver, until the pulley flange will fully engage. When the pulley is fully engaged, the distance from the pulley inner rim to the cast lug on the timing cover should be no more than 5.5 mm.

36 Fit the new pulley retaining bolt and tighten it to the specified torque and through the specified angle while preventing crankshaft rotation using the method employed on removal.

37 Locate a new sealing ring on each camshaft VVT solenoid valve. Lubricate the sealing rings with clean engine oil and carefully insert them into their locations in the timing cover. Position the solenoid valves as shown, then insert the four retaining bolts and tighten them securely **(see illustration 4.32)**.

38 Adjust the valve timing as described in Section 4, paragraphs 18 to 31.

39 Refit the sump as described in Section 11.

40 Refit the coolant pump as described in Chapter 3, Section 8.

41 Locate the auxiliary drivebelt tensioner on the timing cover and secure with the two bolts tightened to the specified torque.

42 Refit the right-hand engine mounting bracket to the timing cover and secure with the three new bolts, tightened to the specified torque, then through the specified angle.

43 Refit the alternator as described in Chapter 5A, Section 8.

44 Refit the air conditioning compressor mounting bracket to the cylinder block and timing cover and secure with the three bolts, tightened to the specified torque.

45 Refit the air conditioning compressor to the mounting bracket, then fit and tighten the mounting bolts to the specified torque. Reconnect the wiring and secure the wiring harness with the cable clips.

46 Locate the auxiliary drivebelt onto the pulleys in the correct routing **(see illustration)**. If the drivebelt is being re-used, make sure it is fitted the correct way around.

47 Turn back the tensioner and remove the locking pin/bolt then release it, making sure that the drivebelt ribs locate correctly on each of the pulley grooves.

48 Refit the right-hand engine mounting as described in Section 16.

49 Refit the camshaft cover as described in Section 3.

6.32 Attach the timing chain guide rail to the cylinder block and tighten the two bolts securely

6.34 Correct location for the longer M6 bolt

50 Refit the air cleaner assembly as described in Chapter 4A, Section 2.

51 Refit the wheel arch liner and the roadwheel and tighten the wheel nuts to the specified torque.

52 Lower the car to the ground and reconnect the battery negative terminal.

53 Refill the engine with fresh oil as described in Chapter 1A, Section 5.

54 Refill the cooling system as described in Chapter 1A, Section 25.

7 Timing chain tensioner – removal, inspection and refitting

Removal

1 Remove the timing cover and chain as described in Section 6.

2 Undo the two tensioner retaining bolts and remove the tensioner from the cylinder head **(see illustration)**.

3 Hold the tensioner plunger, then extract the roll pin used to retain the plunger in the retracted position. Withdraw the tensioner plunger and spring.

Inspection

4 Examine the components for any sign of wear, deformation or damage and, if evident, renew the complete tensioner assembly.

Refitting

5 Lubricate the tensioner spring and plunger,

6.33 Remove the drill bit or roll pin used to secure the timing chain tensioner plunger in the released position

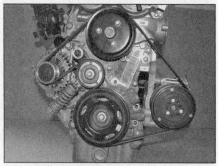

6.46 Auxiliary drivebelt routing

then insert the spring, followed by the plunger into the tensioner body.

6 Fully compress the tensioner plunger and reinsert the retaining roll pin.

7 Refit the tensioner assembly to the cylinder head and tighten the retaining bolts securely.

8 Refit the timing chain and cover as described in Section 6.

8 Camshaft sprockets – removal and refitting

Note: *The special tools described in Section 4 will also be required for this operation. Read through the entire procedure and also the procedures contained in Section 4 to familiarise yourself with the work involved,*

7.2 Timing chain tensioner retaining bolts

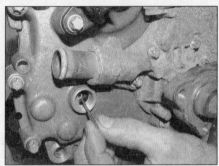

8.3 Unscrew the timing chain tensioner closure bolt from the timing cover

8.6a Insert a 2 mm diameter drill bit or roll pin through the closure bolt aperture...

8.6b ...and into the hole in the tensioner body

then either obtain the manufacturer's special tools, or use the alternatives described. New gaskets and sealing rings will also be required for all disturbed components, together with new camshaft sprocket retaining bolts.

Removal

1 Carry out the operations described in Section 4, paragraphs 1 to 12 to set the engine at TDC for No 1 piston on the compression stroke.

2 Undo the four bolts securing the VVT solenoid valves to the timing cover. Carefully rotate the inlet camshaft solenoid valve anti-clockwise slightly while at the same time pulling it squarely out of the timing cover. Repeat this procedure to remove the exhaust camshaft solenoid valve, but this time, rotate it clockwise during removal **(see illustrations 4.16a and 4.16b)**. Remove the sealing ring from the rear of each solenoid valve, noting that new sealing rings will be required for refitting.

3 Unscrew the timing chain tensioner closure bolt from the front of the timing cover, below the heater hose union on the coolant pump **(see illustration)**. Note that a new closure bolt sealing ring will be required for refitting.

4 Obtain a suitable roll pin or similar of 2 mm diameter and approximately 30 mm long to use as a timing chain tensioner locking tool.

5 Using a suitable spanner engaged with the flats provided on the inlet camshaft, apply tension in a clockwise direction (as viewed from the right-hand side of the car) to the camshaft, to take up any slack in the timing

chain. This will push the timing chain tensioner plunger fully into its bore.

6 Hold the camshaft in this position and retain the tensioner plunger in the released position by inserting the roll pin through the closure bolt aperture and into the hole on the tensioner body **(see illustrations)**.

7 Undo the two bolts and remove the timing chain sliding rail from the top of the cylinder head **(see illustration 4.21)**.

8 Slacken the sprocket retaining bolts for both camshafts, using the spanner to counterhold each camshaft as the bolts are slackened **(see illustrations 4.17a and 4.17b)**.

9 Unscrew and remove both camshaft sprocket retaining bolts together with the camshaft position sensor phase discs.

10 Withdraw both sprockets from the camshaft, then disengage the relevant sprocket from the timing chain and remove it from the engine.

Refitting

11 Engage the relevant sprocket with the timing chain then locate both sprockets on the camshafts. Fit the new retaining bolts together with the phase discs. Tighten the bolts finger tight only at this stage and check that the phase discs can still be turned.

12 Remove the roll pin used to hold the timing chain tensioner plunger in the retracted position.

13 Adjust the valve timing as described in Section 4, paragraphs 20 to 35.

9 Camshafts, hydraulic tappets and rocker arms – removal and refitting

Removal

1 Remove both camshaft sprockets as described in Section 8.

2 Observe the identification numbers and markings on the camshaft bearing caps **(see illustration)**. On the project car used during the compilation of this manual, the bearing caps with odd numbers were fitted to the exhaust camshaft, and the caps with the even numbers were fitted to the inlet camshaft. However, this may not be the case on other engines. Also, as it is possible to fit the caps either way round, it will be necessary to mark the caps with quick-drying paint, or identify them in some way, so that they can be refitted in exactly the same position. On the project car, all the numbers could be read the correct way up, when viewed from the exhaust camshaft side of the engine. Again, this may not always be the case.

3 With the bearing caps correctly identified, initially slacken the bearing cap bolts, one at a time, by half a turn, in the sequence shown **(see illustration)**. When all the bolts have been initially slackened, repeat the procedure, slackening the bolts by a further half a turn. Continue until all the bolts have been fully slackened. The camshaft will rise up under

9.2 Camshaft bearing cap identification numbers

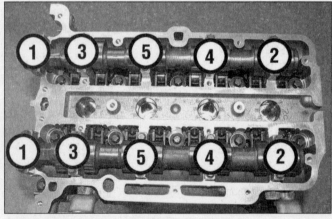

9.3 Camshaft bearing cap retaining bolt slackening sequence

9.4 Remove the retaining bolts and lift off the camshaft bearing caps

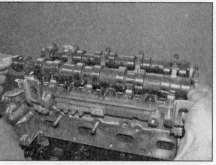

9.5 Carefully lift the camshafts from their locations in the cylinder head

9.6a Withdraw each rocker arm...

the action of the valve springs as the bolts are slackened. Ensure that the camshaft rises uniformly and does not jam in its bearings.

4 When all the bolts have been slackened, remove the bolts and lift off the bearing caps, keeping them in order according to the identification method decided on **(see illustration)**.

5 Carefully lift the camshafts from their locations in the cylinder head. If both camshafts are removed, identify them as exhaust and inlet **(see illustration)**.

6 Obtain sixteen small, clean plastic containers, and number them inlet 1 to 8 and exhaust 1 to 8; alternatively, divide a larger container into sixteen compartments and number each compartment accordingly. Fill each container, or each compartment with clean engine oil so that the hydraulic tappets will be completely submerged. Withdraw each rocker arm and hydraulic tappet in turn, and place them in their respective container **(see illustrations)**. Do not interchange the rocker arms and tappets, or the rate of wear will be much increased.

Inspection

7 Examine the camshaft bearing surfaces and cam lobes for signs of wear ridges and scoring. Renew the camshaft if any of these conditions are apparent. Examine the condition of the bearing surfaces, both on the camshaft journals and in the cylinder head/bearing caps. If the head bearing surfaces are worn excessively, the cylinder head will need to be renewed.

8 Examine the rocker arm and hydraulic tappet bearing surfaces for wear ridges and scoring. Renew any rocker arm or tappet on which these conditions are apparent.

9 If either camshaft is being renewed, it will be necessary to renew all the rocker arms and tappets for that particular camshaft also.

Refitting

10 Before refitting, thoroughly clean all the components and the cylinder head and bearing cap journals.

11 Liberally oil the cylinder head hydraulic tappet bores and the tappets. Carefully refit the tappets to the cylinder head, ensuring that each tappet is refitted to its original bore.

9.6b...and hydraulic tappet in turn...

12 Lay each rocker arm in position over its respective tappet.

13 Liberally oil the camshaft bearings in the cylinder head and the camshaft lobes, then place the camshafts in the cylinder head.

14 Position the camshafts so that the offset slot in the left-hand end of each camshaft is above the camshaft cover mating face of the cylinder head. The indentations between the camshaft lobes should be facing toward the centre of the engine **(see illustration)**.

15 Refit all the bearing caps to their respective locations ensuring they are fitted the correct way round as noted during removal.

16 Initially tighten the bearing cap bolts, one at a time, by half a turn, in the sequence shown **(see illustration)**. When all the bolts have been initially tightened, repeat the

9.6c...and place them in their respective containers

procedure, tightening the bolts by a further half a turn. Continue until all the bearing caps are in contact with the cylinder head and the bolts are lightly tightened.

17 Again, working in the sequence shown, tighten all the bolts to the specified torque.

18 Refit the camshaft sprockets as described in Section 8.

10 Cylinder head – removal and refitting

Note: *The engine must be cold when removing the cylinder head. A new cylinder head gasket, timing cover gasket, cylinder head bolts, camshaft sprocket bolts, right-hand engine mounting bracket bolts, together with seals*

9.14 The indentations between the camshaft lobes should be facing toward the centre of the engine

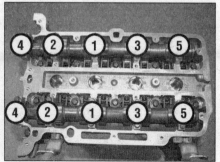

9.16 Camshaft bearing cap retaining bolt tightening sequence

10.8 Extract the retaining spring clip and disconnect the cooling system air bleed hose from the coolant outlet elbow

10.9a Release the retaining clips and disconnect the oil cooler inlet hose and radiator inlet hose...

10.9b ...and the heater matrix inlet hose from the coolant outlet elbow

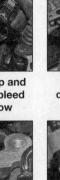

10.11 Undo the retaining bolt and disconnect the cylinder head earth lead

10.12 Extract the retaining spring clip and disconnect the radiator outlet hose from the thermostat housing

and sealing rings will be required for refitting. A suitable sealant will also be needed to seal the timing cover-to-cylinder block joint.

Note: *The special tools described in Section 4 will also be required for this operation. Read through the entire procedure and also the procedures contained in Section 4 to familiarise yourself with the work involved, then either obtain the manufacturer's special tools, or use the alternatives described.*

Removal

1 Disconnect the battery negative lead as described in Chapter 5A Section 4.
2 Remove the air cleaner assembly as described in Chapter 4A, Section 2.
3 Remove the camshaft cover as described in Section 3.

4 Firmly apply the handbrake, then jack up the front of the car and support it securely on axle stands (see *Jacking and vehicle support*). Remove the right-hand front roadwheel then remove the wheel arch liner as described in Chapter 11, Section 21.
5 Drain the cooling system as described in Chapter 1A, Section 25.
6 Remove the inlet manifold as described in Chapter 4A, Section 13.
7 Remove the exhaust manifold as described in Chapter 4A, Section 18.
8 Extract the retaining spring clip and disconnect the cooling system air bleed hose from the coolant outlet elbow on the left-hand side of the cylinder head **(see illustration)**.
9 Release the retaining clips and disconnect the oil cooler inlet hose, radiator inlet hose

and heater matrix inlet hose from the coolant outlet elbow **(see illustrations)**.
10 Disconnect the wiring connector from the coolant temperature sensor on the coolant outlet elbow.
11 Undo the retaining bolt and disconnect the earth lead from the left-hand end of the cylinder head **(see illustration)**.
12 Extract the retaining spring clip and disconnect the radiator outlet hose from the thermostat housing **(see illustration)**.
13 Using a socket or spanner on the pulley centre bolt, turn the auxiliary drivebelt tensioner clockwise against the spring tension. Hold the tensioner in this position by inserting a suitable locking pin/bolt through the special hole provided **(see illustration 5.2)**. Note the routing of the drivebelt, then remove it from the pulleys – mark the drivebelt for fitted direction.
Note: *The drivebelt cannot be removed completely until the right-hand engine mounting has been removed.*
14 The engine must now be supported while the right-hand engine mounting is removed. To do this, use a hoist attached to the top of the engine, or make up a wooden frame to locate beneath the sump and use a trolley jack.
15 Remove the right-hand engine mounting as described in Section 16, then remove the auxiliary drivebelt.
16 Undo the three bolts and remove the right-hand engine mounting bracket from the timing cover. Note that new bolts will be required for refitting.
17 Hold the coolant pump shaft with a suitable spanner, undo the three coolant pump pulley retaining bolts and remove the pulley **(see illustrations)**.
18 Undo the bolts securing the coolant pump and timing cover to the cylinder head **(see illustration)**. Note that it is not necessary to remove all the coolant pump bolts as five are shorter than the rest, and only secure the pump to the timing cover.
19 Remove the camshaft sprockets from the camshafts as described in Section 8. Leave the timing chain engaged and move the chain and both sprockets into the timing cover.

10.17a Hold the coolant pump shaft and undo the three coolant pump pulley retaining bolts...

10.17b...then remove the pulley

20 Working in the specified sequence, progressively slacken the cylinder head retaining bolts half a turn at a time until all the bolts are loose **(see illustration)**. Remove the bolts from their locations noting that new bolts will be required for refitting.

21 Slightly raise the cylinder head so it just clears the cylinder block face and move the head toward the transmission end of the engine. The head will be initially tight due to the sealant on the timing cover gasket and head gasket. Note that the locating dowel holes on the cylinder head are elongated to allow the head to move sideways slightly.

22 As soon as sufficient clearance exists, lift the cylinder head up and off the cylinder block. At the same time, release the timing chain guide rail from the peg on the cylinder head, and guide the chain tensioner clear of the tensioning rail. Check that the tensioner locking pin is not dislodged as the cylinder head is lifted up. Place the cylinder head on wooden blocks after removal to avoid damage to the valves. Recover the cylinder head gasket.

Preparation for refitting

23 The mating faces of the cylinder head and cylinder block must be perfectly clean before refitting the head. Scouring agents are available for this purpose, but acceptable results can be achieved by using a hard plastic or wood scraper to remove all traces of gasket and carbon. The same method can be used to clean the piston crowns. Take particular care to avoid scoring or gouging the cylinder head mating surfaces during the cleaning operations, as aluminium alloy is easily damaged. Make sure that the carbon is not allowed to enter the oil and water passages – this is particularly important for the lubrication system, as carbon could block the oil supply to the engine's components. Using adhesive tape and paper, seal the water, oil and bolt holes in the cylinder block. To prevent carbon entering the gap between the pistons and bores, smear a little grease in the gap. After cleaning each piston, use a small brush to remove all traces of grease and carbon from the gap, then wipe away the remainder with a clean rag.

24 Check the mating surfaces of the cylinder block and the cylinder head for nicks, deep scratches and other damage. If slight, they may be removed carefully with a file, but if excessive, machining may be the only alternative to renewal. If warpage of the cylinder head gasket surface is suspected, use a straight-edge to check it for distortion. Refer to Part E of this Chapter if necessary.

25 Thoroughly clean the threads of the cylinder head bolt holes in the cylinder block. Ensure that the bolts run freely in their threads, and that all traces of oil and water are removed from each bolt hole.

26 Using a sharp knife, partially cut through the timing cover gasket flush with the top of the cylinder block. Release the gasket from the timing cover and bend it in half to break

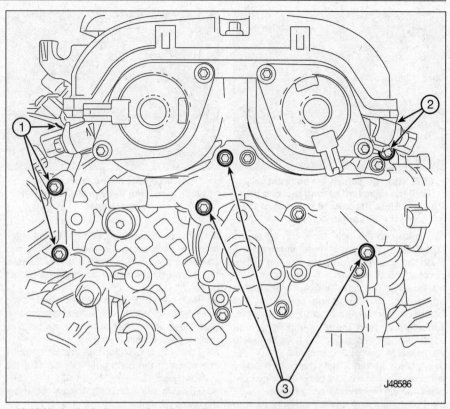

10.18 Undo the bolts securing the coolant pump and timing cover to the cylinder head

1 and 2 Timing cover retaining bolts   *3 Coolant pump retaining bolts*

it off at the cut line. Remove the upper part of the gasket, and thoroughly clean the mating surface, paying particular attention to the cylinder block edge where it contacts the timing cover.

27 Before refitting the cylinder head, using the spanner on the camshaft flats, turn the camshafts slightly, as necessary, until the camshaft setting tool described in Section 4 can be inserted into the slots in the ends of the camshafts **(see illustration)**.

Refitting

28 Apply a 2 mm diameter bead of silicone sealant to the joint between the cylinder block and the timing cover on each side.

29 Check that the locating dowels are in position in the cylinder block, then lay the

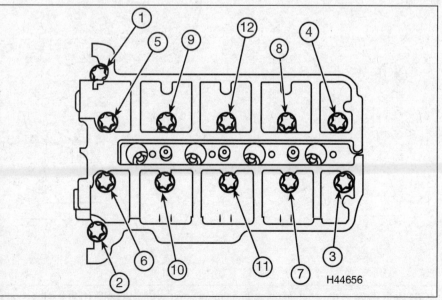

10.20 Cylinder head bolt slackening sequence

10.27 Turn the camshafts as necessary, until the camshaft setting tool can be inserted into the slots in the ends of the camshafts

10.29 Lay the new gasket on the block face, with the words OBEN/TOP uppermost

new gasket on the block face, with the words OBEN/TOP uppermost **(see illustration)**. Push the gasket hard up against the timing cover so that it engages with the sealant.

30 Position the new timing cover gasket upper part on the timing cover, so that its lower ends engage with the sealant. Temporarily insert the left- and right-hand upper timing cover mounting bolts to locate the gasket in the correct position.

31 Apply a further 2 mm diameter bead of silicone sealant to the joint between the cylinder block and the timing cover on each side.

32 Carefully lower the cylinder head into position on the gasket, guiding the chain tensioner past the tensioning rail and guiding the tensioner locking pin into the timing cover

access hole. Check also that the timing chain guide rail engages with cylinder head peg.

33 Once the head is seated on its dowels, tap it towards the timing cover with a rubber mallet.

34 Refit three of the coolant pump and timing cover retaining bolts in the locations shown **(see illustration)**. Tighten the three bolts to the specified torque.

35 Fit the new cylinder head retaining bolts and screw in the bolts until they contact the cylinder head.

36 Remove the camshaft setting tool from the camshaft slots.

37 Working in the reverse order to the loosening sequence shown earlier in this Section, tighten the cylinder head bolts to

the Stage one torque setting given in the Specifications, using a torque wrench. Again working in the correct order, tighten all the bolts through the Stage two angle using an angle measuring gauge.

38 Slacken the three previously-fitted bolts securing the coolant pump and timing cover to the cylinder head.

39 Refit the remaining coolant pump and timing cover retaining bolts and tighten all the bolts to the specified torque.

40 Using the spanner on the camshaft flats, turn the camshafts slightly, as necessary, until the camshaft setting tool can once again be inserted into the camshaft slots.

41 Refit the coolant pump pulley and secure with the three bolts, tightened to the specified torque.

42 Refit the right-hand engine mounting bracket to the timing cover and secure with the three new bolts, tightened to the specified torque, then through the specified angle.

43 Locate the auxiliary drivebelt onto the pulleys in the correct routing **(see illustration 6.46)**. If the drivebelt is being re-used, make sure it is fitted the correct way around.

44 Turn back the tensioner and remove the locking pin/bolt then release it, making sure that the drivebelt ribs locate correctly on each of the pulley grooves.

45 Refit the right-hand engine mounting as described in Section 16.

46 Engage the camshaft sprockets with their respective camshafts and fit the new retaining bolts together with the phase discs. Tighten the bolts finger tight only at this stage and check that the phase discs can still be turned.

47 Remove the roll pin used to hold the timing chain tensioner plunger in the retracted position.

48 Adjust the valve timing as described in Section 4, paragraphs 20 to 32.

49 Connect the radiator outlet hose to the thermostat housing and secure with the retaining spring clip.

50 Refit the earth lead to the left-hand end of the cylinder head and tighten the retaining bolt securely.

51 Reconnect the wiring connector to the coolant temperature sensor on the coolant outlet elbow.

52 Connect the heater matrix inlet hose, radiator inlet hose and oil cooler inlet hose to the coolant outlet elbow and secure with the retaining clips.

53 Connect the cooling system air bleed hose to the coolant outlet elbow and secure with the retaining spring clip.

54 Refit the exhaust manifold as described in Chapter 4A, Section 18.

55 Refit the inlet manifold as described in Chapter 4A, Section 13.

56 Refit the camshaft cover as described in Section 3.

57 Refit the air cleaner assembly as described in Chapter 4A, Section 2, then reconnect the battery negative terminal.

58 Refill the cooling system as described in Chapter 1A, Section 25.

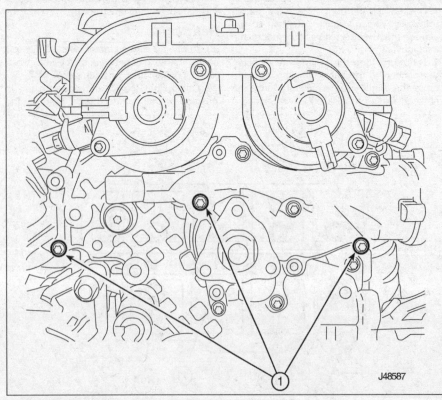

10.34 Refit three of the coolant pump and timing cover retaining bolts (1) in the locations shown

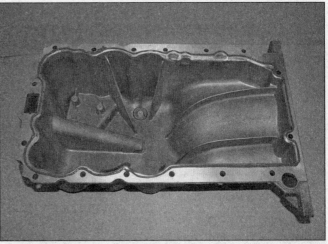

11.12 Screw alignment studs into one of the retaining bolt holes on each side of the cylinder block baseplate

11.13 Apply a bead of silicone sealant to the sump mating face, close to the inner edge of the sump and on the inside of the bolt holes

11 Sump –
removal and refitting

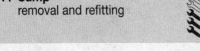

Removal

1 Firmly apply the handbrake, then jack up the front of the car and support it securely on axle stands (see *Jacking and vehicle support*).
2 Drain the engine oil, with reference to Chapter 1A, Section 5 then refit and tighten the sump drain plug.
3 Remove the exhaust system front pipe as described in Chapter 4A, Section 20.
4 On remove the catalytic converter as described in Chapter 4A, Section 20.
5 On 4WD models, remove the transfer gearbox and transfer gearbox adapter as described in Chapter 7C, Section 3.
6 Undo the three bolts securing the sump to the transmission bellhousing.
7 Undo the sixteen bolts securing the sump to the cylinder block baseplate and timing cover.
8 Engage a screwdriver or similar tool in the slot between the sump and cylinder block baseplate. The slot is located on the front facing side of the sump, at the transmission end. Carefully lever the sump away from the baseplate to release it from the sealant, then manoeuvre the sump out from under the car.
9 To remove the oil baffle plate, undo the retaining bolts and remove the baffle plate from the cylinder block baseplate.

Refitting

10 Thoroughly clean the inside and outside of the sump ensuring that all traces of old sealant are removed from the mating face. Also clean the cylinder block baseplate mating face and the groove in the timing cover to remove all traces of old sealant.
11 If removed, refit the oil baffle plate and secure with the retaining bolts tightened securely.

12 Obtain two M6 studs, approximately 30 mm long to act as alignment pins when offering the sump into position. Cut a slot in the end of each pin to allow them to be unscrewed after the sump is in position. Wrap masking tape around the stud so it is a snug fit in the sump retaining bolt hole. Alternatively, use heat-shrink film. Screw each stud into one of the retaining bolt holes on each side of the cylinder block baseplate **(see illustration)**.
13 Apply a 2 mm continuous bead of silicone sealing compound (available from your Vauxhall/Opel dealer) to the sump mating face. The bead should be close to the inner edge of the sump and on the inside of the retaining bolt holes **(see illustration)**.
14 Apply an additional 2 mm bead of sealant to the groove in the timing cover, and around the bolt hole at the rear of the cylinder block baseplate **(see illustrations)**.
15 Locate the sump in position on the cylinder block baseplate, guiding it over the two centring pins.
16 Refit two of the sump retaining bolts to each side of the sump and tighten them sufficiently to hold the sump in place. Remove the two centring pins, then refit the remaining bolts securing the sump to the timing cover and cylinder block baseplate.

17 Refit the three bolts securing the sump to the transmission bellhousing and tighten them to the specified torque. Now tighten the bolts securing the sump to the timing cover and cylinder block baseplate to the specified torque.
18 On 4WD models, refit the transfer gearbox adapter and transfer gearbox as described in Chapter 7C, Section 3.
19 Refit the catalytic converter as described in Chapter 4A, Section 20.
20 Refit the exhaust system front pipe as described in Chapter 4A, Section 20.
21 Refit the roadwheel, then lower the car to the ground and tighten the wheel nuts to the specified torque.
22 Refill the engine with oil as described in Chapter 1A, Section 5.

12 Oil pump –
removal, inspection and refitting

Removal

1 Remove the timing cover and chain as described in Section 6.
2 Remove the securing bolts and withdraw

11.14a Apply an additional bead of sealant to the groove in the timing cover...

11.14b ...and around the bolt hole at the rear of the cylinder block baseplate

12.2 Remove the securing bolts and withdraw the oil pump cover from the rear of the timing cover

12.3a Compress the oil pump slide pin spring...

12.3b...and carefully lever out the pin and spring

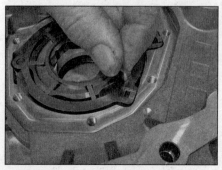

12.4a Lift out the two oil pump slide seal springs...

12.4b...together with the two slide seals

12.5 Lift out the outer oil pump vane ring

12.6a Remove the seven oil pump vanes...

12.6b...then lift out the oil pump vane rotor

the oil pump cover from the rear of the timing cover **(see illustration)**.

3 Using a screwdriver in contact with the oil pump slide pin, compress the slide pin spring and carefully lever out the pin and spring. Use a suitable piece of plastic or similar to protect the edge of the pump housing as the pin and spring are levered out **(see illustrations)**. It is also advisable to cover the spring with a cloth to prevent a sudden and unexpected release!

4 Lift out the two oil pump slide seal springs, together with the two slide seals **(see illustrations)**.

5 Lift out the outer oil pump vane ring **(see illustration)**.

6 Remove the seven oil pump vanes, then lift out the oil pump vane rotor **(see illustrations)**.

7 Lift out the inner oil pump vane ring, followed by the oil pump slide **(see illustrations)**.

8 The oil pressure relief valve components can also be removed from the timing cover by unscrewing the cap. Withdraw the cap, spring and plunger **(see illustrations)**.

Inspection

9 Thoroughly clean all the individual components and the oil pump housing in the timing cover. Blow out all the oil holes and

12.7a Lift out the inner oil pump vane ring...

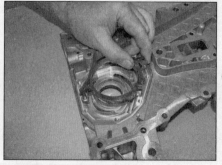

12.7b...followed by the oil pump slide

12.8a Unscrew and remove the oil pressure relief valve cap...

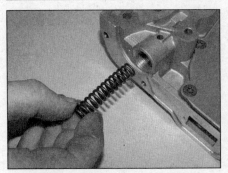

12.8b...then remove the spring...

12.8c...and plunger

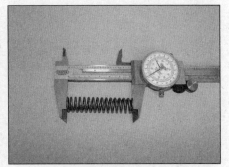

12.11 Measure the free length of the oil pump slide pin spring

galleries in the timing cover using compressed air, if possible.

⚠️ *Warning: Wear eye protection when using compressed air.*

10 Check the components for any signs of scuffing, scoring, wear or distortion. If any abnormalities are found, a pump repair kit is available from Vauxhall/Opel parts stockists. If any wear or damage is visible in the pump housing, a new timing cover will be required.

11 Measure the free length of the oil pump slide pin spring and renew it if it is outside the dimension given in the Specifications **(see illustration)**.

12 Examine the pressure relief valve spring and plunger, and renew if any sign of damage or wear is evident.

13 Checking of the oil pump clearances is carried out after reassembly of the pump.

14 Ensure that all the individual components, and the oil pump housing in the interior of the timing cover, are scrupulously clean before commencing reassembly.

12.17 Refit the oil pump vane rotor with the direction mark facing the oil pump cover

Refitting

15 Thoroughly clean the pressure relief valve components, and lubricate them with clean engine oil before refitting. Insert the plunger, and the spring, then refit the cap. Tighten the cap to the specified torque.

16 Refit the oil pump slide to the pivot pin, then place the inner oil pump vane ring in position.

12.18 Locate the seven oil pump vanes into the slots in the vane rotor

17 Refit the oil pump vane rotor, ensuring that it is fitted with the direction mark facing the oil pump cover **(see illustration)**.

18 Locate the seven oil pump vanes into the slots in the vane rotor, ensuring that they are fitted with the slight scuff marks, caused by the vane rings, facing the vane rotor **(see illustration)**.

19 Place the outer oil pump vane ring in position on the vane rotor.

20 Refit the two oil pump slide seals, followed by the slide seal springs **(see illustration)**.

21 Fit the oil pump slide pin to the slide pin spring, then refit the spring and pin to the pump housing. Using a screwdriver, compress the slide pin as done during removal, and push the slide pin and spring into position. Ensure that the flat side of the slide pin head is facing toward the top of the timing cover **(see illustration)**.

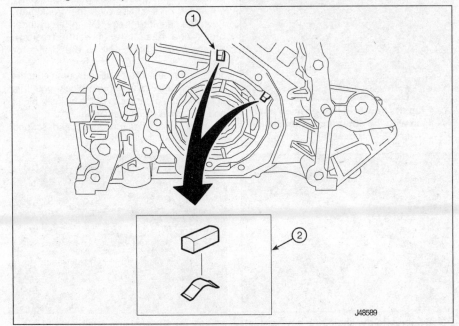

12.20 Refit the two oil pump slide seals and slide seal springs (2) into the locations shown (1)

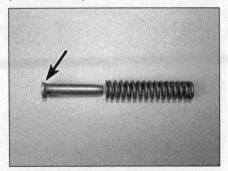

12.21 Ensure that the flat side of the oil pump slide pin head is facing toward the top of the timing cover

12.22a Measure the axial clearances...

12.22b...and radial clearances of the various components using a straight-edge and feeler blades

12.23 Thoroughly lubricate the pump internal components with clean engine oil, then refit the pump cover

22 With the oil pump assembled, measure the axial and radial clearances of the various components using a straight-edge and feeler blades **(see illustrations)**. If any of the measurements exceed the clearances given in the Specifications, a new pump repair kit or a new timing cover will be required.

23 If all is satisfactory, thoroughly lubricate the pump internal components with clean engine oil, then refit the pump cover **(see illustration)**. Progressively tighten the pump cover retaining bolts in a diagonal sequence to the specified torque.

24 Temporarily refit the crankshaft pulley to the timing cover, engaging the pulley shaft with the oil pump vane rotor. Turn the pulley in the normal direction of rotation to check that

the pump operates smoothly and the pulley can be turned easily.

25 Fit a new crankshaft oil seal to the timing cover as described in Section 13.

26 Refit the timing chain and cover to the engine as described in Section 6.

13 Crankshaft oil seals – renewal

Timing chain end oil seal

1 Remove the crankshaft pulley as described in Section 5.

2 The seal can now be carefully prised out with a screwdriver or similar hooked tool **(see illustration)**.

3 Clean the oil seal seat with a wooden or plastic scraper.

4 Tap the new seal into position until it is flush with the outer face of the timing cover, using a wooden block or preferably, an oil seal installer tool **(see illustration)**.

5 Lubricate the sealing lip of the oil seal with clean engine oil, then refit the crankshaft pulley as described in Section 5.

Transmission end oil seal

6 Remove the flywheel/driveplate as described in Section 15.

7 Carefully prise out the old seal from its location using a screwdriver or similar hooked tool.

8 Clean the oil seal seat with a wooden or plastic scraper.

9 If available, fit a protector sleeve over the end of the crankshaft to protect the oil seal lips as the seal is initially fitted **(see illustration)**.

10 Lubricate the lips of the new oil seal and locate it over the protector sleeve. If a protector sleeve is not being used, carefully guide the oil seal lips over the crankshaft. Push the seal into the recess in the cylinder block and baseplate **(see illustration)**. Where applicable, remove the protector sleeve.

11 Tap the seal into position using a suitable socket or tube, or a wooden block, until it is flush with the outer faces of the cylinder block and baseplate **(see illustration)**.

12 Refit the flywheel/driveplate as described in Section 15.

13.2 Carefully prise out the crankshaft oil seal from the timing cover

13.4 Tap the new seal into position until it is flush with the outer face of the timing cover

13.9 If available, fit a protector sleeve over the end of the crankshaft to protect the oil seal lips as the seal is initially fitted

13.10 Lubricate the lips of the new oil seal and locate it over the protector sleeve

13.11 Tap the seal into position until it is flush

14.2 Release the retaining clip and disconnect the oil cooler outlet hose from the thermostat housing

14.3 Undo the bolt securing the oil cooler outlet pipe to the cylinder block

14.4 Release the retaining clip and disconnect the oil cooler inlet hose from the coolant outlet elbow

14 Oil filter housing – removal and refitting

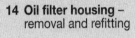

Removal

1 Remove the exhaust manifold as described in Chapter 4A, Section 18.

2 Release the retaining clip and disconnect the oil cooler outlet hose from the thermostat housing **(see illustration)**.

3 Undo the bolt securing the oil cooler outlet pipe to the cylinder block **(see illustration)**.

4 Release the retaining clip and disconnect the oil cooler inlet hose from the coolant outlet elbow **(see illustration)**.

5 Disconnect the wiring connector from the coolant temperature sensor.

6 Undo the three retaining bolts and withdraw the oil filter housing from the cylinder block. Recover the two rubber seals, noting that new seals will be required for refitting **(see illustrations)**.

Refitting

7 Refitting is a reversal of the removal procedure, bearing in mind the following points:

a) Fit a new rubber seal to the filter housing flange, ensuring that is fully seated in the flange groove.

b) Fit a new O-ring seal to the outlet orifice and lubricate the seal with clean engine oil.

14.6a Withdraw the oil filter housing and recover the filter housing flange seal...

c) Tighten the filter housing retaining bolts to the specified torque.

d) Refit the exhaust manifold as described in Chapter 4A, Section 18.

15 Flywheel/driveplate – removal, inspection and refitting

Removal

Note: New flywheel/driveplate securing bolts must be used on refitting.

Flywheel (manual transmission models)

1 Remove the transmission as described in Chapter 7A, Section 8.

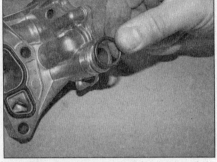

14.6b...and the outlet orifice O-ring seal

2 Remove the clutch assembly as described in Chapter 6, Section 6.

3 Although the flywheel bolt holes are offset so that the flywheel can only be fitted in one position, it will make refitting easier if alignment marks are made between the flywheel and the end of the crankshaft.

4 Prevent the flywheel from turning by jamming the ring gear teeth using a suitable tool **(see illustration)**.

5 Unscrew the retaining bolts, and remove the flywheel **(see illustrations)**.

Caution: Take care, as the flywheel is heavy.

Driveplate (automatic transmission models)

6 Remove the transmission as described in Chapter 7B, Section 9, then remove the driveplate as described in paragraphs 4 and 5.

15.4 Prevent the flywheel from turning by jamming the ring gear teeth using a suitable tool

15.5a Unscrew the retaining bolts...

15.5b...and remove the flywheel

15.18a Tighten the flywheel retaining bolts to the specified torque...

15.18b...then through the specified angles

Inspection

Flywheel

7 A dual-mass flywheel is fitted and has the effect of reducing engine and transmission vibrations and harshness. The flywheel consists of a primary mass and a secondary mass constructed in such a way that the secondary mass is allowed to rotate slightly in relation to the primary mass. Springs within the assembly restrict this movement to set limits.

8 Dual-mass flywheels have earned an unenviable reputation for unreliability and have been known to fail at quite low mileages (sometimes as low as 20 000 miles).

9 Examine the flywheel for wear or chipping of the ring gear teeth. Renewal of the ring gear is not possible and if the wear or chipping is significant, a new flywheel will be required.

10 Examine the flywheel for scoring of the clutch face. If the clutch face is scored significantly, a new flywheel will be required.

11 Look through the bolt hole and inspection openings in the secondary mass and check for any visible damage in the area of the centre bearing.

12 Place your thumbs on the clutch face of the secondary mass at the 3 o'clock and 9 o'clock positions and try to rock it. The maximum movement should not exceed 3 mm. Repeat this check with your thumbs at the 12 o'clock and 6 o'clock positions.

13 Rotate the secondary mass clockwise and anti-clockwise. It should move freely in both directions until spring resistance is felt, with

16.9 Right-hand engine mounting bracket-to-timing cover retaining bolts

no abnormal grating or rattling noises. The maximum rotational movement should not exceed a distance of eight teeth of the ring gear.

14 If there is any doubt about the condition of the flywheel, seek the advice of a Vauxhall/Opel dealer or engine reconditioning specialist. They will be able to advise if the flywheel is an acceptable condition, or whether renewal is necessary.

Driveplate

15 Closely examine the driveplate and ring gear teeth for signs of wear or damage and check the driveplate surface for any signs of cracks.

16 If there is any doubt about the condition of the driveplate, seek the advice of a Vauxhall/Opel dealer or engine reconditioning specialist.

Refitting

Flywheel

17 Offer the flywheel to the end of the crankshaft, and align the previously-made marks on the flywheel and crankshaft (if applicable).

18 Coat the threads of the new flywheel bolts with thread-locking compound (note that new bolts may be supplied ready-coated), then fit the bolts and tighten them to the specified torque, then through the specified angles whilst preventing the flywheel from turning as during removal **(see illustrations)**.

19 Refit the clutch as described in Chapter 6, Section 6 then refit the transmission as described in Chapter 7A, Section 8.

Driveplate

20 Clean the mating surfaces of the driveplate and crankshaft

21 Offer the driveplate and engage it on the crankshaft. Coat the threads of the new retaining bolts with thread-locking compound (note that new bolts may be supplied ready-coated), then fit the bolts and tighten them to the specified torque. Prevent the driveplate from turning as during removal.

22 Remove the locking tool and refit the transmission as described in Chapter 7B, Section 9.

16 Engine/transmission mountings – inspection and renewal

Inspection

1 To improve access, firmly apply the handbrake, then jack up the front of the vehicle and support it on axle stands (see *Jacking and vehicle support*).

2 Check the mounting blocks (rubbers) to see if they are cracked, hardened or separated from the metal at any point. Renew the mounting block if any such damage or deterioration is evident.

3 Check that all the mounting securing nuts and bolts are securely tightened, using a torque wrench to check if possible.

4 Using a large screwdriver, or a similar tool, check for wear in the mounting blocks by carefully levering against them to check for free play. Where this is not possible, enlist the aid of an assistant to move the engine/ transmission unit back-and-forth, and from side-to-side, while you observe the mountings. While some free play is to be expected, even from new components, excessive wear should be obvious. If excessive free play is found, check first to see that the securing nuts and bolts are correctly tightened, then renew any worn components as described in the following paragraphs.

Renewal

Note: *Before slackening any of the engine mounting bolts/nuts, the relative positions of the mountings to their various brackets should be marked to ensure correct alignment upon refitting.*

Right-hand mounting

Note: *New bolts will be required to secure the mounting to the engine mounting bracket and to secure the engine mounting bracket to the timing cover (if removed).*

5 Remove the air cleaner assembly as described in Chapter 4A, Section 2.

6 Support the weight of the engine using a trolley jack with a block of wood placed on its head.

7 Undo the three bolts securing the right-hand engine mounting to the engine mounting bracket. Note that new bolts will be required for refitting.

8 Undo the three nuts securing the mounting to the body, and withdraw the mounting.

9 If necessary, the engine mounting bracket may be unbolted from the timing cover **(see illustration)**. Note that new bolts will be required for refitting.

10 Refitting is a reversal of removal. Tighten all retaining bolts/nut to the specified torque, and where applicable, through the specified angle. Refit the air cleaner assembly as described in Chapter 4A, Section 2.

Left-hand mounting

Note: *New bolts will be required to secure the mounting to the transmission bracket.*

11 Remove the battery and battery tray as described in Chapter 5A, Section 4.

12 Support the weight of the transmission using a trolley jack with a block of wood placed on its head.

13 Unscrew the three bolts securing the mounting to the transmission bracket **(see illustration)**. Note that new bolts will be required for refitting.

14 Undo the two bolts and one nut securing the mounting to the body, and withdraw the mounting.

15 Refitting is a reversal of removal, tightening all retaining bolts/nut to the specified torque. Refit the battery and battery tray as described in Chapter 5A, Section 4.

Rear mounting/torque link

Note: *New bolts will be required at all attachment points.*

FWD models

16 Firmly apply the handbrake, then jack up the front of the vehicle and support it on axle stands (see *Jacking and vehicle support*).

17 Remove the exhaust system front pipe as described in Chapter 4A, Section 20.

18 Using a suitable axle stand or trolley jack, support the rear of the engine/transmission unit.

19 Undo the bolt securing the rear mounting/torque link to the subframe **(see illustration)**. Note that a new bolt will be required for refitting.

20 Undo the three bolts securing the mounting bracket to the side of the transmission **(see illustration)**. Note that new bolts will be required for refitting.

21 Withdraw the mounting and mounting bracket from the subframe and transmission **(see illustration)**, then undo the through bolt and separate the mounting from the mounting bracket. Note that a new through bolt will be required for refitting.

16.13 Left-hand engine mounting-to-transmission bracket retaining bolts

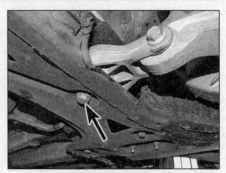

16.19 Undo the bolt securing the rear mounting to the subframe

16.20 Undo the three bolts securing the rear mounting bracket to the transmission

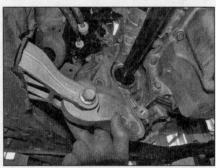

16.21 Withdraw the rear mounting and mounting bracket from the subframe and transmission

22 Refitting is a reversal of removal, tightening all retaining bolts to the specified torque. Refit the exhaust system front pipe as described in Chapter 4A, Section 20.

4WD models

23 Firmly apply the handbrake, then jack up the front of the vehicle and support it on axle stands (see *Jacking and vehicle support*).

24 Using a suitable axle stand or trolley jack, support the rear of the engine/transmission unit.

25 Remove the transfer gearbox and transfer gearbox adapter as described in Chapter 7C, Section 3.

26 Undo the bolt securing the rear mounting/torque link to the subframe **(see illustration 16.19)**. Note that a new bolt will be required for refitting.

27 Withdraw the rear mounting/torque link from the subframe and remove it from under the car.

28 Refitting is a reversal of removal, tightening all retaining bolts to the specified torque. Refit the transfer gearbox adapter and transfer gearbox as described in Chapter 7C, Section 3.

Chapter 2 Part B
1.6 litre petrol engine in-car repair procedures

Contents

Degrees of difficulty

| Easy, suitable for novice with little experience | 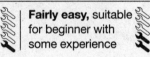 | Fairly easy, suitable for beginner with some experience | 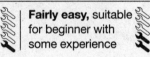 | Fairly difficult, suitable for competent DIY mechanic | 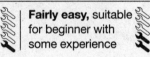 | Difficult, suitable for experienced DIY mechanic | 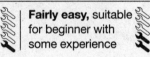 | Very difficult, suitable for expert DIY or professional | |

Specifications

General

Engine type...	Four-cylinder, in-line, water-cooled. Double overhead camshafts, belt-driven
Manufacturer's engine code*.................................	A16XER (LDE)
Bore	79.0 mm
Stroke	81.5 mm
Capacity	1598 cc
Firing order...	1-3-4-2 (No 1 cylinder at timing belt end of engine)
Direction of crankshaft rotation	Clockwise (viewed from timing belt end of engine)
Compression ratio	11.0: 1

*For details of engine code location, see 'Vehicle identification' in Reference

Compression pressures

Maximum difference between any two cylinders..................	1.0 bar

Valve clearances

Engine cold:

Inlet	0.21 to 0.29 mm
Exhaust...	0.26 to 0.34 mm

Lubrication system

Minimum oil pressure at 80°C	1.3 bar at idle speed
Oil pump type.......................................	Rotor-type, driven directly from crankshaft
Rotor-to-housing clearance (endfloat)	0.020 to 0.058 mm

Torque wrench settings

	Nm	lbf ft
Air conditioning compressor mounting bolts	22	16
Auxiliary drivebelt tensioner retaining bolt	55	41
Camshaft bearing cap bolts	8	6
Camshaft cover bolts	8	6
Camshaft sprocket closure bolt	30	22
Camshaft sprocket retaining bolt*:		
Stage 1	50	37
Stage 2	Angle-tighten a further 150°	
Stage 3	Angle-tighten a further 15°	
Connecting rod big-end bearing cap bolt*:		
Stage 1	35	26
Stage 2	Angle-tighten a further 45°	
Stage 3	Angle-tighten a further 15°	
Coolant pipe flange-to-coolant pump	8	6
Crankshaft pulley bolt*:		
Stage 1	95	70
Stage 2	Angle-tighten a further 45°	
Stage 3	Angle-tighten a further 15°	
Cylinder head bolts*:		
Stage 1	25	18
Stage 2	Angle-tighten a further 90°	
Stage 3	Angle-tighten a further 90°	
Stage 4	Angle-tighten a further 90°	
Stage 5	Angle-tighten a further 45°	
Engine mountings:		
Left-hand mounting:		
Mounting-to-body bolts/nut	58	43
Mounting bracket-to-transmission bracket*:		
Stage 1	50	37
Stage 2	Angle-tighten a further 60°	
Rear mounting/torque link:		
Mounting-to-transmission bracket*:		
Stage 1	80	59
Stage 2	Angle-tighten a further 45°	
Mounting-to-subframe*:		
Stage 1	100	74
Stage 2	Angle-tighten a further 100°	
Transmission bracket bolts*:		
Stage 1	100	74
Stage 2	Angle-tighten a further 45°	
Right-hand mounting:		
Mounting bracket-to-engine*:		
Stage 1	60	44
Stage 2	Angle-tighten a further 45°	
Mounting-to-body	62	46
Mounting-to-mounting bracket*:		
Stage 1	50	37
Stage 2	Angle-tighten a further 60°	
Engine-to-transmission bolts:		
M10 bolts	40	32
M12 bolts	60	44
Flywheel bolts*:		
Stage 1	35	26
Stage 2	Angle-tighten a further 30°	
Stage 3	Angle-tighten a further 15°	
Main bearing cap bolts*:		
Stage 1	50	37
Stage 2	Angle-tighten a further 45°	
Stage 3	Angle-tighten a further 15°	
Oil filter housing cap to filter housing	25	18
Oil filter housing to cylinder block	25	18
Oil pump:		
Retaining bolts	20	15
Pump cover screws	8	6
Oil pressure relief valve cap	20	15
Roadwheel nuts	140	103

Torque wrench settings (continued)

	Nm	lbf ft
Sump bolts:		
Sump-to-cylinder block/oil pump bolts	10	7
Sump flange-to-transmission bolts	40	32
Timing belt cover bolts.	6	4
Timing belt idler pulley bolt*:		
Stage 1	20	15
Stage 2	Angle-tighten a further 120°	
Stage 3	Angle-tighten a further 15°	
Timing belt tensioner bolt*:		
Stage 1	20	15
Stage 2	Angle-tighten a further 120°	
Stage 3	Angle-tighten a further 15°	

*Use new fasteners

1 General Information

How to use this Chapter

1 This Part of Chapter 2 describes the repair procedures which can reasonably be carried out on the engine while it remains in the vehicle. If the engine has been removed from the vehicle and is being dismantled as described in Chapter 2E, any preliminary dismantling procedures can be ignored.

2 Note that, while it may be possible physically to overhaul items such as the piston/connecting rod assemblies while the engine is in the vehicle, such tasks are not usually carried out as separate operations, and usually require the execution of several additional procedures (not to mention the cleaning of components and of oil ways); for this reason, all such tasks are classed as major overhaul procedures, and are described in Chapter 2E.

3 Chapter 2E describes the removal of the engine/transmission unit from the vehicle, and the full overhaul procedures which can then be carried out.

Engine description

4 The engine is of four-cylinder in-line type, with double overhead camshafts (DOHC). The engine is mounted transversely at the front of the vehicle.

5 The power unit is known as a 'Twinport' engine due to the design of the inlet manifold and cylinder head combustion chambers. Vacuum operated flap valves located in the inlet manifold are opened or closed according to engine operating conditions, to create a variable venturi manifold arrangement. This system has significant advantages in terms of engine power, fuel economy and reduced exhaust emissions.

6 The crankshaft runs in five shell-type main bearings with crankshaft endfloat being controlled by thrustwashers which are an integral part of No 3 main bearing shells.

7 The connecting rods are attached to the crankshaft by horizontally-split shell-type big-end bearings, and to the pistons by gudgeon pins, which are a sliding fit in the connecting rod small-end eyes and retained by circlips. The aluminium-alloy pistons are fitted with three piston rings – two compression rings and an oil control ring.

8 The camshafts run directly in the cylinder head, and are driven by the crankshaft via a toothed composite rubber timing belt. The camshafts operate each valve via a camshaft follower. One camshaft operates the inlet valves, and the other operates the exhaust valves. The camshaft followers are available in various thicknesses to facilitate valve clearance adjustment.

9 A variable valve timing (VVT) system is employed. The VVT system allows the inlet and exhaust camshaft timing to be varied under the control of the engine management system, to boost both low-speed torque and top-end power, as well as reducing exhaust emissions. The VVT camshaft adjuster is integral with each camshaft timing belt sprocket, and is supplied with two pressurised oil feeds through passages in the camshaft itself. Two electro-magnetic oil control valves, one for each camshaft and operated by the engine management system, are fitted to the cylinder head, and are used to supply the pressurised oil to each camshaft adjuster through the two oil feeds. Each adjuster contains two chambers – depending on which of the two oil feeds is enabled by the control valve, the oil pressure will turn the camshaft clockwise (advance) or anti-clockwise (retard) to adjust the valve timing as required. If pressure is removed from both feeds, this induces a timing 'hold' condition. Thus the valve timing is infinitely variable within a given range.

10 Lubrication is by pressure-feed from a rotor-type oil pump, which is mounted on the right-hand end of the crankshaft. The pump draws oil through a strainer located in the sump, and then forces it through an externally mounted full-flow oil filter. The oil flows into galleries in the cylinder block/crankcase, from where it is distributed to the crankshaft (main bearings) and camshafts. The big-end bearings are supplied with oil via internal drillings in the crankshaft, while the camshaft bearings also receive a pressurised supply. The camshaft lobes and valves are lubricated by splash, as are all other engine components.

11 A semi-closed crankcase ventilation system is employed; crankcase fumes are drawn from camshaft cover, and passed via a hose to the inlet manifold.

12 The coolant pump is located externally on the engine, and is driven by the auxiliary drivebelt.

Repair operations possible with the engine in the car

13 The following operations can be carried out without having to remove the engine from the car.

a) Removal and refitting of the camshaft cover.
b) Adjustment of the valve clearances.
c) Removal and refitting of the VVT oil control valves.
d) Removal and refitting of the cylinder head.
e) Removal and refitting of the timing belt, tensioner and sprockets.
f) Renewal of the camshaft oil seals.
g) Removal and refitting of the camshafts and followers.
h) Removal and refitting of the sump.
i) Removal and refitting of the connecting rods and pistons*.
j) Removal and refitting of the oil pump.
k) Renewal of the crankshaft oil seals.
l) Renewal of the engine mountings.
m) Removal and refitting of the flywheel/driveplate.

*Although the operation marked with an asterisk can be carried out with the engine in the car (after removal of the sump), it is preferable for the engine to be removed, in the interests of cleanliness and improved access. For this reason, the procedure is described in Chapter 2E.

2 Compression test – description and interpretation

1 When engine performance is down, or if misfiring occurs which cannot be attributed to the ignition or fuel systems, a compression test can provide diagnostic clues as to the engine's condition. If the test is performed regularly, it can give warning of trouble before any other symptoms become apparent.

2 The engine must be fully warmed-up to operating temperature, the oil level must be correct and the battery must be fully-charged.

3.5a Align the camshaft timing marks…

3.5b …and the notch on the crankshaft pulley rim with the mark on the timing belt lower cover

3 Open the cover on the engine compartment fuse/relay box and remove the fuel pump relay. Refer to the wiring diagrams at the end of Chapter 12 for information on fuse and relay locations. Now start the engine and allow it to run until it stalls.

4 Remove the throttle housing as described in Chapter 4A, Section 8.

5 Remove the spark plugs as described in Chapter 1A, Section 27.

6 Fit a compression tester to the No 1 cylinder spark plug hole – the type of tester which screws into the spark plug thread is preferable.

7 Arrange for an assistant to crank the engine over for several seconds on the starter motor while you observe the compression gauge reading. The compression will build-up fairly quickly in a healthy engine. Low compression on the first stroke, followed by gradually-increasing pressure on successive strokes, indicates worn piston rings. A low compression on the first stroke which does not rise on successive strokes, indicates leaking valves or a blown head gasket (a cracked cylinder head could also be the cause). Deposits on the underside of the valve heads can also cause low compression. Record the highest gauge reading obtained,

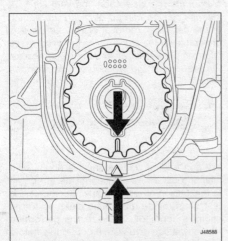

3.6 Crankshaft sprocket and timing belt rear cover timing mark alignment (crankshaft pulley removed)

then repeat the procedure for the remaining cylinders.

8 Due to the variety of testers available, and the fluctuation in starter motor speed when cranking the engine, different readings are often obtained when carrying out the compression test. However, the most important factor is that the compression pressures are uniform in all cylinders, and that is what this test is mainly concerned with.

9 Add some engine oil (about three squirts from a plunger type oil can) to each cylinder through the spark plug holes, and then repeat the test.

10 If the compression increases after the oil is added, the piston rings are probably worn. If the compression does not increase significantly, the leakage is occurring at the valves or the head gasket. Leakage past the valves may be caused by burned valve seats and/or faces, or warped, cracked or bent valves.

11 If two adjacent cylinders have equally low compressions, it is most likely that the head gasket has blown between them. The appearance of coolant in the combustion chambers or on the engine oil dipstick would verify this condition.

12 If one cylinder is about 20 percent lower than the other, and the engine has a slightly rough idle, a worn lobe on the camshaft could be the cause.

13 On completion of the checks, refit the spark plugs, throttle housing and fuel pump relay.

3 Top Dead Centre (TDC) for No 1 piston – locating

1 Top dead centre (TDC) is the highest point in the cylinder that a piston reaches as the crankshaft turns. Each piston reaches TDC at the end of the compression stroke, and again at the end of the exhaust stroke. For the purpose of timing the engine, TDC refers to the position of No 1 piston at the end of its compression stroke. No 1 piston and cylinder are at the timing belt end of the engine.

2 Disconnect the battery negative lead as described in Chapter 5A Section 4. If necessary, remove all the spark plugs as described in Chapter 1A, Section 27 to enable the engine to be easily turned over.

3 Remove the timing belt upper cover as described in Section 6.

4 Apply the handbrake, then jack up the front of the vehicle and support it on axle stands (see *Jacking and vehicle support*). Remove the right-hand front roadwheel, then remove the wheel arch liner as described in Chapter 11, Section 21 for access to the crankshaft pulley.

5 Using a socket and extension bar on the crankshaft pulley bolt, rotate the crankshaft until the timing marks on the camshaft sprockets are facing towards each other, and an imaginary straight line can be drawn through the camshaft sprocket bolts and the timing marks. With the camshaft sprocket marks correctly positioned, align the notch on the crankshaft pulley rim with the mark on the timing belt lower cover **(see illustrations)**. The engine is now positioned with No 1 piston at TDC on its compression stroke.

6 If the crankshaft pulley and lower timing belt cover have been removed, the timing mark on the crankshaft sprocket can be used instead of the mark on the pulley. The mark on the crankshaft sprocket must align with the corresponding mark on the oil pump housing **(see illustration)**.

4 Camshaft cover – removal and refitting

Removal

1 Depressurise the fuel system as described in Chapter 4A, Section 5.

2 Disconnect the fuel feed hose quick-release connector at the fuel rail **(see illustration)**. Be prepared for some loss of fuel. Clamp or plug the open end of the hose, to prevent dirt ingress and further fuel spillage.

3 Release the retaining clips and move the fuel feed hose assembly to one side.

4 Remove the ignition module from the spark plugs as described in Chapter 5B, Section 3.

4.2 Disconnect the fuel feed hose quick-release connector

5 Lift the wiring harness up and out of the support on the right-hand end of the camshaft cover **(see illustration)**.

6 Unclip the wiring harness trough from the rear of the camshaft cover.

7 Pull out the retaining wire clip and disconnect the crankcase ventilation hose from the camshaft cover.

8 Unscrew the eleven bolts securing the camshaft cover to the cylinder head.

9 Lift the camshaft cover away from the cylinder head and recover the rubber seal. Examine the seal for signs of wear or damage and renew if necessary.

Refitting

10 Ensure that the camshaft cover grove and rubber seal are clean and dry with all traces of oil removed. If necessary, de-grease the seal and cover groove with brake cleaner or a similar product.

11 Clean the mating surface of the cylinder head and the area around the camshaft bearing caps at the timing belt end, ensuring that all traces of oil are removed.

12 Locate the rubber seal into the grooves of the camshaft cover, ensuring that it is fully seated, with no chance of it falling out as the cover is fitted **(see illustration)**.

13 Carefully manoeuvre the camshaft cover into position, taking great care to ensure the seal remains correctly seated. Screw in all the cover retaining bolts and tighten them to the specified torque, working in a spiral pattern from the centre outwards.

14 Reconnect the crankcase ventilation hose then clip the wiring harness trough back into position. Engage the wiring harness with the support on the cover.

15 Refit the ignition module to the spark plugs as described in Chapter 5B, Section 3.

16 Reconnect the fuel feed hose quick-release connector at the fuel rail.

---

5 Crankshaft pulley – removal and refitting

Removal

Note: *A new pulley retaining bolt will be required for refitting.*

1 Firmly apply the handbrake, then jack up the front of the car and support it securely on axle stands (see *Jacking and vehicle support*). Remove the right-hand roadwheel, then remove the wheel arch liner as described in Chapter 11, Section 21 for access to the crankshaft pulley.

2 Remove the auxiliary drivebelt as described in Chapter 1A, Section 20. Prior to removal, mark the direction of rotation on the belt to ensure the belt is refitted the same way around.

3 Slacken the crankshaft pulley retaining bolt. To prevent crankshaft rotation, have an assistant select top gear and apply the

4.5 Lift the wiring harness out of the support on the camshaft cover

brakes firmly. Alternatively, remove the starter motor (see Chapter 5A, Section 11), and lock the flywheel ring gear teeth using a large screwdriver or suitable flywheel locking tool.

4 Unscrew the retaining bolt and washer and remove the crankshaft pulley from the end of the crankshaft. Note that a new retaining bolt will be required for refitting.

Refitting

5 Refit the crankshaft pulley, aligning the pulley cut-out with the raised notch on the timing belt sprocket, then fit the washer and new retaining bolt **(see illustration)**.

6 Lock the crankshaft by the method used on removal, and tighten the pulley retaining bolt to the specified Stage 1 torque setting, then angle-tighten the bolt through the specified Stage 2 angle, using a socket and extension bar, and finally through the specified Stage 3 angle. It is recommended that an angle-measuring gauge is used during the final stages of the tightening, to ensure accuracy. If a gauge is not available, use white paint to make alignment marks between the bolt head and pulley prior to tightening; the marks can then be used to check that the bolt has been rotated through the correct angle.

7 Refit the auxiliary drivebelt as described in Chapter 1A, Section 20 using the mark made prior to removal to ensure the belt is fitted the correct way around.

8 Refit the wheel arch liner and the roadwheel. Lower the car to the ground and tighten the wheel nuts to the specified torque.

5.5 Align the pulley cut-out with the raised notch

4.12 Locate the rubber seal into the grooves of the camshaft cover

---

6 Timing belt covers – removal and refitting

Upper cover

Removal

1 Remove the air cleaner assembly and the relevant air ducts as described in Chapter 4A, Section 2.

2 Undo the two retaining bolts then withdraw the upper cover from the rear cover and remove it from the engine compartment **(see illustration)**.

Refitting

3 Refitting is the reverse of removal, tightening the cover retaining bolts to the specified torque.

Lower cover

Removal

4 Remove the crankshaft pulley as described in Section 5.

5 Remove the auxiliary drivebelt tensioner as described in Chapter 5A, Section 7.

6 Undo the four retaining bolts then manoeuvre the lower cover off the engine.

Refitting

7 Refitting is the reverse of removal, tightening the cover retaining bolts to the specified torque.

6.2 Undo the two retaining bolts and withdraw the upper cover from the rear cover

A camshaft locking tool can be made from a steel strip approximately 4.5 mm thick, with two grooves filed in the strip to clear the camshaft sensor trigger lugs

Centre cover

Removal

8 Remove the upper timing belt cover as described previously.

9 Remove the right-hand engine mounting as described in Section 19, then undo the three retaining bolts and remove the mounting bracket bolted to the engine.

10 Release the two clips securing the centre cover to the rear cover and manoeuvre the centre cover off the engine.

Refitting

11 Refitting is the reverse of removal, tightening the engine mounting bracket retaining bolts to the specified torque.

Rear cover

Removal

Note: New cover retaining bolts will be required for refitting.

12 Remove the timing belt as described in Section 7.

13 Remove the camshaft sprockets and timing belt tensioner as described in Section 8.

14 Unclip the wiring harness from the cover, then undo the four retaining bolts and manoeuvre the rear cover off the engine. Discard the four retaining bolts and obtain new bolts for refitting.

8.3 Insert the special tool into the slots on the end of the camshafts to lock the camshafts in the TDC position

Refitting

15 Clean the cover retaining bolt threads in the cylinder head and remove all traces of thread locking compound.

16 If the new retaining bolts are not already pre-coated, apply thread locking compound to the bolt threads. Place the rear cover in position, then fit the four bolts and tighten to the specified torque. Clip the wiring harness back into position on the rear cover.

17 Refit the camshaft sprockets and timing belt tensioner as described in Section 8.

18 Refit the timing belt as described in Section 7.

7 Timing belt – removal and refitting

Note: The timing belt must be removed and refitted with the engine cold.

Removal

1 Position No 1 cylinder at TDC on its compression stroke as described in Section 3.

2 Remove the timing belt centre and lower cover as described in Section 6.

3 Check that the timing marks on the camshaft sprockets are still correctly aligned and facing towards each other, and the timing mark on the crankshaft sprocket is aligned with the corresponding mark on the oil pump housing (Section 3).

4 Using an Allen key inserted in the slot on the face of the timing belt tensioner, rotate the tensioner clockwise to relieve the tension in the timing belt. Insert a small drill bit or similar into the slot on the inner edge of the tensioner body to lock the tensioner in the released position.

5 Slide the timing belt from its sprockets and remove it from the engine. If the belt is to be re-used, use white paint or similar to mark the direction of rotation on the belt. Do not rotate the crankshaft or camshafts until the timing belt has been refitted.

6 Check the timing belt carefully for any signs of uneven wear, splitting or oil contamination, and renew it if there is the slightest doubt about its condition. If the engine is undergoing an overhaul and is approaching the manufacturer's specified interval for belt renewal (see Chapter 1A) renew the belt as a matter of course, regardless of its apparent condition. If signs of oil contamination are found, trace the source of the oil leak and rectify it, then wash down the engine timing belt area and all related components to remove all traces of oil.

Refitting

7 On reassembly, thoroughly clean the timing belt sprockets and tensioner/idler pulleys.

8 Check that the camshaft and crankshaft sprocket timing marks are still correctly aligned as described in Section 3.

9 Fit the timing belt over the crankshaft

and camshaft sprockets and around the idler pulley, ensuring that the belt front run is taut (ie, all slack is on the tensioner side of the belt), then fit the belt over the tensioner pulley. Do not twist the belt sharply while refitting it. Ensure that the belt teeth are correctly seated centrally in the sprockets, and that the timing marks remain in alignment. If a used belt is being refitted, ensure that the arrow mark made on removal points in the normal direction of rotation, as before.

10 Using the Allen key, turn the tensioner clockwise slightly and remove the drill bit or similar tool used to lock the tensioner. Slowly release the tensioner and allow it to turn anti-clockwise and automatically tension the timing belt.

11 Check the sprocket timing marks are still correctly aligned. If adjustment is necessary, release the tensioner again then disengage the belt from the sprockets and make any necessary adjustments.

12 Using a socket on the temporarily refitted crankshaft pulley bolt, rotate the crankshaft smoothly through two complete turns (720°) in the normal direction of rotation to settle the timing belt in position.

13 Set the crankshaft back in the timing position and check that all the sprocket timing marks are still correctly aligned. If this is not the case, repeat the timing belt refitting procedure.

14 If everything is satisfactory, refit the auxiliary drivebelt tensioner and tighten its retaining bolt to the specified torque.

15 Refit the timing belt covers as described in Section 6.

8 Timing belt sprockets, tensioner and idler pulley – removal and refitting

Camshaft sprockets

Removal

Note: Vauxhall/Opel special tools KM-6340 and KM-6628 or suitable alternatives will be required for this procedure.

Note: New sprocket retaining bolt(s) and a new sprocket closure bolt seal will be required for refitting.

1 Remove the camshaft cover as described in Section 4.

2 Remove the timing belt as described in Section 7.

3 Insert Vauxhall/Opel special tool KM-6628 into the slots on the end of the camshafts to lock the camshafts in the TDC position. In the absence of the special tool, a suitable alternative can be fabricated from a steel strip (see **Tool Tip** and **illustration**). It may be necessary to turn the camshafts slightly using an open-ended spanner on the flats provided, to allow the tool to fully engage with the slots.

4 Unscrew the closure bolt from the relevant

camshaft sprocket **(see illustration)**. Note that a new closure bolt seal will be required for refitting.

5 Hold the camshaft using an open-ended spanner on the flats provided, and unscrew the camshaft sprocket retaining bolt. Withdraw the sprocket from the end of the camshaft.

6 If necessary, remove the remaining sprocket using the same method.

Refitting

7 Prior to refitting check the oil seal(s) for signs of damage or leakage. If necessary, renew as described in Section 10.

8 Refit the sprocket to the camshaft end and fit the new retaining bolt. Tighten the bolt finger tight only at this stage. If both sprockets have been removed, ensure each sprocket is fitted to the correct shaft; the exhaust camshaft sprocket has the timing belt guide flange on its inner face and the inlet sprocket has the guide flange on its outer face.

9 Turn the camshaft sprocket(s) until the timing marks are facing towards each other and aligned. It will now be necessary to retain the sprockets in the timing position while the sprocket retaining bolt is tightened. Engage Vauxhall/Opel special tool KM-6340 or a suitable alternative with the teeth on both sprockets, to lock the sprockets together.

10 With the camshafts and sprockets locked in the timing position with the special tools, hold the camshaft using an open-ended spanner on the flats provided and tighten the sprocket retaining bolt to the specified Stage 1 torque setting. Now angle-tighten the bolt through the specified Stage 2 angle, using a socket and extension bar, and finally through the specified Stage 3 angle. It is recommended that an angle-measuring gauge is used during the final stages of the tightening, to ensure accuracy. If a gauge is not available, use white paint to make alignment marks between the bolt head and sprocket prior to tightening; the marks can then be used to check that the bolt has been rotated through the correct angle.

11 Fit a new seal to the camshaft sprocket closure bolt, then refit the closure bolt and tighten to the specified torque.

12 Remove the special tools, then refit the timing belt as described in Section 7, and the camshaft cover as described in Section 4.

Crankshaft sprocket

Removal

13 Remove the timing belt as described in Section 7.

14 Slide the sprocket off from the end of the crankshaft, noting which way around it is fitted.

Refitting

15 Align the sprocket locating key with the crankshaft groove then slide the sprocket into position, making sure its timing mark is facing outwards.

16 Refit the timing belt as described in Section 7.

8.4 Camshaft sprocket closure bolts

Tensioner assembly

Removal

Note: *A new tensioner retaining bolt will be required for refitting.*

17 Remove the timing belt as described in Section 7.

18 Slacken and remove the retaining bolt and remove the tensioner assembly from the engine.

Refitting

19 Clean the tensioner retaining bolt threads in the oil pump housing, ensuring that all traces of sealant, oil or grease are removed.

20 Fit the tensioner to the engine, making sure that the projecting end of the tensioner spring engages with the slot on the oil pump housing. Ensure the tensioner is correctly seated then fit the new retaining bolt.

21 Tighten the tensioner retaining bolt to the specified Stage 1 torque setting then angle-tighten the bolt through the specified Stage 2 angle, using a socket and extension bar, and finally through the specified Stage 3 angle. It is recommended that an angle-measuring gauge is used during the final stages of the tightening, to ensure accuracy. If a gauge is not available, use white paint to make alignment marks between the bolt head and oil pump housing prior to tightening; the marks can then be used to check that the bolt has been rotated through the correct angle.

22 Refit the timing belt as described in Section 7.

9.2 Disconnect the wiring connector from the VVT oil control valve

Idler pulley

Removal

Note: *A new idler pulley retaining bolt will be required for refitting.*

23 Remove the timing belt as described in Section 7.

24 Slacken and remove the retaining bolt and remove the idler pulley from the engine.

Refitting

25 Refit the idler pulley and the new retaining bolt.

26 Tighten the pulley retaining bolt to the specified Stage 1 torque setting then angle-tighten the bolt through the specified Stage 2 angle, using a socket and extension bar, and finally through the specified Stage 3 angle. It is recommended that an angle-measuring gauge is used during the final stages of the tightening, to ensure accuracy. If a gauge is not available, use white paint to make alignment marks between the bolt head and oil pump housing prior to tightening; the marks can then be used to check that the bolt has been rotated through the correct angle.

27 Refit the timing belt as described in Section 7.

---

9 VVT oil control valves –
removal and refitting

Removal

1 The VVT oil control valves are fitted to the camshaft bearing support, adjacent to the camshaft sprockets. Two oil control valves are used, one for each camshaft.

2 Disconnect the wiring connector from the relevant oil control valve **(see illustration)**.

3 Undo the retaining bolt located below the valve, and withdraw the valve from the camshaft bearing support. Be prepared for oil spillage.

Refitting

4 Lubricate the valve sealing rings with clean engine oil and insert the valve into the camshaft bearing support.

5 Refit and tighten the retaining bolt securely, then reconnect the wiring connector.

6 On completion, check and if necessary top-up the engine oil as described in *Weekly checks*.

---

10 Camshaft oil seals –
renewal

---

1 Remove the relevant camshaft sprocket as described in Section 8.

2 Carefully punch or drill two small holes opposite each other in the oil seal. Screw a self-tapping screw into each, and pull on the

10.2 Camshaft oil seal removal method

10.4 Using the old camshaft sprocket bolt and a socket to fit the new camshaft oil seal

screws with pliers to extract the seal **(see illustration)**.

3 Clean the seal housing, and polish off any burrs or raised edges which may have caused the seal to fail in the first place.

4 Press the new seal into position using a suitable tubular drift (such as a socket) which bears only on the hard outer edge of the seal **(see illustration)**. Take care not to damage the seal lips during fitting; note that the seal lips should face inwards.

5 Refit the camshaft sprocket as described in Section.

11 Valve clearances – checking and adjustment

Checking

1 The importance of having the valve clearances correctly adjusted cannot be overstressed, as they vitally affect the performance of the engine. The engine must be cold for the check to be accurate. The clearances are checked as follows.

2 Apply the handbrake, then jack up the front of the car and support it on axle stands (see *Jacking and vehicle support*). Remove the right-hand front roadwheel then remove the wheel arch liner as described in Chapter 11, Section 21 for access to the crankshaft pulley.

3 Remove the camshaft cover as described in Section 4.

11.7a Using feeler blades, measure the clearance between the base of both No 2 cylinder inlet cam lobes...

4 Position No 1 cylinder at TDC on its compression stroke as described in Section 3.

5 With the engine at TDC on compression for No 1 cylinder, the inlet camshaft lobes for No 2 cylinder and the exhaust camshaft lobes for No 3 cylinder are pointing upwards and slightly towards the centre. This indicates that these valves are completely closed, and the clearances can be checked.

6 On a piece of paper, draw the outline of the engine with the cylinders numbered from the timing belt end. Show the position of each valve, together with the specified valve clearance.

7 With the cam lobes positioned as described in paragraph 5, using feeler blades, measure the clearance between the base of both No 2 cylinder inlet cam lobes and No 3 cylinder exhaust cam lobes and their followers. Record the clearances on the paper **(see illustrations)**.

8 Rotate the crankshaft pulley in the normal direction of rotation through a half a turn (180°) to position No 1 cylinder inlet camshaft lobes and No 4 cylinder exhaust camshaft lobes pointing upwards and slightly towards the centre. Measure the clearance between the base of the camshaft lobes and their followers and record the clearances on the paper.

9 Rotate the crankshaft pulley through a half a turn (180°) to position No 3 cylinder inlet camshaft lobes and No 2 cylinder exhaust camshaft lobes pointing upwards and slightly towards the centre. Measure the clearance between the base of the camshaft lobes and their followers and record the clearances on the paper.

11.7b ...and No 3 cylinder exhaust cam lobes and their followers

10 Rotate the crankshaft pulley through a half a turn (180°) to position No 4 cylinder inlet camshaft lobes and No 1 cylinder exhaust camshaft lobes pointing upwards and slightly towards the centre. Measure the clearance between the base of the camshaft lobes and their followers and record the clearances on the paper.

11 If all the clearances are correct, refit the camshaft cover (see Section 4), then refit the wheel arch liner and the roadwheel. Lower the vehicle to the ground and tighten the wheel nuts to the specified torque. If any clearance measured is not correct, adjustment must be carried out as described in the following paragraphs.

Adjustment

12 If adjustment is necessary, remove the relevant camshaft(s) and camshaft followers as described in Section 12.

13 Clean the followers of the valves that require clearance adjustment and note the thickness marking on the follower. The thickness marking is stamped on the underside of each follower. For example, a 3.20 mm thick follower will have a 20 thickness marking, a 3.21 mm thick follower will have a 21 thickness marking etc.

14 Add the measured clearance of the valve to the thickness of the original follower then subtract the specified valve clearance from this figure. This will give you the thickness of the follower required. For example:

Clearance measured of inlet valve	0.31 mm
Plus thickness of original follower	3.20 mm
Equals	3.51 mm
Minus clearance required	0.25 mm
Thickness of follower required	3.26 mm

15 Repeat this procedure on the remaining valves which require adjustment, then obtain the correct thickness of follower(s) required.

16 Refit the camshaft followers and the relevant camshaft(s) as described in Section 12. Rotate the crankshaft a few times to settle all the components, then recheck the valve clearances before refitting the camshaft cover (Section 4).

17 Refit the wheel arch liner and roadwheel then lower the vehicle to the ground and tighten the wheel nuts to the specified torque.

12 Camshafts and followers – removal, inspection and refitting

Note: *New timing belt end oil seals, and a tube of suitable sealant will be required for refitting.*

Removal

1 Remove the camshaft cover as described in Section 4.

2 Remove the timing belt as described in Section 7.

3 Remove the camshaft sprockets as described in Section 8.

4 Remove the timing belt rear cover as described in Section 6.

5 Remove the special tool used to lock the camshafts in the TDC position.

6 Remove the camshaft position sensor(s) as described in Chapter 4A, Section 12.

7 Disconnect the wiring connector at the inlet camshaft and exhaust camshaft VVT oil control valves **(see illustration 9.2)**.

8 Undo the four bolts securing the camshaft bearing support at the timing belt end of the engine **(see illustration)**. Undo the two outer bolts first, followed by the two inner bolts. Using a plastic mallet, gently tap the bearing support free and remove it from the cylinder head.

9 Starting on the inlet camshaft, working in a spiral pattern from the outside inwards, slacken the camshaft bearing cap retaining bolts by half a turn at a time, to relieve the pressure of the valve springs on the bearing caps gradually and evenly (the *reverse* of **illustration 12.21**). Once the valve spring pressure has been relieved, the bolts can be fully unscrewed and removed along with the caps; the bearing caps are numbered inlet camshaft 2 to 5, exhaust camshaft 6 to 9 to ensure the caps are correctly positioned on refitting **(see illustration)**. Take care not to lose the locating dowels (where fitted).

Caution: If the bearing cap bolts are carelessly slackened, the bearing caps might break. If any bearing cap breaks then the complete cylinder head assembly must be renewed; the bearing caps are matched to the head and are not available separately.

10 Lift the camshaft out of the cylinder head and slide off the oil seal.

11 Repeat the operations described in paragraphs 9 and 10 and remove the exhaust camshaft.

12 Obtain sixteen small, clean plastic containers, and label them for identification. Alternatively, divide a larger container into compartments. Using a rubber sucker tool, lift the followers out from the top of the cylinder head and store each one in its respective fitted position **(see illustration)**.

Inspection

13 Examine the camshaft bearing surfaces and cam lobes for signs of wear ridges and scoring. Renew the camshaft if any of these conditions are apparent. Examine the condition of the bearing surfaces both on the camshaft journals and in the cylinder head. If the head bearing surfaces are worn excessively, the cylinder head will need to be renewed.

14 Examine the follower bearing surfaces which contact the camshaft lobes for wear ridges and scoring. Check the followers and their bores in the cylinder head for signs of wear or damage. If any follower is thought to be faulty or is visibly worn it should be renewed.

12.8 Camshaft bearing support retaining bolts

Refitting

15 Commence refitting by turning the crankshaft anti-clockwise by 60°. This will position Nos 1 and 4 pistons a third of the way down the bore, and prevent any chance of the valves touching the piston crowns as the camshafts are being fitted.

16 Thoroughly clean the mating surfaces of the camshaft bearing support and cylinder head, ensuring all traces of old sealant are removed.

17 Where removed, lubricate the followers with clean engine oil and carefully insert each one into its original location in the cylinder head.

18 Lubricate the camshaft followers with molybdenum disulphide paste (or clean engine oil) then lay the camshafts in position.

19 Ensure the mating surfaces of the bearing caps and cylinder head are clean and dry and lubricate the camshaft journals and lobes with clean engine oil.

20 Ensure the locating dowels (where fitted) are in position then refit camshaft bearing caps 2 to 9 and the retaining bolts in their original locations on the cylinder head.

21 Working on the inlet camshaft, tighten the bearing cap bolts by hand only then, working in a spiral pattern from the centre outwards, tighten the bolts by half a turn at a time to gradually impose the pressure of the valve springs on the bearing caps **(see illustration)**. Repeat this sequence until all bearing caps are in contact with the cylinder head then go around and tighten the

12.12 Use a valve lapping tool to remove the cam followers

12.9 Camshaft bearing cap numbers (inlet camshaft shown)

camshaft bearing cap bolts to the specified torque.

Caution: If the bearing cap bolts are carelessly tightened, the bearing caps might break. If any bearing cap breaks then the complete cylinder head assembly must be renewed; the bearing caps are matched to the head and are not available separately.

22 Tighten the exhaust camshaft bearing cap bolts as described in paragraph 21.

23 Apply a smear of sealant to the mating surface of the camshaft bearing support, ensuring that the oil grooves remain free of sealant. Do not apply sealant to the area immediately adjacent to the bearing surface on the inside of the oil groove.

24 Place the camshaft bearing support in position and refit the four retaining bolts. Tighten the bolts to the specified torque, starting with the two inner bolts, then the two outer bolts.

25 If new components have been fitted, the valve clearances should now be checked and, if necessary adjusted, before proceeding with the refitting procedure. Temporarily refit the camshaft sprockets and secure with their retaining bolts, to allow the camshafts to be turned for the check. Refer to the procedure contained in Section 11, but as the timing belt is not fitted, check the clearances of each camshaft individually. Use the sprocket retaining bolt to turn the camshafts as necessary until the cam lobes for each pair of valves are pointing upward, away from the valves.

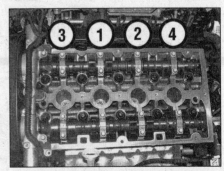

12.21 Camshaft bearing cap tightening sequence (inlet camshaft shown – exhaust identical)

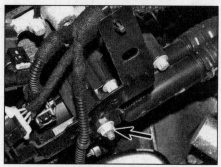

13.17 Undo the nut securing the wiring harness bracket to the thermostat housing

26 Once the valve clearances have been checked and if necessary adjusted, position the camshafts so that their timing slots are parallel with the cylinder head surface and refit the locking tool to the slots. Undo the retaining bolts and remove the camshaft sprockets.

27 Refit the camshaft position sensor(s) as described in Chapter 4A, Section 12.

28 Reconnect the wiring connector at the inlet camshaft and exhaust camshaft VVT oil control valves.

29 Refit the timing belt rear cover as described in Section 6.

30 Fit new camshaft oil seals as described in Section 10.

31 Refit the camshaft sprockets as described in Section 8.

32 Align all the sprocket timing marks to bring the camshafts and crankshaft back to

TDC then refit the timing belt as described in Section 7.

33 Refit the camshaft cover as described in Section 4.

13 Cylinder head – removal and refitting

Note: The engine must be cold when removing the cylinder head. A new cylinder head gasket and new cylinder head bolts must be used for refitting.

Removal

1 Depressurise the fuel system as described in Chapter 4A, Section 5.

2 Disconnect the battery negative lead as described in Chapter 5A Section 4.

3 Apply the handbrake, then jack up the front of the vehicle and support it on axle stands (see *Jacking and vehicle support*). Remove the right-hand front roadwheel, then remove the wheel arch liner as described in Chapter 11, Section 21 for access to the right-hand side of the engine.

4 Drain the cooling system as described in Chapter 1A, Section 25.

5 Remove the spark plugs as described in Chapter 1A, Section 27.

6 Remove the camshaft cover as described in Section 4.

7 Remove the VVT oil control valves as described in Section 9.

8 Remove the timing belt as described in Section 7.

9 Remove the camshaft sprockets, timing belt tensioner, and the timing belt idler pulley, as described in Section 8.

10 Remove the rear timing belt cover as described in Section 6.

11 Refer to Chapter 4A, Section 20 and unbolt the exhaust front pipe from the catalytic converter, taking care to support the flexible section. Release the mounting rubbers and support the front of the exhaust pipe to one side.

Note: Angular movement in excess of 10° can cause permanent damage to the front pipe flexible section.

12 Remove the inlet manifold as described in Chapter 4A, Section 14 and the exhaust manifold as described in Chapter 4A, Section 19.

13 Unclip the coolant hoses from the heater matrix unions on the engine compartment bulkhead to drain the coolant from the cylinder block. Once the flow of coolant has stopped, reconnect both hoses and mop up any spilt coolant.

14 Loosen the clips and remove the upper hose from the radiator and thermostat housing.

15 Loosen the clips and disconnect the heater hoses from the left-hand end of the cylinder head or thermostat housing.

16 Disconnect the wiring connectors from the camshaft position sensor(s).

17 Disconnect the thermostat wiring harness connector, then undo the nut securing the wiring harness bracket to the thermostat housing **(see illustration)**. Move the bracket and wiring harness to one side.

18 Make a final check to ensure that all relevant hoses, pipes and wires have been disconnected.

19 Working in the sequence shown, progressively loosen the cylinder head bolts **(see illustration)**. First loosen all the bolts by quarter of a turn, then loosen all the bolts by half a turn, then finally slacken all the bolts fully and withdraw them from the cylinder head. Recover the washers.

20 Lift the cylinder head from the cylinder block. If necessary, tap the cylinder head gently with a soft-faced mallet to free it from the block, but do not lever at the mating faces. Note that the cylinder head is located on dowels.

21 Recover the cylinder head gasket, and discard it.

Preparation for refitting

22 The mating faces of the cylinder head and block must be perfectly clean before refitting the head. Use a scraper to remove all traces of gasket and carbon, and also clean the tops of the pistons. Take particular care with the aluminium surfaces, as the soft metal is damaged easily. Also, make sure that debris is not allowed to enter the oil and water channels – this is particularly important for the oil circuit, as carbon could block the oil supply to the camshaft or crankshaft bearings. Using adhesive tape and paper, seal the water, oil and bolt holes in the cylinder block. To prevent

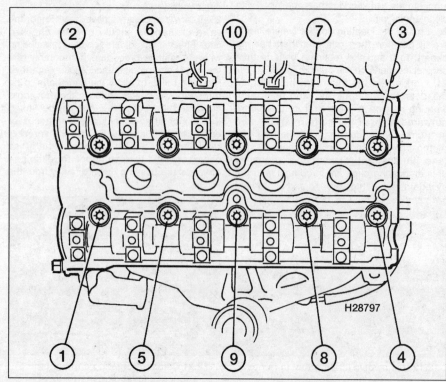

13.19 Cylinder head bolt loosening sequence

13.28 Ensure the head gasket is fitted with its OBEN/TOP marking uppermost

carbon entering the gap between the pistons and bores, smear a little grease in the gap. After cleaning the piston, rotate the crankshaft so that the piston moves down the bore, then wipe out the grease and carbon with a cloth rag. Clean the other piston crowns in the same way.

23 Check the block and head for nicks, deep scratches and other damage. If slight, they may be removed carefully with a file. More serious damage may be repaired by machining, but this is a specialist job.

24 If warpage of the cylinder head is suspected, use a straight-edge to check it for distortion. Refer to Chapter 2E if necessary.

25 Ensure that the cylinder head bolt holes in the crankcase are clean and free of oil. Syringe or soak up any oil left in the bolt holes. This is most important in order that the correct bolt tightening torque can be applied and to prevent the possibility of the block being cracked by hydraulic pressure when the bolts are tightened.

26 Renew the cylinder head bolts regardless of their apparent condition.

Refitting

27 Ensure that the two locating dowels are in position at each end of the cylinder block/crankcase surface.

28 Fit the new cylinder head gasket to the block, making sure it is fitted with the correct way up with its OBEN/TOP mark uppermost **(see illustration)**.

29 Carefully refit the cylinder head, locating it on the dowels.

30 Fit the washers to the new cylinder head bolts then carefully insert them into position (do not drop), tightening them finger-tight only at this stage.

31 Working progressively and in the sequence shown, first tighten all the cylinder head bolts to the Stage 1 torque setting **(see illustrations)**.

32 Once all bolts have been tightened to the Stage 1 torque, again working in the sequence shown, tighten each bolt through its specified Stage 2 angle, using a socket and extension bar. It is recommended that an angle-measuring gauge is used during this stage of the tightening, to ensure accuracy **(see illustration)**.

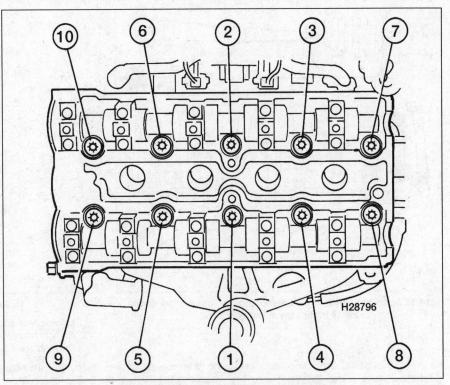

13.31a Cylinder head bolt tightening sequence

33 Working in the specified sequence, go around again and tighten all bolts through the specified Stage 3 angle.

34 Working again in the specified sequence, go around and tighten all bolts through the specified Stage 4 angle.

35 Finally go around in the specified sequence again and tighten all bolts through the specified Stage 5 angle.

36 Refit the wiring harness bracket to the thermostat housing and secure with the retaining nut tightened securely. Reconnect the thermostat wiring harness connector.

37 Reconnect the wiring connectors to the camshaft position sensor(s).

38 Reconnect the heater hoses to the left-hand end of the cylinder head or thermostat housing and tighten the clips.

39 Refit the upper hose to the radiator and thermostat housing and tighten the clips.

40 Refit the exhaust manifold as described in Chapter 4A, Section 19.

41 Refit the inlet manifold as described in Chapter 4A, Section 14.

42 Refer to Chapter 4A, Section 20 and refit the exhaust front pipe to the catalytic converter. Refit the mounting rubbers.

43 Refit the rear timing belt cover as described in Section 6.

44 Refit the camshaft sprockets, timing belt tensioner, and the timing belt idler pulley, with reference to Section 8.

45 Refit the timing belt as described in Section 7.

46 Refit the VVT oil control valves as described in Section 9.

47 Refit the camshaft cover as described in Section 4.

48 Refit the spark plugs as described in Chapter 1A, Section 27.

13.31b Tighten the cylinder head bolts to the specified Stage 1 torque setting...

13.32 ...and then through the various specified angles as described in the text

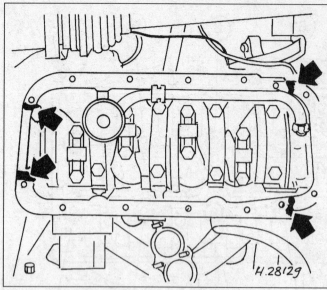

14.11 Apply sealant to the oil pump and rear main bearing cap joints before the sump is refitted

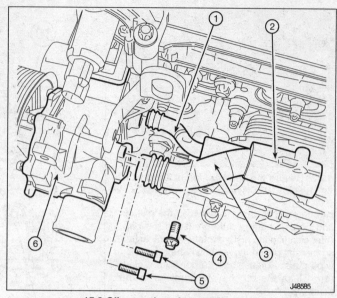

15.9 Oil pump housing attachments

1 Lower metal coolant pipe
2 Oil filter housing
3 Upper metal coolant pipe
4 Support bracket bolt
5 Upper metal coolant pipe bolts

49 Refit the wheel arch liner and front roadwheel, then lower the vehicle to the ground. Tighten the roadwheel nuts to the specified torque.
50 Reconnect the battery negative terminal.
51 Check that all relevant hoses, pipes and wires, etc, have been reconnected. Check the security of the fuel hose connections.
52 Refill and bleed the cooling system as described in Chapter 1A, Section 25.
53 When the engine is started, check for signs of oil or coolant leakage.

14 Sump –
removal and refitting

Removal

1 Disconnect the battery negative lead as described in Chapter 5A Section 4.
2 Apply the handbrake, then jack up the front of the vehicle and support it on axle stands (see *Jacking and vehicle support*). Remove the right-hand front roadwheel then remove the wheel arch liner as described in Chapter 11, Section 21 for access to the right-hand side of the engine.
3 Drain the engine oil as described in Chapter 1A, Section 5, then fit a new seal and refit the drain plug, tightening it to the specified torque (see Chapter 1A).
4 Remove the exhaust system front pipe as described in Chapter 4A, Section 20.
5 Undo the retaining bolt and remove the oil dipstick guide tube.
6 Slacken and remove the three bolts securing the sump flange to the transmission housing.

7 Progressively slacken and remove the bolts securing the sump to the base of the cylinder block/oil pump. Using a wide bladed scraper or similar tool inserted between the sump and cylinder block, carefully break the joint, then remove the sump from under the car.
8 With the sump removed, undo the retaining bolts and remove the baffle plate.

Refitting

9 Thoroughly clean the sump and baffle plate, then remove all traces of sealant and oil from the mating surfaces of the sump and cylinder block. With the sump thoroughly cleaned, refit the baffle plate and tighten the bolts securely.
10 Ensure the sump and cylinder block mating surfaces are clean and dry and remove all traces of locking compound from the sump bolts.
11 Apply a 3.5 mm bead of suitable sealant (available form Vauxhall/Opel dealers) to the areas of the cylinder block mating surface around the areas of the of the oil pump housing and rear main bearing cap joints **(see illustration)**.
12 Apply a 3.5 mm continuous bead of suitable sealant (available from your Vauxhall/Opel dealer) to the sump mating face. The bead should be close to the inner edge of the sump and on the inside of the retaining bolt holes.
13 Offer up the sump, and loosely refit all the retaining bolts. Working out from the centre in a diagonal sequence, progressively tighten the bolts securing the sump to the cylinder block/oil pump to the specified torque setting.
14 Tighten the bolts securing the sump flange to the transmission housing to their specified torque settings.
15 Refit the oil dipstick guide tube.

16 Refit the exhaust system front pipe as described in Chapter 4A, Section 20.
17 Refit the wheel arch liner and front roadwheel, then lower the vehicle to the ground. Tighten the roadwheel nuts to the specified torque.
18 Fill the engine with fresh oil as described in Chapter 1A, Section 5.

15 Oil pump –
removal, overhaul and refitting

Removal

1 Drain the cooling system as described in Chapter 1A, Section 25.
2 Remove the alternator as described in Chapter 5A, Section 8.
3 Disconnect the wiring from the air conditioning compressor and release the wiring harness from the cable clips. Undo the three mounting bolts and support the compressor clear of the engine. Do not disconnect the refrigerant lines from the compressor.
4 Remove the exhaust manifold as described in Chapter 4A, Section 19.
5 Remove the timing belt as described in Section 7.
6 Remove the timing belt tensioner and idler pulley, and the crankshaft sprocket as described in Section 8.
7 Remove the sump as described in Section 14.
8 Release the clamp and disconnect the coolant hose from the coolant pump.
9 Undo the two bolts securing the upper metal coolant pipe to the rear of the coolant pump **(see illustration)**.

10 Undo the support bracket bolt securing the lower metal coolant pipe to the oil filter housing.

11 Release the upper and lower metal coolant pipes from the coolant pump by pushing them into the oil filter housing.

12 Slacken and remove the eight retaining bolts (noting their different lengths) then slide the oil pump housing assembly off of the end of the crankshaft, taking great care not to lose the locating dowels. Remove the housing gasket and discard it.

13 Remove the metal coolant pipe seals.

Overhaul

14 Remove the securing screws and withdraw the oil pump cover from the rear of the oil pump housing (see illustration).

15 Remove the inner and outer rotor from the pump housing, noting which way round they are fitted, and wipe them clean. Also clean the rotor location in the oil pump housing.

16 The oil pressure relief valve components can also be removed from the oil pump housing by unscrewing the cap. Withdraw the cap, spring and plunger (see illustrations).

17 Locate the inner and outer rotor back in the oil pump housing, ensuring they are fitted the right way round as noted during removal.

18 Check the clearance between the end faces of the rotors and the housing (endfloat) using a straight-edge and a feeler gauge (see illustration).

19 If the clearance is outside the specified limits, renew the components as necessary.

20 Examine the pressure relief valve spring and plunger, and renew if any sign of damage or wear is evident.

21 Ensure that the rotor location in the interior of the oil pump housing is scrupulously clean before commencing reassembly.

22 Thoroughly clean the pressure relief valve components, and lubricate them with clean engine oil before refitting. Insert the plunger and spring, then refit the cap and tighten to the specified torque.

23 Ensure that the rotors are clean, then lubricate them with clean engine oil, and refit them to the pump body ensuring they are fitted the right way round as noted during removal.

24 Wipe clean the mating faces of the rear cover and the pump housing, then refit the rear cover. Refit and tighten the securing screws securely. Prime the oil pump by filling it with clean engine oil whilst rotating the inner rotor

Refitting

25 Prior to refitting, carefully lever out the crankshaft oil seal using a flat-bladed screwdriver. Fit the new oil seal, ensuring its sealing lip is facing inwards, and press it squarely into the housing using a tubular drift which bears only on the hard outer edge of the seal. Press the seal into position so that it is flush with the housing.

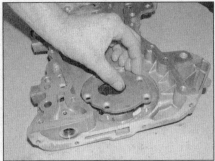

15.14 Remove the securing screws and withdraw the oil pump cover from the rear of the housing

26 Fit new seals to the two metal coolant pipes.

27 Ensure the mating surfaces of the oil pump housing and cylinder block are clean and dry and the locating dowels are in position.

28 Fit a new gasket to the cylinder block.

29 Carefully manoeuvre the oil pump into position and engage the inner rotor with the crankshaft end. Engage the coolant pipes, then locate the pump on the dowels, taking great care not damage the oil seal lip.

30 Refit the pump housing retaining bolts in their original locations and tighten them to the specified torque.

31 Refit the metal coolant pipe retaining bolts and tighten securely.

32 Reconnect the coolant hose to the pump and secure with the retaining clip.

33 Refit the sump as described in Section 14.

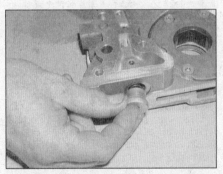

15.16a Unscrew the oil pressure relief valve cap...

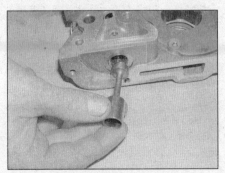

15.16c ...and the plunger

34 Refit the timing belt tensioner, idler pulley and crankshaft sprocket as described in Section 8.

35 Refit the timing belt as described in Section 7.

36 Refit the exhaust manifold as described in Chapter 4A, Section 19.

37 Refit the air conditioning compressor and tighten the mounting bolts to the specified torque. Reconnect the wiring and secure the wiring harness with the cable clips.

38 Refit the alternator as described in Chapter 5A, Section 8.

39 On completion, refer to Chapter 1A, Section 5 and fit a new oil filter and fill the engine with clean oil.

40 Refill the cooling system as described in Chapter 1A, Section 25.

16 Crankshaft oil seals – renewal

Right-hand (timing belt end)

1 Remove the crankshaft sprocket as described in Section 8.

2 Carefully punch or drill two small holes opposite each other in the oil seal. Screw a self-tapping screw into each and pull on the screws with pliers to extract the seal (see illustration).

Caution: Great care must be taken to avoid damage to the oil pump

3 Clean the seal housing and polish off any

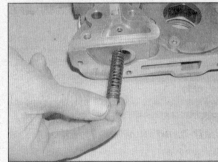

15.16b ...then withdraw the spring...

15.18 Check the oil pump rotor endfloat using a straight-edge and feeler gauge

16.2 Removing the crankshaft oil seal

16.4 Fitting a new crankshaft oil seal

Removal

Note: *New flywheel retaining bolts will be required on refitting.*

1 Remove the transmission as described in Chapter 7A, Section 8 then remove the clutch assembly as described in Chapter 6, Section 6.

2 Prevent the flywheel from turning by locking the ring gear teeth with a similar arrangement to that shown **(see illustration)**.

3 Slacken and remove the retaining bolts and remove the flywheel.

Caution: Take care, as the flywheel is heavy.

Inspection

4 Examine the flywheel for wear or chipping of the ring gear teeth. Renewal of the ring gear is not possible and if the wear or chipping is significant, a new flywheel will be required.

5 Examine the flywheel for scoring of the clutch face. If the clutch face is scored significantly, a new flywheel will be required.

6 If there is any doubt about the condition of the flywheel, seek the advice of a Vauxhall/Opel dealer or engine reconditioning specialist.

Refitting

7 Clean the mating surfaces of the flywheel and crankshaft.

8 Offer up the flywheel and engage it on the crankshaft. Apply a drop of locking compound to the threads of each new flywheel retaining bolt (unless they are already precoated) and install the new bolts.

9 Lock the flywheel by the method used on removal then, working in a diagonal sequence, evenly and progressively tighten the retaining bolts to the specified torque and through the specified angles.

10 Refit the clutch as described in Chapter 6, Section 6, then remove the locking tool and refit the transmission as described in Chapter 7A, Section 8.

16.7 Left-hand crankshaft oil seal – transmission and flywheel removed

18.2 Prevent the flywheel from turning by locking the ring gear teeth with a suitable tool

burrs or raised edges which may have caused the seal to fail in the first place.

4 Carefully ease the new seal into position on the end of the crankshaft. Press the seal squarely into position until it is flush with the housing. If necessary, a suitable tubular drift, such as a socket, which bears only on the hard outer edge of the seal can be used to tap the seal into position **(see illustration)**. Take great care not to damage the seal lips during fitting and ensure that the seal lips face inwards.

5 Wash off any traces of oil, then refit the crankshaft sprocket as described in Section 8.

Left-hand (flywheel end)

6 Remove the flywheel as described in Section 18.

7 Renew the seal as described in paragraphs 2 to 4 **(see illustration)**.

8 Refit the flywheel as described in Section 18.

17 Oil filter housing – removal and refitting

Removal

1 Remove the exhaust manifold as described in Chapter 4A, Section 19.

2 Undo the two bolts securing the upper metal coolant pipe to the rear of the coolant pump **(see illustration 15.9)**.

3 Undo the support bracket bolt securing the lower metal coolant pipe to the oil filter housing.

4 Release the upper and lower metal coolant pipes from the coolant pump by pushing them into the oil filter housing.

5 Undo the two bolts securing the metal coolant pipe to the thermostat housing. Pull the pipe out of the filter housing and collect the O-ring seals. Note that new seals will be required for refitting.

6 Undo the remaining retaining bolts and withdraw the oil filter housing from the cylinder block. Recover the rubber seals, noting that new seals will be required for refitting. Be prepared for oil spillage.

Refitting

7 Refitting is a reversal of the removal procedure, bearing in mind the following points:

a) *Fit new seals and O-rings to all disturbed connections.*

b) *Tighten the filter housing retaining bolts to the specified torque.*

c) *Refit the exhaust manifold as described in Chapter 4A, Section 19.*

d) *Top up the engine oil level as described in 'Weekly checks'.*

19 Engine/transmission mountings – inspection and renewal

1 Refer to Chapter 2A, Section 16, but observe the different torque wrench settings given in the Specifications of this Chapter.

Chapter 2 Part C
1.6 litre diesel engine in-car repair procedures

Contents

Degrees of difficulty

Easy, suitable for novice with little experience	**Fairly easy,** suitable for beginner with some experience	**Fairly difficult,** suitable for competent DIY mechanic	**Difficult,** suitable for experienced DIY mechanic	**Very difficult,** suitable for expert DIY or professional

Specifications

General

Engine type.	Four-cylinder, in-line, water-cooled. Chain-driven double overhead camshafts, 16 valves
Manufacturer's engine code*	B16DTH (LVL) or B16DTN (LVL)
Bore	79.7 mm
Stroke	80.1 mm
Capacity	1598 cc
Firing order.	1-3-4-2 (No 1 cylinder at right-hand end of engine)
Direction of crankshaft rotation	Clockwise (viewed from right-hand end of engine)
Compression ratio	16.0: 1

For details of engine code location, see 'Vehicle identification' in Reference

Compression pressures

Minimum	20.0 bar

Lubrication system

Oil pump type.	Variable displacement vane type with integral vacuum pump
Oil pressure at 80°C (approximate)	1.0 bar at idle speed

Torque wrench settings

	Nm	lbf ft
Auxiliary drivebelt tensioner-to-cylinder block....................	55	41
Big-end bearing cap bolts*:		
Stage 1..	25	18
Stage 2..	Angle-tighten a further 110°	
Camshaft bearing cap bolts (except rear bearing cap)............	10	7
Camshaft bearing cap bolts (rear bearing cap)...................	25	18
Camshaft cover bolts...................................	10	7
Camshaft sprocket bolts	25	18
Crankshaft pulley bolts*:		
Stage 1..	50	37
Stage 2..	Angle-tighten a further 45°	
Cylinder block baseplate bolts:		
M10 bolts (bolts 1 to 10)*:		
Stage 1..	50	37
Stage 2..	Angle-tighten a further 90°	
Stage 3..	Angle-tighten a further 90°	
M8 bolts (bolts 11 to 21)...........................	34	25
Cylinder head bolts (except two outer bolts)*:		
Stage 1..	50	37
Stage 2..	Angle-tighten a further 90°	
Stage 3..	Angle-tighten a further 90°	
Stage 4..	Angle-tighten a further 90°	
Cylinder head bolts (two outer bolts)*:		
Stage 1..	20	15
Stage 2..	Angle-tighten a further 45°	
Engine lifting bracket bolts................................	58	43
Engine mountings:		
Left-hand mounting:		
Mounting-to-body bolts/nut...........................	58	43
Mounting bracket-to-transmission bracket*:		
Stage 1..	50	37
Stage 2..	Angle-tighten a further 60°	
Rear mounting/torque link:		
Mounting-to-transmission bracket*:		
Stage 1..	80	59
Stage 2..	Angle-tighten a further 60°	
Mounting-to-subframe*:		
Stage 1..	100	74
Stage 2..	Angle-tighten a further 100°	
Right-hand mounting:		
Mounting bracket-to-engine...........................	62	46
Mounting-to-body...................................	62	46
Mounting-to-mounting bracket*:		
Stage 1..	50	37
Stage 2..	Angle-tighten a further 60°	
Engine-to-transmission bolts...............................	58	43
Flywheel/driveplate bolts*:		
Stage 1..	60	44
Stage 2..	Angle-tighten a further 45°	
Fuel pump sprocket retaining nut	64	47
Idler sprocket retaining bolts	25	18
Main vacuum pipe assembly retaining bolts	10	7
Oil cooler retaining bolts.................................	25	18
Oil pump chain tensioner bolts:		
Lower bolt....................................	10	7
Upper bolt....................................	25	18
Oil pump retaining bolts.................................	25	18
Oil pump sprocket bolt*:		
Stage 1..	25	18
Stage 2..	Angle-tighten a further 60°	
Roadwheel nuts..	140	103
Sump-to-cylinder block baseplate and timing cover...............	25	18
Sump-to-transmission...................................	58	43
Timing chain guide rail bolts..............................	10	7
Timing chain tensioner bolts	10	7
Timing chain tensioner rail pivot bolt.........................	25	18
Timing cover retaining bolts...............................	25	18
Timing cover sleeve bolt.................................	25	18

Use new fasteners

1 General Information

How to use this Chapter

1 This Part of Chapter 2 describes the repair procedures which can reasonably be carried out on the engine while it remains in the vehicle. If the engine has been removed from the vehicle and is being dismantled as described in Chapter 2E, any preliminary dismantling procedures can be ignored.

2 Note that, while it may be possible physically to overhaul items such as the piston/connecting rod assemblies while the engine is in the vehicle, such tasks are not usually carried out as separate operations, and usually require the execution of several additional procedures (not to mention the cleaning of components and of oil ways); for this reason, all such tasks are classed as major overhaul procedures, and are described in Chapter 2E.

3 Chapter 2E describes the removal of the engine/transmission unit from the vehicle, and the full overhaul procedures which can then be carried out.

Engine description

4 The 1.6 litre common-rail diesel engine is of sixteen-valve, in-line four-cylinder, double overhead camshaft (DOHC) type, mounted transversely at the front of the car with the transmission attached to its left-hand end.

5 The crankshaft runs in five main bearings. Thrustwashers are fitted to No 3 main bearing shell (upper half) to control crankshaft endfloat.

6 The cylinder block is made of aluminium incorporating cast iron dry liners. An aluminium baseplate is bolted to the cylinder block and forms the lower half of the crankcase.

7 The connecting rods rotate on horizontally-split bearing shells at their big-ends. The pistons are attached to the connecting rods by gudgeon pins, which are a sliding fit in the connecting rod small-end eyes and retained by circlips. The aluminium-alloy pistons are fitted with three piston rings – two compression rings and an oil control ring.

8 The camshafts run directly in the cylinder head. The inlet camshaft is driven by the crankshaft by a hydraulically tensioned timing chain and drives the exhaust camshaft via a spur gear. Each cylinder has four valves (two inlet and two exhaust), operated via rocker arms which are supported at their pivot ends by hydraulic self-adjusting valve lifters (tappets). One camshaft operates the inlet valves, and the other operates the exhaust valves. In what is an unusual configuration, the timing chain and timing gear components are situated at the left-hand (flywheel/driveplate) end of the engine

9 The inlet and exhaust valves are each closed by a single valve spring, and operate in guides pressed into the cylinder head.

10 A variable displacement vane-type oil pump is located below the cylinder block baseplate, and is driven by a chain from the left-hand end of the crankshaft. The brake servo vacuum pump is integral with the oil pump.

11 The coolant pump is located externally at the right-hand end of the engine, and is driven by the auxiliary drivebelt.

Operations with engine in place

12 The following operations can be carried out without having to remove the engine from the vehicle.

a) Removal and refitting of the cylinder head.
b) Removal and refitting of the timing covers.
c) Removal and refitting of the timing chain, tensioner, sprockets and guide rails.
d) Removal and refitting of the hydraulic tappets and rocker arms.
e) Removal and refitting of the camshafts.
f) Removal and refitting of the sump.
g) Removal and refitting of the big-end bearings, connecting rods, and pistons*.
h) Removal and refitting of the oil pump.
i) Removal and refitting of the oil cooler.
j) Renewal of the crankshaft oil seals.
k) Renewal of the engine mountings.
l) Removal and refitting of the flywheel/driveplate.

*Although the operation marked with an asterisk can be carried out with the engine in the vehicle (after removal of the sump), it is preferable for the engine to be removed, in the interests of cleanliness and improved access. For this reason, the procedure is described in Chapter 2E.

2 Compression test – description and interpretation

Compression test

Note: *A compression tester specifically designed for diesel engines must be used for this test.*

1 When engine performance is down, or if misfiring occurs which cannot be attributed to the fuel system, a compression test can provide diagnostic clues as to the engine's condition. If the test is performed regularly, it can give warning of trouble before any other symptoms become apparent.

2 A compression tester specifically intended for diesel engines must be used, because of the higher pressures involved. The tester is connected to an adapter which screws into the glow plug or injector hole. On these models, an adapter suitable for use in the injector holes will be required. It is unlikely to be worthwhile buying such a tester for occasional use, but it may be possible to borrow or hire one – if not, have the test performed by a garage.

3 Unless specific instructions to the contrary are supplied with the tester, observe the following points:

a) The battery must be in a good state of charge, the air filter must be clean, and the engine should be at normal operating temperature.
b) All the fuel injectors must be removed before starting the test (see Chapter 4B, Section 16).
c) Open the cover on the engine compartment fuse/relay box and remove the fuel pump relay. Refer to the wiring diagrams in Chapter 14 for information on fuse and relay locations.

4 Screw the compression tester and adapter in to the fuel injector hole of No 1 cylinder.

5 With the help of an assistant, crank the engine on the starter motor; after one or two revolutions, the compression pressure should build-up to a maximum figure, and then stabilise. Record the highest reading obtained.

6 Repeat the test on the remaining cylinders, recording the pressure in each.

7 All cylinders should produce very similar pressures; any difference greater than that specified indicates the existence of a fault. Note that the compression should build-up quickly in a healthy engine; low compression on the first stroke, followed by gradually-increasing pressure on successive strokes, indicates worn piston rings. A low compression reading on the first stroke, which does not build-up during successive strokes, indicates leaking valves or a blown head gasket (a cracked head could also be the cause).

Note: *The cause of poor compression is less easy to establish on a diesel engine than on a petrol one. The effect of introducing oil into the cylinders ('wet' testing) is not conclusive, because there is a risk that the oil will sit in the recess on the piston crown instead of passing to the rings.*

8 On completion of the test, refit the fuel pump relay, then refit the fuel injectors as described in Chapter 4B, Section 16.

Leakdown test

9 A leakdown test measures the rate at which compressed air fed into the cylinder is lost. It is an alternative to a compression test, and in many ways it is better, since the escaping air provides easy identification of where pressure loss is occurring (piston rings, valves or head gasket).

10 The equipment needed for leakdown testing is unlikely to be available to the home mechanic. If poor compression is suspected, have the test performed by a Vauxhall/Opel dealer or suitably-equipped garage.

3 Engine assembly/valve timing tools – general information and usage

Note: *Do not attempt to rotate the engine whilst the camshafts and crankshaft are locked in position. If the engine is to be left in this state for a long period of time, it is a good*

3.6 Camshaft locking tool inserted through the bearing cap and into the exhaust camshaft timing gear

3.7 Crankshaft locking tool securing the crankshaft in the TDC position

4 Crankshaft pulley – removal and refitting

Removal

1 Firmly apply the handbrake, then jack up the front of the car and support it securely on axle stands (see *Jacking and vehicle support*), Section. Remove the right-hand roadwheel, then remove the wheel arch liner as described in Chapter 11, Section 21.

2 Remove the auxiliary drivebelt as described in Chapter 1B, Section 30.

3 It will now be necessary to hold the crankshaft pulley to enable the retaining bolts to be removed. Vauxhall/Opel special tools EN-49979 and EN-49979-100 are available for this purpose, however, a home-made tool can easily be fabricated **(see Tool Tip)**.

4 Using the holding tool to prevent rotation of the crankshaft, slacken the four pulley retaining bolts. Remove the holding tool, unscrew the retaining bolts and remove the pulley **(see illustration)**. Note that new pulley retaining bolts will be required for refitting.

Refitting

5 Locate the pulley in position on the crankshaft.

6 Fit the new pulley retaining bolts and tighten them to the specified torque, then through the specified angle while preventing crankshaft rotation using the method employed on removal.

7 Refit the auxiliary drivebelt as described in Chapter 1B, Section 30.

8 Refit the wheel arch liner as described in Chapter 11, Section 21, then refit the roadwheel and lower the car to the ground. Tighten the wheel nuts to the specified torque.

5 Camshaft cover – removal and refitting

Removal

1 Disconnect the battery negative lead as described in Chapter 5A Section 4.

2 Remove the fuel injectors as described in Chapter 4B, Section 15.

3 Disconnect the wiring connector at the fuel pressure sensor on the fuel rail **(see illustration)**.

4 Remove the oxygen sensor from the top of the diesel particulate filter as described in Chapter 4C, Section 3.

5 Undo the nut and four bolts securing the upper heat shield to the turbocharger. Remove the washer from the particulate filter stud, then lift off the heat shield **(see illustrations)**.

6 Undo the two bolts and remove the turbocharger inner heat shield **(see illustration)**.

7 Release the two retaining clips and

idea to place suitable warning notices inside the car, and in the engine compartment. This will reduce the possibility of the engine being accidentally cranked on the starter motor, which is likely to cause damage with the locking tools in place.

1 To accurately set the valve timing for all operations requiring removal and refitting of the timing chain, a hole is machined in the exhaust camshaft timing gear to allow a locking tool to be inserted. When the hole (and locking tool) are aligned with a corresponding hole in the camshaft bearing cap, No 1 piston will be at TDC on its compression stroke. An additional tool is used to lock the crankshaft in the TDC position. With the crankshaft pulley removed, the locking tool is bolted to the crankshaft using the pulley bolts, and a projection on the tool engages with a timing hole in the cylinder block. This arrangement ensures that the correct valve timing can be obtained. The design of the engine is such

that there are no conventional timing marks on the crankshaft or camshafts to indicate the normal TDC position. Therefore, for any work on the timing chain, camshafts or cylinder head, the locking tools must be used.

2 The Vauxhall/Opel tool numbers are as follows:

Camshaft locking tool	EN-51143
Crankshaft locking tool	EN-51140

3 To position the engine at TDC for No 1 piston on compression, first remove the camshaft cover as described in Section 5.

4 Remove the crankshaft pulley as described in Section 4. Refit the crankshaft pulley retaining bolts and screw them in a few turns to allow the crankshaft to be turned.

5 Using a large screwdriver or pry bar engaged with the crankshaft pulley bolts, turn the crankshaft in the normal direction of rotation (clockwise as viewed from the right-hand side of the engine) until the hole in the exhaust camshaft timing gear is aligned with the corresponding hole in the camshaft bearing cap.

6 Insert the camshaft locking tool through the hole in the bearing cap and into the hole in the gear **(see illustration)**.

7 Remove the crankshaft pulley retaining bolts and place the crankshaft locking tool in position. Ensure that the projection on the tool engages with the timing hole in the cylinder block. Refit two of the crankshaft pulley retaining bolts to secure the tool in position and tighten them securely **(see illustration)**.

To make a pulley/sprocket holding tool, obtain two lengths of steel strip about 6 mm thick by about 30 mm wide or similar, one 600 mm long, the other 200 mm long (all dimensions are approximate). Bolt the two strips together to form a forked end, leaving the bolt slack so that the shorter strip can pivot freely. At the other end of each 'prong' of the fork, drill a suitable hole and fit a nut and bolt to allow the tool to engage with the holes in the pulley.

4.4 Undo the four retaining bolts and remove the crankshaft pulley

5.3 Lift the retaining catch and disconnect the fuel pressure sensor wiring connector

5.5a Undo the nut and four bolts...

5.5b ...and lift off the turbocharger upper heat shield

5.6 Undo the two bolts and remove the inner heat shield

5.7 Release the clips and remove the crankcase ventilation hose

5.8a Undo the turbocharger coolant pipe bracket bolt...

remove the crankcase ventilation hose from the camshaft cover and turbocharger (see illustration).

8 Undo the bolts securing the turbocharger wastegate actuator pipe and turbocharger

coolant pipe support brackets to their attachments. Disconnect the pipe hose ends from the vacuum pipe and wastegate actuator and remove the pipe (see illustrations).

9 Undo the retaining bolt and remove

the left-hand engine lifting bracket (see illustration).

10 Undo the bolt securing the right-hand engine lifting bracket to the cylinder head (see illustration).

5.8b ...and wastegate actuator pipe bracket bolt at the left-hand end of the engine...

5.8c ...at the centre of the engine...

5.8d ...and at the right-hand end of the engine...

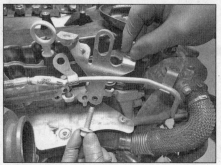

5.8e ...then disconnect the hose ends and remove the wastegate actuator pipe

5.9 Unbolt and remove the left-hand engine lifting bracket

5.10 Undo the bolt securing the right-hand engine lifting bracket to the cylinder head

5.11 Undo the two bolts and move the wiring harness and engine lifting bracket to one side

5.16 Locate the new rubber seal into the grooves of the camshaft cover

11 Undo the two bolts securing the wiring harness support bracket to the camshaft cover **(see illustration)**. Move the wiring harness and the engine lifting bracket to one side.

12 Remove the engine oil level dipstick.

13 Progressively unscrew the fifteen outer bolts followed by the five inner bolts securing the camshaft cover to the cylinder head. Carefully lift up the cover and remove it from the engine.

14 Remove the camshaft cover rubber seal and the washer and rubber seal from each of the retaining bolts, noting the larger diameter of the washers and seals on the inner bolts. Check the condition of the retaining bolt seals and renew as necessary. It is recommended that the cover rubber seal is renewed as a matter of course.

Refitting

15 Ensure that the camshaft cover grove and rubber seal are clean and dry with all traces of oil removed. If necessary, de-grease the seal and cover groove with brake cleaner or a similar product.

16 Locate the rubber seal into the grooves of the camshaft cover, ensuring that it is fully seated, with no chance of it falling out as the cover is fitted **(see illustration)**.

17 Insert the retaining bolts into the camshaft cover ensuring that the bushing and rubber seal are correctly seated **(see illustration)**.

18 Clean the mating surface of the cylinder head ensuring that all traces of oil are removed, then lower the camshaft cover into position. Initially tighten all the bolts lightly to draw the cover down into contact with the cylinder head, then tighten them in the sequence shown to the specified torque **(see illustration)**.

19 Refit the engine oil level dipstick.

20 Locate the wiring harness support bracket back in position on the camshaft cover, then refit and securely tighten the two retaining bolts.

21 Refit the bolt securing the right-hand engine lifting bracket to the cylinder head and tighten the bolt to the specified torque.

22 Refit the left-hand engine lifting bracket and tighten the retaining bolt to the specified torque.

23 Attach the wastegate actuator pipe ends to the vacuum pipe and actuator, then refit the support brackets and tighten the bolts securely.

24 Refit the crankcase ventilation hose to the camshaft cover and turbocharger.

25 Refit the turbocharger inner heat shield

and upper heat shield, securely tightening the retaining bolts.

26 Refit the oxygen sensor to the diesel particulate filter as described in Chapter 4C, Section 3.

27 Reconnect the wiring connector to the fuel pressure sensor on the fuel rail.

28 Refit the fuel injectors as described in Chapter 4B, Section 15.

29 On completion, reconnect the battery negative terminal as described in Chapter 5A Section 4.

6 Timing covers – removal and refitting

Removal

Upper cover

1 Disconnect the battery negative lead as described in Chapter 5A Section 4.

2 Lift off the plastic cover over the top of the engine.

3 Remove the throttle housing as described in Chapter 4B, Section 9.

4 Remove the EGR valve cooler as described in Chapter 4C, Section 3.

5 Remove the diesel particulate filter as described in Chapter 4B, Section 26.

6 Disconnect the brake servo unit vacuum hose at the quick-release connector on the main vacuum pipe assembly.

7 Disconnect the turbocharger wastegate actuator pipe hose end from the main vacuum pipe assembly **(see illustration 5.8e)**.

8 Disconnect the turbocharger wastegate actuator solenoid vacuum hose from the main vacuum pipe assembly **(see illustration)**.

9 Undo the main vacuum pipe assembly retaining bolts at the lower flange, at the centre of the upper timing cover, and at the top, then remove the main vacuum pipe assembly **(see illustrations)**.

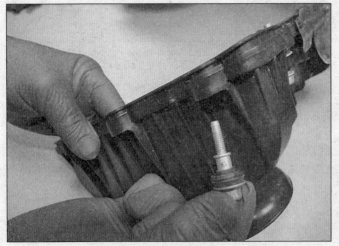

5.17 Insert the bolts into the camshaft cover ensuring the bushing and seal are correctly seated

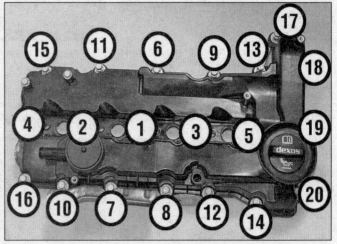

5.18 Camshaft cover bolt tightening sequence

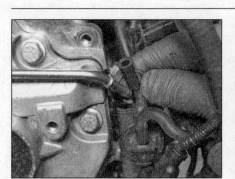

6.8 Disconnect the wastegate actuator solenoid vacuum hose from the main vacuum pipe assembly

6.9a Undo the main vacuum pipe assembly retaining bolts at the lower flange…

6.9b …at the upper timing cover…

6.9c …and at the top…

6.9d …then remove the main vacuum pipe assembly

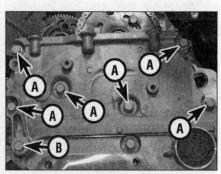

6.11 Upper timing cover retaining bolts (A) and spreading sleeve bolt (B)

10 Remove the camshaft cover as described in Section 5.

11 Undo the seven bolts securing the upper timing cover to the cylinder head **(see illustration)**.

12 Using an Allen key or suitable socket bit, screw in the spreading sleeve to push the upper timing cover from its location. Once the sealant has been initially released, use a knife or spatula to cut through the remaining sealant and allow the upper timing cover to be removed **(see illustrations)**.

13 Once the upper timing cover has been removed, unscrew the spreading sleeve until it is flush with the outer face of the timing cover **(see illustration)**.

14 Thoroughly clean the timing cover and remove all traces of sealant from all the mating surfaces. Similarly clean the lower timing cover, cylinder block and cylinder head mating surfaces. Ensure that all traces of old sealant are removed, particularly from the joint areas between these components.

Lower cover

15 Remove the timing chain upper cover as described previously in this Section.

16 Remove the crankshaft speed/position sensor as described in Chapter 4B, Section 9.

17 Remove the sump as described in Section 12.

18 Remove the transmission as described in Chapter 7A, Section 8 (manual transmission) or Chapter 7B, Section 9 (automatic transmission).

19 Remove the flywheel/driveplate as described in Section 17.

20 Carefully withdraw the crankshaft speed/

6.12a Screw in the spreading sleeve…

6.13 With the upper timing cover removed, unscrew the spreading sleeve until it is flush with the cover outer face

position sensor reluctor ring from the left-hand end of the crankshaft **(see illustration)**.

21 Undo the eleven retaining bolts securing

6.12b …to push the upper timing cover from its location

6.20 Withdraw the crankshaft speed/ position sensor reluctor ring

6.21 Lower timing cover retaining bolt locations

6.25 Fit two new seals to the timing cover recesses

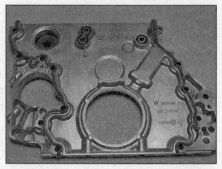

6.26 Apply a bead of sealant to each side of the lower timing cover

6.27a Refit the lower timing cover...

6.27b ...and tighten the retaining bolts in the sequence shown

the lower timing cover to the cylinder block **(see illustration)**.

22 Use a knife or spatula to cut through the sealant, then carefully prise the cover free and remove it from the engine. Recover the two rubber seals from the rear of the cover.

23 Carefully prise out the old crankshaft oil seal with a screwdriver or similar hooked tool. Clean the oil seal seat with a wooden or plastic scraper.

24 Thoroughly clean the timing cover and cylinder block to remove all traces of sealant from all the mating surfaces.

Refitting

Lower cover

25 Fit the two new rubber seals to the recesses on the rear of the lower timing cover **(see illustration)**.

26 Apply a 2 to 3 mm bead of silicone sealant to the mating surfaces on each side of the lower timing cover. The sealant bead should be around the inside of the bolt holes on the sides of the cover **(see illustration)**.

27 Place the lower cover in position on the cylinder block and refit the eleven retaining bolts. Progressively tighten the retaining bolts to the specified torque in the sequence shown **(see illustrations)**.

28 Fit a new crankshaft left-hand oil seal to the lower timing cover as described in Section 15.

29 Refit the crankshaft speed/position sensor reluctor ring.

30 Refit the flywheel/driveplate as described in Section 17.

31 Refit the transmission as described in Chapter 7A, Section 8 (manual transmission) or Chapter 7B, Section 9 (automatic transmission).

32 Refit the sump as described in Section 12.

33 Refit the crankshaft speed/position sensor as described in Chapter 4B, Section 9.

34 Refit the timing chain upper cover as described later in this Section.

Upper cover

35 Apply a 2 to 3 mm bead of silicone sealant to the mating surfaces of the upper timing cover. The sealant bead should be around the inside of the bolt holes on the sides of the cover and there should be two beads, one each side of the groove along the lower face of the cover. Ensure that a bead of sealant is also applied around the two central bolt holes **(see illustrations)**.

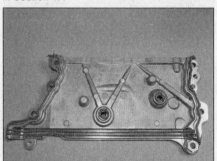

6.35a Apply a bead of sealant to the timing cover mating surfaces...

6.35b ...in the areas shown

6.36 Upper timing cover retaining bolt tightening sequence

6.37 Refit the sleeve bolt and tighten it to the specified torque

36 Place the upper timing cover in position on the engine and refit the six retaining bolts (don't fit the sleeve bolt at this stage). Tighten the bolts to the specified torque in the sequence shown **(see illustration)**.

37 Screw in the spreading sleeve until it just makes contact with the lower timing cover. Refit the sleeve bolt and tighten it to the specified torque **(see illustration)**.

38 Refit the camshaft cover as described in Section 5.

39 Place the main vacuum pipe assembly in position and refit the retaining bolts at the lower flange, at the centre of the upper timing cover, and at the top. Tighten the bolts to the specified torque.

40 Reconnect the turbocharger wastegate actuator pipe hose end and actuator solenoid hose to the main vacuum pipe assembly

41 Resconnect the brake servo unit vacuum hose to the quick-release connector on the main vacuum pipe assembly.

42 Refit the diesel particulate filter as described in Chapter 4B, Section 26.

43 Refit the EGR valve cooler as described in Chapter 4C, Section 3.

44 Refit the throttle housing as described in Chapter 4B, Section 9.

45 Refit the plastic cover over the top of the engine, then reconnect the battery negative terminal.

7 Timing chain – removal and refitting

Removal

1 Remove the lower timing cover as described in Section 6.

2 Refer to Section 3 and set the engine at TDC for No. 1 piston. Insert the camshaft and crankshaft locking tools to retain the engine in the TDC position.

3 Push the timing chain tensioner plunger back into its bore and insert a suitable drill bit or similar tool to retain it in the released position **(see illustration)**.

4 Hold the camshaft using a spanner on the camshaft hexagon, then undo the three bolts and withdraw the sprocket and timing chain from the inlet camshaft **(see illustration)**. Disengage the sprocket from the chain and remove the sprocket.

5 Undo the two bolts and remove the timing chain upper guide rail **(see illustration)**.

6 Disengage the timing chain from the idler sprocket and fuel pump sprocket and remove the chain from the engine **(see illustration)**.

7 Inspect the timing chain, sprockets, tensioner rail and guide rails for any sign of wear or deformation, and renew any suspect components as necessary.

8 Push the timing chain tensioner plunger into the tensioner body and remove the locking drill bit. Check that the tensioner plunger is free to move in out of the tensioner body with no trace of binding. If any binding or sticking of the plunger is felt, renew the tensioner assembly. On completion of the check, or if a new tensioner is being fitted, compress the plunger and refit the locking drill bit.

Refitting

9 Remove all traces of thread locking compound from the timing chain upper guide rail retaining bolt holes by running a suitable tap through the bolt hole threads. Also, suitably clean the threads of the retaining bolts.

10 Hold the timing chain so that the two coloured links, separated by a plain link are uppermost. Half way down the right-hand side of the chain is another coloured link. Engage the chain with the fuel pump sprocket so that the coloured link is adjacent to the dot on the sprocket **(see illustration)**.

11 Feed the chain around the idler sprocket aligning the coloured link with the mark on the sprocket **(see illustration)**.

7.3 Push the timing chain tensioner plunger back into its bore and retain it using a suitable drill bit

7.4 Undo the three bolts and withdraw the sprocket and timing chain from the camshaft

7.5 Remove the timing chain upper guide rail

7.6 Disengage the chain from the sprockets and remove the chain

7.10 Engage the chain with the fuel pump sprocket so that the coloured link is adjacent to the dot

7.11 Feed the chain around the idler sprocket so that the coloured link is aligned with the sprocket mark

7.13a Locate the camshaft sprocket in the chain so that the plain link between the two coloured links is aligned with the dot on the sprocket…

7.13b …then fit the retaining bolts and tighten them to the specified torque

7.14 Withdraw the locking drill bit and release the tensioner plunger

12 Apply thread locking compound to the two upper guide rail retaining bolts. Refit the timing chain upper guide rail and tighten the retaining bolts to the specified torque.

13 Locate the camshaft sprocket in the chain so that the plain link between the two coloured links is aligned with the dot on the sprocket. Fit the sprocket to the inlet camshaft, insert the three bolts and tighten them to the specified torque **(see illustrations)**. Hold the camshaft using a spanner on the camshaft hexagon as the bolts are tightened.

14 Depress the timing chain tensioner plunger, withdraw the locking drill bit and release the plunger to apply tension to the chain **(see illustration)**.

15 Remove the tools used to lock the camshaft and crankshaft in the TDC position. Refit the crankshaft pulley retaining bolts and rotate the crankshaft through two complete revolutions.

16 Remove the crankshaft pulley retaining bolts, refit the tool used to lock the crankshaft in the TDC position and secure with the pulley retaining bolts. Check that it is possible to insert the camshaft sprocket locking tool. If the tool cannot be inserted, slacken the inlet camshaft sprocket retaining bolts. Using a spanner on the camshaft hexagon, rotate the camshaft slightly until the tool can be inserted. When all is correct, tighten the camshaft sprocket retaining bolts to the specified torque, then remove the locking tools.

17 Refit the crankshaft pulley with reference to Section 4.

18 Refit the lower timing cover as described in Section 6.

8 Timing chain sprockets, tensioner and guide rails – removal and refitting

Camshaft sprocket

1 Removal and refitting of the camshaft sprocket is part of the timing chain removal and refitting procedures contained in Section 7.

Fuel pump sprocket

Removal

Note: *Vauxhall/Opel special tool EN-50885-2 or suitable equivalent will be required for this operation.*

2 Remove the timing chain as described in Section 7.

3 Locate the body of the special tool on the fuel pump sprocket, engaging the three lugs on the tool with the slots on the sprocket.

4 Feed the special tool insert into the tool body and screw it into the fuel pump sprocket.

5 Use a socket and extension bar to unscrew the fuel pump sprocket retaining nut whilst counter-holding the special tool body with a 1/2 inch square drive extension bar engaged with the lug on the tool body.

6 Insert the special tool pressure spindle into the tool insert. Screw in the pressure spindle to press out the fuel pump sprocket, then remove the tool and sprocket from the fuel pump. If necessary, separate the special tool from the sprocket.

Refitting

7 Clean the fuel pump shaft and the sprocket hub ensuring that all traces of oil or grease are removed.

8 Fit the retaining nut to the pump shaft, then assemble the special tool to the sprocket.

9 Using the tool, hold the sprocket stationary and tighten the retaining nut to the specified torque.

10 Refit the timing chain as described in Section 8.

Idler sprocket

Removal

Note: *Vauxhall/Opel special tool EN-51141 or suitable equivalent will be required for this operation.*

11 Remove the timing chain as described in Section 7.

12 The idler sprocket incorporates a backlash compensating gear. This must now be locked to the fixed gear by attaching Vauxhall/Opel special tool EN-51141 (or suitable equivalent) to the gear teeth **(see illustration)**. This prevents the spring preload of the compensating gear being lost when the sprocket is removed.

13 Undo the three bolts and remove the sprocket and special tool from the cylinder block **(see illustrations)**. Do not remove the tool if the existing sprocket is to be refitted.

8.12 Engage the special tool with the teeth of the idler sprocket

8.13a Undo the three bolts…

8.13b …and remove the idler sprocket and special tool

Refitting

14 Engage the idler sprocket gear teeth with the crankshaft sprocket gear teeth so that the timing mark on the idler sprocket tooth is between the two dots on the crankshaft sprocket teeth **(see illustration)**.

15 Refit the sprocket retaining bolts, remove the special tool, and tighten the bolts to the specified torque.

16 Refit the timing chain as described in Section 7.

Tensioner

Removal

17 Remove the upper timing cover as described in Section 6.

18 Refer to Section 3 and set the engine at TDC for No. 1 piston. Insert the camshaft and crankshaft locking tools to retain the engine in the TDC position.

19 Push the timing chain tensioner plunger back into its bore and insert a suitable drill bit or similar tool to retain it in the released position **(see illustration 7.3)**.

20 Undo the two retaining bolts and remove the tensioner from the cylinder head.

21 Push the timing chain tensioner plunger into the tensioner body and remove the locking drill bit. Check that the tensioner plunger is free to move in out of the tensioner body with no trace of binding. If any binding or sticking of the plunger is felt, renew the tensioner assembly. On completion of the check, or if a new tensioner is being fitted, Compress the plunger and refit the locking drill bit.

Refitting

22 Refit the timing chain tensioner and tighten the two retaining bolts to the specified torque. Depress the tensioner plunger, withdraw the locking drill bit and release the plunger.

23 Remove the TDC locking tools, then refit the upper timing cover as described in Section 6.

Upper guide rail

Removal

24 Remove the upper timing cover as described in Section 6.

8.14 The mark on the idler sprocket tooth must be between the two dots on the crankshaft sprocket teeth

25 Refer to Section 3 and set the engine at TDC for No. 1 piston. Insert the camshaft and crankshaft locking tools to retain the engine in the TDC position.

26 Push the timing chain tensioner plunger back into its bore and insert a suitable drill bit or similar tool to retain it in the released position **(see illustration 7.3)**.

27 Undo the two bolts and remove the timing chain upper guide rail **(see illustration)**.

Refitting

28 Remove all traces of thread locking compound from the upper guide rail retaining bolt holes by running a suitable tap through the bolt hole threads. Also, suitably clean the threads of the retaining bolts.

29 Apply thread locking compound to the two upper guide rail retaining bolts. Refit the upper guide rail and tighten the retaining bolts to the specified torque.

30 Depress the timing chain tensioner plunger, withdraw the locking drill bit and release the plunger to apply tension to the chain.

31 Remove the TDC locking tools, then refit the upper timing cover as described in Section 6.

Lower guide rail

Removal

32 Remove the lower timing cover as described in Section 6.

33 Refer to Section 3 and set the engine at

TDC for No. 1 piston. Insert the camshaft and crankshaft locking tools to retain the engine in the TDC position.

34 Push the timing chain tensioner plunger back into its bore and insert a suitable drill bit or similar tool to retain it in the released position **(see illustration 7.3)**.

35 Undo the two retaining bolts and remove the lower guide rail from the cylinder block **(see illustration)**.

Refitting

36 Locate the lower guide rail in position, refit the two bolts and tighten them to the specified torque.

37 Depress the timing chain tensioner plunger, withdraw the locking drill bit and release the plunger to apply tension to the chain.

38 Remove the TDC locking tools, then refit the lower timing cover as described in Section 6.

Tensioner rail

Removal

39 Remove the lower timing cover as described in Section 6.

40 Refer to Section 3 and set the engine at TDC for No. 1 piston. Insert the camshaft and crankshaft locking tools to retain the engine in the TDC position.

41 Push the timing chain tensioner plunger back into its bore and insert a suitable drill bit or similar tool to retain it in the released position **(see illustration 7.3)**.

42 Undo the lower pivot bolt and remove the tensioner rail from the cylinder block **(see illustration)**.

Refitting

43 Locate the tensioner rail in position, refit the pivot bolt and tighten it to the specified torque.

44 Depress the timing chain tensioner plunger, withdraw the locking drill bit and release the plunger to apply tension to the chain.

45 Remove the TDC locking tools, then refit the lower timing cover as described in Section 6.

8.27 Undo the bolts and remove the upper guide rail

8.35 Lower guide rail retaining bolts

8.42 Tensioner rail lower pivot bolt

9.3a Undo the four bolts...

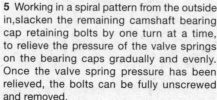

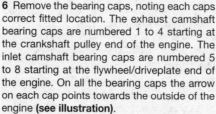

9.3b ...and lift off the camshaft rear bearing cap

9.6 The camshaft bearing caps are numbered 1 to 4 (exhaust) and 5 to 8 (inlet) and the arrow points toward the outside of the engine

9 Camshafts –
removal, inspection and refitting

Removal

1 Remove the timing chain (see Section 7).
2 Remove the tool used to lock the exhaust camshaft in the TDC position.
3 Undo the four bolts and lift off the camshaft rear bearing cap **(see illustrations)**.
4 The exhaust camshaft gear incorporates a backlash compensating gear. This must now be locked to the fixed exhaust camshaft gear by inserting a suitably-sized bolt/rod into the hole on the inboard face of the fixed gear, and through into the backlash compensating gear. This prevents the spring preload of the compensating gear being lost when either camshaft is removed.

5 Working in a spiral pattern from the outside in, slacken the remaining camshaft bearing cap retaining bolts by one turn at a time, to relieve the pressure of the valve springs on the bearing caps gradually and evenly. Once the valve spring pressure has been relieved, the bolts can be fully unscrewed and removed.
6 Remove the bearing caps, noting each caps correct fitted location. The exhaust camshaft bearing caps are numbered 1 to 4 starting at the crankshaft pulley end of the engine. The inlet camshaft bearing caps are numbered 5 to 8 starting at the flywheel/driveplate end of the engine. On all the bearing caps the arrow on each cap points towards the outside of the engine **(see illustration)**.
7 Carefully lift the camshafts from their locations in the cylinder head **(see illustration)**. If both camshafts are removed, identify them as exhaust and inlet.

8 With the camshafts removed, lift out the rear bearing base **(see illustration)**.
9 If required, remove the rocker arms and hydraulic tappets from the cylinder head as described in Section 10.

Inspection

10 Examine the camshaft bearing surfaces and cam lobes for signs of wear ridges and scoring. Renew the camshaft if any of these conditions are apparent. Examine the condition of the bearing surfaces in the cylinder head. If the any wear or scoring is evident, the cylinder head will need to be renewed.
11 If either camshaft is being renewed, it will be necessary to renew all the rocker arms and tappets for that particular camshaft also (see Section 10).
12 Check the condition of the camshaft drive gears for chipped or damaged teeth, wear ridges and scoring. Renew any components as necessary.

Refitting

13 Before refitting, thoroughly clean all the components and the cylinder head and bearing cap journals.
14 Liberally oil the cylinder head hydraulic tappet bores and the tappets. Carefully refit the tappets and rocker arms to the cylinder head, ensuring that each tappet is refitted to its original bore **(see illustrations)**.
15 Refit the camshaft rear bearing base to the cylinder head, ensuring that it locates correctly over the positioning dowels **(see illustration)**.

9.7 Lift the camshafts out of the cylinder head

9.8 Lift out the camshaft rear bearing base

9.14a Liberally oil the cylinder head tappet bores and tappets...

9.14b ...then refit the tappets and rocker arms to their original locations

9.15 Refit the camshaft rear bearing base

9.16a Liberally oil the exhaust camshaft bearings in the cylinder head...

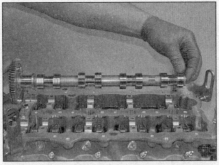

9.16b ...then place the exhaust camshaft in position

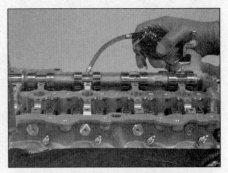

9.17a Liberally oil the camshaft bearings and bearing caps...

9.17b ...then refit the bearing caps to their respective locations

9.20a Refit the inlet camshaft so that the timing marks on the camshaft gears are aligned...

9.20b ...noting that it may be necessary to align the backlash compensating gear to allow the gear teeth to engage

16 Liberally oil the exhaust camshaft bearings in the cylinder head, then place the exhaust camshaft in position **(see illustrations)**.

17 Liberally oil the exhaust camshaft bearing journals and the bearing caps, Refit the bearing caps to their respective locations ensuring they are fitted the correct way round as noted during removal **(see illustrations)**.

18 Refit the bearing cap bolts and initially tighten them, one at a time, by half a turn, working from the centre outward in a spiral pattern. When all the bolts have been initially tightened, repeat the procedure, tightening the bolts by a further half a turn. Continue until all the bearing caps are in contact with the cylinder head and the bolts are lightly tightened.

19 Again, working in the same sequence, tighten all the bolts to the specified torque.

20 Liberally oil the inlet camshaft bearings in the cylinder head. Place the inlet camshaft in position, ensuring that the timing marks on the camshaft gears are aligned. Note that it may be necessary to use a large screwdriver to align the teeth of the backlash compensating gear to allow the gear teeth to engage **(see illustrations)**.

21 Liberally oil the rear camshaft bearing journals and the camshaft rear bearing cap, then place the cap in position. Refit the four retaining bolts and tighten them to the specified torque starting with the two outer bolts, then the two inner bolts **(see illustrations)**.

22 Refit the timing chain as described in Section 7.

9.21a Place the camshaft rear bearing cap in position...

9.21b ...then refit the retaining bolts and tighten them to the specified torque

10 Hydraulic tappets and rocker arms – removal, inspection and refitting

Removal

1 Remove the camshafts as described in Section 9.

2 Obtain sixteen small, clean plastic containers, and number them inlet 1 to 8 and exhaust 1 to 8; alternatively, divide a larger container into sixteen compartments and number each compartment accordingly.

3 Withdraw each rocker arm and hydraulic tappet in turn, unclip the rocker arm from the tappet, and place them in their respective container **(see illustrations)**. Do not interchange

the rocker arms and tappets, or the rate of wear will be much increased.

10.3a Withdraw each rocker arm and hydraulic tappet in turn...

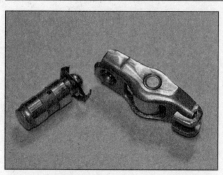

10.3b ...then unclip the rocker arm from the tappet

Inspection

4 Examine the rocker arm and hydraulic tappet bearing surfaces for wear ridges and scoring. Renew any rocker arm or tappet on which these conditions are apparent.

Refitting

5 Liberally oil the cylinder head hydraulic tappet bores and the tappets. Working on one assembly at a time, clip the rocker arm back onto the tappet, then refit the tappet to the cylinder head, ensuring that it is refitted to its original bore. Lay the rocker arm over its respective valve.

6 Refit the remaining tappets and rocker arms in the same way.

7 With all the tappets and rocker arms in place, refit the camshafts as described in Section 9.

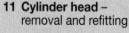

11 Cylinder head – removal and refitting

Removal

Note: *New cylinder head retaining bolts will be required for refitting.*

1 Remove the camshafts as described in Section 9, and the hydraulic tappets and rocker arms as described in Section 10.

2 Remove the inlet manifold as described in Chapter 4B, Section 17.

3 Remove the exhaust manifold as described in Chapter 4B, Section 24.

11.22 Using a dial test indicator to measure piston protrusion

4 Remove the fuel rail as described in Chapter 4B, Section 13.

5 The engine must now be supported while the right-hand engine mounting is removed. To do this, use a hoist attached to the top of the engine, or make up a wooden frame to locate beneath the sump and use a trolley jack.

6 Remove the right-hand engine mounting as described in Section 18.

7 Undo the four bolts and remove the right-hand engine mounting bracket.

8 Undo and remove the two outer cylinder head retaining bolts at the timing chain end of the engine.

9 Working in the reverse of the tightening sequence **(see illustration 11.28)**, progressively slacken the cylinder head bolts by half a turn at a time, until all bolts can be unscrewed by hand. Remove the cylinder head bolts.

10 Engage the help of an assistant and lift the cylinder head from the cylinder block. Take care as it is a bulky and heavy assembly.

11 Remove the gasket and keep it for identification purposes (see paragraph 18).

12 If the cylinder head is to be dismantled for overhaul, then refer to Part E of this Chapter.

Preparation for refitting

13 The mating faces of the cylinder head and cylinder block/crankcase must be perfectly clean before refitting the head. Use a hard plastic or wood scraper to remove all traces of gasket and carbon; also clean the piston crowns. Take particular care, as the surfaces are damaged easily. Also, make sure that the carbon is not allowed to enter the oil and water passages – this is particularly important for the lubrication system, as carbon could block the oil supply to any of the engine's components. Using adhesive tape and paper, seal the water, oil and bolt holes in the cylinder block/crankcase. To prevent carbon entering the gap between the pistons and bores, smear a little grease in the gap. After cleaning each piston, use a small brush to remove all traces of grease and carbon from the gap, then wipe away the remainder with a clean rag. Clean all the pistons in the same way.

14 Check the mating surfaces of the cylinder block/crankcase and the cylinder head for nicks, deep scratches and other damage. If slight, they may be removed carefully with a file, but if excessive, machining may be the only alternative to renewal.

15 Ensure that the cylinder head bolt holes in the crankcase are clean and free of oil. Syringe or soak up any oil left in the bolt holes. This is most important in order that the correct bolt tightening torque can be applied and to prevent the possibility of the block being cracked by hydraulic pressure when the bolts are tightened.

16 The cylinder head bolts must be discarded and renewed, regardless of their apparent condition.

17 If warpage of the cylinder head gasket surface is suspected, use a straight-edge to check it for distortion. Refer to Part E of this Chapter if necessary.

18 On this engine, the cylinder head-to-piston clearance is controlled by fitting different thickness head gaskets. The gasket thickness can be determined by looking at the tab located adjacent to No. 2 cylinder, on the inlet manifold side, and checking the position of the identification hole.

Identification hole position	Gasket thickness
Left-hand side of tab	*1.25 mm*
Centre of tab	*1.35 mm*
Right-hand side of tab	*1.45 mm*

19 The correct thickness of gasket required is selected by measuring the piston protrusions as follows.

20 Remove the crankshaft locking tool and temporarily refit the four crankshaft pulley retaining bolts to enable the crankshaft to be turned.

21 Mount a dial test indicator securely on the block so that its pointer can be easily pivoted between the piston crown and block mating surface. Turn the crankshaft to bring No 1 piston roughly to the TDC position. Move the dial test indicator probe over and in contact with No 1 piston. Turn the crankshaft back and forth slightly until the highest reading is shown on the gauge, indicating that the piston is at TDC.

22 Zero the dial test indicator on the gasket surface of the cylinder block then carefully move the indicator over No 1 piston. Measure its protrusion at the highest point between the valve cut-outs, and then again at its highest point between the valve cut-outs at 90° to the first measurement **(see illustration)**. Repeat this procedure with No 4 piston.

23 Rotate the crankshaft half a turn (180°) to bring No 2 and 3 pistons to TDC. Ensure the crankshaft is accurately positioned then measure the protrusions of No 2 and 3 pistons at the specified points. Once all pistons have been measured, rotate the crankshaft to return No 1 piston to the TDC position and refit the crankshaft locking tool.

24 Select the correct thickness of head gasket required by determining the largest amount of piston protrusion, and using the following table.

Piston protrusion measurement	Gasket thickness required
0.440 to 0.535 mm	*1.25 mm (hole on left-hand side of tab)*
0.535 to 0.630 mm	*1.35 mm (hole in centre of tab)*
0.630 to 0.725 mm	*1.45 mm (hole on right-hand side of tab)*

Refitting

25 Wipe clean the mating surfaces of the cylinder head and cylinder block/crankcase. Place the new gasket in position with the words ALTO/TOP uppermost.

11.28 Cylinder head bolt tightening sequence

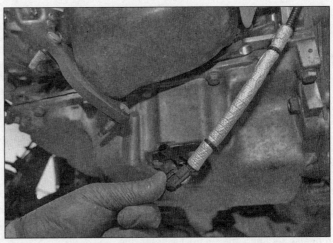

12.8 Disconnect the oil level sensor wiring connector

26 With the aid of an assistant, carefully refit the cylinder head assembly to the block, aligning it with the locating dowels.

27 Carefully enter each new cylinder head bolt into its relevant hole (do not drop them in). Screw all bolts in, by hand only, until finger-tight.

28 Working progressively in the sequence shown, tighten the cylinder head bolts to their Stage 1 torque setting, using a torque wrench and suitable socket **(see illustration)**.

29 Once all bolts have been tightened to the Stage 1 torque, working again in the same sequence, go around and tighten all bolts through the specified Stage 2 angle, then through the specified Stage 3 and Stage 4 angles using an angle-measuring gauge.

30 Fit two new outer cylinder head retaining bolts at the timing chain end of the engine and tighten them to the specified torque.

31 Refit the right-hand engine mounting bracket and engine mounting as described in Section 18.

32 Refit the fuel rail as described in Chapter 4B, Section 13.

33 Refit the exhaust manifold as described in Chapter 4B, Section 24.

34 Refit the inlet manifold as described in Chapter 4B, Section 17.

35 Refit the hydraulic tappets and rocker arms as described in Section 10, and the camshafts as described in Section 9.

12 Sump –
removal and refitting

Removal

1 Disconnect the battery negative lead as described in Chapter 5A Section 4.

2 Apply the handbrake, then jack up the front of the vehicle and support it on axle stands (see *Jacking and vehicle support*).

3 Remove the engine undertray as described in Chapter 11, Section 21.

4 Drain the engine oil and remove the oil filter as described in Chapter 1B, Section 5.

5 Remove the exhaust system front pipe as described in Chapter 4B, Section 26.

6 On 4WD vehicles, remove the transfer gearbox as described in Chapter 7C, Section 3.

7 Remove the engine oil dipstick.

8 Disconnect the wiring connector from the oil level sensor **(see illustration)**.

9 Undo the three bolts (manual transmission models) or two bolts (automatic transmission models) securing the sump flange to the transmission casing.

10 Undo the fifteen bolts securing the sump to the cylinder block baseplate and lower timing cover, noting the locations of the

different length bolts. Using a wide bladed scraper or similar tool inserted between the sump and baseplate, carefully break the joint, then remove the sump from under the car. Recover the oil filter housing seal from the sump upper face **(see illustrations)**. Obtain a new seal for refitting.

11 If required, the oil level sensor can be removed as described in Chapter 5A, Section 15.

Refitting

12 Thoroughly clean the inside and outside of the sump ensuring that all traces of old sealant are removed from the mating face. Preferably, undo the six bolts and remove the baffle plate so the inside of the sump can be completely cleaned. Also clean the cylinder block baseplate mating face to remove all traces of old sealant. Refit the baffle plate on completion.

13 If removed, refit the oil level sensor as described in Chapter 5A, Section 15.

14 Fit a new oil filter housing seal to the sump upper face, ensuring that it seats fully in the housing groove.

15 Apply a 3 to 4 mm bead of silicone sealant to the sump mating face, ensuring the sealant bead runs around the inside of the bolt holes **(see illustration)**. Position the sump on the cylinder block baseplate, then refit the retaining bolts. Progressively tighten the bolts to the specified torque.

12.10a Undo the fifteen sump retaining bolts...

12.10b ...remove the sump from the baseplate...

12.10c ...and recover the oil filter housing seal

12.15 Apply a bead of sealant to the sump mating face, ensuring the bead runs around the inside of the bolt holes

16 Refit the three bolts (manual transmission models) or two bolts (automatic transmission models) securing the sump flange to the transmission casing, and tighten the bolts to the specified torque.

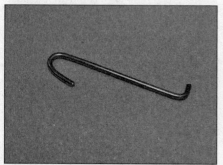

13.2a Make up a tool out of welding rod or stiff wire…

13.3a Use a suitable forked tool to hold the sprocket…

13.5b …and withdraw the pump from the cylinder block baseplate

17 Reconnect the oil level sensor wiring connector.
18 Refit the engine oil dipstick.
19 On 4WD vehicles, refit the transfer gearbox as described in Chapter 7C, Section 3.
20 Refit the exhaust system front pipe as described in Chapter 4B, Section 26.
21 Fit a new oil filter, then fill the engine with oil as described in Chapter 1B, Section 5.
22 Refit the engine undertray as described in Chapter 11, Section 21, then lower the vehicle to the ground.
23 Reconnect the battery negative terminal.

13 Oil pump – removal and refitting

Note: *The oil pump also incorporates the brake servo vacuum pump as an integral assembly.*

13.2b …to hold the chain tensioner arm in the released position

13.3b …then undo and remove the sprocket retaining bolt

13.8a Fit a new rubber seal…

Removal

Note: *The following procedure assumes that the transmission and flywheel/driveplate have been removed. Although it is possible to remove the oil pump with the transmission and flywheel/driveplate installed, it is an awkward operation and clearance is extremely limited.*

1 Remove the sump as described in Section 12.
2 Screw one of the sump retaining bolts into the outer front bolt hole on the lower timing cover. Push the oil pump chain tensioner forward to remove the tension in the chain and retain it in the released position using a short length of stiff wire or similar, suitably bent to shape. Engage the wire with the tensioner arm, then wrap it around the sump bolt **(see illustrations)**.
3 Undo the oil pump sprocket retaining bolt, while preventing the sprocket from rotating with a suitable forked tool engaged with the sprocket holes **(see illustrations)**. Note that a new bolt will be required for refitting.
4 Using a suitable screwdriver, ease the sprocket off the pump shaft.
5 Undo the four bolts securing the oil pump to the cylinder block baseplate. Withdraw the pump from its location and recover the two rubber seals **(see illustrations)**. Note that new seals will be required for refitting.
6 The oil pump and integral vacuum pump is a sealed assembly. No internal components are available and no repair is possible.

Refitting

7 Thoroughly clean the pump and cylinder block baseplate contact areas, paying particular attention to the oil seal locations.

13.5a Undo the four oil pump retaining bolts…

13.8b …to each side of the oil pump

8 Fit a new rubber seal to each side of the pump, then refit the pump to the cylinder block baseplate **(see illustrations)**. Refit the four retaining bolts and tighten them to the specified torque.

9 Turn the oil pump shaft as necessary until the flat on the shaft aligns with the flat on the sprocket, then fit the sprocket to the pump shaft.

10 Fit a new sprocket retaining bolt and tighten it to the specified torque, then through the specified angle. Hold the sprocket stationary as the bolt is tightened using the method employed on removal.

11 Undo the sump retaining bolt used to retain the tensioner in the released position and allow the tensioner to contact and tension the chain.

12 Refit the sump as described in Section 12.

---

14 Oil pump sprocket, chain and tensioner – removal and refitting

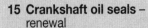

Removal

1 Remove the lower timing cover as described in Section 6.

2 Undo the oil pump sprocket retaining bolt, while preventing the sprocket from rotating with a suitable forked tool engaged with the sprocket holes **(see illustrations 13.03a and 13.03b)**. Note that a new bolt will be required for refitting.

3 Using a suitable screwdriver, ease the sprocket off the pump shaft. Push the tensioner arm away from the oil pump chain and disengage the sprocket from the chain.

4 Disengage the chain from the crankshaft and remove the chain from the engine **(see illustration)**.

5 To remove the tensioner, undo the two retaining bolts and remove it from the cylinder block baseplate.

Refitting

6 Refitting is a reversal of removal, bearing in mind the following points:

a) *Tighten all retaining bolts to the specified torque and, where applicable, through the specified angle.*

b) *When refitting the oil pump sprocket, turn the oil pump shaft as necessary until the flat on the shaft aligns with the flat on the sprocket, then fit the sprocket to the pump shaft.*

c) *Refit the lower timing cover as described in Section 6.*

---

15 Crankshaft oil seals – renewal

Front (crankshaft pulley end) oil seal

1 Remove the crankshaft pulley as described in Section 4.

2 Carefully punch or drill a small hole in the seal. Screw in a self-tapping screw and pull on the screw with pliers to extract the seal.

3 Clean the oil seal seat with a wooden or plastic scraper.

4 Tap the new seal into position until it is flush with the outer face of the housing, using a wooden block or preferably, an oil seal installer tool.

5 Refit the crankshaft pulley as described in Section 4.

Rear (transmission end) oil seal

6 Remove the flywheel/driveplate as described in Section 17.

7 Carefully withdraw the crankshaft speed/position sensor reluctor ring from the end of the crankshaft **(see illustration)**.

15.7 Withdraw the crankshaft speed/position sensor reluctor ring

14.4 Disengage the chain from the crankshaft

8 Carefully punch or drill a small hole in the seal. Screw in a self-tapping screw and pull on the screw with pliers to extract the seal **(see illustration)**.

9 Clean the oil seal seat with a wooden or plastic scraper.

10 Fit the new seal to the fitting tool supplied with the seal, so that the seal lip is spread open toward the crankshaft side **(see illustration)**.

11 Position the seal, together with the fitting tool over the end of the crankshaft. Remove the fitting tool then tap the seal into position using a suitable socket or tube, or a wooden block, until it is flush with the outer face of the lower timing cover **(see illustrations)**.

12 Refit the crankshaft speed/position sensor reluctor ring to the end of the crankshaft.

13 Refit the flywheel/driveplate as described in Section 17.

15.8 Screw in a self-tapping screw and pull on the screw with pliers to extract the seal

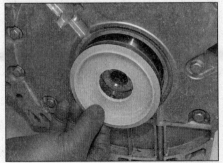

15.10 Fit the oil seal to the fitting tool, so that the seal lip is spread open toward the crankshaft side

15.11a Position the seal, together with the fitting tool over the end of the crankshaft, then remove the tool

15.11b Tap the seal into position using a suitable wooden block or similar

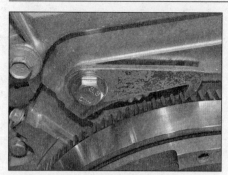

17.4 Prevent the flywheel from turning by jamming the ring gear teeth using a suitable tool

16 Oil cooler – removal and refitting

Removal

1 Disconnect the battery negative lead as described in Chapter 5A Section 4.
2 Drain the cooling system as described in Chapter 1B, Section 27.
3 Remove the driveshaft intermediate shaft and bracket as described in Chapter 8, Section 3.
4 Remove the exhaust system front pipe as described in Chapter 4B, Section 26.
5 Release the retaining spring clip and disconnect the coolant hose from the oil cooler.
6 Undo the six retaining bolts and remove the oil cooler from the cylinder block. Recover the three rubber seals from the rear of the housing. Note that new seals will be required for refitting.

Refitting

7 Fit the three new rubber seals to the rear of the oil cooler.
8 Position the oil cooler on the cylinder block and fit the six retaining bolts. Tighten the bolts to the specified torque.
9 Connect the coolant hose to the oil cooler and secure with the retaining spring clip.
10 Refit the exhaust system front pipe as described in Chapter 4B, Section 26.

17.5a Unscrew the retaining bolts...

11 Refit the driveshaft intermediate shaft and bracket as described in Chapter 8, Section 3.
12 Lower the vehicle to the ground, then reconnect the battery negative terminal.
13 Refill the cooling system as described in Chapter 1B, Section 27.

17 Flywheel/driveplate – removal, inspection and refitting

Removal

Note: *New flywheel/driveplate securing bolts must be used on refitting.*

Flywheel (manual transmission models)

1 Remove the transmission as described in Chapter 7A, Section 8.
2 Remove the clutch assembly as described in Chapter 6, Section 6.
3 Although the flywheel bolt holes are offset so that the flywheel can only be fitted in one position, it will make refitting easier if alignment marks are made between the flywheel and the end of the crankshaft.
4 Prevent the flywheel from turning by jamming the ring gear teeth using a suitable tool **(see illustration)**.
5 Unscrew the eight retaining bolts, and remove the flywheel **(see illustrations)**.
Caution: Take care, as the flywheel is heavy.

Driveplate (automatic transmission models)

6 Remove the transmission as described in Chapter 7B, Section 9, then remove the driveplate as described in paragraphs 4 and 5.

Inspection

Flywheel

7 A dual-mass flywheel is fitted and has the effect of reducing engine and transmission vibrations and harshness. The flywheel consists of a primary mass and a secondary mass constructed in such a way that the secondary mass is allowed to rotate slightly in relation to the primary mass. Springs within the assembly restrict this movement to set limits.

17.5b ...and remove the flywheel

8 Dual-mass flywheels have earned an unenviable reputation for unreliability and have been known to fail at quite low mileages (sometimes as low as 20 000 miles).
9 Examine the flywheel for wear or chipping of the ring gear teeth. Renewal of the ring gear is not possible and if the wear or chipping is significant, a new flywheel will be required.
10 Examine the flywheel for scoring of the clutch face. If the clutch face is scored significantly, a new flywheel will be required.
11 Look through the bolt hole and inspection openings in the secondary mass and check for any visible damage in the area of the centre bearing.
12 Place your thumbs on the clutch face of the secondary mass at the 3 o'clock and 9 o'clock positions and try to rock it. The maximum movement should not exceed 3 mm. Repeat this check with your thumbs at the 12 o'clock and 6 o'clock positions.
13 Rotate the secondary mass clockwise and anti-clockwise. It should move freely in both directions until spring resistance is felt, with no abnormal grating or rattling noises. The maximum rotational movement should not exceed a distance of eight teeth of the ring gear.
14 If there is any doubt about the condition of the flywheel, seek the advice of a Vauxhall/Opel dealer or engine reconditioning specialist. They will be able to advise if the flywheel is an acceptable condition, or whether renewal is necessary.

Driveplate

15 Closely examine the driveplate and ring gear teeth for signs of wear or damage and check the driveplate surface for any signs of cracks.
16 If there is any doubt about the condition of the driveplate, seek the advice of a Vauxhall/Opel dealer or engine reconditioning specialist.

Refitting

Flywheel

17 Offer the flywheel to the end of the crankshaft, and align the previously-made marks on the flywheel and crankshaft (if applicable).
18 Coat the threads of the new flywheel bolts with thread-locking compound (note that new bolts may be supplied ready-coated), then fit the bolts and tighten them to the specified torque, then through the specified angle whilst preventing the flywheel from turning as during removal **(see illustrations)**.
19 Refit the clutch as described in Chapter 6, Section 6 then refit the transmission as described in Chapter 7A, Section 8.

Driveplate

20 Clean the mating surfaces of the driveplate and crankshaft
21 Offer the driveplate and engage it on the crankshaft. Coat the threads of the new retaining bolts with thread-locking compound

(note that new bolts may be supplied ready-coated), then fit the bolts and tighten them to the specified torque. Prevent the driveplate from turning as during removal.

22 Remove the locking tool and refit the transmission as described in Chapter 7B, Section 9.

17.18a Tighten the flywheel retaining bolts to the specified torque...

17.18b ...then through the specified angle

18 Engine/transmission mountings – inspection and renewal

Inspection

1 To improve access, firmly apply the handbrake, then jack up the front of the vehicle and support it on axle stands (see *Jacking and vehicle support*).

2 Check the mounting blocks (rubbers) to see if they are cracked, hardened or separated from the metal at any point. Renew the mounting block if any such damage or deterioration is evident.

3 Check that all the mounting securing nuts and bolts are securely tightened, using a torque wrench to check if possible.

4 Using a large screwdriver, or a similar tool, check for wear in the mounting blocks by carefully levering against them to check for free play. Where this is not possible, enlist the aid of an assistant to move the engine/ transmission unit back-and-forth, and from side-to-side, while you observe the mountings. While some free play is to be expected, even from new components, excessive wear should be obvious. If excessive free play is found, check first to see that the securing nuts and bolts are correctly tightened, then renew any worn components as described in the following paragraphs.

Renewal

Note: *Before slackening any of the engine mounting bolts/nuts, the relative positions of the mountings to their various brackets should be marked to ensure correct alignment upon refitting.*

Right-hand mounting

Note: *New bolts will be required to secure the mounting to the engine mounting bracket.*

5 Remove the air cleaner assembly as described in Chapter 4B, Section 3.

6 Support the weight of the engine using a trolley jack with a block of wood placed on its head.

7 Undo the three bolts and remove the washers securing the right-hand engine mounting to the engine mounting bracket **(see illustration)**. Note that new bolts will be required for refitting.

8 Undo the three nuts securing the mounting to the body, and withdraw the mounting.

9 If necessary, the engine mounting bracket may be unbolted from the cylinder block **(see illustration)**.

10 Refitting is a reversal of removal. Tighten all retaining bolts/nuts to the specified torque,

and where applicable, through the specified angle. Refit the air cleaner assembly as described inChapter 4B, Section 3.

Left-hand mounting

Note: *New bolts will be required to secure the mounting to the transmission bracket.*

11 Remove the battery and battery tray as described in Chapter 5A, Section 4.

12 Support the weight of the transmission using a trolley jack with a block of wood placed on its head.

13 Unscrew the bolts securing the mounting to the transmission bracket **(see illustration)**. Note that new bolts will be required for refitting.

14 Undo the two bolts and one nut securing the mounting to the body, and withdraw the mounting.

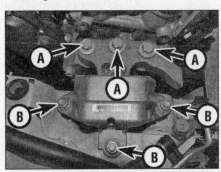

18.7 Right-hand mounting-to-engine mounting bracket bolts (A) and mounting-to-body nuts (B)

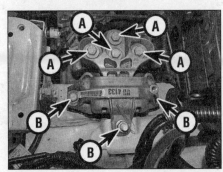

18.13 Left-hand mounting-to-transmission bracket bolts (A) and mounting-to-body bolts/nut (B) – manual transmission shown

15 Refitting is a reversal of removal, tightening all retaining bolts/nut to the specified torque. Refit the battery and battery tray as described in Chapter 5A, Section 4.

Rear mounting/torque link

Note: *New bolts will be required at all attachment points.*

FWD models

16 Firmly apply the handbrake, then jack up the front of the vehicle and support it on axle stands (see *Jacking and vehicle support*).

17 Remove the exhaust system front pipe as described in Chapter 4B, Section 26.

18 Using a suitable axle stand or trolley jack, support the rear of the engine/transmission unit.

19 Undo the bolt securing the rear mounting/ torque link to the subframe **(see illustration)**.

18.9 Right-hand engine mounting bracket-to-cylinder block retaining bolts

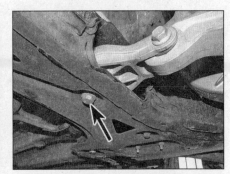

18.19 Undo the bolt securing the rear mounting to the subframe

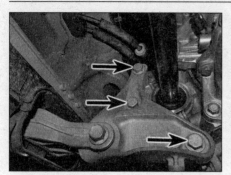

18.20 Undo the three bolts securing the rear mounting bracket to the transmission

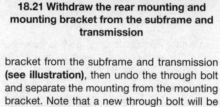

18.21 Withdraw the rear mounting and mounting bracket from the subframe and transmission

Note that a new bolt will be required for refitting.

20 Undo the three bolts securing the mounting bracket to the side of the transmission **(see illustration)**. Note that new bolts will be required for refitting.

21 Withdraw the mounting and mounting bracket from the subframe and transmission **(see illustration)**, then undo the through bolt and separate the mounting from the mounting bracket. Note that a new through bolt will be required for refitting.

22 Refitting is a reversal of removal, tightening all retaining bolts to the specified torque. Refit the exhaust system front pipe as described in Chapter 4B, Section 26.

4WD models

23 Firmly apply the handbrake, then jack up the front of the vehicle and support it on axle stands (see *Jacking and vehicle support*).

24 Using a suitable axle stand or trolley jack, support the rear of the engine/transmission unit.

25 Remove the transfer gearbox and transfer gearbox adapter as described in Chapter 7C, Section 3.

26 Undo the bolt securing the rear mounting/torque link to the subframe **(see illustration 18.19)**. Note that a new bolt will be required for refitting.

27 Withdraw the rear mounting/torque link from the subframe and remove it from under the car.

28 Refitting is a reversal of removal, tightening all retaining bolts to the specified torque. Refit the transfer gearbox adapter and transfer gearbox as described in Chapter 7C, Section 3.

Chapter 2 Part D
1.7 litre diesel engine in-car repair procedures

Contents

Degrees of difficulty

Easy, suitable for novice with little experience	**Fairly easy,** suitable for beginner with some experience	**Fairly difficult,** suitable for competent DIY mechanic	**Difficult,** suitable for experienced DIY mechanic	**Very difficult,** suitable for expert DIY or professional

Specifications

General

Engine type.	Four-cylinder, in-line, water-cooled. Belt-driven double overhead camshafts, 16 valves
Manufacturer's engine code*	A17DTS (LUD)
Bore	79.0 mm
Stroke	86.0 mm
Capacity	1686 cc
Firing order.	1-3-4-2 (No 1 cylinder at timing belt end of engine)
Direction of crankshaft rotation	Clockwise (viewed from timing belt end of engine)
Compression ratio	16.5 : 1

For details of engine code location, see 'Vehicle identification' in Reference

Compression pressures

Maximum difference between any two cylinders.	1.5 bar

Valve clearances

Engine cold:	
Inlet and exhaust	0.40 ± 0.05 mm

Lubrication system

Oil pump type.	Rotor-type, driven by timing belt	
Oil pressure at 80°C (approximate)	0.8 bar at idle speed	
Oil pump clearances:	**Standard**	**Service limit**
Outer rotor-to-body clearance	0.24 to 0.36 mm	0.40 mm
Inner-to-outer rotor clearance	0.13 to 0.15 mm	0.20 mm
Rotor endfloat.	0.035 to 0.100 mm	0.150 mm

Torque wrench settings

	Nm	lbf ft
Auxiliary drive belt tensioner bolt	50	37
Auxiliary drivebelt guide roller bolt	50	37
Baffle plate-to-cylinder block bolts	19	14
Camshaft bearing cap bolts/nuts.	27	20
Camshaft cover bolts.	10	7
Camshaft housing-to-cylinder head:		
M8 bolts	22	16
M10 bolts	27	20
Camshaft sprocket bolt	111	82
Coolant pump pulley bolts	12	9

Torque wrench settings (continued)

	Nm	lbf ft
Connecting rod big-end bearing cap nuts*:		
Stage 1	25	18
Stage 2	Angle-tighten a further 100°	
Stage 3	Angle-tighten a further 15°	
Crankshaft pulley bolts	20	15
Crankshaft oil seal housing bolts	10	7
Crankshaft sprocket bolt	196	145
Cylinder head bolts*:		
Stage 1	50	37
Stage 2	Angle-tighten a further 90°	
Stage 3	Angle-tighten a further 90°	
Driveshaft support bearing housing bolts	58	43
Engine mountings:		
Left-hand mounting:		
Mounting-to-body bolts/nut	58	43
Mounting bracket-to-transmission bracket*:		
Stage 1	50	37
Stage 2	Angle-tighten a further 60°	
Rear mounting/torque link:		
Mounting-to-transmission bracket:		
Stage 1	80	59
Stage 2	Angle-tighten a further 45°	
Mounting-to-subframe:		
Stage1	100	74
Stage 2	Angle-tighten a further 100°	
Transmission bracket bolts*:		
Stage 1	100	74
Stage 2	Angle-tighten a further 45°	
Right-hand mounting:		
Mounting bracket-to-engine*:		
Stage 1	60	44
Stage 2	Angle-tighten a further 60°	
Mounting-to-body	62	46
Mounting-to-mounting bracket*:		
Stage 1	50	37
Stage 2	Angle-tighten a further 60°	
Engine-to-transmission bolts:		
M10 bolts	40	32
M12 bolts	60	44
Flywheel/driveplate bolts*:		
Stage 1	85	63
Stage 2	Angle-tighten a further 30°	
Stage 3	Angle-tighten a further 15°	
High-pressure fuel pipe union nuts	25	18
High-pressure fuel pump sprocket nut	69	51
Main bearing cap bolts*:		
Stage 1	39	29
Stage 2	Angle-tighten a further 60°	
Oil dipstick guide tube bolts	10	7
Oil filter housing to cylinder block:		
Exhaust pressure sensor bracket bolt	20	15
Filter housing flange bolt	46	34
Filter housing hollow bolt	110	81
Oil pump cover retaining bolts	10	7
Oil pump sprocket nut	60	44
Oil pump pick-up/strainer bolts	26	19
Sump bolts:		
Drain plug	20	15
Temperature sensor bracket bolt	46	34
Main casting-to-block bolts	10	7
Main casting-to-transmission bolts	40	32
Sump pan-to-main casting bolts	10	7
Roadwheel nuts	140	103
Timing belt cover bolts	10	7
Timing belt idler pulley bolt	80	59
Timing belt tensioner pulley bolt	49	36

*Use new fasteners

1 General Information

How to use this Chapter

1 This Part of Chapter 2 describes the repair procedures which can reasonably be carried out on the engine while it remains in the vehicle. If the engine has been removed from the vehicle and is being dismantled as described in Chapter 2E, any preliminary dismantling procedures can be ignored.

2 Note that, while it may be possible physically to overhaul items such as the piston/connecting rod assemblies while the engine is in the vehicle, such tasks are not usually carried out as separate operations, and usually require the execution of several additional procedures (not to mention the cleaning of components and of oil ways); for this reason, all such tasks are classed as major overhaul procedures, and are described in Chapter 2E.

3 Chapter 2E describes the removal of the engine/transmission unit from the vehicle, and the full overhaul procedures which can then be carried out.

Engine description

4 The 1.7 litre common-rail diesel engine is of the sixteen-valve, in-line four-cylinder, double overhead camshaft (DOHC) type, mounted transversely at the front of the car with the transmission attached to its left-hand end.

5 The crankshaft runs in five main bearings. Thrustwashers are fitted to No 2 main bearing shell (upper half) to control crankshaft endfloat.

6 The connecting rods rotate on horizontally-split bearing shells at their big-ends. The pistons are attached to the connecting rods by gudgeon pins, which are a sliding fit in the connecting rod small-end eyes and retained by circlips. The aluminium-alloy pistons are fitted with three piston rings – two compression rings and an oil control ring.

7 The cylinder block is made of cast iron and the cylinder bores are an integral part of the block. On this type of engine the cylinder bores are sometimes referred to as having dry liners.

8 The inlet and exhaust valves are each closed by coil springs, and operate in guides pressed into the cylinder head.

9 The inlet camshaft is driven by the crankshaft by a timing belt and rotates directly in the camshaft housing. The exhaust camshaft is driven by the inlet camshaft via a spur gear. The camshafts operate the valves via followers, which are situated directly below the camshafts. Valve clearances are adjusted using camshaft followers of different thickness.

10 Lubrication is by means of an oil pump, which is driven by the timing belt. It draws oil through a strainer located in the sump, and then forces it through an externally-mounted filter into galleries in the cylinder block/crankcase. From there, the oil is distributed to the crankshaft (main bearings) and camshaft. The big-end bearings are supplied with oil via internal drillings in the crankshaft, while the camshaft bearings also receive a pressurised supply. The camshaft lobes and valves are lubricated by splash, as are all other engine components. An oil cooler is fitted to keep the oil temperature stable under arduous operating conditions.

Operations with engine in place

11 The following operations can be carried out without having to remove the engine from the vehicle.

a) Compression pressure testing.
b) Camshaft cover – removal and refitting.
c) Timing belt cover – removal and refitting.
d) Timing belt – removal and refitting.
e) Timing belt tensioner and sprockets – removal and refitting.
f) Valve clearances – checking and adjustment.
g) Camshaft and followers – removal, inspection and refitting.
h) Cylinder head – removal and refitting.
i) Connecting rods and pistons – removal and refitting*.
j) Sump – removal and refitting.
k) Oil pump – removal, overhaul and refitting.
l) Oil cooler – removal and refitting.
m) Crankshaft oil seals – renewal.
n) Engine/transmission mountings – inspection and renewal.
o) Flywheel – removal, inspection and refitting.
p) Camshaft housing – removal and refitting.

* Although the operation marked with an asterisk can be carried out with the engine in the car after removal of the sump, it is better for the engine to be removed, in the interests of cleanliness and improved access. For this reason, the procedure is described in Chapter 2E.

2 Compression test – description and interpretation

Compression test

Note: A compression tester specifically designed for diesel engines must be used for this test.

1 When engine performance is down, or if misfiring occurs which cannot be attributed to the fuel system, a compression test can provide diagnostic clues as to the engine's condition. If the test is performed regularly, it can give warning of trouble before any other symptoms become apparent.

2 A compression tester specifically intended for diesel engines must be used, because of the higher pressures involved. The tester is connected to an adapter which screws into the glow plug or injector hole. On these models, an adapter suitable for use in the injector holes will be required. It is unlikely to be worthwhile buying such a tester for occasional use, but it may be possible to borrow or hire one – if not, have the test performed by a garage.

3 Unless specific instructions to the contrary are supplied with the tester, observe the following points:

a) The battery must be in a good state of charge, the air filter must be clean, and the engine should be at normal operating temperature.
b) All the fuel injectors must be removed before starting the test (see Chapter 4B, Section 16).
c) Open the cover on the engine compartment fuse/relay box and remove the fuel pump relay. Refer to the wiring diagrams in Chapter 14 for information on fuse and relay locations.
d) Disconnect the wiring connector from the crankshaft position sensor as described in Chapter 4B, Section 10.

4 Screw the compression tester and adapter in to the fuel injector hole of No 1 cylinder.

5 With the help of an assistant, crank the engine on the starter motor; after one or two revolutions, the compression pressure should build-up to a maximum figure, and then stabilise. Record the highest reading obtained.

6 Repeat the test on the remaining cylinders, recording the pressure in each.

7 All cylinders should produce very similar pressures; any difference greater than that specified indicates the existence of a fault. Note that the compression should build-up quickly in a healthy engine; low compression on the first stroke, followed by gradually-increasing pressure on successive strokes, indicates worn piston rings. A low compression reading on the first stroke, which does not build-up during successive strokes, indicates leaking valves or a blown head gasket (a cracked head could also be the cause).

Note: The cause of poor compression is less easy to establish on a diesel engine than on a petrol one. The effect of introducing oil into the cylinders ('wet' testing) is not conclusive, because there is a risk that the oil will sit in the recess on the piston crown instead of passing to the rings.

8 On completion of the test, refit the fuel pump relay, reconnect the crankshaft position sensor wiring connector, then refit the fuel injectors as described in Chapter 4B, Section 16.

Leakdown test

9 A leakdown test measures the rate at which compressed air fed into the cylinder is lost. It is an alternative to a compression test, and in many ways it is better, since the escaping air provides easy identification of where pressure loss is occurring (piston rings, valves or head gasket).

10 The equipment needed for leakdown testing is unlikely to be available to the home mechanic. If poor compression is suspected, have the test performed by a Vauxhall/Opel dealer or suitably-equipped garage.

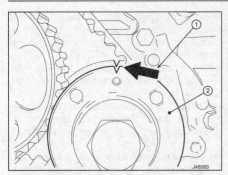

3.6 Rotate the crankshaft until the notch (1) on the sprocket (2) is aligned with the mark on the oil pump housing

3.8a The camshaft sprocket can be locked using a 6 mm bolt...

3.8b ...and the high-pressure fuel pump sprocket by an 8 mm bolt

3 Top Dead Centre (TDC) for No 1 piston – locating

Note: *If the engine is to be locked in position with No 1 piston at TDC on its compression stroke then two M6 bolts will be required.*

1 In its travel up and down its cylinder bore, Top Dead Centre (TDC) is the highest point that each piston reaches as the crankshaft rotates. While each piston reaches TDC both at the top of the compression stroke and again at the top of the exhaust stroke, for the purpose of timing the engine, TDC refers to the piston position (usually number 1) at the top of its compression stroke.

2 Number 1 piston (and cylinder) is at the right-hand (timing belt) end of the engine, and its TDC position is located as follows. Note that the crankshaft rotates clockwise when viewed from the right-hand side of the car.

3 Disconnect the battery negative lead as described in Chapter 5A Section 4.

4 Firmly apply the handbrake, then jack up the front of the car and support it securely on axle stands (see *Jacking and vehicle support*). Remove the right-hand front roadwheel.

5 Remove the timing belt upper and lower covers as described in Section 6.

6 Using a socket and extension bar on the crankshaft sprocket bolt, rotate the crankshaft until the notch on the sprocket is aligned with the mark on the oil pump housing **(see illustration)**. Once the notch is correctly aligned, No 1 and 4 pistons are at TDC.

7 To determine which piston is at TDC on its compression stroke, check the position of the timing holes in the camshaft and high-pressure fuel pump sprockets. When No 1 piston is at TDC on its compression stroke, both sprocket holes will be aligned with the threaded holes in the cylinder head/block, and both exhaust camshaft lobes for cylinder No 1 are pointing upwards if viewed through the oil filler hole. If the timing holes are out of alignment, continue rotating the crankshaft until they align. Note that because the sprockets are of different diameters, it may take up to six revolutions of the crankshaft to achieve alignment.

8 With No 1 piston at TDC on its compression stroke, if necessary, the camshaft and high-pressure fuel pump sprockets can be locked in position. Secure the camshaft sprocket in position by screwing an M6 bolt into the hole in the cylinder head and lock the high-pressure fuel pump sprocket in position by screwing an M8 bolt into the cylinder block **(see illustrations)**.

4 Camshaft cover – removal and refitting

Removal

⚠️ **Warning: Refer to the information contained in Chapter 4B, Section 2 before proceeding.**

1 Disconnect the battery negative lead as described in Chapter 5A Section 4.

2 Remove the engine oil filler cap, then lift off the plastic cover over the top of the engine **(see illustrations)**. Refit the oil filler cap.

3 Undo the two bolts and move the cooling system expansion tank to one side.

4 Squeeze together the legs of the retaining clip and disconnect the crankcase ventilation hose from the air cleaner outlet air duct **(see illustration)**.

5 Undo the bolt securing the outlet air duct to the support bracket **(see illustration)**. Slacken the retaining clip securing the outlet air duct to the airflow meter and detach the duct.

6 Undo the bolt securing the outlet air duct to the camshaft cover **(see illustration)**.

4.2a Remove the engine oil filler cap...

4.2b ...then lift off the plastic cover over the top of the engine

4.4 Squeeze together the retaining clip legs and disconnect the crankcase ventilation hose from the outlet air duct

4.5 Undo the bolt securing the outlet air duct to the support bracket

4.6 Undo the bolt securing the outlet air duct to the camshaft cover

4.7 Disconnect the vacuum hose from the EGR vacuum valve

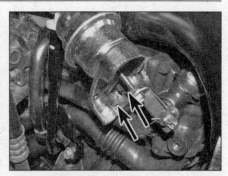

4.8a Undo the two nuts...

7 Disconnect the vacuum hose from the EGR vacuum valve **(see illustration)**.
8 Undo the two nuts securing the vacuum valve to the EGR valve cooler. Withdraw the valve from the two studs and lay aside **(see illustrations)**.
9 Spread the sides of the wire retaining clip and disconnect the intercooler outlet duct from the throttle housing **(see illustration)**.
10 Slacken the clip securing the lower end of the outlet air duct to the turbocharger. Pull the duct off the turbocharger, disengage the radiator hose from the support clip and withdraw the duct up and off the engine **(see illustrations)**.
11 Undo the three bolts securing the wiring harness trough to the top of the camshaft cover **(see illustration)**.
12 Disconnect the wiring connectors from the following components:
a) *Coolant temperature sensor.*
b) *EGR valve.*
c) *Throttle housing.*
d) *Fuel injectors.*
e) *Glow plugs.*
f) *Camshaft sensor.*
g) *Mass airflow sensor.*
h) *Inlet manifold vacuum solenoid valves.*
i) *Charge (boost) pressure sensor.*
j) *Engine management electronic control unit.*
k) *Wiring harness connection plug.*
13 Release the wiring harness from the cable ties and retaining clips and move the harness to one side, clear of the camshaft cover.
14 Using a small screwdriver, carefully prise

4.8b ...then withdraw the vacuum valve from the two studs and lay aside

off the retaining clips securing the fuel leak-off hose connection to the top of each fuel injector. Lift the leak-off hose off the injectors

4.10a Slacken the clip securing the lower end of the outlet air duct to the turbocharger...

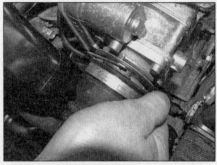

4.9 Spread the sides of the wire clip and disconnect the intercooler outlet duct from the throttle housing

and suitably plug or cover the open injector unions to prevent dirt entry **(see illustrations)**.
15 Open the retaining clip securing the fuel

4.10b ...pull the duct off the turbocharger...

4.10c ...disengage the radiator hose from the support clip and withdraw the duct up and off the engine

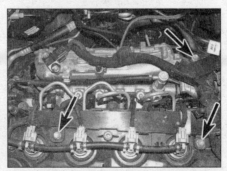

4.11 Undo the three bolts securing the wiring harness trough to the top of the camshaft cover

4.14a Carefully prise off the retaining clips securing the fuel leak-off hose connection to the top of each fuel injector...

4.14b ...lift the leak-off hose off the injectors...

4.14c ...and suitably plug or cover the open injector unions

4.15a Open the retaining clip securing the fuel leak-off hose to the banjo union on the fuel rail...

4.15b ...then disconnect the hose from the banjo union

4.16 Undo the bolt securing the fuel leak-off hose support clip to the inlet manifold

17 Thoroughly clean the fuel pipe unions on the fuel injectors and fuel rail. Using an open-ended spanner, unscrew the union nuts securing the high-pressure fuel pipes to the injectors and fuel rail. Withdraw the high-pressure fuel pipes and plug or cover the open unions to prevent dirt entry (see illustrations).
18 Using a suitable screwdriver, prise free the four fuel injector seal rings and remove them from the camshaft cover (see illustrations).
19 Undo the two bolts securing the timing belt upper cover to the camshaft cover (see illustration).
20 Unscrew the ten retaining bolts and lift away the camshaft cover, complete with rubber seal (see illustration).
21 Prior to refitting, obtain four new high-pressure fuel pipes, four new fuel injector seal rings, fuel leak-off hose retaining clips,

leak-off hose to the banjo union on the fuel rail, then disconnect the hose from the banjo union (see illustrations).
16 Undo the bolt securing the fuel leak-off hose support clip to the inlet manifold.

Release the leak-off hose from the guides on the timing belt cover and move the hose assembly to one side (see illustration). Slip a plastic bag over the disconnected leak off hose to prevent dirt entry.

4.17a Unscrew the union nuts...

4.17b ...and remove the high-pressure fuel pipes

4.18a Prise free the four fuel injector seal rings...

4.18b ...and remove them from the camshaft cover

4.19 Undo the two bolts securing the timing belt upper cover to the camshaft cover

4.20 Unscrew the retaining bolts and lift away the camshaft cover

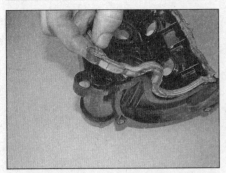

4.23 Fit the new rubber seal to the camshaft cover ensuring it is fully seated in the cover groove

4.24a Apply sealant to the camshaft housing mating surfaces at the left-hand end…

4.24b …and right-hand end of each camshaft…

4.24c …ensuring that the oil borehole at the right-hand end of the exhaust camshaft is not covered in sealant

4.25 Camshaft cover retaining bolt tightening sequence

4.26 Fit the four new fuel injector seal rings, ensuring they are pushed fully home in the camshaft cover

a tube of Vauxhall/Opel sealant and a new camshaft cover rubber seal.

Refitting

22 Thoroughly clean the camshaft cover and the mating surfaces of the cover and camshaft housing ensuring that all traces of oil and old sealant are removed from the mating surfaces.

23 Fit the new rubber seal to the camshaft cover ensuring it is fully seated in the cover groove **(see illustration)**.

24 Apply Vauxhall/Opel sealant to the camshaft housing mating surfaces at the end of each camshaft. Ensure that the oil borehole at the right-hand end of the exhaust camshaft is not covered in sealant **(see illustrations)**.

25 Carefully lower the cover into position, ensuring the seal remains correctly seated. Refit the timing belt upper cover and camshaft cover retaining bolts and tighten them finger tight only at this stage. With all the bolts installed, tighten the camshaft cover retaining bolts to the specified torque in the sequence shown **(see illustration)**. When all the camshaft cover bolts have been tightened, tighten the two timing belt cover bolts to the specified torque.

26 Fit the four new fuel injector seal rings, ensuring they are pushed fully home in the camshaft cover **(see illustration)**.

27 Working on one fuel injector at a time, remove the blanking plugs from the fuel pipe unions on the fuel rail and the relevant injector. Locate the new high-pressure fuel pipe over the unions and screw on the union nuts finger

tight. Tighten the union nuts to the specified torque using a torque wrench and crow-foot adaptor **(see illustrations)**. Repeat this operation for the remaining three injectors.

28 Place the fuel leak-off hose in position, locating it in the guides in the timing belt cover. Refit the bolt securing the leak-off hose support clip to the inlet manifold and tighten it securely.

29 Reconnect the leak-off hose to the banjo union on the fuel rail and secure the hose with a new retaining clip.

30 Engage the fuel-leak-off hose connectors with the four injectors and secure using new retaining clips.

31 Place the wiring harness trough in position over the camshaft cover and reconnect the wiring connectors to the components listed in

paragraph 12. Secure the wiring harness with the three retaining bolts, cable clips and cable ties.

32 Place the air cleaner outlet air duct in position, attach the lower end to the turbocharger and secure with the retaining clip.

33 Reconnect the intercooler outlet duct to the throttle housing, ensuring that the wire retaining clip is correctly seated in its slots.

34 Refit the EGR vacuum valve to the studs on the EGR cooler. Refit the retaining nuts and tighten securely. Reconnect the vacuum hose to the valve and secure the hose with the retaining clips.

35 Refit and tighten the bolt securing the outlet air duct to the camshaft housing.

36 Engage the outlet air duct with the airflow meter and tighten the retaining clip. Refit and

4.27a Locate the new high-pressure fuel pipes over the unions and screw on the union nuts finger tight…

4.27b …then tighten the union nuts using a torque wrench and crow-foot adaptor

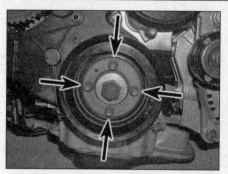

5.3 Remove the bolts securing the crankshaft pulley to the sprocket

5.4a Refit the pulley to the crankshaft sprocket, aligning the pulley hole…

5.4b …with the sprocket locating pin

tighten the bolt securing the outlet air duct to the support bracket.

37 Reconnect the crankcase ventilation hose to the outlet air duct.

38 Place the cooling system expansion tank back in position and secure with the two bolts.

39 Reconnect the battery negative terminal.

40 Observing the precautions listed in Chapter 4B, Section 2, prime the fuel system as described in Chapter 4B, Section 5, then start the engine and allow it to idle. Check for leaks at the high-pressure fuel pipe unions with the engine idling. If satisfactory, increase the engine speed to 4000 rpm and check again for leaks. Take the car for a short road test and check for leaks once again on return. If any leaks are detected, obtain and fit a new high-pressure fuel pipe(s).

41 Refit the engine cover on completion.

5 Crankshaft pulley – removal and refitting

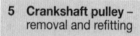

Removal

1 Firmly apply the handbrake, then jack up the front of the car and support it securely on axle stands (see *Jacking and vehicle support*). Remove the right-hand roadwheel.

2 Remove the auxiliary drivebelt as described in Chapter 1B, Section 22. Prior to removal, mark the direction of rotation on the belt to ensure the belt is refitted the same way around.

3 Slacken and remove the four small retaining

bolts securing the pulley to the crankshaft sprocket and remove the pulley from the engine **(see illustration)**. If necessary, prevent crankshaft rotation by holding the sprocket retaining bolt with a suitable socket.

Refitting

4 Refit the pulley to the crankshaft sprocket, aligning the pulley hole with the sprocket locating pin **(see illustrations)**. Refit the pulley retaining bolts, tightening them to the specified torque.

5 Refit the auxiliary drivebelt as described in Chapter 1B, Section 22, using the mark made prior to removal to ensure the belt is fitted the correct way around.

6 Refit the roadwheel, then lower the car to the ground and tighten the roadwheel nuts to the specified torque.

6 Timing belt covers – removal and refitting

Removal

Upper cover

1 Disconnect the battery negative lead as described in Chapter 5A Section 4.

2 Remove the engine oil filler cap, then lift off the plastic cover over the top of the engine **(see illustrations 4.2a and 4.2b)**. Refit the oil filler cap.

3 Remove the air cleaner assembly and outlet duct as described in Chapter 4B, Section 3.

4 Disconnect the wiring connector from the camshaft position sensor, then undo the retaining bolt and remove the sensor **(see illustration)**.

5 Undo the retaining nut and bolt and remove the protecting cover plate located below the camshaft position sensor location.

6 Support the weight of the engine using a trolley jack with a block of wood placed on its head.

7 Undo the three bolts securing the right-hand engine mounting to the engine mounting support bracket. Note that new bolts will be required for refitting.

8 Undo the two bolts and one nut securing the mounting to the body, and withdraw the mounting.

9 Undo the three retaining bolts and separate the engine mounting support bracket from the engine block mounting adapter **(see illustration)**. Note that new bolts will be required for refitting.

10 Undo the eight upper cover retaining bolts and withdraw the cover **(see illustration)**. Note that the cover bolts are of different lengths.

Lower cover

11 Firmly apply the handbrake, then jack up the front of the car and support it securely on axle stands (see *Jacking and vehicle support*). Remove the right-hand roadwheel.

12 Remove the auxiliary drivebelt as described in Chapter 1B, Section 22. Prior to removal, mark the direction of rotation on the belt to ensure the belt is refitted the same way around.

6.4 Disconnect the wiring connector from the camshaft position sensor

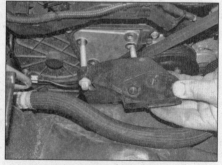

6.9 Undo the bolts and separate the engine mounting support bracket from the mounting adapter

6.10 Undo the eight upper cover retaining bolts and withdraw the cover

13 Remove the timing belt upper cover as described previously.

14 Slacken and remove the four small retaining bolts securing the crankshaft pulley to the crankshaft sprocket and remove the pulley from the engine **(see illustration 5.3)**. If necessary, prevent crankshaft rotation by holding the sprocket retaining bolt with a suitable socket.

15 Undo the three retaining bolts and remove the coolant pump pulley.

16 Undo the three retaining bolts, and remove the lower cover from the oil pump housing, then manoeuvre the engine mounting adapter out from the timing belt area **(see illustrations)**.

Rear cover

17 Remove timing belt as described in Section 7.

18 Remove the timing belt idler pulley, camshaft sprocket and high-pressure fuel pump sprocket as described in Section 8.

19 Undo the six retaining bolts, and remove the rear timing belt cover.

Refitting

20 Refitting is the reverse of removal, ensuring all retaining bolts are tightened to the specified torque, where given.

7 Timing belt – removal and refitting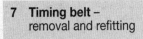

Note: *The timing belt must be removed and refitted with the engine cold.*

Removal

1 Position No 1 cylinder at TDC on its compression stroke as described in Section 3. Lock the camshaft and high-pressure fuel pump sprockets in position by screwing the bolts into the threaded holes in the cylinder head/block.

2 Slacken the timing belt tensioner retaining bolt, then turn the tensioner anti-clockwise to relieve the belt tension, using a suitable hexagon bit or Allen key engaged with the hole provided in the tensioner front plate **(see illustration)**. Hold the tensioner in this position and tighten the retaining bolt.

3 Slide the timing belt off its sprockets and

7.2 Slacken the timing belt tensioner retaining bolt and turn the tensioner anti-clockwise to relieve the belt tension

6.16a Undo the three retaining bolts, and remove the lower cover from the oil pump housing...

remove it from the engine. If the belt is to be re-used, use white paint or similar to mark the direction of rotation on the belt. Do not rotate the crankshaft until the timing belt has been refitted.

4 Check the timing belt carefully for any signs of uneven wear, splitting or oil contamination, and renew it if there is the slightest doubt about its condition. If the engine is undergoing an overhaul and is approaching the specified interval for belt renewal (see Chapter 1B) renew the belt as a matter of course, regardless of its apparent condition. If signs of oil contamination are found, trace the source of the oil leak and rectify it, then wash down the engine timing belt area and all related components to remove all traces of oil.

Refitting

5 On reassembly, thoroughly clean the timing belt sprockets and ensure that No 1 cylinder is still positioned at TDC on its compression stroke, with the camshaft and high-pressure fuel pump sprockets locked correctly in position (see Section 3).

6 Fit the timing belt over the crankshaft, oil pump, high-pressure fuel pump and camshaft sprockets, then guide it over the tensioner pulley. Ensure that the belt rear run is taut (ie, all slack is on the tensioner pulley side of the belt). Do not twist the belt sharply while refitting it. Ensure that the belt teeth are correctly seated centrally in the sprockets, and that the timing mark remains in alignment. If a used belt is being refitted, ensure that the arrow mark made on removal points in the normal direction of rotation, as before.

7 Tension the belt by slackening the tensioner retaining bolt and allowing the tensioner to automatically take up the slack in the belt. Once the slack is taken up, tighten the tensioner retaining bolt to the specified torque.

8 Check that the crankshaft sprocket timing mark is still correctly positioned then unscrew the locking bolts from the high-pressure fuel pump and camshaft sprockets.

9 Rotate the crankshaft through two complete turns in the normal direction of rotation to settle the timing belt in position. Realign the crankshaft sprocket timing mark and check that the camshaft and high-pressure fuel

6.16b ...then manoeuvre the engine mounting adapter out from the timing belt area

pump sprocket locking bolts can be refitted. If it is not possible to fit the locking bolts with the crankshaft sprocket timing mark aligned, repeat the belt refitting procedure.

10 Refit the timing belt covers as described in Section 6.

8 Timing belt tensioner and sprockets – removal and refitting

Camshaft sprocket

Removal

1 Remove the timing belt as described in Section 7.

2 Remove the locking bolt securing the camshaft sprocket in the TDC position.

3 It will now be necessary to hold the camshaft sprocket to enable the retaining bolt to be removed. Vauxhall/Opel special tools EN-956-1 and EN-6347 are available for this purpose, however, a home-made tool can easily be fabricated **(see Tool Tip)**.

To make a sprocket holding tool, obtain two lengths of steel strip about 6 mm thick by about 30 mm wide or similar, one 600 mm long, the other 200 mm long (all dimensions are approximate). Bolt the two strips together to form a forked end, leaving the bolt slack so that the shorter strip can pivot freely. At the other end of each 'prong' of the fork, drill a suitable hole and fit a nut and bolt to allow the tool to engage with the spokes in the sprocket.

8.4 Using the holding tool to prevent rotation of the camshaft, slacken the sprocket retaining bolt

8.11 Using a sprocket holding tool to prevent rotation as the high-pressure fuel pump sprocket nut is slackened

4 Using the holding tool to prevent rotation of the camshaft, slacken the sprocket retaining bolt. Remove the holding tool, unscrew the retaining bolt and remove the sprocket, noting which way round it is fitted **(see illustration)**. If the sprocket locating pin is a loose fit, remove it from the camshaft end and store it with the sprocket for safe-keeping.

Refitting

5 Ensure the locating pin is in position then refit the sprocket to the camshaft end aligning its locating hole with the pin.
6 Refit the sprocket retaining bolt, then tighten the bolt to the specified torque using the holding tool to prevent rotation.
7 Turn the camshaft sprocket slightly, as necessary, then refit the locking bolt to secure the sprocket in the TDC position.
8 Refit the timing belt as described in Section 7.

High-pressure fuel pump sprocket

Removal

9 Remove the timing belt as described in Section 7.
10 Remove the locking bolt securing the high-pressure fuel pump sprocket in the TDC position.
11 Refer to the information contained in paragraph 3 and suitably hold the sprocket, then slacken and remove the sprocket retaining nut **(see illustration)**.
12 Remove the sprocket from the fuel pump shaft, noting which way around it is fitted. If the Woodruff key is a loose fit in the pump shaft, remove it and store it with the sprocket for safe-keeping. The sprocket is a tapered-fit on the fuel pump shaft and in some cases a suitable puller may be needed to free it from the shaft **(see illustration)**.

Refitting

13 Ensure the Woodruff key is correctly fitted to the pump shaft then refit the sprocket, aligning the sprocket groove with the key **(see illustration)**.
14 Refit the retaining nut and tighten it to the specified torque whilst using the holding tool to prevent rotation.
15 If not already done, align the sprocket timing hole with the threaded hole in the cylinder block and screw in the locking bolt.
16 Refit the timing belt as described in Section 7.

Crankshaft sprocket

Removal

17 Remove the timing belt as described in Section 7.
18 Slacken the crankshaft sprocket retaining bolt. To prevent crankshaft rotation, have an assistant select top gear and apply the brakes firmly. If the engine is removed from the vehicle it will be necessary to lock the flywheel (see Section 18).
19 Unscrew the retaining bolt and washer and remove the crankshaft sprocket from the end of the crankshaft **(see illustration)**. If the sprocket is a tight fit, draw it off of the crankshaft using a suitable puller. If the Woodruff key is a loose fit in the crankshaft, remove it and store it with the sprocket for safe-keeping.
20 Slide the flanged spacer off of the crankshaft, noting which way around it is fitted.

Refitting

21 Refit the flanged spacer to the crankshaft with its convex surface facing away from the oil pump housing **(see illustration)**.
22 Ensure the Woodruff key is correctly fitted then slide on the crankshaft sprocket aligning its groove with the key.
23 Refit the retaining bolt and washer then lock the crankshaft by the method used on removal, and tighten the sprocket retaining bolt to the specified torque setting **(see illustration)**.
24 Refit the timing belt as described in Section 7.

8.12 Using a puller to remove the high-pressure fuel pump sprocket

8.13 Align the keyway in the sprocket with the Woodruff key in the shaft

8.19 Slide the sprocket from the shaft

8.21 Fit the flanged spacer with the convex side away from the oil pump cover

8.23 Refit the sprocket retaining bolt and washer

Oil pump sprocket

Removal

25 Remove the timing belt as described in Section 7.

26 Prevent the oil pump sprocket from rotating using a socket and extension bar fitted to one of the oil pump cover bolts then slacken and remove the sprocket retaining nut **(see illustration)**.

27 Remove the sprocket from the oil pump shaft, noting which way around it is fitted.

Refitting

28 Refit the sprocket, aligning it with the flat on the pump shaft, and fit the retaining nut. Tighten the sprocket retaining nut to the specified torque, using the socket and extension bar to prevent rotation.

29 Refit the timing belt as described in Section 7.

Tensioner assembly

Removal

30 Remove the timing belt as described in Section 7.

31 Unscrew the retaining bolt and remove the tensioner assembly from the engine. Disengage the tension spring from its locating stud as the tensioner is removed.

Refitting

32 Fit the tensioner assembly to the engine, engaging the tension spring over the stud. Screw in the tensioner retaining bolt, then turn the tensioner anti-clockwise using a suitable hexagon bit or Allen key engaged with the hole provided in the tensioner front plate. Hold the tensioner in this position and tighten the retaining bolt.

33 Refit the timing belt as described in Section 7.

Idler pulley

Removal

34 Remove the timing belt as described in Section 7.

35 Slacken and remove the retaining bolt and remove the idler pulley from the engine **(see illustration)**.

Refitting

36 Refit the idler pulley and tighten the retaining bolt to the specified torque.

37 Refit the timing belt as described in Section 7.

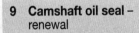

9 Camshaft oil seal – renewal

1 Remove the camshaft sprocket as described in Section 8.

2 Carefully punch or drill a small hole in the oil seal. Screw in a self-tapping screw, and pull on the screw with pliers to extract the seal **(see illustration)**.

8.26 Use a socket and extension bar on one of the oil pump cover bolts to prevent the sprocket from turning

9.2 Pull the screw to extract the seal

3 Clean the seal housing, and polish off any burrs or raised edges which may have caused the seal to fail in the first place.

4 Press the new seal into position using a suitable tubular drift (such as a socket) which bears only on the hard outer edge of the seal. Take care not to damage the seal lips during fitting; note that the seal lips should face inwards **(see illustration)**.

5 Refit the camshaft sprocket as described in Section 8.

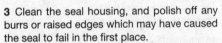

10 Valve clearances – checking and adjustment

Checking

1 The importance of having the valve clearances correctly adjusted cannot be overstressed, as they vitally affect the performance of the engine. The engine must be cold for the check to be accurate. The clearances are checked as follows.

2 Firmly apply the handbrake, then jack up the front of the car and support it securely on axle stands (see *Jacking and vehicle support*). Remove the right-hand roadwheel.

3 Remove the camshaft cover as described in Section 4, then remove the fuel injectors as described in Chapter 4B, Section 16.

4 Using a socket and extension on the crankshaft sprocket bolt, rotate the crankshaft in the normal direction (clockwise) until the

8.35 Timing belt idler pulley retaining bolt

9.4 Press the oil seal into position using a tubular drift such as a socket

inlet camshaft lobes for No 1 cylinder (nearest the timing belt end of the engine) and the exhaust camshaft lobes for No 3 cylinder are pointing away from the followers. This indicates that these valves are completely closed, and the clearances can be checked.

5 On a piece of paper, draw the outline of the engine with the cylinders numbered from the timing belt end. Show the position of each valve, together with the specified valve clearance. Note that the clearance for both the inlet and exhaust valves is the same.

6 With the cam lobes positioned as described in paragraph 4, using feeler blades, measure the clearance between the base of both No 1 cylinder inlet cam lobes and No 3 cylinder exhaust cam lobes and their followers. Record the clearances on the paper **(see illustration)**.

10.6 With the camshaft lobe pointing away from the follower measure the clearance

11.6 Insert a screw through the fixed gear and into the backlash compensating gear

11.8 The camshaft bearing caps are numbered 1 to 5 and the arrow on each cap points towards the timing belt end of the engine

7 Rotate the crankshaft pulley through a half a turn (180º) to position No 3 cylinder inlet camshaft lobes and No 4 cylinder exhaust camshaft lobes pointing away from their followers. Measure the clearance between the base of the camshaft lobes and their followers and record the clearances on the paper.

8 Rotate the crankshaft pulley through a half a turn (180º) to position No 2 cylinder exhaust camshaft lobes and No 4 cylinder inlet camshaft lobes pointing away from their followers. Measure the clearance between the base of the camshaft lobes and their followers and record the clearances on the paper.

9 Rotate the crankshaft pulley through a half a turn (180º) to position No 1 cylinder exhaust camshaft lobes and No 2 cylinder inlet camshaft lobes pointing away from their followers. Measure the clearance between the base of the camshaft lobes and their followers and record the clearances on the paper.

10 If all the clearances are correct, refit the fuel injectors as described in Chapter 4B, Section 16, and the camshaft cover as described in Section 4. Refit the roadwheel, lower the vehicle to the ground and tighten the roadwheel nuts to the specified torque. If any clearance measured is not correct, adjustment must be carried out as described in the following paragraphs.

Adjustment

11 If adjustment is necessary, remove the relevant camshaft(s) and camshaft followers as described in Section 11.

12 Clean the followers of the valves that require clearance adjustment and note

11.10 Lift the cam followers from the cylinder head

the thickness marking on the follower. The thickness marking is stamped on the underside of each follower. For example, a 3.46 mm thick follower will have a 46 thickness marking, a 3.48 mm thick follower will have a 48 thickness marking etc. New followers are available in thicknesses ranging from 3.46 mm to 4.14 mm in increments of 0.02 mm.

13 Add the measured clearance of the valve to the thickness of the original follower then subtract the specified valve clearance from this figure. This will give you the thickness of the follower required. For example:

Clearance measured of valve	*0.48 mm*
Plus thickness of the original follower	*3.46 mm*
Equals	*3.94 mm*
Minus clearance required	*0.40 mm*
Thickness of follower required	*3.54 mm*

14 Repeat this procedure on the remaining valves which require adjustment, then obtain the correct thickness of follower(s) required.

15 Refit the camshaft followers and the relevant camshaft(s) as described in Section 11. Rotate the crankshaft a few times to settle all the components, then recheck the valve clearances.

16 If all the clearances are correct, refit the fuel injectors as described in Chapter 4B, Section 16, and the camshaft cover as described in Section 4. Refit the roadwheel, lower the vehicle to the ground and tighten the roadwheel nuts to the specified torque.

---

11 Camshafts and followers – removal, inspection and refitting

Removal

1 Remove the camshaft cover as described in Section 4.

2 Remove the fuel injectors as described in Chapter 4B, Section 16.

3 Remove the vacuum pump as described in Chapter 9, Section 19.

4 Remove the camshaft sprocket as described in Section 8.

5 Undo the bolt and two nuts and remove the No 5 bearing cap from the left-hand (transmission) end of the camshafts.

6 The exhaust camshaft gear incorporates a backlash compensating gear. This must now be locked to the fixed exhaust camshaft gear by inserting a suitably-sized screw/bolt into the hole on the inboard face of the fixed gear, and through into the backlash compensating gear. This prevents the spring preload of the compensating gear being lost when either camshaft is removed **(see illustration)**.

7 Working in a spiral pattern from the outside in, slacken the remaining camshaft bearing cap retaining bolts and nuts by one turn at a time, to relieve the pressure of the valve springs on the bearing caps gradually and evenly. Once the valve spring pressure has been relieved, the bolts and nuts can be fully unscrewed and removed.

Caution: If the bearing cap nuts/bolts are carelessly slackened, the bearing caps might break. If any bearing cap breaks then the complete camshaft housing assembly must be renewed; the bearing caps are matched to the housing and are not available separately.

8 Remove the bearing caps, noting each caps correct fitted location. The bearing caps are numbered 1 to 5 and the arrow on each cap points towards the timing belt end of the engine **(see illustration)**.

9 Lift the camshafts out of the cylinder head.

10 Obtain sixteen small, clean plastic containers, and label them for identification. Alternatively, divide a larger container into compartments. Using a rubber sucker tool, lift the followers out from the top of the cylinder head and store each one in its respective fitted position. Make sure the followers are not mixed to ensure the valve clearances remain correctly adjusted on refitting **(see illustration)**.

Inspection

11 Examine the camshaft bearing surfaces and camshaft lobes for signs of wear ridges and scoring. Renew the camshaft if any of these conditions are apparent.

12 Examine the condition of the bearing surfaces both on the camshaft journals and in the cylinder head. If the head bearing surfaces are worn excessively, the cylinder head will need to be renewed.

13 Examine the followers and their bores in the cylinder head for signs of wear or damage. If any follower is visibly worn it should be renewed.

Refitting

14 Lubricate the followers with clean engine oil and carefully insert each one into its original location in the cylinder head.

15 Rotate the crankshaft approximately 60º backwards (anti-clockwise) as a precaution against accidental piston-to-valve contact. Lubricate the camshaft followers with clean engine oil then lay the camshafts in position. Check that the exhaust camshaft backlash compensating gear is still locked to the fixed gear. Ensure that the mark on the outer face exhaust camshaft gear lies between the two marks on the outer face of the inlet camshaft gear, and that the marks are approximately level with the upper edge of the camshaft housing **(see illustration)**. If the exhaust camshaft is being renewed, it will be necessary to obtain Vauxhall/Opel tool No KM 6092, and pre-tension the backlash compensating gear prior to installation.

16 Ensure the mating surfaces of the bearing caps and camshaft housing are clean and dry and lubricate the camshaft journals and lobes with clean engine oil.

17 Refit the No 1 to 4 camshaft bearing caps in their original locations on the cylinder head. The caps are numbered 1 to 5 (No 1 cap being at the timing belt end of the engine) and the arrow cast onto the top of each cap should point towards the timing belt end of the engine **(see illustration 11.8)**.

18 Refit the No 1 to 4 bearing cap bolts and nuts, tightening them by hand only.

19 Remove the screw/bolt locking the backlash compensating gear to the exhaust camshaft fixed gear, and refit the No 5 camshaft bearing cap.

20 Working in a spiral pattern from the inside out, tighten the bolts and nuts by one turn at a time to gradually impose the pressure of the valve springs evenly on the bearing caps. Repeat this sequence until all bearing caps are in contact with the cylinder head then go around again, in a spiral pattern from the inside out, and tighten them to the specified torque.

Caution: If the bearing cap nuts/bolts are carelessly tightened, the bearing caps might break. If any bearing cap breaks then the complete camshaft housing assembly must be renewed; the bearing caps are matched to the housing and are not available separately.

21 Fit a new camshaft oil seal as described in Section 9.

22 Refit the camshaft sprocket as described in Section 8 and the timing belt as described in Section 7.

23 Check the valve clearances as described in Section 10.

24 Refit the vacuum pump as described in Chapter 9, Section 19.

25 Refit the fuel injectors as described in Chapter 4B, Section 16.

26 Refit the camshaft cover as described in Section 4.

12 Camshaft housing – removal and refitting

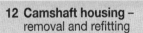

Removal

1 Remove the camshaft cover as described in Section 4.

2 Remove the fuel injectors as described in Chapter 4B, Section 16.

3 Remove the vacuum pump as described in Chapter 9, Section 19.

11.15 Align the camshaft gear marks

4 Remove the camshaft sprocket as described in Section 8.

5 Remove the EGR valve cooler as described in Chapter 4C, Section 3.

6 Undo the two bolts securing the rear timing belt cover to the camshaft housing **(see illustration)**.

7 Slacken the camshaft housing retaining bolts 1/2 turn at a time working in a spiral pattern from inside to outside.

8 Undo the bolts completely and remove the camshaft housing. Remove and discard the gasket.

Refitting

9 Ensure all mating surfaces are clean and free from any gasket or sealant residue.

10 Prior to refitting the housing, check that the camshaft gears are correctly aligned, with the mark on the outer face of the exhaust gear between the marks on the outer face of the inlet gear **(see illustration 11.15)**.

11 With a new gasket in place, position the camshaft housing on to the cylinder head, and tighten the housing bolts evenly and gradually to the specified torque, in a spiral pattern from outside to inside.

12 Refit the two bolts securing the rear timing belt cover to the camshaft housing and tighten the bolts to the specified torque.

13 Refit the EGR valve cooler as described in Chapter 4C, Section 3.

14 Refit the camshaft sprocket as described in Section 8 and the timing belt as described in Section 7.

15 Check the valve clearances as described in Section 10.

12.6 Remove the two bolts securing the rear timing belt cover to the camshaft housing

16 Refit the vacuum pump as described in Chapter 9, Section 19.

17 Refit the fuel injectors as described in Chapter 4B, Section 16.

18 Refit the camshaft cover as described in Section 4.

13 Cylinder head – removal and refitting

Caution: Be careful not to allow dirt into the fuel injection pump or injector pipes during this procedure.

Note: *New cylinder head bolts and a new high-pressure fuel pipe will be required for refitting.*

Removal

1 Drain the engine oil as described in Chapter 1B, Section 5.

2 Drain the cooling system as described in Chapter 1B, Section 27.

3 Remove the exhaust manifold as described in Chapter 4B, Section 25.

4 Remove the camshaft housing as described in Section 12.

5 As a precaution, obtain sixteen small, clean plastic containers, and label them for identification. Alternatively, divide a larger container into compartments. Lift the followers out from the top of the cylinder head and store each one in its respective fitted position **(see illustration)**.

6 Thoroughly clean the fuel pipe unions on the fuel pump and fuel rail. Using an open-ended spanner, unscrew the union nuts securing the high-pressure fuel pipe to the fuel pump and fuel rail. Counterhold the union on the pump with a second spanner, while unscrewing the union nut. Undo the bolt securing the high-pressure fuel pipe to the support bracket and withdraw the pipe. Plug or cover the open unions to prevent dirt entry.

7 Release the retaining clips and disconnect the remaining coolant hoses from the thermostat housing and cylinder head.

8 Disconnect the two vacuum hoses from the solenoid valves at the rear of the inlet manifold. Release the hoses from their retaining clips.

13.5 Store the followers the correct way up

13.11 Lift the cylinder head off the block

9 Undo the bolt securing the rear timing belt cover to the cylinder head.

10 Working in the reverse of the tightening sequence **(see illustration 13.29)**, progressively slacken the cylinder head bolts by half a turn at a time, until all bolts can be unscrewed by hand. Lift out the cylinder head bolts.

11 Lift the cylinder head away; seek assistance if possible, as it is a heavy assembly **(see illustration)**. Remove the gasket, noting the two locating dowels fitted to the top of the cylinder block. If they are a loose fit, remove the locating dowels and store them with the head for safe-keeping. Keep the head gasket for identification purposes (see paragraph 18).

12 If the cylinder head is to be dismantled for overhaul, refer to Chapter 2E, Section 6.

Preparation for refitting

13 The mating faces of the cylinder head and cylinder block/crankcase must be perfectly clean before refitting the head. Use a hard plastic or wood scraper to remove all traces of gasket and carbon; also clean the piston crowns. Take particular care, as the surfaces are damaged easily. Also, make sure that the carbon is not allowed to enter the oil and water passages – this is particularly important for the lubrication system, as carbon could block the oil supply to any of the engine's components. Using adhesive tape and paper, seal the water, oil and bolt holes in the cylinder block/crankcase. To prevent carbon entering the gap between the pistons and bores, smear a little grease in the gap. After

13.21 Measure the piston projection at the highest points between the valve cut-outs

cleaning each piston, use a small brush to remove all traces of grease and carbon from the gap, then wipe away the remainder with a clean rag. Clean all the pistons in the same way.

14 Check the mating surfaces of the cylinder block/crankcase and the cylinder head for nicks, deep scratches and other damage. If slight, they may be removed carefully with a file, but if excessive, machining may be the only alternative to renewal.

15 Ensure that the cylinder head bolt holes in the crankcase are clean and free of oil. Syringe or soak up any oil left in the bolt holes. This is most important in order that the correct bolt tightening torque can be applied and to prevent the possibility of the block being cracked by hydraulic pressure when the bolts are tightened.

16 The cylinder head bolts must be discarded and renewed, regardless of their apparent condition.

17 If warpage of the cylinder head gasket surface is suspected, use a straight-edge to check it for distortion. Refer to Chapter 2E, Section 7 if necessary.

18 On this engine, the cylinder head-to-piston clearance is controlled by fitting different thickness head gaskets. The gasket thickness can be determined by looking at the front centre of the gasket and checking on the number of holes.

Holes in gasket	Gasket thickness
One hole	0.95 mm
Two holes	1.00 mm
Three holes	1.05 mm

19 The correct thickness of gasket required is selected by measuring the piston protrusions as follows.

20 Mount a dial test indicator securely on the block so that its pointer can be easily pivoted between the piston crown and block mating surface. Turn the crankshaft to bring No 1 piston roughly to the TDC position. Move the dial test indicator probe over and in contact with No 1 piston. Turn the crankshaft back and forth slightly until the highest reading is shown on the gauge, indicating that the piston is at TDC.

21 Zero the dial test indicator on the gasket surface of the cylinder block then carefully

move the indicator over No 1 piston. Measure its protrusion at the highest point between the valve cut-outs, and then again at its highest point between the valve cut-outs at 90° to the first measurement **(see illustration)**. Repeat this procedure with No 4 piston.

22 Rotate the crankshaft half a turn (180º) to bring No 2 and 3 pistons to TDC. Ensure the crankshaft is accurately positioned then measure the protrusions of No 2 and 3 pistons at the specified points. Once both pistons have been measured, rotate the crankshaft through a further one and a half turns (540º) to bring No 1 and 4 pistons back to TDC.

23 Select the correct thickness of head gasket required by determining the largest amount of piston protrusion, and using the following table.

Piston protrusion measurement	Gasket thickness required
0.230 to 0.296 mm	0.95 mm
0.297 to 0.362 mm	1.00 mm
0.363 to 0.429 mm	1.05 mm

Refitting

24 Wipe clean the mating surfaces of the cylinder head and cylinder block/crankcase.

25 Check that the two locating dowels are in position then fit a new gasket to the cylinder block **(see illustration)**.

26 If not already positioned at TDC, rotate the crankshaft so that No 1 piston is at its highest point in the cylinder. Now turn the crankshaft 60° backwards (anti-clockwise). This is to ensure that whist the cylinder head and camshafts are being refitted, there is little chance of accidental piston-to-valve contact.

27 With the aid of an assistant, carefully refit the cylinder head assembly to the block, aligning it with the locating dowels.

28 Carefully enter each new cylinder head bolt into its relevant hole (do not drop them in). Screw all bolts in, by hand only, until finger-tight.

29 Working progressively in the sequence shown, tighten the cylinder head bolts to their Stage 1 torque setting, using a torque wrench and suitable socket **(see illustration)**.

30 Once all bolts have been tightened to the Stage 1 torque, working again in the same sequence, go around and tighten all bolts through the specified Stage 2 angle, then through the specified Stage 3 angle using an angle-measuring gauge.

31 Refit the bolt securing the rear timing belt cover to the cylinder head and tighten the bolt to the specified torque.

32 Reconnect the two vacuum hoses to the solenoid valves at the rear of the inlet manifold. Refit the hoses to their retaining clips.

33 Reconnect the coolant hoses to the thermostat housing and cylinder head.

34 Remove the blanking plugs from the fuel pipe unions on the fuel rail and fuel pump. Locate the new high-pressure fuel pipe over the unions and screw on the union nuts finger

13.25 Check the locating dowels are in place then fit the new gasket

tight. Tighten the union nuts to the specified torque using a torque wrench and crow-foot adaptor. Counterhold the union on the pump with an open-ended spanner, while tightening the union nut. Refit and tighten the bolt securing the high-pressure fuel pipe to the support bracket.

35 Lubricate the camshaft followers with clean engine oil and carefully insert each one into its original location in the cylinder head.

36 Refit the camshaft housing as described in Section 12.

37 Refit the exhaust manifold as described in Chapter 4B, Section 25.

38 Refit the plastic cover over the top of the engine.

39 Refit the wheel arch liner and roadwheel, then lower the car to the ground and tighten the roadwheel nuts to the specified torque.

40 Refill the engine with fresh oil as described in Chapter 1B, Section 5, then reconnect the battery negative terminal.

41 Refill the cooling system as described in Chapter 1B, Section 27.

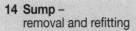

14 Sump –
removal and refitting

Removal

1 Disconnect the battery negative lead as described in Chapter 5A Section 4.

2 Firmly apply the handbrake, then jack up the front of the car and support it securely on axle stands (see *Jacking and vehicle support*).

3 Drain the engine oil as described in Chapter 1B, Section 5, then fit a new sealing washer and refit the drain plug, tightening it to the specified torque.

4 Slacken and remove the bolts securing the sump lower pan to the main casting. Using a wide bladed scraper or similar tool inserted between the sump and main casting, carefully break the joint, then remove the sump from under the car **(see illustration)**.

5 To remove the main casting from the engine, remove the exhaust system front pipe as described in Chapter 4B, Section 26.

6 Undo the two bolts securing the oil dipstick guide tube to the main sump casting and the bolt securing the guide tube to the exhaust manifold. Remove the guide tube and collect the O-ring seal. Note that a new seal will be required for refitting.

7 Undo the two bolts securing the driveshaft support bearing housing to the sump main casting.

8 Disconnect the wiring connector from the engine oil level sensor.

9 Release the exhaust temperature sensor wiring connector from the support bracket, then undo the retaining bolt securing the support bracket to the main casting.

10 Undo the two retaining nuts and remove the wiring harness trough from the studs at the front of the main casting.

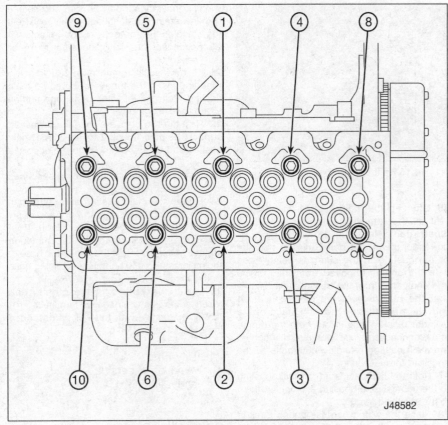

13.29 Cylinder head bolt tightening sequence

11 Progressively slacken and remove the nuts and bolts securing the main casting to the base of the cylinder block/oil pump cover and transmission. Break the joint using a wide scraper or similar tool carefully inserted in the joint between the main casting and cylinder block. Lower the casting away from the engine and withdraw it from underneath the vehicle.

12 While the sump main casting is removed, take the opportunity to check the oil pump pick-up/strainer for signs of clogging or splitting. If necessary, unbolt the pick-up/strainer and remove it from the engine along with its sealing rings **(see illustration)**. The strainer can then be cleaned easily in solvent or renewed.

Refitting

13 Remove all traces of dirt, oil and sealant from the mating surfaces of the sump main casting and pan, the cylinder block and (where removed) the pick-up/strainer.

14 Where necessary, fit new sealing rings to the oil pump pick-up/strainer and fit the strainer to the base of the cylinder block. Refit the strainer retaining bolt and tighten it to the specified torque.

15 Ensure the main casting and cylinder block mating surfaces are clean and dry, and apply a coat of suitable sealant (available from Vauxhall/Opel dealers) to the upper mating surface of the casting.

14.4 If the lower pan is stuck to the main casting, carefully ease it away using a wide-bladed scraper

14.12 Remove the oil pick-up pipe and strainer

14.16 Refit the sump main casting

16 Offer up the main casting and loosely refit all the retaining nuts and bolts **(see illustration)**. Note that the four long bolts correspond with the bolts holes at the rear of the casting. If the sump is being fitted with the engine removed from the vehicle and separated from the transmission, ensure that the rear face of the casting is flush with the transmission mounting face of the cylinder block. Working out from the centre in a diagonal sequence, progressively tighten the main casting retaining nuts/bolts to the specified torque setting.

17 Refit the bolts securing the main casting to the transmission housing and tighten them to the specified torque.

18 Place the wiring harness trough back in position over the two studs and secure with the two nuts tightened securely.

19 Refit the exhaust temperature sensor

wiring connector support bracket and secure with the retaining bolt tightened securely. Refit the wiring connector to the support bracket.

20 Reconnect the wiring connector to the oil level sensor.

21 Fit new seals to the oil dipstick guide tube and insert the tube into the sump main casting. Refit the retaining bolts, and tighten to the specified torque.

22 Refit the two bolts securing the driveshaft support bearing housing to the sump main casting and tighten the bolts to the specified torque.

23 Refit the exhaust front pipe as described in Chapter 4B, Section 26.

24 Ensure the main casting and sump pan mating surfaces are clean and dry and apply a coat of suitable sealant (available from Vauxhall/Opel dealers) to the upper mating surface of the pan. Refit the pan to the base of the main casting and tighten its retaining bolts to the specified torque.

25 Lower the vehicle to the ground and reconnect the battery negative terminal.

26 Fill the engine with fresh oil as described in Chapter 1B, Section 5.

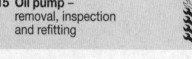

15 Oil pump –
   removal, inspection
   and refitting

Removal

1 Remove the timing belt as described in Section 7.

2 Remove the oil pump and crankshaft timing belt sprockets as described in Section 8.

3 Remove the sump main casting as described in Section 14. If the engine was supported on a trolley jack for removal of the timing belt, temporarily refit the right-hand engine mounting components, then remove the trolley jack from under the sump.

4 Slacken and remove the retaining bolts then slide the oil pump cover off of the end of the crankshaft, taking great care not to lose the locating dowels. Remove the sealing ring, which is fitted around the oil pump housing section of the cover, and discard it.

5 Using a suitable marker pen, mark the surface of the pump outer rotor; the mark can then be used to ensure the rotor is refitted the correct way around.

6 Remove the oil pump inner and outer rotors from the cylinder block **(see illustrations)**.

Inspection

7 Clean the components, and carefully examine the rotors, pump housing and cover for any signs of scoring or wear. Renew any component which shows signs of wear or damage. If the pump housing in the cylinder block is marked then seek the advice of a Vauxhall/Opel dealer on the best course of action.

8 If the components appear serviceable, fit the rotors into the housing and measure the clearance between the outer rotor and pump housing, and the inner rotor tip-to-outer rotor clearance using feeler blades **(see illustration)**. Also measure the rotor endfloat, and check the flatness of the end cover. If the clearances exceed the specified tolerances, renew the worn components.

Refitting

9 Lubricate the pump rotors with clean engine oil and refit them to the pump housing, using the mark made prior to removal to ensure the outer rotor is fitted the correct way around.

10 Prior to refitting, carefully lever out the crankshaft and oil pump oil seals using a flat-bladed screwdriver. Fit the new oil seals, ensuring that each seals sealing lip is facing inwards, and press them squarely into the housing using a tubular drift which bears only on the hard outer edge of the seal. Press each seal into position so that it is flush with the housing **(see illustration)**.

11 Ensure the mating surfaces of the oil pump and cylinder block are clean and dry and the locating dowels are in position. Remove all traces of sealant from the threads of the pump cover bolts.

12 Fit a new seal into the groove around the oil pump housing section of the cover, and apply a bead of suitable sealant (available from Vauxhall/Opel dealers) to the pump cover mating surface **(see illustration)**.

13 Carefully manoeuvre the oil pump cover into position, taking great care not to damage

15.6a Remove the oil pump inner rotor...

15.6b ...and outer rotor

15.8 Measure the inner rotor tip-to-outer rotor clearance

15.10 The seal fits flush with the housing

the oil seal lips on the crankshaft and inner rotor shaft. Locate the cover on the dowels making sure the pump sealing ring remains correctly positioned.

14 Apply a smear of sealant to the threads of each cover retaining bolt then refit all bolts and tighten them to the specified torque. Note that the longer bolt corresponds to the lower left-hand bolt hole in the cover.

15 Refit the sump as described in Section 14.

16 Refit the timing belt sprockets as described in Section 8, then refit the timing belt as described in Section 7.

17 On completion refill the engine with fresh oil as described in Chapter 1B, Section 5.

16 Oil pump seal – renewal

1 Remove the oil pump sprocket as described in Section 8.

2 Carefully punch or drill a small hole in the oil seal. Screw a self-tapping screw into the seal, and pull on the screw with pliers to extract the seal **(see illustration)**.

Caution: Great care must be taken to avoid damage to the oil pump.

3 Clean the seal housing, and polish off any burrs or raised edges which may have caused the seal to fail in the first place.

4 Press the new seal into position using a suitable tubular drift (such as a socket) which bears only on the hard outer edge of the seal. Take care not to damage the seal lips during fitting; note that the seal lips should face inwards.

5 Refit the oil pump sprocket as described in Section 8.

17 Oil filter housing – removal and refitting

Removal

1 The oil filter housing and integral oil cooler is situated on the rear left-hand side of the cylinder block. Firmly apply the handbrake, then jack up the front of the car and support it securely on axle stands (see *Jacking and vehicle support*).

2 Drain the cooling system as described in Chapter 1B, Section 27.

3 Position a suitable container beneath the oil filter. Release the retaining clip and disconnect the coolant return hose from the oil cooler.

4 Release the retaining clip and disconnect the oil return hose from the filter housing.

5 Undo the retaining bolt and remove the exhaust pressure sensor bracket from the side of the filter housing.

6 Disconnect the wiring connector from the oil pressure warning light switch on the side of the filter housing.

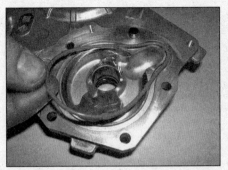

15.12 Fit a new seal to the oil pump cover

17.10a Fit new O-ring seals to the hollow bolt...

7 Undo the small retaining bolt securing the oil filter housing flange to the cylinder block.

8 Undo the large hollow bolt securing the oil filter housing to the cylinder block and remove the assembly from the engine.

9 Remove and discard the filter housing sealing ring and the two O-ring seals on the hollow bolt. Obtain new seals for refitting.

Refitting

10 Refitting is a reversal of removal, bearing in mind the following points:

l) Fit new O-ring seals to the hollow bolt and a new sealing ring to the filter housing **(see illustrations)**.

m) Tighten all retaining bolts to the specified torque, where given.

n) Check and if necessary top-up the engine oil as described in 'Weekly checks'.

o) Refill the cooling system as described in Chapter 1B, Section 27.

18 Flywheel/driveplate – removal, inspection and refitting

Removal

Note: *New flywheel/driveplate securing bolts must be used on refitting.*

Flywheel (manual transmission models)

1 Remove the transmission as described in Chapter 7A, Section 8.

16.2 Pull the screw to extract the seal

17.10b ...and a new sealing ring to the filter housing

2 Remove the clutch assembly as described in Chapter 6, Section 6.

3 Although the flywheel bolt holes are offset so that the flywheel can only be fitted in one position, it will make refitting easier if alignment marks are made between the flywheel and the end of the crankshaft.

4 Prevent the flywheel from turning by jamming the ring gear teeth using a suitable tool **(see illustration)**.

5 Unscrew the retaining bolts, and remove the flywheel.

Caution: Take care, as the flywheel is heavy.

Driveplate (automatic transmission models)

6 Remove the transmission as described in Chapter 7B, Section 9, then remove the driveplate as described in paragraphs 4 and 5.

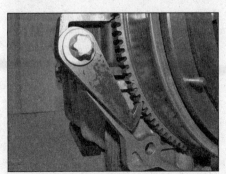

18.4 Prevent the flywheel from turning by jamming the ring gear teeth using a suitable tool

Inspection

Flywheel

7 On certain models a dual-mass flywheel is fitted and this has the effect of reducing engine and transmission vibrations and harshness. The flywheel consists of a primary mass and a secondary mass constructed in such a way that the secondary mass is allowed to rotate slightly in relation to the primary mass. Springs within the assembly restrict this movement to set limits.

8 Dual-mass flywheels have earned an unenviable reputation for unreliability and have been known to fail at quite low mileages (sometimes as low as 20 000 miles).

9 Examine the flywheel for wear or chipping of the ring gear teeth. Renewal of the ring gear is not possible and if the wear or chipping is significant, a new flywheel will be required.

10 Examine the flywheel for scoring of the clutch face. If the clutch face is scored significantly, a new flywheel will be required.

11 Look through the bolt hole and inspection openings in the secondary mass and check for any visible damage in the area of the centre bearing.

12 Place your thumbs on the clutch face of the secondary mass at the 3 o'clock and 9 o'clock positions and try to rock it. The maximum movement should not exceed 3 mm. Repeat this check with your thumbs at the 12 o'clock and 6 o'clock positions.

13 Rotate the secondary mass clockwise and anti-clockwise. It should move freely in both directions until spring resistance is felt, with no abnormal grating or rattling noises. The maximum rotational movement should not exceed a distance of eight teeth of the ring gear.

14 If there is any doubt about the condition of the flywheel, seek the advice of a Vauxhall/ Opel dealer or engine reconditioning specialist. They will be able to advise if the flywheel is an acceptable condition, or whether renewal is necessary.

Driveplate

15 Closely examine the driveplate and ring gear teeth for signs of wear or damage and check the driveplate surface for any signs of cracks.

16 If there is any doubt about the condition of the driveplate, seek the advice of a Vauxhall/Opel dealer or engine reconditioning specialist.

Refitting

Flywheel

17 Offer the flywheel to the end of the crankshaft, and align the previously-made marks on the flywheel and crankshaft (if applicable).

18 Coat the threads of the new flywheel bolts with thread-locking compound (note that new bolts may be supplied ready-coated), then fit the bolts and tighten them to the specified torque, then through the specified angles whilst preventing the flywheel from turning as during removal.

19 Refit the clutch as described in Chapter 6, Section 6 then refit the transmission as described in Chapter 7A, Section 8.

Driveplate

20 Clean the mating surfaces of the driveplate and crankshaft

21 Offer the driveplate and engage it on the crankshaft. Coat the threads of the new retaining bolts with thread-locking compound (note that new bolts may be supplied ready-coated), then fit the bolts and tighten them to the specified torque. Prevent the driveplate from turning as during removal.

22 Remove the locking tool and refit the transmission as described in Chapter 7B, Section 9.

19 Crankshaft oil seals – renewal

Right-hand (timing belt end) oil seal

1 Remove the crankshaft sprocket as described in Section 8.

2 Carefully punch or drill two small holes opposite each other in the oil seal. Screw a self-tapping screw into each and pull on the screws with pliers to extract the seal.

3 Clean the seal housing and polish off any burrs or raised edges which may have caused the seal to fail in the first place.

4 Ease the new seal into position on the end of the crankshaft. Press the seal squarely into position until it is flush with the housing. If necessary, a suitable tubular drift which bears only on the hard outer edge of the seal can be used to tap the seal into position. Take great care not to damage the seal lips during fitting and ensure that the seal lips face inwards.

5 Wash off any traces of oil, then refit the crankshaft sprocket as described in Section 8.

Left-hand (flywheel/driveplate end) oil seal

6 Remove the flywheel/driveplate as described in Section 18.

7 Renew the seal as described in paragraphs 2 to 4.

8 Refit the flywheel/driveplate as described in Section 18.

20 Engine/transmission mountings – inspection and renewal

1 Refer to Chapter 2C, Section 18, but observe the different torque wrench settings given in the Specifications of this Chapter.

Chapter 2 Part E
Engine removal and overhaul procedures

Contents

Degrees of difficulty

Easy, suitable for novice with little experience	**Fairly easy,** suitable for beginner with some experience	**Fairly difficult,** suitable for competent DIY mechanic	**Difficult,** suitable for experienced DIY mechanic	**Very difficult,** suitable for expert DIY or professional

Specifications

Engine identification

Engine type	Manufacturer's engine code*
1.4 litre (1364 cc) DOHC 16-valve turbocharged petrol engine	A14NET (LUJ)
1.6 litre (1598 cc) DOHC 16-valve normally aspirated petrol engine . . .	A16XER (LDE)
1.6 litre (1598 cc) DOHC 16-valve turbocharged diesel engine	B16DTH (LVL) and B16DTN (LVL)
1.7 litre (1686 cc) DOHC 16-valve turbocharged diesel engine	A17DTS (LUD)

* For details of engine code location, see 'Vehicle identification' in Reference

1.4 litre petrol engines – A14NET (LUJ)

Note: *Where specifications are given as N/A, no information was available at the time of writing. Refer to your Vauxhall/Opel dealer for the latest information available.*

Cylinder head

Maximum gasket face distortion	0.050 mm
Cylinder head height	N/A
Valve seat angle in cylinder head	90° 30'
Valve seat width in cylinder head:	
Inlet valve	1.40 to 1.80 mm
Exhaust valve	1.00 to 1.40 mm

Valves and guides

	Inlet	Exhaust
Stem diameter (standard size.	4.950 to 4.965 mm	4.930 to 4.945 mm
Valve head diameter.	27.900 to 28.100 mm	24.900 to 25.100 mm
Valve length (standard size)	92.90 mm	92.70 mm
Maximum permissible valve stem play in guide.	0.026 to 0.057 mm	0.046 to 0.077 mm.
Valve clearances.	Automatic adjustment by hydraulic cam followers	

1.4 litre petrol engines – A14NET (LUJ) (continued)

Cylinder block
Maximum gasket face distortion . . . . 0.050 mm
Cylinder bore diameter. . . . . 72.492 to 72.508 mm (nominal)
Maximum cylinder bore ovality and taper . . . . N/A

Crankshaft and bearings
Number of main bearings. 5
Main bearing journal diameter:
    Standard size . . . . 50.004 to 50.017 mm (nominal)
Big-end bearing journal diameter:
    Standard size . . . . 42.971 to 42.987 mm (nominal)
Crankshaft endfloat . . . . 0.100 to 0.202 mm

Pistons
Piston diameter . . . . 72.453 to 72.467 mm (nominal)

Piston rings
Number of rings (per piston). . . . . 2 compression, 1 oil control
Ring end gap:
    Upper compression . . . . 0.25 to 0.40 mm
    Lower compression . . . . 0.40 to 0.60 mm
    Oil control . . . . 0.25 to 0.75 mm

Torque wrench settings
Refer to Chapter 2A

1.6 litre petrol engines – A16XER (LDE)

Note: *Where specifications are given as N/A, no information was available at the time of writing. Refer to your Vauxhall/Opel dealer for the latest information available.*

Cylinder head
Maximum gasket face distortion . . . . 0.050 mm
Cylinder head height . . . . N/A
Valve seat angle in cylinder head. . . . . 90° 30'
Valve seat width in cylinder head:
    Inlet valve . . . . 1.00 to 1.40 mm
    Exhaust valve . . . . 1.40 to 1.80 mm

Valves and guides

	Inlet	Exhaust
Stem diameter (standard size)	4.965 to 4.980 mm	4.950 to 4.965 mm
Valve head diameter	31.10 to 31.30 mm	27.40 to 27.60 mm
Valve length (standard size)	117.00 to 117.40 mm	116.16 to 116.36 mm
Maximum permissible valve stem play in guide	0.020 to 0.051 mm	0.035 to 0.066 mm
Valve clearances (cold)	0.21 to 0.29 mm	0.26 to 0.34 mm

Cylinder block
Maximum gasket face distortion . . . . 0.05 mm
Cylinder bore diameter. . . . . 78.995 to 79.005 mm (nominal)
Maximum cylinder bore ovality and taper . . . . N/A

Crankshaft and bearings
Number of main bearings. 5
Main bearing journal diameter:
    Standard size . . . . 54.980 to 54.997 mm
Big-end bearing journal diameter. . . . . N/A
Crankshaft endfloat . . . . N/A

Pistons
Piston diameter . . . . 78.943 to 78.957 mm (nominal)

Piston rings
Number of rings (per piston). . . . . 2 compression, 1 oil control
Ring end gap:
    Upper compression . . . . 0.15 to 0.30 mm
    Lower compression . . . . 0.30 to 0.50 mm
    Oil control . . . . 0.20 to 0.70 mm

Torque wrench settings
Refer to Chapter 2B

1.6 litre diesel engines – B16DTH (LVL) and B16DTN (LVL)

Note: *Where specifications are given as N/A, no information was available at the time of writing. Refer to your Vauxhall/Opel dealer for the latest information available.*

Cylinder head

Maximum gasket face distortion	0.10 mm
Cylinder head height	129.95 to 130.05 mm
Valve seat angle in cylinder head	90º 30'
Valve seat width in cylinder head	N/A

Valves and guides

	Inlet	Exhaust
Stem diameter	4.969 to 4.985 mm	4.959 to 4.975 mm
Valve head diameter	26.54 to 26.80 mm	24.24 to 24.50 mm
Valve length	N/A	
Maximum permissible valve stem play in guide	N/A	
Valve clearances	Automatic adjustment by hydraulic cam followers	

Cylinder block

Maximum gasket face distortion	N/A
Cylinder bore diameter	79.692 to 79.708 mm (nominal)
Maximum cylinder bore ovality and taper	0.013 mm

Crankshaft and bearings

Number of main bearings	5
Main bearing journal diameter:	
Standard size	52.982 to 52.998 mm (nominal)
Big-end bearing journal diameter:	
Standard size	49.992 to 50.008 mm (nominal)
Crankshaft endfloat	N/A

Pistons

Piston diameter	79.648 to 79.662 mm (nominal)

Piston rings

Number of rings (per piston)	2 compression, 1 oil control
Ring end gap:	
Upper compression	0.20 to 0.35 mm
Lower compression	0.50 to 0.70 mm
Oil control	0.25 to 0.50 mm

Torque wrench settings

Refer to Chapter 2C

1.7 litre diesel engines – A17DTS (LUD)

Note: *Where specifications are given as N/A, no information was available at the time of writing. Refer to your Vauxhall/Opel dealer for the latest information available.*

Cylinder head

Maximum gasket face distortion	0.10 mm
Cylinder head height	94.95 to 95.05 mm
Valve seat angle	89.5º
Valve seat width:	
Inlet	1.6 to 1.8 mm
Exhaust	1.4 to 1.6 mm

Valves and guides

	Inlet	Exhaust
Stem diameter	5.96 to 5.97 mm	5.96 to 5.97 mm
Valve head diameter	27.5 mm	25.5 mm
Valve length	102.5 mm	102.2 mm
Maximum permissible valve stem play in guide	0.019 mm	0.021 mm
Valve clearances (cold)	0.40 ± 0.05 mm	0.40 ± 0.05 mm

Cylinder block

Maximum gasket face distortion	N/A
Cylinder bore diameter	79.000 to 79.010 mm (nominal)
Maximum cylinder bore ovality	N/A
Maximum cylinder bore taper	N/A

1.7 litre diesel engines – A17DTS (LUD) (continued)

Crankshaft and bearings

Number of main bearings.   5
Main bearing journal diameter:
   Standard size . . . .   51.928 to 51.938 mm (nominal)
Big-end bearing journal (crankpin) diameter . . . .   N/A
Crankshaft endfloat . . . .   0.030 to 0.120 mm

Pistons

Piston diameter . . . .   78.930 to 78.939 mm (nominal)

Piston rings

Number of rings (per piston). . . . .   2 compression, 1 oil control
Ring end gap:
   Upper compression . . . .   0.25 to 0.50 mm
   Lower compression . . . .   0.50 to 0.70 mm
   Oil control . . . .   0.25 to 0.50 mm

Torque wrench settings

Refer to Chapter 2D

1 General Information

1 Included in this Part of Chapter 2 are details of removing the engine/transmission from the car and general overhaul procedures for the cylinder head, cylinder block/crankcase and all other engine internal components.

2 The information given ranges from advice concerning preparation for an overhaul and the purchase of replacement parts, to detailed step-by-step procedures covering removal, inspection, renovation and refitting of engine internal components.

3 After Section 5, all instructions are based on the assumption that the engine has been removed from the car. For information concerning in-car engine repair, as well as the removal and refitting of those external components necessary for full overhaul, refer to Part A, B, C or D of this Chapter (as applicable) and to Section 5. Ignore any preliminary dismantling operations described in Part A, B, C or D that are no longer relevant once the engine has been removed from the car.

4 Apart from torque wrench settings, which are given at the beginning of Part A, B, C or D (as applicable), all specifications relating to engine overhaul are at the beginning of this Part of Chapter 2.

2 Engine overhaul – general information

1 It is not always easy to determine when, or if, an engine should be completely overhauled, as a number of factors must be considered.

2 High mileage is not necessarily an indication that an overhaul is needed, while low mileage does not preclude the need for an overhaul. Frequency of servicing is probably the most important consideration. An engine which has had regular and frequent oil and filter changes, as well as other required maintenance, should give many thousands of miles of reliable service. Conversely, a neglected engine may require an overhaul very early in its life.

3 Excessive oil consumption is an indication that piston rings, valve seals and/or valve guides are in need of attention. Make sure that oil leaks are not responsible before deciding that the rings and/or guides are worn. Have a compression test performed to determine the likely cause of the problem.

4 Check the oil pressure with a gauge fitted in place of the oil pressure switch, and compare it with that specified. If it is extremely low, the main and big-end bearings, and/or the oil pump, are probably worn out.

5 Loss of power, rough running, knocking or metallic engine noises, excessive valve gear noise, and high fuel consumption may also point to the need for an overhaul, especially if they are all present at the same time. If a complete service does not cure the situation, major mechanical work is the only solution.

6 A full engine overhaul involves restoring all internal parts to the specification of a new engine. During a complete overhaul, the pistons and the piston rings are renewed, and the cylinder bores are reconditioned. New main and big-end bearings are generally fitted; if necessary, the crankshaft may be reground, to compensate for wear in the journals. The valves are also serviced as well, since they are usually in less-than-perfect condition at this point. Always pay careful attention to the condition of the oil pump when overhauling the engine, and renew it if there is any doubt as to its serviceability. The end result should be an as-new engine that will give many trouble-free miles.

7 Critical cooling system components such as the hoses, thermostat and coolant pump should be renewed when an engine is overhauled. The radiator should also be checked carefully, to ensure that it is not clogged or leaking.

8 Before beginning the engine overhaul, read through the entire procedure, to familiarise yourself with the scope and requirements of the job. Check on the availability of parts and make sure that any necessary special tools and equipment are obtained in advance. Most work can be done with typical hand tools, although a number of precision measuring tools are required for inspecting parts to determine if they must be renewed.

9 The services provided by an engineering machine shop or engine reconditioning specialist will almost certainly be required, particularly if major repairs such as crankshaft regrinding or cylinder reboring are necessary. Apart from carrying out machining operations, these establishments will normally handle the inspection of parts, offer advice concerning reconditioning or renewal and supply new components such as pistons, piston rings and bearing shells. It is recommended that the establishment used is a member of the Federation of Engine Re-Manufacturers, or a similar society.

10 Always wait until the engine has been completely dismantled, and until all components (especially the cylinder block/crankcase and the crankshaft) have been inspected, before deciding what service and repair operations must be performed by an engineering works. The condition of these components will be the major factor to consider when determining whether to overhaul the original engine, or to buy a reconditioned unit. Do not, therefore, purchase parts or have overhaul work done on other components until they have been thoroughly inspected. As a general rule, time is the primary cost of an overhaul, so it does not pay to fit worn or sub-standard parts.

11 As a final note, to ensure maximum life and minimum trouble from a reconditioned engine, everything must be assembled with care, in a spotlessly-clean environment.

3 Engine removal – methods and precautions

1 If you have decided that the engine must be removed for overhaul or major repair work, several preliminary steps should be taken.

2 Engine/transmission removal is extremely complicated and involved on these vehicles. It must be stated, that unless the vehicle can be positioned on a ramp, or raised and supported on axle stands over an inspection pit, it will be very difficult to carry out the work involved.

3 Cleaning the engine compartment and engine/transmission before beginning the removal procedure will help keep tools clean and organised.

4 An engine hoist will also be necessary. Make sure the equipment is rated in excess of the combined weight of the engine and transmission. Safety is of primary importance, considering the potential hazards involved in removing the engine/transmission from the car.

5 The help of an assistant is essential. Apart from the safety aspects involved, there are many instances when one person cannot simultaneously perform all of the operations required during engine/transmission removal.

6 Plan the operation ahead of time. Before starting work, arrange for the hire of or obtain all of the tools and equipment you will need. Some of the equipment necessary to perform engine/transmission removal and installation safely (in addition to an engine hoist) is as follows: a heavy duty trolley jack, complete sets of spanners and sockets as described in the rear of this manual, wooden blocks, and plenty of rags and cleaning solvent for mopping up spilled oil, coolant and fuel. If the hoist must be hired, make sure that you arrange for it in advance, and perform all of the operations possible without it beforehand. This will save you money and time.

7 Plan for the car to be out of use for quite a while. An engineering machine shop or engine reconditioning specialist will be required to perform some of the work which cannot be accomplished without special equipment. These places often have a busy schedule, so it would be a good idea to consult them before

removing the engine, in order to accurately estimate the amount of time required to rebuild or repair components that may need work.

8 During the engine/transmission removal procedure, it is advisable to make notes of the locations of all brackets, cable ties, earthing points, etc, as well as how the wiring harnesses, hoses and electrical connections are attached and routed around the engine and engine compartment. An effective way of doing this is to take a series of photographs of the various components before they are disconnected or removed; the resulting photographs will prove invaluable when the engine/transmission is refitted.

9 Always be extremely careful when removing and refitting the engine/transmission. Serious injury can result from careless actions. Plan ahead and take your time, and a job of this nature, although major, can be accomplished successfully.

10 On all Mokka models, the engine must be removed complete with the transmission as an assembly. There is insufficient clearance in the engine compartment to remove the engine leaving the transmission in the vehicle. The assembly is removed by raising the front of the vehicle, and lowering the assembly from the engine compartment.

4 Engine and transmission unit – removal, separation and refitting 🔧

Note: *The engine can be removed from the car only as a complete unit with the transmission; the two are then separated for overhaul. The engine/transmission unit is lowered out of position, and withdrawn from under the vehicle. Bearing this in mind, and also bearing in mind the information contained in Section 3, ensure the vehicle is raised sufficiently so that there is enough clearance between the front of the vehicle and the floor to allow the engine/transmission unit to be slid out once it has been lowered out of position.*
Note: *Such is the complexity of the power unit arrangement on these vehicles, and the variations that may be encountered according to model and optional equipment fitted, that the following should be regarded as a guide to*

the work involved, rather than a step-by-step procedure. Where differences are encountered, or additional component disconnection or removal is necessary, make notes of the work involved as an aid to refitting.

Removal

1 Have the air conditioning system fully discharged by an air conditioning specialist.

2 Position the vehicle as described in Section 3, paragraph 2 and remove both front roadwheels. Remove the engine undertray and both front wheel arch liners as described in Chapter 11, Section 21.

3 Disconnect the battery negative lead as described in Chapter 5A Section 4.

4 Remove the front bumper as described in Chapter 11, Section 6.

5 Where fitted, remove the plastic cover from the top of the engine.

6 Remove the battery and battery tray as described in Chapter 5A, Section 4.

7 Lift off the cover from the engine compartment fuse/relay box, and unscrew the bolts securing the upper section of the fuse/relay box to the lower section **(see illustration)**.

8 Unscrew the nut from the stud at the rear of the fuse/relay box, then unscrew the nut and disconnect the battery positive feed cable from the fuse/relay box.

9 Release the retaining catches and lift the upper section of the fuse/relay box off the lower section, while at the same time disconnecting the engine wiring harness block connectors. Ensure that the block connectors remain in the lower section.

10 Lift the engine management wiring harness block connector(s) out of the fuse/relay box lower section **(see illustration)**.

11 Disconnect the engine management wiring harness connector and earth lead(s) from beneath the battery tray location, then release the retaining clips so the wiring harness is free to be removed with the engine **(see illustrations)**.

12 Carry out the following operations as described in Chapter 1A or Chapter 1B, as applicable:
a) Drain the engine oil.
b) Drain the cooling system.
c) Remove the auxiliary drivebelt.

4.7 Unscrew the bolts securing the upper section of the fuse/relay box to the lower section

4.10 Lift the wiring harness block connectors out of the fuse/relay box lower section

4.11a Disconnect the engine management wiring harness connector...

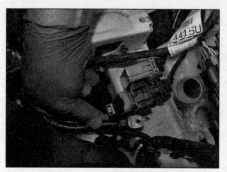

4.11b ...and earth lead...

4.11c ...then release the clips so the harness is free to be removed with the engine

4.13 Undo the two bolts and remove the coolant expansion tank

13 Release the retaining clips and disconnect the hoses at the coolant expansion tank. Undo the two retaining bolts and remove the expansion tank from the engine compartment **(see illustration)**.

14 Remove the air cleaner assembly, outlet air ducts and, where applicable, intercooler air ducts as described in Chapter 4A, Section 2 (petrol engine models) or Chapter 4B, Section 3 (diesel engine models).

15 Disconnect the reversing light switch wiring connector and release the wiring harness from the transmission.

16 Disconnect the wiring connector(s) from the radiator cooling fan housing. Release the wiring harness from the retaining clips so that it is free to be removed with the engine.

17 Remove the air conditioning system compressor as described in Chapter 3, Section 13.

Petrol engine models

18 Depressurise the fuel system with reference to Chapter 4A, Section 5, then disconnect the fuel supply pipe from the fuel rail and support bracket **(see illustration)**. Where applicable, also disconnect the fuel return pipe. Be prepared for fuel spillage, and take adequate precautions. Clamp or plug the open unions, to minimise further fuel loss.

19 Disconnect the brake vacuum servo hose, and fuel evaporation purge hose. Release the hoses from their clips and support brackets.

Diesel engine models

20 Disconnect the brake vacuum servo hose from the vacuum pump.

21 Disconnect the fuel supply and return hose quick-release fittings from the fuel pump or fuel rail. Suitably plug or cover the open unions to prevent dirt entry.

22 Disconnect the vacuum hoses and wiring connector from the turbocharger wastegate (charge pressure) solenoid valve.

All models

23 Release the retaining clips and remove the upper and lower radiator hoses.

24 Release the retaining clips and disconnect the coolant hoses at the thermostat housing and coolant outlet housing.

25 Using a small screwdriver, lift up the wire clips securing the two heater hoses to the heater matrix pipe stubs, and disconnect the hoses from the stubs **(see illustration)**.

26 Using a large screwdriver, prise the inner cable end fitting(s) from the transmission selector lever ballpin(s) **(see illustration)**.

27 Push the retaining sleeve(s) toward the rear of the car and detach the outer cable(s) from the mounting bracket on the transmission **(see illustration)**.

28 Drain the manual transmission oil or automatic transmission fluid as described in Chapter 7A, Section 2 (manual transmission) or Chapter 7B, Section 2 (automatic transmission).

29 Remove both driveshafts as described in Chapter 8, Section 2.

30 Position the steering with the front roadwheels straight-ahead, and lock the steering by removing the ignition key.

31 In the driver's footwell, unscrew the bolt securing the bottom of the steering column intermediate shaft universal joint to the steering gear pinion **(see illustration)**. Use paint or a suitable marker pen to make

4.18 Disconnect the fuel supply pipe from the fuel rail and support bracket

4.25 Lift up the wire clips and disconnect the heater hoses from the heater matrix pipe stubs

4.26 On manual transmission models, prise the selector cable end fittings from the transmission selector leverballpins

4.27 Push the retaining sleeves rearward and detach the outer cables from the transmission bracket

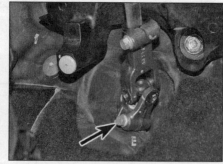

4.31 Unscrew the bolt securing the steering column intermediate shaft universal joint to the steering gear pinion

alignment marks between the universal joint and the steering gear pinion, then pull the joint from the pinion and position to one side.

Caution: To prevent damage to the airbag wiring contact unit, the steering lock must remain locked until the intermediate shaft is re-attached to the pinion shaft.

32 On manual transmission models, remove the filler cap from the brake/clutch fluid reservoir on the bulkhead, then tighten it onto a piece of polythene. This will reduce the loss of fluid when the clutch hydraulic hose is disconnected. Alternatively, fit a hose clamp to the flexible hose next to the clutch hydraulic connection on the transmission housing.

33 Place some cloth rags beneath the hose, then prise out the retaining clip securing the clutch hydraulic hose to the end fitting on the transmission. Detach the hose from the end fitting **(see illustration)**. Plug/cover both the end fitting and hose end to minimise fluid loss and prevent the entry of dirt into the hydraulic system. Whilst the hose is disconnected, do not depress the clutch pedal.

34 On automatic transmission models, unscrew the central retaining bolt (or nut) and detach the fluid cooler pipes from the transmission. Suitably cover the pipe ends and plug the transmission orifices to prevent dirt entry.

35 Remove the front subframe as described in Chapter 10, Section 7.

36 Attach a suitable hoist and lifting tackle to the engine lifting brackets on the cylinder head, and support the weight of the engine/transmission.

37 Mark the position of the three bolts securing the right-hand engine mounting bracket to the engine bracket and undo the bolts. There is no need to remove the mounting, since the engine/transmission is lowered from the engine compartment.

38 Mark the position of the three bolts securing the left-hand engine mounting to the transmission bracket and undo the bolts.

39 Make a final check to ensure that all relevant pipes, hoses, wires, etc, have been disconnected, and that they are positioned clear of the engine and transmission.

40 With the help of an assistant, carefully lower the engine/transmission assembly to the ground. Make sure that the surrounding components in the engine compartment are not damaged. Ideally, the assembly should be lowered onto a trolley jack or low platform with castors, so that it can easily be withdrawn from under the car.

41 Ensure that the assembly is adequately supported, then disconnect the engine hoist and lifting tackle, and withdraw the engine/transmission assembly from under the front of the vehicle.

42 Clean away any external dirt using paraffin or a water-soluble solvent and a stiff brush.

43 On manual transmission models remove the right-hand driveshaft intermediate shaft as described in Chapter 8, Section 3.

44 On 4WD models remove the transfer

4.33 Prise out the retaining clip then detach the clutch hydraulic hose from the end fitting on the transmission bellhousing

gearbox as described in Chapter 7C, Section 3.

45 On automatic transmission models, remove the starter motor as described in Chapter 5A, Section 11. Mark the position of the torque converter to the driveplate for reassembly, then turn the crankshaft until one of the torque converter retaining bolts becomes accessible through the starter motor aperture. Undo the retaining bolt, then turn the crankshaft until the next bolt becomes accessible. Undo the bolt and repeat the procedure until all the torque converter retaining bolts have been removed.

46 With reference to Chapter 7A, Section 8 or Chapter 7B, Section 9, unbolt the transmission from the engine. Carefully withdraw the transmission from the engine. On manual transmission models, ensure that its weight is not allowed to hang on the input shaft while engaged with the clutch friction disc. On automatic transmission models, ensure that the torque converter is removed together with the transmission so that it remains engaged with the oil pump. Note that the transmission locates on dowels positioned in the rear of the cylinder block.

Refitting

47 With reference to Chapter 7A, Section 8 or Chapter 7B, Section 9, refit the transmission to the engine and tighten the bolts to the specified torque.

48 On automatic transmission models, align the torque converter and driveplate using the marks made on removal, then fit the retaining bolts, turning the crankshaft as necessary. Tighten the bolts to the specified torque, then refit the starter motor.

49 On 4WD models refit the transfer gearbox as described in Chapter 7C, Section 3.

50 On manual transmission models refit the right-hand driveshaft intermediate shaft as described in Chapter 8, Section 3.

51 With the front of the vehicle raised and supported on axle stands, move the engine/transmission assembly under the vehicle, ensuring that the assembly is adequately supported.

52 Reconnect the hoist and lifting tackle to the engine lifting brackets, and carefully raise

the engine/transmission assembly up into the engine compartment with the help of an assistant.

53 Reconnect the right- and left-hand engine/transmission mountings and tighten the bolts to the specified torque given in the relevant Part of this Chapter. Ensure that the marks made on removal are correctly aligned when tightening the retaining bolts.

54 Refit the front subframe as described in Chapter 10, Section 7.

55 Disconnect the hoist and lifting tackle from the engine lifting brackets.

56 Refit the driveshafts as described in Chapter 8, Section 2.

57 The remainder of refitting is a reversal of removal, noting the following additional points:

a) Make sure that all mating faces are clean, and use new gaskets where necessary.

b) Tighten all nuts and bolts to the specified torque setting, where given.

c) Check and if necessary adjust the gearchange/selector cables as described in Chapter 7A, Section 3 or Chapter 7B, Section 3.

d) Refill the transmission with lubricant as described in Chapter 7A, Section 2 or Chapter 7B, Section 2.

e) Top-up and bleed the clutch hydraulic system as described in Chapter 6, Section 2.

f) Refill the engine with oil and coolant as described in Chapter 1A or Chapter 1B.

g) Have the air conditioning system evacuated, charged and leak-tested by the specialist who discharged it.

5 Engine overhaul – dismantling sequence

1 It is much easier to dismantle and work on the engine if it is mounted on a portable engine stand. These stands can often be hired from a tool hire shop. Before the engine is mounted on a stand, the flywheel/driveplate should be removed, so that the stand bolts can be tightened into the end of the cylinder block/crankcase.

2 If a stand is not available, it is possible to dismantle the engine with it blocked up on a sturdy workbench, or on the floor. Be extra-careful not to tip or drop the engine when working without a stand.

3 If you are going to obtain a reconditioned engine, all the external components must be removed first, to be transferred to the replacement engine (just as they will if you are doing a complete engine overhaul yourself). These components include the following:

a) Engine wiring harness and supports.

b) Alternator and air conditioning compressor mounting brackets (as applicable).

c) Coolant pump (where applicable), thermostat and inlet/outlet housings.

d) Dipstick tube.

e) Fuel system components.

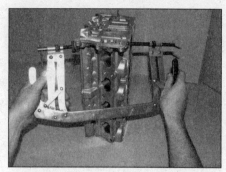

6.3 Using a valve spring compressor, compress the valve spring to relieve the pressure on the collets

6.5 Extract the two split collets by hooking them out using a small screwdriver

f) All electrical switches and sensors.
g) Inlet and exhaust manifolds and, where fitted, the turbocharger.
h) Oil filter housing.
i) Flywheel/driveplate.

Note: *When removing the external components from the engine, pay close attention to details that may be helpful or important during refitting. Note the fitted position of gaskets, seals, spacers, pins, washers, bolts, and other small items.*

4 If you are obtaining a 'short' engine (which consists of the engine cylinder block/crankcase, crankshaft, pistons and connecting rods all assembled), then the cylinder head, sump, oil pump, and timing chain/belt will have to be removed also.

5 If you are planning a complete overhaul, the engine can be dismantled, and the internal components removed, in the order given below.

Petrol engines:

a) Inlet manifold (see Chapter 4A).
b) Exhaust manifold (see Chapter 4A).
c) Timing cover, chain, sprockets and tensioner (see Chapter 2A).
d) Timing belt, sprockets, tensioner and idler pulley (see Chapter 2B).
e) Coolant pump (see Chapter 3).
f) Cylinder head (see Chapter 2A or 2B).
g) Flywheel/driveplate (see Chapter 2A or 2B).
h) Sump (see Chapter 2A or 2B).
i) Oil pump (see Chapter 2A or 2B).
j) Pistons/connecting rod assemblies (see Section 9).
k) Crankshaft (see Section 10).

Diesel engines:

a) Inlet manifold (see Chapter 4B).
b) Exhaust manifold (see Chapter 4B).

c) Timing cover, chain, sprockets and tensioner (see Chapter 2C).
d) Timing belt, sprockets, tensioner and idler pulleys (see Chapter 2D).
e) Coolant pump (see Chapter 3).
f) Cylinder head (see Chapter 2C or 2D).
g) Flywheel/driveplate (see Chapter 2C or 2D).
h) Sump (see Chapter 2C or 2D).
i) Oil pump (see Chapter 2C or 2D).
j) Piston/connecting rod assemblies (see Section 9).
k) Crankshaft (see Section 10).

6 Before beginning the dismantling and overhaul procedures, make sure that you have all of the correct tools necessary. See *Tools and working facilities* in Reference for further information.

6 Cylinder head – dismantling

Note: *New and reconditioned cylinder heads are available from the manufacturer, and from engine overhaul specialists. Due to the fact that some specialist tools are required for the dismantling and inspection procedures, and new components may not be readily available, it may be more practical and economical for the home mechanic to purchase a reconditioned head rather than to dismantle, inspect and recondition the original head. A valve spring compressor tool will be required for this operation.*

1 Remove the cylinder head with reference to the following Chapters and Sections:

1.4 litre petrol engines – Chapter 2A, Section 10
1.6 litre petrol engines – Chapter 2B, Section 13
1.6 litre diesel engines – Chapter 2C, Section 11
1.7 litre diesel engines – Chapter 2D, Section 13

2 Remove any remaining ancillary components from the cylinder head as necessary, referring to the relevant Chapters and Sections of this manual.

3 To remove a valve, fit a valve spring compressor tool. Ensure that the arms of the compressor tool are securely positioned on the head of the valve and the spring cap **(see illustration)**. The valves are deeply-recessed and a suitable extension piece may be required for the spring compressor.

4 Compress the valve spring to relieve the pressure of the spring cap acting on the collets.

 HAYNES HINT *If the spring cap sticks to the valve stem, support the compressor tool, and give the end a light tap with a soft-faced mallet to help free the spring cap.*

5 Extract the two split collets by hooking them out using a small screwdriver, then slowly release the compressor tool **(see illustration)**.

6 Remove the valve spring cap and the spring, then withdraw the valve through the combustion chamber. Using pliers or a valve stem oil seal remover tool, remove the valve stem oil seal **(see illustrations)**.

6.6a Remove the valve spring cap...

6.6b ...and the spring...

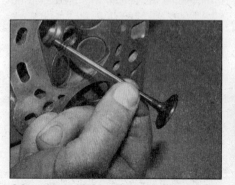

6.6c ...then withdraw the valve through the combustion chamber

6.6d Using pliers, remove the valve stem oil seal

7 Repeat the procedure for the remaining valves, keeping all components in strict order so that they can be refitted in their original positions, unless all the components are to be renewed. If the components are to be kept and used again, place each valve assembly in a labelled polythene bag or a similar small container **(see illustration)**. Note that as with cylinder numbering, the valves are numbered from the timing chain/belt end of the engine. Make sure that the valve components are identified as inlet and exhaust, as well as numbered.

7 Cylinder head and valves – cleaning and inspection

1 Thorough cleaning of the cylinder head and valve components, followed by a detailed inspection, will enable you to decide how much valve service work must be carried out during the engine overhaul.
Note: *If the engine has been severely overheated, it is best to assume that the cylinder head is warped – check carefully for signs of this.*

Cleaning

2 Scrape away all traces of old gasket material from the cylinder head.
3 Scrape away the carbon from the combustion chambers and ports, then wash the cylinder head thoroughly with paraffin or a suitable solvent.
4 Scrape off any heavy carbon deposits that may have formed on the valves, then use a power-operated wire brush to remove deposits from the valve heads and stems.

Inspection

Note: *Be sure to perform all the following inspection procedures before concluding that the services of a machine shop or engine overhaul specialist are required. Make a list of all items that require attention.*

Cylinder head

5 Inspect the head very carefully for cracks, evidence of coolant leakage, and other damage. If cracks are found, a new cylinder head should be obtained.

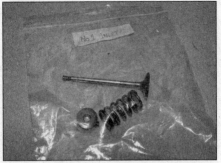

6.7 Place each valve assembly in a labelled polythene bag or similar container

6 Use a straight-edge and feeler gauge blade to check that the cylinder head surface is not distorted **(see illustration)**. If it is, it will be necessary to obtain a new or reconditioned cylinder head.
7 Examine the valve seats in each of the combustion chambers. If they are severely pitted, cracked or burned, then they will need to be recut by an engine overhaul specialist. If they are only slightly pitted, this can be removed by grinding-in the valve heads and seats with fine valve-grinding compound, as described below.
8 If the valve guides are worn, indicated by a side-to-side motion of the valve, oversize valve guides are available, and valves with oversize stems can be fitted. This work is best carried out by an engine overhaul specialist. A dial gauge may be used to determine whether the amount of side play of a valve exceeds the specified maximum.
9 Check the tappet bores in the cylinder head for wear. If excessive wear is evident, the cylinder head must be renewed. Also check the tappet oil holes in the cylinder head for obstructions.

Valves

10 Examine the head of each valve for pitting, burning, cracks and general wear, and check the valve stem for scoring and wear ridges. Rotate the valve, and check for any obvious indication that it is bent. Look for pitting and excessive wear on the tip of each valve stem. Renew any valve that shows any such signs of wear or damage.

11 If the valve appears satisfactory at this stage, measure the valve stem diameter at several points using a micrometer **(see illustration)**. Any significant difference in the readings obtained indicates wear of the valve stem. Should any of these conditions be apparent, the valve(s) must be renewed.
12 If the valves are in satisfactory condition, they should be ground (lapped) into their respective seats, to ensure a smooth gas-tight seal. If the seat is only lightly pitted, or if it has been recut, fine grinding compound only should be used to produce the required finish. Coarse valve-grinding compound should not be used unless a seat is badly burned or deeply pitted; if this is the case, the cylinder head and valves should be inspected by an expert to decide whether seat recutting, or even the renewal of the valve or seat insert, is required.
13 Valve grinding is carried out as follows. Place the cylinder head upside-down on a bench, with a block of wood at each end to give clearance for the valve stems.
14 Smear a trace of the appropriate grade of valve-grinding compound on the seat face, and press a suction grinding tool onto the valve head. With a semi-rotary action, grind the valve head to its seat, lifting the valve occasionally to redistribute the grinding compound **(see illustration)**. A light spring placed under the valve head will greatly ease this operation.
15 If coarse grinding compound is being used, work only until a dull, matt even surface is produced on both the valve seat and the valve, then wipe off the used compound and repeat the process with fine compound. When a smooth unbroken ring of light grey matt finish is produced on both the valve and seat, the grinding operation is complete. Do not grind in the valves any further than absolutely necessary, or the seat will be prematurely sunk into the cylinder head.
16 When all the valves have been ground-in, carefully wash off all traces of grinding compound using paraffin or a suitable solvent before reassembly of the cylinder head.

Valve components

17 Examine the valve springs for signs of damage and discoloration; if possible; also

7.6 Using a straight-edge and feeler gauge to check cylinder head surface distortion

7.11 Using a micrometer to measure valve stem diameter

7.14 Grinding-in a valve

8.1 Lubricate the valve stem with engine oil and insert the valve into the correct guide

8.2a If available, fit an oil seal protector over the valve stem

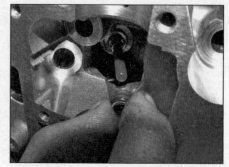

8.2b Fit the valve stem oil seal...

8.2c ...pressing it onto the valve guide with a suitable socket

8.3a Locate the spring in position...

8.3b ...and fit the spring cap

compare the existing spring free length with new components.

18 Stand each spring on a flat surface, and check it for squareness. If any of the springs are damaged, distorted or have lost their tension, obtain a complete new set of springs.

8 Cylinder head – reassembly

1 Lubricate the stems of the valves, and insert them into their original locations **(see illustration)**. If new valves are being fitted, insert them into the locations to which they have been ground.

2 Working on the first valve, dip the new

HAYNES HiNT

Use a little dab of grease to hold the collets in position on the valve stem while the spring compressor is released.

valve stem seal in fresh engine oil, then carefully locate it over the valve and onto the guide. Take care not to damage the seal as it is passed over the valve stem. Use a suitable socket or metal tube to press the seal firmly onto the guide. If genuine seals are being fitted, use the oil seal protector which is supplied with the seals; the protector fits over the valve stem and prevents the oil seal lip being damaged on the valve **(see illustrations)**.

3 Locate the spring in position and fit the spring cap **(see illustrations)**.

4 Compress the valve spring, and locate the split collets in the recess in the valve stem **(see illustration** and **Haynes Hint)**. Release the compressor, then repeat the procedure on the remaining valves.

5 With all the valves installed, support the cylinder head on blocks on the bench and,

8.4 Compress the valve and locate the collets in the recess on the valve stem

using a hammer and interposed block of wood, tap the end of each valve stem to settle the components.

6 Refit the components removed during cylinder head dismantling, with reference to the relevant Chapters and Sections of this manual.

9 Pistons/connecting rods – removal

Note: *New connecting rod big-end cap bolts will be needed on refitting.*

1 On petrol engine models, referring to Chapter 2A or 2B, remove the cylinder head and sump. On 1.4 litre turbocharged engines, undo the retaining bolts and lift off the oil baffle plate **(see illustration)**.

9.1 On 1.4 litre turbocharged engines, undo the bolts and lift off the oil baffle plate

2 On diesel engine models, referring to Chapter 2C or 2D, remove the cylinder head and sump. Where fitted, undo the retaining bolts and lift off the oil baffle plate.

3 On all models, if there is a pronounced wear ridge at the top of any bore, it may be necessary to remove it with a scraper or ridge reamer, to avoid piston damage during removal. Such a ridge indicates excessive wear of the cylinder bore.

4 If the connecting rods and big-end caps are not marked to indicate their positions in the cylinder block (ie, marked with cylinder numbers), suitably mark both the rod and cap with quick-drying paint or similar **(see illustration)**. Note which side of the engine the marks face and accurately record this also. There may not be any other way of identifying which way round the cap fits on the rod, when refitting.

5 Turn the crankshaft to bring pistons 1 and 4 to BDC (bottom dead centre).

6 Unscrew the bolts or nuts from No 1 piston big-end bearing cap, then take off the cap and recover the bottom half-bearing shell **(see illustration)**. If the bearing shells are to be re-used, tape the cap and the shell together.

Caution: On some engines, the connecting rod/bearing cap mating surfaces are not machined flat; the big-end bearing caps are 'cracked' off from the rod during production and left untouched to ensure the cap and rod mate perfectly. Where this type of connecting rod is fitted, great care must be taken to ensure the mating surfaces of the cap and rod are not marked or damaged in anyway. Any damage to the mating surfaces will adversely affect the strength of the connecting rod and could lead to premature failure.

7 Using a hammer handle, push the piston up through the bore, and remove it from the top of the cylinder block. Recover the bearing shell, and tape it to the connecting rod for safe-keeping.

8 Loosely refit the big-end cap to the connecting rod, and secure with the bolts/nuts – this will help to keep the components in their correct order.

9.4 Suitably mark both the connecting rod and cap with their cylinder number

9 Remove No 4 piston assembly in the same way.

10 Turn the crankshaft through 180° to bring pistons 2 and 3 to BDC, and remove them in the same way.

10 Crankshaft – removal

1.4 litre petrol engines

Note: *New cylinder block baseplate retaining bolts will be required for refitting.*

1 Remove the flywheel/driveplate as described in Chapter 2A, Section 15.

2 Remove the pistons and connecting rods, as described in Section 9.

3 Invert the engine so that the crankshaft is uppermost.

4 Before removing the crankshaft, check the endfloat using a dial gauge in contact with the end of the crankshaft. Push the crankshaft fully one way, and then zero the gauge. Push the crankshaft fully the other way, and check the endfloat **(see illustration)**. The result should be compared with the specified limit, and will give an indication as to the size of the main bearing shell thrust journal width which will be required for reassembly.

5 If a dial gauge is not available, a feeler gauge can be used to measure crankshaft

9.6 Unscrew the bolts or nuts from the big-end bearing cap, then take off the cap and recover the bottom half-bearing shell

endfloat. Push the crankshaft fully towards one end of the crankcase, and insert a feeler gauge between the thrust flange of the main bearing shell and the machined surface of the crankshaft web **(see illustration)**. Before measuring, ensure that the crankshaft is fully forced towards one end of the crankcase, to give the widest possible gap at the measuring location. The measurement should be taken at the bearing with the thrustwasher (see Section 17).

6 Working in a diagonal sequence, progressively slacken the outer (M6) bolts securing the cylinder block baseplate to the cylinder block.

7 When all the outer bolts have been slackened, repeat the procedure on the inner (M8) retaining bolts.

8 Remove all the bolts and lift the cylinder block baseplate off the cylinder block **(see illustration)**. If the baseplate is initially tight to remove, carefully tap it free using a soft-faced mallet.

9 As the baseplate is withdrawn check that the lower main bearing shells come away with the baseplate. If they remain on the crankshaft journals, lift them off and refit them to their respective locations in the baseplate.

10 Lift the crankshaft from the cylinder block and remove the crankshaft oil seal **(see illustration)**.

11 Extract the upper bearing shells, and identify them for position if they are to be re-used.

10.4 Check the crankshaft endfloat using a dial gauge...

10.5 ...or a feeler gauge

10.8 Remove all the bolts and lift the cylinder block baseplate off the cylinder block – 1.4 litre petrol engines

**10.10 Removing the crankshaft –
1.4 litre petrol engines**

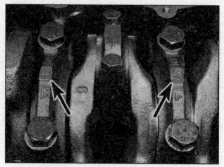

**10.15 Main bearing cap identification
markings – 1.6 litre petrol engines**

1.6 litre petrol engines

Note: *New main bearing cap retaining bolts will be required for refitting.*

12 Remove the flywheel and the oil pump as described in Chapter 2B, Section 18, and Chapter 2B, Section 15.

13 Remove the pistons and connecting rods, as described in Section 9.

14 Before removing the crankshaft, check the endfloat as described in paragraphs 4 and 5.

15 The main bearing caps should be numbered 1 to 5 from the timing belt end of the engine and all identification numbers should be the right way up when read from the rear of the cylinder block **(see illustration)**. If the bearing caps are not marked, using a hammer and punch or a suitable marker pen, number the caps from 1 to 5 from the timing belt end of the engine and mark each cap to indicate its correct fitted direction to avoid confusion on refitting.

16 Working in a diagonal sequence, evenly and progressively slacken the ten main bearing cap retaining bolts by half a turn at a time until all bolts are loose. Remove all the bolts.

17 Carefully remove each cap from the cylinder block, ensuring that the lower main bearing shell remains in position in the cap.

18 Carefully lift out the crankshaft, taking care not to displace the upper main bearing shells. Remove the oil seal and discard it.

19 Recover the upper bearing shells from the cylinder block, and tape them to their respective caps for safe-keeping.

1.6 litre diesel engines

Note: *New cylinder block baseplate (M10) retaining bolts will be required for refitting.*

20 Remove the flywheel/driveplate as described in Chapter 2C Section 17.

21 Remove the pistons and connecting rods, as described in Section 9.

22 Before removing the crankshaft, check the endfloat as described in paragraphs 4 and 5.

23 Working in a diagonal sequence, progressively slacken the outer (M8) bolts securing the cylinder block baseplate to the cylinder block.

24 When all the outer bolts have been slackened, repeat the procedure on the inner (M10) retaining bolts. Note that new bolts will be required for refitting.

25 Remove all the bolts and lift the cylinder block baseplate off the cylinder block. If the baseplate is initially tight to remove, carefully tap it free using a soft-faced mallet. Alternatively, use a spatula to break the seal between the baseplate and cylinder block.

26 As the baseplate is withdrawn check that the lower main bearing shells come away with the baseplate. If they remain on the crankshaft journals, lift them off and refit them to their respective locations in the baseplate.

27 Lift the crankshaft from the cylinder block and remove the crankshaft oil seal.

28 Extract the upper bearing shells, and identify them for position if they are to be re-used.

1.7 litre diesel engines

Note: *New main bearing cap retaining bolts will be required for refitting.*

29 Remove the flywheel/driveplate, main sump casting, oil pump cover and oil pick-up pipe as described in Chapter 2D, Section 18, 2D Section 14 and 2D Section 15.

30 Unbolt the crankshaft rear oil seal housing and remove it from the cylinder block. If the housing locating dowels are a loose fit, remove them and store them with the housing for safe-keeping.

31 Remove the piston and connecting rod assemblies as described in Section 9.

32 Before removing the crankshaft, check the endfloat as described in paragraphs 4 and 5.

33 Unscrew the main bearing cap bolts and remove the bearing caps. Note that the caps should are numbered 1 to 5, Number 1 cap being at the timing belt end, and the arrow on each cap should point towards the timing belt end of the engine. Ensure that the lower main bearing shell remains in position in the cap

34 Carefully remove the crankshaft, and recover the thrustwasher halves from the sides of Number 2 main bearing.

11 Cylinder block – cleaning and inspection

Cleaning

1 For complete cleaning, remove all external components (senders, sensors, brackets, oil pipes, coolant pipes, etc) from the cylinder block.

2 Scrape all traces of gasket and/or sealant from the cylinder block and cylinder block baseplate (where applicable), taking particular care not to damage the cylinder head and sump mating faces.

3 Remove all oil gallery plugs, where fitted. The plugs are usually very tight – they may have to be drilled out and the holes retapped. Use new plugs when the engine is reassembled. Where fitted, undo the retaining bolts and remove the piston oil spray nozzles from inside the cylinder block **(see illustrations)**.

4 If the block is extremely dirty, it should be steam-cleaned.

5 If the block has been steam-cleaned, clean all oil holes and oil galleries one more time on completion. Flush all internal passages with warm water until the water runs clear. Dry the block thoroughly and wipe all machined surfaces with a light oil. If you have access to compressed air, use it to speed-up the drying process, and to blow out all the oil holes and galleries.

⚠️ *Warning: Wear eye protection when using compressed air.*

6 If the block is relatively clean, an adequate cleaning job can be achieved with hot soapy water and a stiff brush. Take plenty of time, and do a thorough job. Regardless of the cleaning method used, be sure to clean all oil holes and galleries very thoroughly, dry

11.3a Unscrew the retaining bolts...

11.3b ...and remove the piston oil spray nozzles from the cylinder block

everything completely, and coat all cast-iron machined surfaces with light oil.

7 The threaded holes in the cylinder block must be clean, to ensure accurate torque readings when tightening fixings during reassembly. Run the correct-size tap (which can be determined from the size of the relevant bolt) into each of the holes to remove rust, corrosion, thread sealant or other contamination, and to restore damaged threads. If possible, use compressed air to clear the holes of debris produced by this operation. Do not forget to clean the threads of all bolts and nuts which are to be re-used, as well.

8 Where applicable, apply suitable sealant to the new oil gallery plugs, and insert them into the relevant holes in the cylinder block. Tighten the plugs securely. Refit the oil spray nozzles into the block and secure with the retaining bolts tightened securely.

9 If the engine is to be left dismantled for some time, cover the cylinder block with a large plastic bag to keep it clean and prevent corrosion.

Inspection

10 Visually check the block for cracks, rust and corrosion. Look for stripped threads in the threaded holes. It's also a good idea to have the block checked for hidden cracks by an engine reconditioning specialist that has the equipment to do this type of work, especially if the vehicle had a history of overheating or using coolant. If defects are found, have the block repaired, if possible, or renewed.

11 If in any doubt as to the condition of the cylinder block, have it inspected and measured by an engine reconditioning specialist. If the bores are worn or damaged, they will be able to carry out any necessary reboring (where possible), and supply appropriate oversized pistons, etc.

12 Pistons/connecting rods – inspection

1 Before the inspection process can begin, the piston/connecting rod assemblies must be cleaned, and the original piston rings removed from the pistons.

Note: *Always use new piston rings when the engine is reassembled.*

2 Carefully expand the old rings over the top of the pistons. The use of two or three old feeler gauges will be helpful in preventing the rings dropping into empty grooves **(see illustration)**. Take care, however, as piston rings are sharp.

3 Scrape away all traces of carbon from the top of the piston. A hand-held wire brush, or a piece of fine emery cloth, can be used once the majority of the deposits have been scraped away.

4 Remove the carbon from the ring grooves in the piston, using an old ring. Break the ring in half to do this (be careful not to cut your

fingers – piston rings are sharp). Be very careful to remove only the carbon deposits – do not remove any metal, and do not nick or scratch the sides of the ring grooves.

5 Once the deposits have been removed, clean the piston/connecting rod assembly with paraffin or a suitable solvent, and dry thoroughly. Make sure that the oil return holes in the ring grooves are clear.

6 If the pistons and cylinder bores are not damaged or worn excessively, and if the cylinder block does not need to be rebored, the original pistons can be refitted. Normal piston wear shows up as even vertical wear on the piston thrust surfaces, and slight looseness of the top ring in its groove. New piston rings should always be used when the engine is reassembled.

7 Carefully inspect each piston for cracks around the skirt, at the gudgeon pin bosses, and at the piston ring lands (between the ring grooves).

8 Look for scoring and scuffing on the thrust faces of the piston skirt, holes in the piston crown, and burned areas at the edge of the crown. If the skirt is scored or scuffed, the engine may have been suffering from overheating, and/or abnormal combustion ('pinking') which caused excessively-high operating temperatures. The cooling and lubrication systems should be checked thoroughly. A hole in the piston crown, or burned areas at the edge of the piston crown indicates that abnormal combustion (pre-ignition, 'pinking', knocking or detonation) has been occurring. If any of the above problems exist, the causes must be investigated and corrected, or the damage will occur again.

9 Corrosion of the piston, in the form of pitting, indicates that coolant has been leaking into the combustion chamber and/or the crankcase. Again, the cause must be corrected, or the problem may persist in the rebuilt engine.

10 If in any doubt as to the condition of the pistons and connecting rods, have them inspected and measured by an engine reconditioning specialist. If new parts are required, they will be able to supply and fit appropriate-sized pistons/rings, and rebore (where possible) or hone the cylinder block.

13 Crankshaft – inspection

1 Clean the crankshaft using paraffin or a suitable solvent, and dry it, preferably with compressed air if available. Be sure to clean the oil holes with a pipe cleaner or similar probe, to ensure that they are not obstructed.

⚠️ **Warning: Wear eye protection when using compressed air.**

2 Check the main and big-end bearing journals for uneven wear, scoring, pitting and cracking.

3 Big-end bearing wear is accompanied by

12.2 Using a feeler blade to remove a piston ring

distinct metallic knocking when the engine is running (particularly noticeable when the engine is pulling from low revs), and some loss of oil pressure.

4 Main bearing wear is accompanied by severe engine vibration and rumble – getting progressively worse as engine revs increase – and again by loss of oil pressure.

5 Check the bearing journal for roughness by running a finger lightly over the bearing surface. Any roughness (which will be accompanied by obvious bearing wear) indicates that the crankshaft requires regrinding.

6 If the crankshaft has been reground, check for burrs around the crankshaft oil holes (the holes are usually chamfered, so burrs should not be a problem unless regrinding has been carried out carelessly). Remove any burrs with a fine file or scraper, and thoroughly clean the oil holes as described previously.

7 Have the crankshaft journals measured by an engine reconditioning specialist. If the crankshaft is worn or damaged, they may be able to regrind the journals and supply suitable undersize bearing shells. If no undersize shells are available and the crankshaft has worn beyond the specified limits, it will have to be renewed. Consult your Vauxhall/Opel dealer or engine reconditioning specialist for further information on parts availability.

8 If a new crankshaft is to be fitted, undo the screws securing the crankshaft speed/position sensor pulse pick-up ring to the crankshaft, and transfer the ring to the new crankshaft **(see illustration)**.

13.8 Transfer the crankshaft speed/position sensor pulse pick-up ring to the new crankshaft

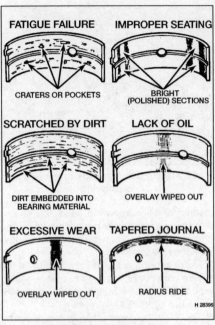

FATIGUE FAILURE

CRATERS OR POCKETS

IMPROPER SEATING

BRIGHT (POLISHED) SECTIONS

SCRATCHED BY DIRT

DIRT EMBEDDED INTO BEARING MATERIAL

LACK OF OIL

OVERLAY WIPED OUT

EXCESSIVE WEAR

OVERLAY WIPED OUT

TAPERED JOURNAL

RADIUS RIDE

H 28395

14.2 Typical bearing failures

14 Main and big-end bearings – inspection

1 Even though the main and big-end bearings should be renewed during the engine overhaul, the old bearings should be retained for close examination, as they may reveal valuable information about the condition of the engine.

2 Bearing failure occurs because of lack of lubrication, the presence of dirt or other foreign particles, overloading the engine, or corrosion **(see illustration)**. If a bearing fails, the cause must be found and eliminated before the engine is reassembled, to prevent the failure from happening again.

3 To examine the bearing shells, remove them from the cylinder block, the main bearing caps, the connecting rods and the big-end bearing caps, and lay them out on a clean surface in the same order as they were fitted to the engine. This will enable any bearing problems to be matched with the corresponding crankshaft journal.

4 Dirt and other foreign particles can enter the engine in a variety of ways. Contamination may be left in the engine during assembly, or it may pass through filters or the crankcase ventilation system. Normal engine wear produces small particles of metal, which can eventually cause problems. If particles find their way into the lubrication system, it is likely that they will eventually be carried to the bearings. Whatever the source, these foreign particles often end up embedded in the soft bearing material, and are easily

recognised. Large particles will not embed in the bearing, and will score or gouge the bearing and journal. To prevent possible contamination, clean all parts thoroughly, and keep everything spotlessly-clean during engine assembly. Once the engine has been installed in the vehicle, ensure that engine oil and filter changes are carried out at the recommended intervals.

5 Lack of lubrication (or lubrication breakdown) has a number of interrelated causes. Excessive heat (which thins the oil), overloading (which squeezes the oil from the bearing face), and oil leakage (from excessive bearing clearances, worn oil pump or high engine speeds) all contribute to lubrication breakdown. Blocked oil passages, which may be the result of misaligned oil holes in a bearing shell, will also starve a bearing of oil and destroy it. When lack of lubrication is the cause of bearing failure, the bearing material is wiped or extruded from the steel backing of the bearing. Temperatures may increase to the point where the steel backing turns blue from overheating.

6 Driving habits can have a definite effect on bearing life. Full-throttle, low-speed operation (labouring the engine) puts very high loads on bearings, which tends to squeeze out the oil film. These loads cause the bearings to flex, which produces fine cracks in the bearing face (fatigue failure). Eventually the bearing material will loosen in places, and tear away from the steel backing. Regular short journeys can lead to corrosion of bearings, because insufficient engine heat is produced to drive off the condensed water and corrosive gases which form inside the engine. These products collect in the engine oil, forming acid and sludge. As the oil is carried to the bearings, the acid attacks and corrodes the bearing material.

7 Incorrect bearing installation during engine assembly will also lead to bearing failure. Tight-fitting bearings leave insufficient bearing lubrication clearance, and will result in oil starvation. Dirt or foreign particles trapped behind a bearing shell results in high spots on the bearing which can lead to failure.

8 Do not touch any shell's bearing surface with your fingers during reassembly; there is a risk of scratching the delicate surface, or of depositing particles of dirt on it.

9 As mentioned at the beginning of this Section, the bearing shells should be renewed as a matter of course during engine overhaul; to do otherwise is false economy.

15 Engine overhaul – reassembly sequence

1 Before reassembly begins, ensure that all necessary new parts have been obtained

(particularly gaskets, and various bolts which must be renewed), and that all the tools required are available. Read through the entire procedure to familiarise yourself with the work involved, and to ensure that all items necessary for reassembly of the engine are to hand. In addition to all normal tools and materials, a thread-locking compound will be required. A tube of suitable sealant will be required to seal joint faces which are not fitted with gaskets.

2 In order to save time and avoid problems, engine reassembly can be carried out in the following order:

Petrol engines:

a) *Piston rings (see Section 16).*
b) *Crankshaft (see Section 17).*
c) *Pistons/connecting rod assemblies (see Section 18).*
d) *Oil pump (see Chapter 2A, Section 12 or Chapter 2B, Section 15).*
e) *Sump (see Chapter 2A, Section 11 or Chapter 2B, Section 14).*
f) *Flywheel/driveplate (see Chapter 2A, Section 15 or Chapter 2B, Section 18).*
g) *Cylinder head (see Chapter 2A, Section 10 or Chapter 2B, Section 13).*
h) *Coolant pump (see Chapter 3, Section 8).*
i) *Timing belt, sprockets, tensioner and idler pulley (see Chapter 2B, Section 8).*
j) *Timing cover, chain, sprockets and tensioner (see Chapter 2A, Section 6).*
k) *Exhaust manifold (see Chapter 4A, Section 18 or Chapter 4A, Section 19).*
l) *Inlet manifold (see Chapter 4A Section 13 or Chapter 4A, Section 14).*

Diesel engines:

a) *Piston rings (see Section 16).*
b) *Crankshaft (see Section 17).*
c) *Pistons/connecting rod assemblies (see Section 18).*
d) *Oil pump (see Chapter 2C, Section 13 or Chapter 2D, Section 15).*
e) *Sump (see Chapter 2C, Section 12 or Chapter 2D, Section 14).*
f) *Flywheel/driveplate (see, Section or Chapter 2D, Section 18).*
g) *Cylinder head (see Chapter 2C, Section 11 or Chapter 2D, Section 13).*
h) *Coolant pump (see Chapter 3, Section 9).*
i) *Timing belt, sprockets, tensioner and idler pulleys (see Chapter 2D, Section 7 and Chapter 2D, Section 8).*
j) *Timing cover, chain, sprockets and tensioner (see Chapter 2C, Section 8).*
k) *Exhaust manifold (see Chapter 4B, Section 24 or Chapter 4B, Section 25).*
l) *Inlet manifold (see Chapter 4B, Section 17 or Chapter 4B, Section 18).*

3 At this stage, all engine components should be absolutely clean and dry, with all faults repaired. The components should be laid out (or in individual containers) on a completely clean work surface.

16 Piston rings – refitting

1 Before refitting the new piston rings, the ring end gaps must be checked as follows.
2 Lay out the piston/connecting rod assemblies and the new piston ring sets, so that the ring sets will be matched with the same piston and cylinder during the end gap measurement and subsequent engine reassembly.
3 Insert the top ring into the first cylinder, and push it down the bore slightly using the top of the piston. This will ensure that the ring remains square with the cylinder walls. Push the ring down into the bore until it is positioned 15 to 20 mm down from the top edge of the bore, then withdraw the piston.
4 Measure the end gap using feeler gauges, and compare the measurements with the figures given in the Specifications **(see illustration)**.
5 If the gap is too small (unlikely if genuine Vauxhall/Opel parts are used), it must be enlarged or the ring ends may contact each other during engine operation, causing serious damage. Ideally, new piston rings providing the correct end gap should be fitted, but as a last resort, the end gap can be increased by filing the ring ends very carefully with a fine file. Mount the file in a vice equipped with soft jaws, slip the ring over the file with the ends contacting the file face, and slowly move the ring to remove material from the ends – take care, as piston rings are sharp, and are easily broken.
6 With new piston rings, it is unlikely that the end gap will be too large. If they are too large, check that you have the correct rings for your engine and for the particular cylinder bore size.
7 Repeat the checking procedure for each ring in the first cylinder, and then for the rings in the remaining cylinders. Remember to keep rings, pistons and cylinders matched up.
8 Once the ring end gaps have been checked and if necessary corrected, the rings can be fitted to the pistons.
9 The oil control ring (lowest one on the piston) is composed of three sections, and should be installed first. Fit the lower steel ring, then the spreader ring, followed by the upper steel ring **(see illustration)**.

16.4 Measuring a piston ring end gap using a feeler gauge

10 With the oil control ring components installed, the second (middle) ring can be fitted. It is usually stamped with a mark (TOP) which must face up, towards the top of the piston. Using two or three old feeler blades, as for removal of the old rings, carefully slip the ring into place in the middle groove.
Note: *Always follow the instructions supplied with the new piston ring sets – different manufacturers may specify different procedures. Do not mix up the top and middle rings, as they have different cross-sections.*
11 Fit the top ring in the same manner, ensuring that, where applicable, the mark on the ring is facing up. If a stepped ring is being fitted, fit the ring with the smaller diameter of the step uppermost.
12 Repeat the procedure for the remaining pistons and rings.

17 Crankshaft – refitting

Note: *It is recommended that new main bearing shells are fitted regardless of the condition of the original ones.*
1 Refitting the crankshaft is the first step in the engine reassembly procedure. It is assumed at this point that the cylinder block, cylinder block baseplate (where applicable) and crankshaft have been cleaned, inspected and repaired or reconditioned as necessary.
2 Position the cylinder block with the sump mating face uppermost.

16.9 Fitting the oil control spreader ring

3 Clean the bearing shells and the bearing recesses in both the cylinder block and the bearing caps. If new shells are being fitted, ensure that all traces of the protective grease are cleaned off using paraffin. Wipe the shells dry with a clean lint-free cloth.
4 Note that the crankshaft endfloat is controlled by thrustwashers located on one of the main bearing shells. The thrustwashers may be separate, incorporated into, or attached to, the bearing shells themselves.
5 If the original bearing shells are being re-used, they must be refitted to their original locations in the block or caps.
6 Fit the upper main bearing shells in place in the cylinder block, ensuring that the tab on each shell engages in the notch in the cylinder block **(see illustration)**. Where separate thrustwashers are fitted, use a little grease to stick them to each side of their respective bearing upper location; ensure that the oilway grooves on each thrustwasher face outwards (away from the block).

1.4 litre petrol engines

Note: *New cylinder block baseplate retaining bolts must be used when refitting the crankshaft. A tube of sealant (Loctite 5900 or equivalent) will be required when fitting the baseplate to the cylinder block.*
7 Liberally lubricate each bearing shell in the cylinder block, and lower the crankshaft into position **(see illustration)**.
8 Fit the bearing shells in the baseplate **(see illustration)**.
9 Lubricate the crankshaft journals, and

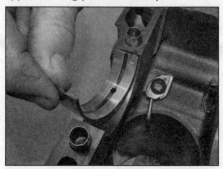

17.6 Fitting a main bearing shell to the cylinder block

17.7 Lubricate the upper bearing shells with clean engine oil then fit the crankshaft – 1.4 litre petrol engines

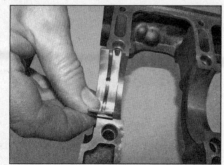

17.8 Fit the bearing shells in the baseplate – 1.4 litre petrol engines

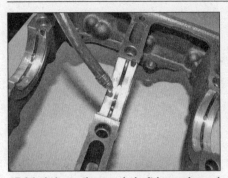

17.9 Lubricate the crankshaft journals, and the bearing shells in the baseplate – 1.4 litre petrol engines

17.10 Apply a 2 mm diameter bead of sealant to the outside of the groove in the baseplate – 1.4 litre petrol engines

17.12a Tighten the inner (M8) baseplate retaining bolts to the specified torque...

17.12b ...then through the specified angles – 1.4 litre petrol engines

17.12c Tighten the outer (M6) baseplate retaining bolts to the specified torque...

17.12d ...then through the specified angles – 1.4 litre petrol engines

the bearing shells in the baseplate **(see illustration)**.

10 Ensure that the cylinder block and baseplate mating surfaces are clean and dry, then apply a 2 mm diameter bead of sealant to the outside of the groove (not in the groove itself) in the baseplate **(see illustration)**.

11 Locate the baseplate over the crankshaft and onto the cylinder block.

12 Fit the new baseplate retaining bolts, then working progressively and in a diagonal sequence, tighten the inner (M8) bolts to the specified torque, then through the specified angles, in the three stages given in the Specifications. Now similarly tighten the outer (M6) retaining bolts **(see illustrations)**.

13 Check that the crankshaft is free to rotate smoothly; if excessive pressure is required

to turn the crankshaft, investigate the cause before proceeding further.

14 Check the crankshaft endfloat with reference to Section 10.

15 Refit the pistons and connecting rods as described in Section 18.

16 Referring to Chapter 2A, fit a new left-hand crankshaft oil seal, then refit the flywheel, timing chain and cover components, sump and cylinder head.

1.6 litre petrol engines

Note: *New main bearing cap retaining bolts must be used when refitting the crankshaft. A tube of sealing compound (Vauxhall/Opel recommend the use of sealant, part no 93165267, availablefrom your dealer) will be required when fitting No5 main bearing cap.*

17 Liberally lubricate each bearing shell in the cylinder block, and lower the crankshaft into position **(see illustration)**.

18 Fit the bearing shells into the bearing caps.

19 Lubricate the bearing shells in the bearing caps, and the crankshaft journals, then fit Nos 1, 2, 3 and 4 bearing caps, and tighten the new bolts as far as possible by hand **(see illustration)**.

20 Ensure the rear (No 5) bearing cap is clean and dry then fill the groove on each side of the cap with sealing compound **(see illustration)**. Fit the bearing cap to the engine, ensuring it is fitted the correct way around, and tighten the new bolts as far as possible by hand.

21 Working in a diagonal sequence from the centre outwards, tighten the main bearing cap

17.17 Lubricate the upper bearing shells with clean engine oil then fit the crankshaft – 1.6 litre petrol engines

17.19 Lubricate the crankshaft journals then refit bearing caps Nos 1 to 4 – 1.6 litre petrol engines

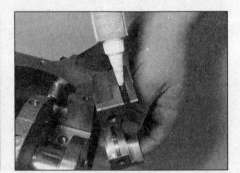

17.20 Fill the side grooves of the rear (No 5) bearing cap with sealant prior to refitting it to the engine – 1.6 litre petrol engines

17.21 Tighten the bolts to the specified Stage 1 torque setting...

**17.22 ...and then through the specified Stage 2 and 3 angles –
1.6 litre petrol engines**

bolts to the specified Stage 1 torque setting **(see illustration)**.

22 Once all bolts are tightened to the specified Stage 1 torque, go around again and tighten all bolts through the specified Stage 2 angle then go around once more and tighten all bolts through the specified Stage 3 angle. It is recommended that an angle-measuring gauge is used during the final stages of the tightening, to ensure accuracy **(see illustration)**.

23 Once all the bolts have been tightened, inject more sealant down the grooves in the rear main bearing cap until sealant is seen to be escaping through the joints. Once you are sure the cap grooves are full of sealant, wipe off all excess sealant using a clean cloth.

24 Check that the crankshaft is free to rotate smoothly; if excessive pressure is required to turn the crankshaft, investigate the cause before proceeding further.

25 Check the crankshaft endfloat with reference to Section 10.

26 Refit the pistons and connecting rods as described in Section 18.

27 Referring to Chapter 2B, fit a new left-hand crankshaft oil seal, then refit the oil pump, sump, flywheel, cylinder head, timing belt sprocket(s) and fit a new timing belt.

1.6 litre diesel engines

Note: *New cylinder block baseplate (M10) retaining bolts must be used when refitting the crankshaft. A tube of silicone sealant will be required when fitting the baseplate to the cylinder block.*

28 Liberally lubricate each bearing shell in the cylinder block, and lower the crankshaft into position.

29 Fit the bearing shells in the baseplate.

30 Lubricate the crankshaft journals, and the bearing shells in the baseplate.

31 Ensure that the cylinder block and baseplate mating surfaces are clean and dry, then apply a 2.5 mm diameter bead of silicone sealant to the cylinder block face.

32 Locate the baseplate over the crankshaft and onto the cylinder block.

33 Fit the new baseplate retaining bolts, then

working progressively in the sequence shown, tighten the inner (M10) bolts to the specified torque, then through the specified angles, in the stages given in the Specifications **(see illustration)**. Now similarly tighten the outer (M8) retaining bolts, in the sequence shown, to the specified torque.

34 Check that the crankshaft is free to rotate smoothly; if excessive force is required to turn the crankshaft, investigate the cause before proceeding further.

35 Check the crankshaft endfloat with reference to Section 10.

36 Refit the pistons and connecting rods as described in Section 18.

37 Referring to Chapter 2C, fit a new left-hand crankshaft oil seal, then refit the flywheel/driveplate.

1.7 litre diesel engines

38 Using a little grease, stick the thrustwashers to each side of the Number 2 main bearing upper location; ensure that the oilway grooves on each thrustwasher face outwards **(see illustration)**.

**17.33 Tightening sequence for the cylinder block baseplate bolts –
1.6 litre diesel engines**

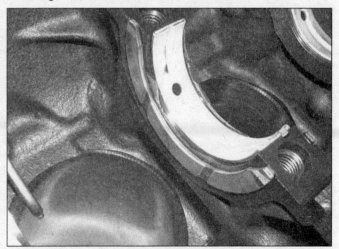

**17.38 Fit the thrust washers to the No 2 main bearing with the oil
grooves facing outwards – 1.7 litre diesel engines**

17.39 Lower the crankshaft into position –
1.7 litre diesel engines

17.40 Fit the bearing caps with the arrows
towards the timing belt end of the engine –
1.7 litre diesel engines

17.45 Apply sealant to the oil seal housing
mating surface – 1.7 litre diesel engines

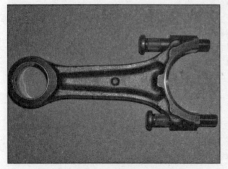

18.1 Replace the big-end bolts –
1.7 litre diesel engines

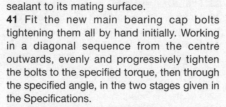

18.3 Fit the bearing shells making sure
their tabs are correctly located in the
connecting rod/cap groove

18 Pistons/connecting rods – refitting

Note: *It is recommended that new big-end bearing shells are fitted regardless of the condition of the original ones.*

1 On 1.7 litre diesel engines, prior to refitting, carefully tap the original bolts out from the connecting rods and install new bolts **(see illustration)**.

2 Clean the backs of the big-end bearing shells and the recesses in the connecting rods and big-end caps. If new shells are being fitted, ensure that all traces of the protective grease are cleaned off using paraffin. Wipe the shells, caps and connecting rods dry with a lint-free cloth.

3 Press the bearing shells into their locations, ensuring that the tab on each shell engages in the notch in the connecting rod and cap **(see illustration)**. If there is no tab on the bearing shell (and no notch in the rod or cap) position the shell equidistant from each side of the rod and cap. If the original bearing shells are being used ensure they are refitted in their original locations.

4 Lubricate the bores, the pistons and piston rings then lay out each piston/connecting rod assembly in its respective position.

Petrol engines

5 Lubricate No 1 piston and piston rings, and check that the ring gaps are correctly positioned. The gaps in the upper and lower steel rings of the oil control ring should be offset by 25 to 50 mm to the right and left of the spreader ring gap. The two upper compression ring gaps should be offset by 180° to each other **(see illustration)**.

6 Fit a ring compressor to No 1 piston, then insert the piston and connecting rod into the cylinder bore so that the base of the compressor stands on the block. With the crankshaft big-end bearing journal positioned at its lowest point, tap the piston carefully into the cylinder bore with the wooden handle of a hammer, and at the same time guide the connecting rod onto the bearing journal.

39 Lubricate the upper shells with clean engine oil then lower the crankshaft into position **(see illustration)**.

40 Fit the bearing shells to the bearing caps, then lubricate the crankshaft journals and bearing shells in the caps. Refit the caps to the cylinder block **(see illustration)**. Ensure the caps are fitted in their correct locations, with Number 1 cap at the timing belt end, and are fitted the correct way

around so that the arrows all point towards the timing belt end of the engine. Prior to refitting Number 1 cap, apply a smear of sealant to its mating surface.

41 Fit the new main bearing cap bolts tightening them all by hand initially. Working in a diagonal sequence from the centre outwards, evenly and progressively tighten the bolts to the specified torque, then through the specified angle, in the two stages given in the Specifications.

42 Check that the crankshaft is free to rotate smoothly; if excessive force is required to turn the crankshaft, investigate the cause before proceeding further.

43 Check the crankshaft endfloat with reference to Section 10.

44 Ensure that the mating surfaces of the oil seal housing and cylinder block are clean and dry. Note the correct fitted depth of the oil seal then tap/lever the seal out of the housing.

45 Apply a bead of sealant to the oil seal housing mating surface, and make sure that the locating dowels are in position **(see illustration)**. Slide the housing over the end of the crankshaft, and into position on the cylinder block. Put a drop of locking compound on the threads, then tighten the retaining bolts to the specified torque.

46 Refit the pistons and connecting rods as described in Section 18.

47 Referring to Chapter 2D, fit a new left-hand crankshaft oil seal, then refit the flywheel/driveplate.

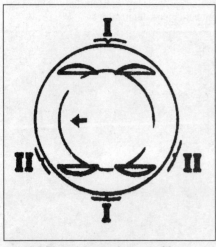

18.5 Piston ring end gap positions –
petrol engines

I Top and second compression rings
II Oil control ring side rails

Note that the arrow on the piston crown must point towards the timing chain/belt end of the engine **(see illustrations)**.

7 Liberally lubricate the bearing journals and bearing shells, and fit the bearing cap in its original location **(see illustrations)**.

8 Screw in the new bearing cap retaining bolts, and tighten both bolts to the specified Stage 1 torque setting then tighten them through the specified Stage 2 angle and finally, on 1.6 litre engines, through the specified Stage 3 angle.

9 Refit the remaining three piston and connecting rod assemblies in the same way.

10 Rotate the crankshaft, and check that it turns freely, with no signs of binding or tight spots.

11 Refit the oil baffle plate (where fitted), sump and the cylinder head as described in Chapter 2A or Chapter 2B.

Diesel engines

12 Lubricate No 1 piston and piston rings, and space the ring gaps uniformly around the piston at 120° intervals **(see illustration)**.

13 Fit a ring compressor to No 1 piston, then insert the piston and connecting rod into the cylinder bore so that the base of the compressor stands on the block. Ensure that the arrow or dot on the piston crown is toward the timing chain/belt end of the engine **(see illustration)**.

14 With the crankshaft big-end bearing journal positioned at its lowest point, tap the piston carefully into the cylinder bore with the wooden handle of a hammer, and at the same time guide the connecting rod onto the bearing journal.

15 Liberally lubricate the bearing journals and bearing shells, and fit the bearing cap in its original location.

16 Screw in the new bearing cap retaining bolts/nuts. Tighten both bolts/nuts to the specified Stage 1 torque setting then tighten them through the specified Stage 2 angle and finally, on 1.7 litre engines, through the specified Stage 3 angle.

17 Refit the remaining three piston and connecting rod assemblies in the same way.

18 Rotate the crankshaft, and check that it turns freely, with no signs of binding or tight spots.

19 Refit the oil baffle plate (where fitted), sump and the cylinder head as described in Chapter 2C or Chapter 2D.

19 Engine – initial start-up after overhaul

1 With the engine refitted in the vehicle, double-check the engine oil and coolant levels. Make a final check that everything has been reconnected, and that there are no tools or rags left in the engine compartment.

2 On diesel engines, prime and bleed the fuel system as described in Chapter 4B, Section 5.

18.6a Tap the piston gently into the bore using the handle of a hammer – petrol engines

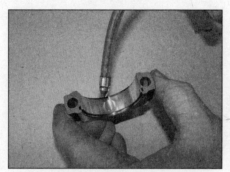
18.7a Liberally lubricate the bearing journals and bearing shells...

3 Start the engine, noting that this may take a little longer than usual. Make sure that the oil pressure warning light goes out.

4 While the engine is idling, check for fuel, water and oil leaks. Don't be alarmed if there are some odd smells and smoke from parts getting hot and burning off oil deposits.

5 Assuming all is well, run the engine until it reaches normal operating temperature, then switch off the engine.

6 After a few minutes, recheck the oil and

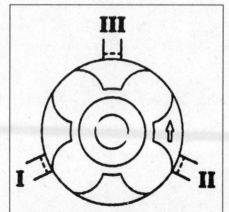
18.12 Piston ring end gap positions – diesel engines

I Top compression ring
II Second compression ring
III Oil control ring

18.6b Ensuring the arrow on the piston crown (circled) is pointing towards the timing chain/belt end of the engine – petrol engines

18.7b ...and fit the bearing cap in its original location – petrol engines

coolant levels as described in *Weekly checks* and top-up as necessary.

7 Note that there is no need to retighten the cylinder head bolts once the engine has first run after reassembly.

8 If new pistons, rings or crankshaft bearings have been fitted, the engine must be treated as new, and run-in for the first 600 miles. Do not operate the engine at full-throttle, or allow it to labour at low engine speeds in any gear. It is recommended that the oil and filter be changed at the end of this period.

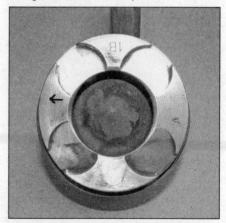
18.13 Position the piston with the arrow or notch (as applicable) pointing towards the timing chain/belt end of the engine – diesel engines

Chapter 3
Cooling, heating and air conditioning systems

Contents

Degrees of difficulty

Easy, suitable for novice with little experience	**Fairly easy,** suitable for beginner with some experience	**Fairly difficult,** suitable for competent DIY mechanic	**Difficult,** suitable for experienced DIY mechanic	**Very difficult,** suitable for expert DIY or professional

Specifications

Thermostat

Opening temperatures:

Petrol engines:

Electric element	90°C
Wax capsule	105°C
Diesel engines	91°C

Air conditioning system

Refrigerant	R134a or R1234yf
Refrigerant quantity	570g

Torque wrench settings

	Nm	lbf ft
Petrol engine models		
Air conditioning compressor mounting bolts	22	16
Air conditioning refrigerant pipe block connections	22	16
Auxiliary drivebelt tensioner retaining bolt (1.6 litre engines)	55	41
Bumper lower impact bar retaining bolts:		
Inner bolts	22	16
Outer bolts	58	43
Bumper upper impact bar retaining bolts	32	24
Coolant pump pulley bolts	22	16
Coolant pump retaining bolts	8	6
Radiator upper mounting bracket retaining bolts	18	13
Thermostat housing:		
1.4 litre engines	8	6
1.6 litre engines	10	7
Diesel engine models		
Air conditioning compressor mounting bolts	22	16
Air conditioning refrigerant pipe block connections	22	16
Bumper lower impact bar retaining bolts:		
Inner bolts	22	16
Outer bolts	58	43
Bumper upper impact bar retaining bolts	32	24
Coolant pump pulley bolts	12	9
Coolant pump retaining bolts	25	18
Coolant temperature sensor	22	16
Radiator upper mounting bracket retaining bolts	18	13
Thermostat bypass pipe bolts (1.6 litre engines)	10	7
Thermostat housing/cover:		
1.6 litre engines	10	7
1.7 litre engines	25	18
Turbocharger coolant pipe banjo union bolt	25	18

1 General information and precautions

General information

1 The cooling system is of pressurised type, comprising a coolant pump, a crossflow radiator, electric cooling fan, and thermostat.
2 The coolant pump is driven by the auxiliary drivebelt on all engines.
3 The system functions as follows. Cold coolant from the radiator passes through the bottom hose to the coolant pump, where it is pumped around the cylinder block, head passages and heater matrix. After cooling the cylinder bores, combustion surfaces and valve seats, the coolant reaches the underside of the thermostat, which is initially closed. The coolant passes through the heater, and is returned to the coolant pump.
4 When the engine is cold, the coolant circulates only through the cylinder block, cylinder head and heater. When the coolant reaches a predetermined temperature, the thermostat opens and the coolant also passes through to the radiator. As the coolant circulates through the radiator, it is cooled by the inrush of air when the car is in forward motion. Airflow is supplemented by the action of the electric cooling fan when necessary. Once the coolant has passed through the radiator, and has cooled, the cycle is repeated.
5 On petrol engines, an electrically assisted thermostat is fitted. Engine coolant temperature is monitored by the engine management system electronic control unit, via the coolant temperature sensor. In conjunction with information received from various other engine sensors, the thermostat opening temperature can be controlled according to engine speed and load. During normal engine operation the thermostat operates conventionally. Under conditions of high engine speed and load, an electric heating element within the thermostat is energised, to cause the thermostat to open at a lower temperature (typically 90°).
6 The electric cooling fan, mounted on the rear of the radiator, is controlled by the engine management system electronic control unit, in conjunction with a cooling fan module. At a predetermined coolant temperature, the fan is actuated.
7 An expansion tank is fitted to the front left-hand side of the engine compartment to accommodate expansion of the coolant when hot.

Precautions

⚠ **Warning: Do not attempt to remove the expansion tank filler cap, or disturb any part of the cooling system, while the engine is hot; there is a high risk of scalding. If the cap must be removed before the engine and radiator have fully cooled (even though this** *is not recommended) the pressure in the cooling system must first be relieved. Cover the cap with a thick layer of cloth, to avoid scalding, and slowly unscrew the filler cap until a hissing sound can be heard. When the hissing has stopped, indicating that the pressure has reduced, slowly unscrew the filler cap until it can be removed; if more hissing sounds are heard, wait until they have stopped before unscrewing the cap completely. At all times, keep well away from the filler cap opening.*

⚠ **Warning: Do not allow antifreeze to come into contact with the skin, or with the painted surfaces of the vehicle. Rinse off spills immediately, with plenty of water. Never leave antifreeze lying around in an open container, or in a puddle on the driveway or garage floor. Children and pets are attracted by its sweet smell, but antifreeze can be fatal if ingested.**

⚠ **Warning: If the engine is hot, the electric cooling fan may start rotating even if the engine is not running; be careful to keep hands, hair and loose clothing well clear when working in the engine compartment.**

⚠ **Warning: Refer to Section 12 for precautions to be observed when working on models equipped with air conditioning.**

2 Cooling system hoses – disconnection and renewal

Note: *Refer to the warnings given in Section 1 of this Chapter before proceeding. Do not attempt to disconnect any hose while the system is still hot.*

1 If the checks described in Chapter 1A, Section 6 or Chapter 1B, Section 6 reveal a faulty hose, it must be renewed as follows.
2 First drain the cooling system (see Chapter 1A, Section 25 or Chapter 1B, Section 27). If the coolant is not due for renewal, it may be re-used if it is collected in a clean container.
3 Before disconnecting a hose, first note its routing in the engine compartment, and whether it is secured by any additional retaining clips or cable ties. Use a pair of pliers to release the clamp-type clips, or a screwdriver to slacken the screw-type clips, then move the clips along the hose, clear of the relevant inlet/outlet union. Carefully work the hose free.
4 Depending on engine, some of the hose attachments may be of the quick-release type. Where this type of hose is encountered, lift the ends of the wire retaining clip, to spread the clip, then withdraw the hose from the inlet/outlet union.
5 Note that the radiator inlet and outlet unions are fragile; do not use excessive force when attempting to remove the hoses. If a hose proves to be difficult to remove, try to release it by rotating the hose ends before attempting to free it. It may be beneficial to spray an penetrating aerosol lubricant (WD-40 or equivalent) onto the end of the hose to aid its release.
6 When fitting a hose, first slide the clips onto the hose, then work the hose into position. If clamp-type clips were originally fitted, it is a good idea to use screw-type clips when refitting the hose. If the hose is stiff, use a little soapy water (washing-up liquid is ideal) as a lubricant, or soften the hose by soaking it in hot water.
7 Work the hose into position, checking that it is correctly routed and secured. Slide each clip along the hose until it passes over the flared end of the relevant inlet/outlet union, before tightening the clips securely.
8 Refill the cooling system with reference to Chapter 1A, Section 25 or Chapter 1B, Section 27.
9 Check thoroughly for leaks as soon as possible after disturbing any part of the cooling system.

3 Radiator – removal, inspection and refitting

Removal

1 Have the air conditioning system refrigerant discharged at a dealer service department or an automotive air conditioning repair facility.
2 Disconnect the battery negative lead as described in Chapter 5A Section 4.
3 Apply the handbrake, then jack up the front of the vehicle and support it on axle stands (see *Jacking and vehicle support*).
4 Remove the front bumper as described in Chapter 11, Section 6.
5 Drain the cooling system as described in Chapter 1A, Section 25 (petrol engines) or Chapter 1B, Section 27 (diesel engines).
6 On automatic transmission models, drain the automatic transmission fluid as described in Chapter 7B, Section 2.
7 On 1.4 litre petrol engine models and all diesel engine models, remove the intercooler as described in Chapter 4A, Section 15 (petrol engines) or Chapter 4B, Section 19 (diesel engines).
8 Remove the air conditioning system condenser as described in Section 13.
9 On automatic transmission models equipped with an auxiliary transmission fluid cooler, disconnect the fluid pipes from the fluid cooler. Undo the two retaining bolts and remove the auxiliary fluid cooler. Suitably cover the open unions after disconnection. Note that new seals will be required.
10 Disconnect the top and bottom coolant hoses from the radiator.
11 On models with automatic transmission, detach the protective ring (where fitted) over the fluid cooler pipe unions at the radiator. Using a small screwdriver, release the quick-release fitting retaining lugs and disconnect the fitting from the radiator.

Suitably cover the open unions after disconnection. Note that new seals will be required for each disconnected fitting. Unclip the pipes from the cooling fan shroud.

12 Disconnect the wiring connector from the electric cooling fan **(see illustration)**.

13 Undo the radiator upper mounting bracket retaining bolts and remove the upper mounting bracket on each side **(see illustrations)**.

14 Check that all hoses, and connections are released from the radiator in the engine compartment.

15 Lift the radiator up to disengage it from its lower mounting rubbers and remove it from the car **(see illustration)**.

Inspection

16 If the radiator has been removed due to suspected blockage, reverse-flush it as described in Chapter 1A, Section 25 or Chapter 1B, Section 27.

17 Clean dirt and debris from the radiator fins, using an air line (in which case, wear eye protection) or a soft brush.

Caution: Be careful, as the fins are easily damaged, and are sharp.

18 If necessary, a radiator specialist can perform a 'flow test' on the radiator, to establish whether an internal blockage exists.

19 A leaking radiator must be referred to a specialist for permanent repair. Do not attempt to weld or solder a leaking radiator.

20 In an emergency, minor leaks from the radiator can be cured by using a suitable radiator sealant (in accordance with its manufacturer's instructions) with the radiator fitted in the vehicle.

21 Inspect the radiator mounting rubbers, and renew them if necessary.

Refitting

22 Refitting is a reversal of removal, bearing in mind the following points.

a) *Ensure that the radiator lower mounting pegs are correctly engaged with the mounting bracket rubber bushes when refitting.*

b) *Ensure that all hoses are correctly reconnected, and their retaining clips securely tightened.*

c) *On automatic transmission models, renew the seals on all disturbed fluid cooler pipe unions.*

3.12 Disconnect the cooling fan wiring connector

3.13b ...and remove the upper mounting bracket on each side

d) *Refill the cooling system as described in Chapter 1A, Section 25 or Chapter 1B, Section 27.*

e) *On models with automatic transmission, refill the transmission with fresh fluid as described in Chapter 7B, Section 2.*

f) *Have the air conditioning system evacuated, charged and leak-tested by the specialist that discharged it.*

4 Thermostat (petrol engine models) – removal and refitting

Removal

1.4 litre engines

1 The thermostat housing is attached to

3.13a Undo the radiator upper mounting bracket retaining bolts...

3.15 Carefully lift the radiator up and remove it from the car

the coolant pump at the right-hand side of the engine. The thermostat is integral with the housing and cannot be renewed separately.

2 Disconnect the battery negative lead as described in Chapter 5A Section 4.

3 Apply the handbrake, then jack up the front of the vehicle and support it on axle stands (see *Jacking and vehicle support*).

4 Drain the cooling system as described in Chapter 1A, Section 25.

5 Slacken the retaining clips and remove the air cleaner outlet air duct from the air cleaner and turbocharger **(see illustrations)**. Suitably cover the turbocharger air inlet to prevent the possible entry of foreign material.

6 Open the two crankcase ventilation hose retaining clips at the rear of the camshaft cover **(see illustration)**.

7 Pull out the retaining spring clip and

4.5a Slacken the retaining clips at the air cleaner...

4.5b ...and turbocharger and remove the outlet air duct – 1.4 litre engines

4.6 Open the two crankcase ventilation hose retaining clips at the rear of the camshaft cover – 1.4 litre engines

4.7 Disconnect the crankcase ventilation hose from the inlet manifold – 1.4 litre engines

4.8 Disconnect the other end of the crankcase ventilation hose from the turbocharger air inlet – 1.4 litre engines

4.9 Release the retaining clips and disconnect the vacuum hose from the turbocharger – 1.4 litre engines

disconnect the crankcase ventilation hose from the inlet manifold **(see illustration)**.

8 Open the quick-release connector and disconnect the other end of the crankcase ventilation hose from the turbocharger air inlet **(see illustration)**. Suitably cover the turbocharger inlet to prevent the possible entry of foreign material.

9 Release the retaining clips and disconnect the vacuum hose from the turbocharger and from the charge air bypass regulator solenoid on the underside of the inlet manifold **(see illustration)**.

10 Lift the crankcase ventilation hose, together with the vacuum hose out of the support clips and remove the two hoses from the engine.

11 Pull out the retaining spring clip and

disconnect the radiator inlet hose from the thermostat housing **(see illustration)**.

12 Disconnect the thermostat wiring connector.

13 Release the retaining clip and disconnect the engine oil cooler outlet hose from the thermostat housing **(see illustration)**.

14 Slacken and remove the three retaining bolts, and remove the thermostat housing **(see illustrations)**.

15 Remove the sealing ring from the housing and discard it; a new one should be used on refitting **(see illustration)**. Thoroughly clean the housing contact surfaces.

1.6 litre engines

16 The thermostat is located in a housing attached to the left-hand side of the cylinder

head. The thermostat is integral with the housing cover and cannot be renewed separately.

17 Disconnect the battery negative lead as described in Chapter 5A Section 4.

18 Apply the handbrake, then jack up the front of the vehicle and support it on axle stands (see *Jacking and vehicle support*).

19 Drain the cooling system as described in Chapter 1A, Section 25.

20 Disconnect the adjacent wiring harness connectors, then undo the retaining nut and remove the wiring harness support bracket.

21 Disconnect the thermostat wiring connector.

22 Release the clips and disconnect the radiator hose from the thermostat housing cover.

23 Unscrew the four bolts and remove the thermostat cover from the housing.

24 Remove the sealing ring from the cover and discard it; a new one should be used on refitting. Thoroughly clean the housing and cover contact surfaces.

Refitting

25 Refitting is a reversal of removal, bearing in mind the following points:

a) Fit a new sealing ring and tighten the mounting bolts to the specified torque.

b) Refill the cooling system as described in Chapter 1A, Section 25.

4.11 Pull out the retaining spring clip and disconnect the radiator hose from the thermostat housing

4.13 Release the retaining clip and disconnect the oil cooler outlet hose from the thermostat housing – 1.4 litre engines

4.14a Slacken and remove the three retaining bolts...

4.14b ...and remove the thermostat housing – 1.4 litre engines

4.15 Remove the sealing ring from the housing – 1.4 litre engines

5 Thermostat (diesel engine models) – removal and refitting

Removal

1.6 litre engines

1 The thermostat housing is attached to the rear of the coolant pump. The thermostat is integral with the housing.

2 Remove the coolant pump as described in Section 9.

3 Undo and remove the four retaining bolts and remove the thermostat housing from the rear of the coolant pump.

4 Remove the sealing ring from the housing and discard it; a new one should be used on refitting. Thoroughly clean the housing and coolant pump contact surfaces.

1.7 litre engines

5 The thermostat is located on the left-hand end of the cylinder head.

6 Remove the battery and battery tray as described in Chapter 5A, Section 4.

7 Apply the handbrake, then jack up the front of the vehicle and support it on axle stands (see *Jacking and vehicle support*).

8 Drain the cooling system as described in Chapter 1B, Section 27.

9 Release the retaining clip, and disconnect the coolant hose from the thermostat housing cover.

10 Unscrew the two bolts and remove the thermostat cover from the housing.

11 Lift the thermostat from its housing and recover the gasket **(see illustration)**. Discard the gasket; a new one should be used on refitting. Thoroughly clean the cover and housing contact surfaces.

Refitting

12 Refitting is a reversal of removal, bearing in mind the following points:
a) Fit a new sealing ring/gasket and tighten the mounting bolts to the specified torque.
b) On 1.6 litre engines, refit the coolant pump as described in Section 9.
c) On 1.7 litre engines, refit the battery tray and battery as described in Chapter 5A, Section 4.
d) Refill the cooling system as described in Chapter 1B, Section 27.

6 Electric cooling fan – removal and refitting

 Warning: If the engine is hot, the cooling fan may start up at any time. Take extra precautions when working in the vicinity of the fan.

Removal

1 Remove the radiator as described in Section 3.

5.11 Lift the thermostat from its housing – 1.7 litre engines

2 Undo the two (three on 1.6 litre petrol engine models) retaining bolts securing the fan shroud to the radiator **(see illustration)**.

3 Lift the fan shroud upwards to disengage the lower retaining lugs and remove the shroud from the radiator **(see illustration)**.

4 To remove the fan motor, undo the fan motor resistor retaining screw (if fitted), then unclip the resistor and unclip the wiring harness from the fan shroud. Undo the three retaining bolts and remove the fan motor from the shroud **(see illustrations)**. Note that the fan motor and fan blades are a complete assembly and are not available separately.

Refitting

5 Refitting is a reversal of removal. Refit the radiator as described in Section 3.

6.2 Undo the fan shroud retaining bolts...

6.4a Unclip the fan motor resistor...

7 Coolant temperature sensor – testing, removal and refitting

Testing

1 Testing of the coolant temperature sensor must be entrusted to a Vauxhall/Opel dealer or suitably equipped repairer, who will have the necessary specialist diagnostic equipment.

Removal

2 Partially drain the cooling system with reference to Chapter 1A, Section 25 (petrol engines) or Chapter 1B, Section 27 (diesel engines). Alternatively, it is possible to change the sensor quickly with minimal loss of coolant by first releasing any pressure from the cooling system. With the engine cold, temporarily remove the expansion tank cap.

3 Where fitted, remove the plastic cover over the top of the engine.

Petrol engines

4 Two temperature sensors are fitted; one located in the side of the coolant outlet housing at the left-hand end of the cylinder head, and one located in the radiator right-hand side tank **(see illustrations)**. The removal and refitting procedures are the same for both sensors.

5 To gain access to the sensor in the coolant outlet housing on 1.4 litre engines, remove the intercooler outlet air duct as described in Chapter 4A, Section 2.

6.3 ...and lift the shroud up and off the radiator

6.4b ...then undo the three retaining bolts and remove the fan motor from the shroud

7.4a Coolant temperature sensor located in the coolant outlet housing – 1.4 litre petrol engine shown

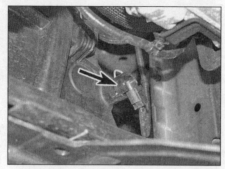

7.4b Coolant temperature sensor located in the radiator right-hand side tank – 1.6 litre petrol engine shown

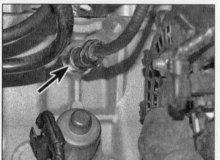

7.8 Coolant temperature sensor location – 1.6 litre engines

7.10 Coolant temperature sensor location – 1.7 litre engines

6 To gain access to the sensor in the radiator on 1.6 litre engines, remove the engine undertray as described in Chapter 11, Section 21.

7 Disconnect the wiring connector, then extract the wire retaining clip. Remove the sensor from the outlet housing or radiator. If the cooling system has not been drained, either insert the new sensor or fit a blanking plug to prevent further loss of coolant.

1.6 litre diesel engines

8 The coolant temperature sensor is located on the rear facing side of the cylinder block, adjacent to the alternator **(see illustration)**.

9 Disconnect the wiring connector, then unscrew the sensor and remove it from the cylinder block. If the cooling system has not been drained, either insert the new sensor or

fit a blanking plug to prevent further loss of coolant.

1.7 litre diesel engines

10 The coolant temperature sensor is located in the thermostat housing on the left-hand end of the cylinder head **(see illustration)**.

11 Remove the thermostat as described in Section 5.

12 Undo the retaining nut and bolt and remove the inlet manifold support bracket.

13 Disconnect the wiring connector, then unscrew the sensor and remove it from the thermostat housing. Recover the seal noting that a new seal will be required for refitting. If the cooling system has not been drained, either insert the new sensor or fit a blanking plug to prevent further loss of coolant.

Refitting

14 Fit the new sensor using a reversal of the removal procedure. Tighten the sensor to the specified torque (diesel engines) and refill/top-up the cooling system with reference to Chapter 1A, Section 25 (petrol engines) or Chapter 1B, Section 27 (diesel engines) and *Weekly checks*.

8 Coolant pump (petrol engine models) – removal and refitting

Removal

1 Disconnect the battery negative lead as described in Chapter 5A Section 4.

2 Where fitted, remove the plastic cover over the top of the engine.

3 Remove the air cleaner assembly and air outlet duct as described in Chapter 4A, Section 2.

4 Apply the handbrake, then jack up the front of the vehicle and support it on axle stands (see *Jacking and vehicle support*).

5 Drain the cooling system as described in Chapter 1A, Section 25.

6 Remove the auxiliary drivebelt as described in Chapter 1A, Section 29.

1.4 litre engines

7 Pull out the retaining spring clip and disconnect the radiator inlet hose from the thermostat housing **(see illustration 4.11)**.

8 Disconnect the thermostat wiring connector.

9 Release the retaining clip and disconnect the engine oil cooler outlet hose from the thermostat housing **(see illustration 4.13)**.

10 Release the retaining clip and disconnect the heater outlet hose from the coolant pump **(see illustration)**.

11 Hold the coolant pump pulley hub using a spanner on the flats provided. Undo the three pulley retaining bolts and remove the pulley **(see illustrations)**.

12 Unscrew the ten coolant pump retaining bolts, noting the locations of the different length bolts **(see illustration)**. The short bolts secure the pump to the timing cover, and the

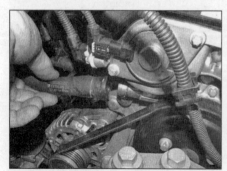

8.10 Release the retaining clip and disconnect the heater outlet hose from the coolant pump – 1.4 litre engines

8.11a Hold the coolant pump pulley hub using a spanner on the flats provided and undo the three pulley retaining bolts...

8.11b ...remove the pulley from the pump – 1.4 litre engines

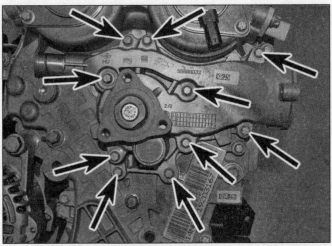

8.12 Unscrew the coolant pump retaining bolts noting the locations of the different length bolts – 1.4 litre engines

8.13 Withdraw the coolant pump from the timing cover – 1.4 litre engines

long bolts secure the pump and the timing cover to the cylinder block and cylinder head.

13 Withdraw the coolant pump from the timing cover, noting that it may be necessary to tap the pump lightly with a soft-faced hammer to free it from the locating dowels **(see illustration)**.

14 Recover the pump sealing gasket, and discard it; a new one must be used on refitting.

15 Note that it is not possible to overhaul the pump. If it is faulty, the unit must be renewed complete.

16 If the pump is being renewed, remove the thermostat (referring to the procedures in Section 4) and transfer it to the new pump.

1.6 litre engines

17 Remove the right-hand engine mounting as described in Chapter 2B, Section 19.

18 Unscrew the retaining bolt and remove the auxiliary drivebelt tensioner.

19 Lower the right-hand side of the engine slightly on the jack to provide access to the coolant pump pulley retaining bolts. Unscrew the pulley retaining bolts and remove the pulley from the pump flange.

20 Undo the five retaining bolts and remove the pump from the oil pump housing. Recover the pump sealing ring, and discard it; a new one must be used on refitting.

21 Note that it is not possible to overhaul the pump. If it is faulty, the unit must be renewed complete.

Refitting

1.4 litre engines

22 Ensure that the pump and timing cover mating surfaces are clean and dry.

23 Place a new gasket on the pump and insert two pump retaining bolts to hold the gasket in position **(see illustration)**.

24 Place the pump in position and refit the retaining bolts in their correct locations **(see illustration 8.12)**. Tighten the bolts to the specified torque.

25 Refit the coolant pump pulley and secure with the three retaining bolts tightened to the specified torque.

26 Reconnect the heater outlet hose to the coolant pump.

27 Reconnect the engine oil cooler outlet hose to the thermostat housing.

28 Reconnect the thermostat wiring connector.

29 Reconnect the radiator inlet hose to the thermostat housing.

1.6 litre engines

30 Ensure that the pump and timing cover mating surfaces are clean and dry.

31 Place a new sealing ring on the pump, then locate the pump in position on the oil pump housing. Refit the retaining bolts and tighten them to the specified torque.

32 Refit the coolant pump pulley and the retaining bolts. Tighten the bolts to the specified torque.

33 Refit the auxiliary drivebelt tensioner and tighten the retaining bolt to the specified torque.

34 Refit the right-hand engine mounting as described in Chapter 2B, Section 19.

All engines

35 Refit the auxiliary drivebelt as described in Chapter 1A, Section 29.

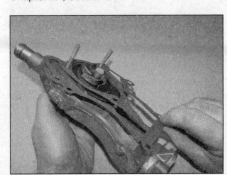

8.23 Place a new gasket on the pump – 1.4 litre engines

36 Refit the air cleaner assembly and air outlet duct as described in Chapter 4A, Section 2.

37 Reconnect the battery negative terminal, then refill the cooling system as described in Chapter 1A, Section 25.

38 Where applicable, refit the plastic cover over the top of the engine.

9 Coolant pump (diesel engine models) – removal and refitting

Removal

1 Disconnect the battery negative lead as described in Chapter 5A Section 4.

2 Where fitted, remove the plastic cover over the top of the engine.

3 Remove the air cleaner assembly and air outlet duct as described in Chapter 4B, Section 3.

4 Apply the handbrake, then jack up the front of the vehicle and support it on axle stands (see *Jacking and vehicle support*).

5 Drain the cooling system as described in Chapter 1B, Section 27.

6 Remove the auxiliary drivebelt as described in Chapter 1B, Section 30.

1.6 litre engines

7 Remove the diesel particulate filter as described in Chapter 4B, Section 26.

8 Extract the wire spring clip and disconnect the intercooler intake air duct from the turbocharger.

9 Release the retaining clip, and disconnect the coolant hose from the thermostat housing.

10 Undo the retaining bolt and release the wiring harness support bracket from the coolant pump.

11 Unscrew the turbocharger coolant pipe banjo union bolt from the side of the coolant pump and collect the two sealing washers **(see illustration)**.

12 Disconnect the wiring connectors at the

9.11 Unscrew the turbocharger coolant pipe banjo union bolt and collect the sealing washers – 1.6 litre engines

9.13 Undo the thermostat bypass pipe retaining bolts – 1.6 litre engines

9.14a Undo the turbocharger coolant pipe bracket bolt...

9.14b ...and wastegate actuator pipe bracket bolt at the left-hand end of the engine...

9.14c ...at the centre of the engine...

9.14d ...and at the right-hand end of the engine...

air conditioning compressor clutch and the coolant pump.

13 Undo the two bolts securing the thermostat bypass pipe to the rear of the coolant pump and slacken the two bolts

securing the bypass pipe bracket to the cylinder block **(see illustration)**. Recover the gasket and discard it; a new one must be used on refitting

14 Undo the bolts securing the turbocharger

wastegate actuator pipe and turbocharger coolant pipe support brackets to their attachments. Disconnect the pipe hose ends from the vacuum pipe and wastegate actuator and remove the pipe **(see illustrations)**.

15 Undo the air conditioning compressor upper mounting bolt.

16 Undo the four bolts securing the coolant pump to the cylinder block noting the locations of the different length bolts **(see illustration)**. Withdraw the pump from the engine and recover the seal. Note that a new seal must be used on refitting.

17 If required, remove the thermostat assembly from the rear of the pump with reference to Section 5.

18 Note that it is not possible to overhaul the pump. If it is faulty, the unit must be renewed complete.

1.7 litre engines

19 Hold the pump pulley stationary (if necessary using an old auxiliary drivebelt or an oil filter strap wrench), then unscrew and remove the bolts and remove the pulley from the drive flange on the coolant pump **(see illustration)**.

20 Unscrew and remove the coolant pump retaining bolts **(see illustration)**.

21 Withdraw the coolant pump from the cylinder block, noting that it may be necessary to tap the pump lightly with a soft-faced mallet to free it.

22 Recover the gasket and discard it; a new one must be used on refitting **(see illustration)**.

9.14e ...then disconnect the hose ends and remove the wastegate actuator pipe – 1.6 litre engines

9.16 Coolant pump retaining bolts – 1.6 litre engines

9.19 Remove the pulley from the drive flange on the coolant pump – 1.7 litre engines

9.20 Unscrew and remove the coolant pump retaining bolts – 1.7 litre engines

23 Note that it is not possible to overhaul the pump. If it is faulty, the unit must be renewed complete.

Refitting

24 Refitting is a reversal of the removal procedure, bearing in mind the following points:

a) *Ensure that all mating surfaces are clean and dry and renew all disturbed gaskets and seals.*

b) *Tighten all retaining bolts/nuts to the specified torque.*

c) *On 1.6 litre engines, refit the diesel particulate filter as described in Chapter 4B, Section 26.*

d) *Refit the auxiliary drivebelt as described in Chapter 1B, Section 30.*

e) *Refill the cooling system as described in Chapter 1B, Section 27.*

10 Heating and ventilation system – general information

1 The heater/ventilation system consists of a multi-speed blower motor (housed behind the facia), face-level vents in the centre and at each end of the facia, and air ducts to the front and rear footwells.

2 The heater controls are located in the centre of the facia, and the controls operate flap valves to deflect and mix the air flowing through the various parts of the heater/ventilation system. The flap valves are contained in the air distribution housing, which

9.22 Withdraw the coolant pump and remove the gasket – 1.7 litre engines

acts as a central distribution unit, passing air to the various ducts and vents.

3 Cold air enters the system through the grille at the rear of the engine compartment. A pollen filter is fitted to the ventilation intake, to filter out dust, soot, pollen and spores from the air entering the vehicle.

4 The air (boosted by the blower fan if required) then flows through the various ducts, according to the settings of the controls. Stale air is expelled through ducts at the rear of the vehicle. If warm air is required, the cold air is passed through the heater matrix, which is heated by the engine coolant.

5 A recirculation switch enables the outside air supply to be closed off, while the air inside the vehicle is recirculated. This can be useful to prevent unpleasant odours entering from outside the vehicle, but should only be used briefly, as the recirculated air inside the vehicle will soon deteriorate.

11 Heater/ventilation system components – removal and refitting

Air vents

Removal

1 To remove the vents from the side of the facia, carefully prise free the relevant vent using a plastic spatula or similar tool to release the retaining tabs. Note that the vent will be damaged during removal and a new vent will be required for refitting.

2 To remove the vent panel from the centre of the facia, carefully prise free the panel using a plastic spatula or similar tool. Withdraw the panel from the facia, disconnect the switch wiring connector and remove the panel **(see illustrations)**.

3 If necessary, the switch block can be removed from the centre vent panel as described in Chapter 12, Section 4.

Refitting

4 Refitting is a reversal of removal.

Heater blower motor

Removal

5 Undo the three retaining screws (where fitted) and remove the facia lower panel on the passenger's side **(see illustration)**.

6 Remove the facia left-hand lower trim panel by carefully pulling it away from the facia and centre console to release the internal retaining clips, then sliding it rearward to disengage the upper locating peg **(see illustrations)**.

11.2a Prise free the centre vent panel with a plastic spatula or similar tool…

11.2b …withdraw the switch and vent panel from the facia…

11.2c …then disconnect the wiring connector and remove the panel

11.5 Remove the facia lower panel on the passenger's side

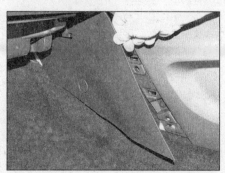

11.6a Carefully pull the lower trim panel away from the facia and centre console…

11.6b …then slide it rearward to disengage the upper locating peg

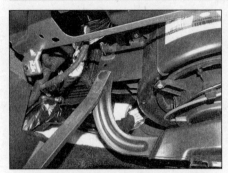

11.7 Extract the expanding rivet and remove the footwell air duct

11.8 Disconnect the wiring connector at the blower motor resistor

11.9a Undo the four screws...

11.9b ...then remove the motor down and out from under the facia

11.12 Disconnect the two wiring connectors at the blower motor resistor

11.13 Undo the two screws and remove the blower motor resistor

7 Extract the expanding rivet securing the footwell air duct to the blower motor housing. Detach the air duct from the air distribution housing and remove it from the car **(see illustration)**.

8 Disconnect the wiring connector at the blower motor resistor **(see illustration)**.
9 Undo the four screws (there may only be three on some versions) securing the blower motor to the housing, then remove the motor

down and out from under the facia **(see illustrations)**.

Refitting

10 Refitting is a reversal of removal.

Heater blower motor resistor

Removal

11 Undo the three retaining screws (where fitted) and remove the facia lower panel on the passenger's side **(see illustration 11.5)**.
12 Disconnect the two wiring connectors at the blower motor resistor **(see illustration)**.
13 Undo the two screws (if fitted) and remove the blower motor resistor **(see illustration)**.

Refitting

14 Refitting is a reversal of removal.

Heater matrix

Removal

15 Remove the air distribution housing as described later in this Section.
16 Remove the bulkhead rubber seal from the expansion valve and heater matrix pipe stubs **(see illustration)**.
17 Undo the two screws and remove the heater matrix pipe support bracket from the side of the air distribution housing **(see illustrations)**.
18 Undo the two screws on the underside of the air distribution housing and remove the heater matrix cover **(see illustrations)**.
19 Disconnect the wiring connector, undo the three retaining screws and remove the servo motor above the heater matrix upper pipe **(see illustration)**.

11.16 Remove the bulkhead rubber seal

11.17a Undo the two screws...

11.17b ...and remove the matrix pipe support bracket

11.18a Undo the two screws...

11.18b ...and remove the heater matrix cover

11.19 Remove the servo motor for access to the matrix pipe attachment

11.20a Open the two clips securing the heater pipes to the matrix...

11.20b ...remove the clips...

11.20c ...then rewmove the pipes and recover the seals

11.21 Pull the heater matrix out of the air distribution housing and remove it

20 Extract the two retaining clips securing the heater pipes to the matrix. Pull out the heater pipes and recover the seals on the pipes **(see illustrations)**.

21 Undo the retaining screw, then pull the heater matrix out of the air distribution housing and remove it from the car **(see illustration)**.

Refitting

22 Refitting is a reversal of removal, bearing in mind the following:

a) Use new seals when refitting the heater pipes to the matrix.

b) Refit the air distribution housing as described later in this Section.

Heater/air conditioning control unit

Removal

23 Remove the facia lower centre panel as described in Chapter 11, Section 27.

24 Using a small screwdriver, release the retaining tabs, then carefully prise free the control unit and remove it from the facia lower centre panel **(see illustrations)**.

Refitting

25 Refitting is a reversal of removal.

Heater/air conditioning control module

Removal

26 Remove the facia centre air vent panel as described earlier in this Section.

27 Lift the module from its location, disconnect the wiring connectors and remove the control module **(see illustration)**.

Refitting

28 Refitting is a reversal of removal.

Air distribution housing

Note: *It is not possible to remove the air distribution housing without opening the refrigerant circuit (see Sections 12 and 13). Have the refrigerant discharged at a dealer service department or an automotive air conditioning repair facility before proceeding.*

Note: *This is an involved and complex operation and it is suggested that the contents of this Section, and the relevant Sections in Chapter 11 are studied carefully to gain an understanding of the work involved, before proceeding.*

29 Drain the cooling system as described in Chapter 1A, Section 25 (petrol engines), Chapter 1B, Section 27 (diesel engines).

30 Remove the complete facia assembly and the facia crossmember as described in Chapter 11, Section 27 and Chapter 11, Section 28 respectively.

11.24a Release the retaining tabs...

11.24b ...then carefully prise free the control unit and remove it from the facia lower centre panel

11.27 Disconnect the wiring connectors and remove the heater/air conditioning control module

11.31 Disconnect the coolant hoses from the heater matrix pipe stubs

11.32 Refrigerant pipe block connection retaining nut

11.34 Unscrew the nut securing the air distribution housing to the bulkhead

31 Using a small screwdriver, lift up the wire clip securing the heater hoses to the heater matrix pipe stubs, and disconnect the hoses from the stubs **(see illustration)**. Be prepared for some loss of coolant as the hoses are released, by placing cloth rags beneath them.

32 Undo the nut securing the refrigerant pipe block connection to the expansion valve, then unscrew the retaining stud from the valve. Withdraw the refrigerant pipes from the valve **(see illustration)**. Note that new seals for the refrigerant pipes will be required for refitting. Suitably plug or cover the disconnected pipes.

33 Release the expansion valve rubber seal from the bulkhead by pushing it inward (the seal remains on the valve).

34 From within the engine compartment, unscrew the nut securing the air distribution housing to the right-hand side of the engine compartment bulkhead **(see illustration)**.

35 Working inside the vehicle, disconnect the condensation drain tube from the base of the housing, then withdraw the air distribution housing from the bulkhead.

Note: *Keep the matrix unions uppermost as the housing is removed, to prevent coolant spillage. Mop up any spilt coolant immediately, and wipe the affected area with a damp cloth to prevent staining.*

Refitting

36 Refitting is the reverse of removal. On completion, refill the cooling system as described in Chapter 1A, Section 25 or Chapter 1B, Section 27. Have the air conditioning system evacuated, charged and leak-tested by the specialist who discharged it.

13.5 Compressor wiring connectors (A) and refrigerant pipe block connectors (B)

12 Air conditioning system – general information and precautions

General information

1 Air conditioning is standard equipment on all Mokka models. It enables the temperature of incoming air to be lowered, and also dehumidifies the air, which makes for rapid demisting and increased comfort.

2 The cooling side of the system works in the same way as a domestic refrigerator. Refrigerant gas is drawn into a belt-driven compressor, and passes into a condenser mounted in front of the radiator, where it loses heat and becomes liquid. The liquid passes through an expansion valve to an evaporator, where it changes from liquid under high pressure to gas under low pressure. This change is accompanied by a drop in temperature, which cools the evaporator. The refrigerant returns to the compressor, and the cycle begins again.

3 Air blown through the evaporator passes to the heater assembly, where it is mixed with hot air blown through the heater matrix, to achieve the desired temperature in the passenger compartment.

4 The heating side of the system works in the same way as on models without air conditioning (see Section 10).

5 The operation of the system is controlled electronically. Any problems with the system should be referred to a Vauxhall/Opel dealer or an air conditioning specialist.

Precautions

6 It is necessary to observe special precautions whenever dealing with any part of the system, its associated components, and any items which necessitate disconnection of the system.

⚠ *Warning: The refrigeration circuit contains a liquid refrigerant. This refrigerant is potentially dangerous, and should only be handled by qualified persons. If it is splashed onto the skin, it can cause frostbite. It is not itself poisonous, but in the presence of a naked flame it forms a*

poisonous gas; inhalation of the vapour through a lighted cigarette could prove fatal. Uncontrolled discharging of the refrigerant is dangerous, and potentially damaging to the environment. Do not disconnect any part of the system unless it has been discharged by a Vauxhall/Opel dealer or an air conditioning specialist. Caution: Do not operate the air conditioning system if it is known to be short of refrigerant, as this may damage the compressor.

13 Air conditioning system components – removal and refitting

⚠ *Warning: The air conditioning system is under high pressure. Do not loosen any fittings or remove any components until after the system has been discharged. Air conditioning refrigerant should be properly discharged into an approved type of container at a dealer service department or an automotive air conditioning repair facility capable of handling R134a or R1234yf refrigerant. Cap or plug the pipe lines as soon as they are disconnected, to prevent the entry of moisture. Always wear eye protection when disconnecting air conditioning system fittings.*

Note: *This Section refers to the components of the air conditioning system itself – refer to Sections 10 and 11 for details of components common to the heating/ventilation system.*

Compressor

Removal

1 Have the refrigerant discharged at a dealer service department or an automotive air conditioning repair facility.

2 Disconnect the battery negative lead as described in Chapter 5A Section 4.

3 Firmly apply the handbrake, then jack up the front of the car and support it securely on axle stands (see *Jacking and vehicle support*).

4 Where fitted, remove the engine undertray as described in Chapter 11, Section 21.

5 Disconnect the compressor wiring connector and the refrigerant pressure sensor wiring connector **(see illustration)**.

13.8 Air conditioning compressor mounting bolts

13.15 Disconnect the servo motor wiring connector

13.16a Undo the lower screws...

13.16b ...and upper screws securing the air distributor case to the blower motor housing

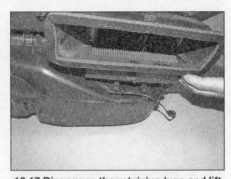

13.17 Disengage the retaining lugs and lift off the air distributor case

13.18 Slip the servo motor wiring out of the guides on the housing

6 Remove the auxiliary drivebelt as described in Chapter 1A, Section 29 (petrol engines), or Chapter 1B, Section 30 (diesel engines).

7 With the system discharged, undo the two retaining bolts and disconnect the refrigerant pipe block connectors from the compressor. Discard the O-ring seals – new ones must be used when refitting. Suitably cap the open fittings immediately to keep moisture and contamination out of the system.

8 Unbolt the compressor from the cylinder block/crankcase/sump, then withdraw the compressor downwards from under the vehicle (see illustration).

Refitting

9 Refit the compressor in the reverse order of removal; renew all seals disturbed.

10 If you are installing a new compressor, refer to the compressor manufacturer's

instructions for adding refrigerant oil to the system.

11 Have the system evacuated, charged and leak-tested by the specialist that discharged it.

Evaporator

Removal

12 Have the refrigerant discharged at a dealer service department or an automotive air conditioning repair facility.

13 Remove the air distribution housing as described in Section 11.

14 The air distribution housing itself consists of a number of sub-housings, all of which must be removed to gain access to the evaporator. Begin by removing the heater matrix pipes as described in Section 11 paragraphs 16 to 20.

15 Disconnect the wiring connector at the servo motor on the air distributor case (see illustration).

16 Work around the air distributor case and undo all the screws securing it to the blower motor housing (see illustrations).

17 Disengage the retaining lugs and lift the air distributor case off the blower motor housing (see illustration).

18 Note the routing of the servo motor wiring harness and slip the harness out of the guides on the blower motor housing (see illustration).

19 Undo the screws securing the blower motor housing to the heater matrix housing (see illustration).

20 Disconnect all relevant wiring connectors to allow the two housings to be separated.

21 Ease the two housings apart, disengage the lower retaining lugs and remove the heater matrix housing from the blower motor housing (see illustrations).

22 Undo the retaining screws and lift off the

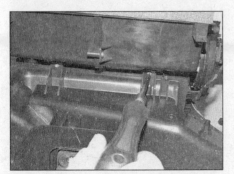

13.19 Undo the screws securing the blower motor housing to the heater matrix housing

13.21a Ease the two housings apart...

13.21b ...then disengage the lower lugs and remove the heater matrix housing

13.22a Undo the retaining screws...

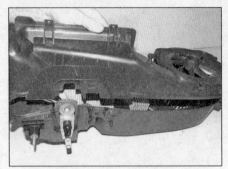

13.22b ...and lift off the upper half of the blower motor housing

13.23 Carefully lift the evaporator out of the housing

upper half of the blower motor housing **(see illustrations)**.

23 Carefully lift the evaporator out of the lower half of the blower motor housing **(see illustration)**.

Refitting

24 Refitting is a reversal of removal, bearing in mind the following points:

a) *Ensure that all disturbed pipe seals are renewed.*

b) *Refit the air distribution housing as described in Section 11.*

c) *Have the system evacuated, charged and leak-tested by the specialist who discharged it.*

Condenser

Removal

25 Have the refrigerant discharged at a dealer service department or an automotive air conditioning repair facility.

26 Disconnect the battery negative lead as described in Chapter 5A Section 4.

27 Apply the handbrake, then jack up the front of the vehicle and support it on axle stands (see *Jacking and vehicle support*).

28 Remove the front bumper as described in Chapter 11, Section 6.

29 Remove the foam energy absorbers from the bumper upper and lower impact bars **(see illustration)**.

30 Undo the five retaining bolts and remove the bumper upper support **(see illustrations)**.

31 Pull out the centre pins and extract the four plastic expanding rivets, then remove the radiator upper air baffle **(see illustrations)**.

32 Undo the three bolts each side securing the bumper lower impact bar to the subframe and withdraw the bar from under the car **(see illustrations)**.

33 Pull out the centre pins and extract the four plastic expanding rivets, then remove the radiator lower air baffle **(see illustrations)**. Note that there are five expanding rivets on 1.6 litre petrol engine models.

34 Undo the four bolts each side and

13.29 Remove the energy absorbers from the bumper impact bars

13.30a Undo the five bolts...

13.30b ...and remove the bumper upper support

13.31a Extract the four expanding rivets...

13.31b ...and remove the radiator upper air baffle

13.32a Undo the three bolts each side...

13.32b ...and remove the bumper lower impact bar

13.33a Extract the four expanding rivets...

13.33b ...and remove the radiator lower air baffle

13.34a Undo the four bolts each side...

13.34b ...and remove the bumper upper impact bar

13.35 Undo the retaining nut and disconnect the refrigerant pipe block connector

13.36 Undo the three condenser retaining bolts

remove the bumper upper impact bar **(see illustrations)**.

35 Undo the retaining nut and disconnect the refrigerant pipe block connector from the side of the condenser **(see illustration)**. Discard the O-ring seals – new ones must be used when refitting. Suitably cap the open fittings immediately to keep moisture and contamination out of the system.

36 On all except 1.6 litre petrol engine models, undo the three bolts securing the con/denser to the radiator **(see illustration)**.

37 Move the condenser sideways to disengage the refrigerant pipes, then lift the condenser from its location **(see illustrations)**.

Refitting

38 Refitting is a reverse of the removal procedure, bearing in mind the following points:

a) Ensure that all disturbed pipe seals are renewed.

b) Tighten all retaining bolts to the specified torque (where given).

c) Refit the front bumper as described in Chapter 11, Section 6.

13.37a Disengage the refrigerant pipes...

d) Have the system evacuated, charged and leak-tested by the specialist who discharged it.

Receiver-dryer

Removal

39 Remove the condenser as described previously in this Section.

40 Remove the retaining plug from the base

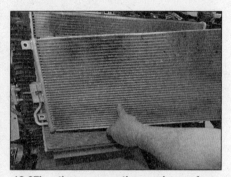

13.37b ...then remove the condenser from the car

of the condenser on the left-hand side and collect the O-ring seal(s).

41 Remove the receiver-dryer from the condenser.

Refitting

42 Refitting is the reverse of removal, using a new O-ring seal.

43 On completion, refit the condenser as described previously in this Section.

Chapter 4 Part A
Fuel and exhaust systems – petrol engines

Contents

Degrees of difficulty

Easy, suitable for novice with little experience	**Fairly easy,** suitable for beginner with some experience	**Fairly difficult,** suitable for competent DIY mechanic	**Difficult,** suitable for experienced DIY mechanic	**Very difficult,** suitable for expert DIY or professional

Specifications

System type
1.4 litre engines	Bosch E78
1.6 litre engines	AC Delco/Hitachi E83 sequential multi-point fuel injection

Fuel system data
Fuel supply pump type	Electric, immersed in tank
Fuel pump regulated constant pressure	3.8 bar
Specified idle speed	Not adjustable – controlled by ECU
Idle mixture CO content	Not adjustable – controlled by ECU

Recommended fuel
Minimum octane rating	95 RON unleaded (UK premium unleaded) Leaded fuel or LRP must NOT be used

Torque wrench settings
	Nm	lbf ft
Bumper lower impact bar retaining bolts:		
Inner bolts	22	16
Outer bolts	58	43
Camshaft sensor	8	6
Catalytic converter clamp bolt nut	13	9
Catalytic converter support bracket nuts/bolts	22	16
Crankshaft position sensor bolt	8	6
Exhaust front pipe-to-main section nuts*	17	13
Exhaust front pipe-to-catalytic converter nuts*	22	16
Exhaust manifold nuts: *		
1.4 litre engines	8	6
1.6 litre engines	20	15
Exhaust manifold heat shield bolts	8	6
Inlet manifold bolts:		
1.4 litre engines	10	7
1.6 litre engines	20	15
Intercooler bolts	10	7
Knock sensor bolt	25	18
Throttle housing bolts	8	6
Turbocharger coolant supply pipe banjo union bolt	30	22
Turbocharger oil return pipe-to-turbocharger bolts	8	6
Turbocharger oil supply pipe banjo union bolt	30	22
Turbocharger oil supply pipe-to-cylinder block bolt	10	7

*Use new fasteners

1 General information and precautions

1 The fuel system consists of a fuel tank (which is mounted under the rear of the car, with an electric fuel pump immersed in it), and the fuel feed lines. The fuel pump supplies fuel to the fuel rail, which acts as a reservoir for the four fuel injectors which inject fuel into the inlet tracts.

2 The electronic control unit controls both the fuel injection system and the ignition system, integrating the two into a complete engine management system. Refer to Section 9 for further information on the operation of the fuel system and to Chapter 5B for details of the ignition side of the system.

Precautions

Note: *Refer to Part C of this Chapter for general information and precautions relating to the catalytic converter.*

3 Before disconnecting any fuel lines, or working on any part of the fuel system, the system must be depressurised as described in Section 5.

4 Care must be taken when disconnecting the fuel lines. When disconnecting a fuel union or hose, loosen the union or clamp screw slowly, to avoid sudden uncontrolled fuel spillage. Take adequate fire precautions.

5 When working on fuel system components, scrupulous cleanliness must be observed, and care must be taken not to introduce any foreign matter into fuel lines or components.

6 After carrying out any work involving disconnection of fuel lines, it is advisable to check the connections for leaks; pressurise the system by switching the ignition on and off several times.

7 Electronic control units are very sensitive components, and certain precautions must be taken to avoid damage to these units as follows.

a) *When carrying out welding operations on the vehicle using electric welding equipment, the battery and alternator should be disconnected.*

b) *Although the underbonnet-mounted control units will tolerate normal underbonnet conditions, they can be adversely affected by excess heat or moisture. If using welding equipment or pressure-washing equipment in the vicinity of an electronic control unit, take care not to direct heat, or jets of water or steam, at the unit. If this cannot be avoided, remove the control unit from the vehicle, and protect its wiring plug with a plastic bag.*

c) *Before disconnecting any wiring, or removing components, always ensure that the ignition is switched off.*

d) *After working on fuel injection/engine management system components, ensure that all wiring is correctly reconnected before reconnecting the battery or switching on the ignition.*

 Warning: Many of the procedures in this Chapter require the removal of fuel lines and connections, which may result in some fuel spillage. Before carrying out any operation on the fuel system, refer to the precautions given in 'Safety first!', and follow them implicitly. Petrol is a highly-dangerous and volatile liquid, and the precautions necessary when handling it cannot be overstressed.

Note: *Residual pressure will remain in the fuel lines long after the vehicle was last used. Before disconnecting any fuel line, first depressurise the fuel system as described in Section 5.*

2 Air cleaner assembly and air ducts – removal and refitting

Removal

Air cleaner assembly

1 Remove the air cleaner inlet air duct by pulling upward to release the duct from the rubber grommets and the air cleaner assembly (see illustrations).

2 On 1.4 litre engines, release the clip securing the mass airflow sensor wiring harness to the rear of the air cleaner (see illustration).

3 Disconnect the wiring connector at the mass airflow sensor (see illustration).

4 Slacken the retaining clip securing the outlet air duct to the turbocharger or throttle housing and detach the duct (see illustration).

5 Slacken the retaining clip securing the outlet air duct to the air cleaner and remove the duct (see illustration).

2.1a Pull the inlet air duct upward to release it from the grommets…

2.1b …then disengage the duct from the air cleaner elbow and remove it

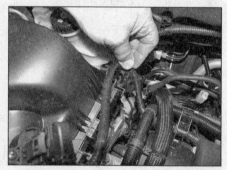

2.2 Release the wiring harness retaining clip from the rear of the air cleaner

2.3 Disconnect the mass airflow sensor wiring connector

2.4 Slacken the retaining clip and detach the outlet air duct from the turbocharger or throttle housing

2.5 Slacken the retaining clip and remove the outlet duct

2.6a Lift up the air cleaner housing...

2.6b ...to disengage it from the mounting rubbers

2.15 Slacken the retaining clip and disconnect the intercooler inlet air duct from the turbocharger

2.16 Extract the wire spring clip using a small screwdriver and disconnect the inlet air duct from the intercooler

2.23a Extract the wire spring clip...

2.23b ...and disconnect the outlet air duct from the intercooler

6 Lift up the air cleaner housing to disengage it from the mounting rubbers, then lift the air cleaner housing from the engine compartment **(see illustrations)**.

Outlet air duct – 1.4 litre engines

7 Slacken the retaining clip securing the outlet air duct to the air cleaner and detach the duct **(see illustration 2.5)**.

8 Slacken the retaining clip securing the outlet air duct to the turbocharger, then detach the duct and remove it from the engine **(see illustration 2.4)**. Suitably cover the turbocharger air inlet to prevent the possible entry of foreign material.

Outlet air duct – 1.6 litre engines

9 Slacken the retaining clip securing the outlet air duct to the air cleaner and detach the duct.

10 Slacken the retaining clip securing the outlet air duct to the throttle housing, then detach the duct and remove it from the engine.

Inlet air duct

11 Remove the air cleaner inlet air duct by pulling upward to release the duct from the rubber grommets and the air cleaner assembly **(see illustrations 2.1a and 2.1b)**.

Intercooler inlet air duct – 1.4 litre engines

12 Apply the handbrake, then jack up the front of the vehicle and support it on axle stands (see *Jacking and vehicle support*).

13 Where applicable, remove the engine undertray as described in Chapter 11, Section 21.

14 Remove the front bumper as described in Chapter 11, Section 6.

15 Slacken the retaining clip and disconnect the inlet air duct from the turbocharger **(see illustration)**.

16 Extract the wire spring clip using a small screwdriver and disconnect the inlet air duct from the intercooler, then manipulate the air duct out from the engine compartment **(see illustration)**.

Intercooler outlet air duct – 1.4 litre engines

17 Apply the handbrake, then jack up the front of the vehicle and support it on axle stands (see *Jacking and vehicle support*).

18 Where applicable, remove the engine undertray as described in Chapter 11, Section 21.

19 Remove the front bumper as described in Chapter 11, Section 6.

20 Disconnect the wiring connector from the charge (boost) pressure sensor located in the intercooler outlet air duct.

21 Undo the screw securing the outlet air duct support bracket to the subframe.

22 Slacken the retaining clip and disconnect the outlet air duct from the throttle housing.

23 Extract the wire spring clip using a small screwdriver and disconnect the outlet air duct from the intercooler, then manipulate the air duct out from under the car **(see illustrations)**.

Refitting

24 Refitting is the reverse of removal, making sure all air ducts are securely reconnected. Where applicable, refit the front bumper as

described in Chapter 11, Section 6, and the engine undertray as described in Chapter 11, Section 21.

3 Accelerator pedal/ position sensor – removal and refitting

Removal

1 Disconnect the battery negative lead as described in Chapter 5A Section 4.

2 Working in the driver's footwell under the facia, disconnect the wiring connector from the side of the accelerator pedal/position sensor.

3 Unscrew the retaining bolt and withdraw the sensor from the brake pedal mounting bracket **(see illustration)**.

Refitting

4 Refitting is a reversal of removal.

3.3 Unscrew the retaining bolt and withdraw the sensor from the brake pedal mounting bracket

5.2 Fuel pressure connection valve

4 Unleaded petrol – general information and usage

Note: *The information given in this Chapter is correct at the time of writing. If updated information is thought to be required, check with a Vauxhall/Opel dealer. If travelling abroad, consult one of the motoring organisations (or a similar authority) for advice on the fuel available.*

1 All petrol engine models are designed to run on fuel with a minimum octane rating of 95 RON. However, if unavailable, 91 octane may be used although a reduction in engine power and torque will be noticed.

2 All models have a catalytic converter, and so must be run on unleaded fuel only. Under no circumstances should leaded fuel or LRP be used, as this will damage the converter.

6.3a Disconnect the fuel supply and return pipes...

6.5a Using a suitable forked tool to unscrew the locking ring...

3 Super unleaded petrol (98 octane) can also be used in all models if wished, though there is no advantage in doing so.

5 Fuel injection system – depressurisation

⚠️ *Warning: Refer to the warning in Section 1 before proceeding. The following procedure will merely relieve the pressure in the fuel system – remember that fuel will still be present in the system components, and take precautions accordingly before disconnecting any of them.*

1 The fuel system referred to in this Section is defined as the tank-mounted fuel pump, the fuel injectors, and the metal pipes and flexible hoses of the fuel lines between these components. All these contain fuel which will be under pressure while the engine is running, and/or while the ignition is switched on. The pressure will remain for some time after the ignition has been switched off, and it must be relieved in a controlled fashion when any of these components are disturbed for servicing work.

2 Locate the fuel pressure connection valve which is fitted to the top of the fuel rail **(see illustration)**.

3 Unscrew the cap from the valve and position a container beneath the valve. Hold a wad of rag over the valve and relieve the pressure in the fuel system by depressing the valve core with a suitable screwdriver.

6.3b ...and evaporative emission system pipes from the pump and fuel gauge sender unit

6.5b ...and remove it from the tank

Be prepared for the squirt of fuel as the valve core is depressed and catch it with the rag. Hold the valve core down until no more fuel is expelled from the valve.

4 Once all pressure is relieved, securely refit the valve cap.

6 Fuel pump/fuel gauge sender unit – removal and refitting

FWD models

Removal

Note: *Refer to the warning in Section 1 before proceeding. Vauxhall/Opel specify the use of their service tool EN-48279 to remove and refit the fuel pump/fuel gauge sender unit locking ring. Suitable alternatives to this tool are readily available from accessory stores and motor factor outlets.*

1 A combined fuel pump and fuel gauge sender unit is located in the top face of the fuel tank. The combined unit can only be detached and withdrawn from the tank after the tank is released and lowered from under the car. Refer to Section 7 and remove the fuel tank, then proceed as follows.

2 Disconnect the wiring connector from the fuel pump/fuel gauge sender unit.

3 Disconnect the fuel supply pipe, fuel return pipe (where fitted) and evaporative emission system pipes from the stubs by squeezing the quick-release lugs **(see illustrations)**. A Vauxhall/Opel special tool is available to release the fuel line connectors, but provided care is taken, the connectors can be released using a pair of long-nosed pliers, or a similar tool, to depress the retaining tangs. Suitably cap or plug the disconnected unions.

4 For improved access, open the fuel pipe retaining clip on the side of the tank and remove the pipes from the tank.

5 Unscrew and remove the fuel pump/fuel gauge sender unit locking ring, by unscrewing it with the Vauxhall/Opel tool or a suitable alternative **(see illustrations)**.

6 Carefully lift out the fuel pump/gauge sender unit from the tank **(see illustration)**. Take care that the sender unit float and arm are not damaged as the unit is removed.

6.6 Carefully lift out the fuel pump/gauge sender unit from the tank

6.7 Renew the O-ring seal

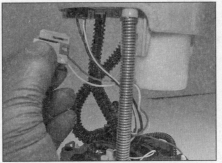

6.8a Disconnect the wiring connector…

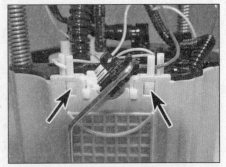

6.8b …release the two retaining tabs…

6.8c …and lift the sender unit off the side of the pump

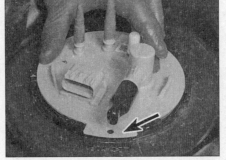

6.10a Position the sender unit so that the protruding tab…

6.10b …aligns with the word LOCATOR on the tank

7 Remove the O-ring seal and obtain a new seal for refitting **(see illustration)**.

8 To remove the fuel gauge sender unit, disconnect the wiring connector, then release the two retaining tabs and lift the unit off of the side of the pump **(see illustrations)**.

Refitting

9 Locate the gauge sender unit in position on the side of the pump and reconnect the wiring connector.

10 Place a new O-ring seal on the fuel tank and carefully lower the sender unit into the tank, taking care not to damage the float arm. Position the sender unit so that the protruding tab aligns with the word LOCATOR stamped on the tank **(see illustrations)**.

11 Refit the locking ring and tighten it with the tool used for removal. Ensure that the ring is tightened fully so that the projections on the ring engage with the centre tab at each fastening point.

Note: *The aid of an assistant will be helpful to steady the tank as the locking ring is tightened. Considerable force will be necessary.*

12 Refit the fuel pipes to the tank and reconnect the quick-release fittings.

13 Reconnect the wiring connector to the fuel pump/fuel gauge sender unit.

14 Refit the fuel tank as described in Section 7.

4WD models

Removal

Note: *Refer to the warning in Section 1 before proceeding. Vauxhall/Opel specify the use of their service tool EN-48482 to remove and*

refit the fuel tank sender unit locking ring and EN-48279 to remove and refit the fuel pump/ fuel gauge sender unit locking ring. Suitable alternatives to these tools are generally available from accessory stores and motor factor outlets.

15 There are two fuel level sensors on 4WD models. The one fitted to the left-hand side of the tank is the fuel tank sender unit. This is fitted due to the fuel tank having a tunnel moulded down the centre, which allows room for the propeller shaft to the rear axle.

16 A combined fuel pump and fuel gauge sender unit is located in the top face of the fuel tank on the right-hand side. The combined unit can only be detached and withdrawn from the tank after the tank is released and lowered from under the car. Refer to Section 6 and remove the fuel tank, then proceed as follows.

17 Disconnect the wiring connector from the fuel pump/fuel gauge sender unit.

18 Disconnect the fuel supply pipe, fuel return pipe (where fitted) and evaporative emission system pipes from the stubs by squeezing the quick-release lugs **(see illustrations 6.3a to 6.3c)**. Suitably cap or plug the disconnected unions.

19 For improved access, open the fuel pipe retaining clip on the side of the tank and remove the pipes from the tank.

20 Unscrew and remove the fuel tank sender unit locking ring, by unscrewing it with the Vauxhall/Opel tool or a suitable alternative.

21 Raise the fuel tank sender unit enough to gain access to the fuel pump/gauge sender

unit fuel pickup and disconnect the fuel tank sender from the fuel pickup. Carefully lift out the fuel tank sender unit from the tank. Take care that the sender unit float and arm are not damaged as the unit is removed.

22 Remove the sender unit O-ring seal and obtain a new seal for refitting.

23 Continue with the fuel pump/fuel gauge sender unit removal procedure as described previously in paragraphs 5 to 8 of this Section. As the unit is being lifted from the tank, disconnect the fuel tank vent pipe from the underside of the unit **(see illustration)**.

Refitting

24 Refit the fuel pump/fuel gauge sender unit as described previously in paragraphs 9 to 13 of this Section. As the unit is being inserted into the tank, remember to reconnect the fuel tank vent pipe.

6.23 Disconnect the tank vent pipe as the fuel pump/fuel gauge sender unit is withdrawn

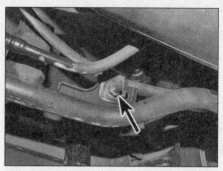

7.9 Undo the retaining nut and release the handbrake cable from the fuel tank guard

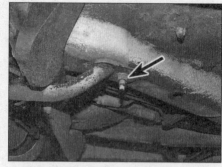

7.10a Undo the front retaining nut...

7.10b ...centre retaining nut...

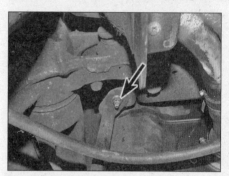

7.10c ...and rear retaining nut...

7.10d ...then remove the fuel tank guard on the right-hand side

7.11 Fuel supply line and evaporative vent line quick-release connectors

25 Place a new O-ring seal on the fuel tank and carefully lower the fuel tank sender unit into the tank, taking care not to damage the float arm. Connect the tank sender unit to the fuel pickup, then position the sender so that the protruding tab aligns with the word LOCATOR stamped on the tank.

26 Refit the locking ring and tighten it with the tool used for removal. Ensure that the ring is tightened fully so that the projections on the ring engage with the centre tab at each fastening point.

Note: *The aid of an assistant will be helpful to steady the tank as the locking ring is tightened. Considerable force will be necessary.*

27 Refit the fuel tank as described in Section 7.

7 Fuel tank – removal and refitting

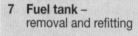

⚠ *Warning: Refer to the warning in Section 1 before proceeding.*

Removal

1 Before removing the fuel tank, all fuel must be drained from the tank. Since a fuel tank drain plug is not provided, it is therefore preferable to carry out the removal operation when the tank is nearly empty. The remaining fuel can then be syphoned or hand-pumped from the tank.

2 Disconnect the battery negative lead as described in Chapter 5A Section 4.

3 Chock the front wheels, then jack up the rear of the vehicle, and support it securely on axle stands (see *Jacking and vehicle support*). Remove the right-hand rear roadwheel.

4 Remove the rear wheel arch liner on the right-hand side as described in Chapter 11, Section 21.

5 Depressurise the fuel system as described in Section 5.

6 Remove the exhaust system main section and rear silencer as described in Section 20.

7 On 4WD models, remove the propeller shaft as described in Chapter 8, Section 6.

8 Disconnect the fuel pump and fuel gauge sender unit wiring harness at the connector under the right-hand wheel arch.

9 Undo the retaining nut and release the handbrake cable from the fuel tank guard on the right-hand side **(see illustration)**.

10 Undo the three nuts and remove the fuel tank guard on the right-hand side **(see illustrations)**.

11 Disconnect the fuel supply and return line or evaporative vent line at the underbody quick-release connectors on the right-hand side of the tank **(see illustration)**. Be prepared for some loss of fuel. A Vauxhall/Opel special tool is available to release the fuel line connectors, but provided care is taken, the connectors can be released using a pair of long-nosed pliers, or a similar tool, to depress the retaining tangs. Suitably plug the disconnected fuel and vent hoses to prevent entry of dust and dirt.

12 Place a suitable container under the tank, then slacken the clip and disconnect the fuel filler hose from the fuel tank **(see illustration)**. Collect the escaping fuel in the container.

13 Disconnect the fuel tank filler vent pipe at the quick-release connector **(see illustration)**.

14 Support the weight of the fuel tank on a jack with interposed block of wood.

15 Undo the bolts securing the front of each fuel tank retaining strap to the underbody.

7.12 Slacken the clip and disconnect the fuel filler hose from the fuel tank

7.13 Disconnect the filler vent pipe at the quick-release connector

Disengage the straps from the underbody at the rear and remove the straps **(see illustrations)**.

16 Taking care not to damage the charcoal canister, slowly lower the tank and move it forwards.

17 Continue to lower the tank until it can be removed from under the vehicle.

18 If necessary, remove the charcoal canister, fuel lines and hoses, heat shield and wiring from the tank for transfer to the new tank.

19 If the tank contains sediment or water, it may cleaned out with two or three rinses of clean fuel. Remove the fuel pump/fuel gauge sender unit as described in Section 6. Shake the tank vigorously, and change the fuel as necessary to remove all contamination from the tank.

Caution: This procedure should be carried out in a well-ventilated area, and it is vital to take adequate fire precautions.

20 Any repairs to the fuel tank should be carried out by a professional. Do not under any circumstances attempt any form of DIY repair to a fuel tank.

Refitting

21 Refitting is the reverse of the removal procedure, noting the following points:

a) When lifting the tank back into position, take care to ensure that none of the hoses become trapped between the tank and vehicle body. Refit the retaining straps and tighten the bolts securely.

b) Ensure all pipes and hoses are correctly routed and all hose unions are securely joined.

c) On 4WD models, refit the propeller shaft as described in Chapter 8, Section 6.

d) Refit the exhaust system main section and rear silencer as described in Section 20.

e) Refit the rear wheel arch liner as described in Chapter 11, Section 21.

f) On completion, refill the tank with a small amount of fuel, and check for signs of leakage prior to taking the vehicle out on the road.

8 Throttle housing – removal and refitting

Removal

1 Disconnect the battery negative lead as described in Chapter 5A Section 4.

1.4 litre engines

2 Slacken the retaining clip and disconnect the intercooler outlet air duct from the throttle housing.

3 Disconnect the wiring connector from the throttle housing and release the wiring from the retaining clip.

4 Undo the four bolts and lift the throttle housing off the inlet manifold. Recover the O-ring seal.

5 It is not possible to obtain the throttle valve

7.15a Undo the bolt securing each fuel tank retaining strap to the underbody…

control motor or throttle valve position sensor separately, so if either is faulty, the complete throttle housing must be renewed.

1.6 litre engines

6 Remove the air cleaner outlet air duct as described in Section 2.

7 Disconnect the crankcase ventilation hose at the quick-release connector on the throttle housing. Clamp the coolant hoses to minimise coolant loss, then release the retaining clip and disconnect the two coolant hoses from the throttle housing.

8 Undo the four bolts and lift the throttle housing off the inlet manifold **(see illustration)**. Recover the O-ring seal.

9 It is not possible to obtain the throttle valve control motor or throttle valve position sensor separately, so if either is faulty, the complete throttle housing must be renewed.

Refitting

10 Refitting is a reversal of removal, but thoroughly clean the mating faces and use a new O-ring seal. Tighten the bolts progressively and securely. On 1.6 litre engines, top-up the coolant level as described in *Weekly checks*.

9 Fuel Injection system – general information

1 The engine management (fuel injection/ ignition) systems incorporate a closed-loop catalytic converter, and an evaporative emission control system. The fuel injection side of the systems operate as follows; refer to Chapter 5B, Section 1 for information on the ignition system.

2 The fuel supply pump, immersed in the fuel tank, pumps fuel from the fuel tank to the fuel rail. Fuel supply pressure is controlled by the pressure regulator located either in the fuel tank, or on the fuel rail.

3 The electrical control system consists of the ECU, along with the following sensors.

a) Throttle potentiometer (integral with the throttle housing) – informs the ECU of the throttle position, and confirms the signals received from the accelerator pedal position sensor.

7.15b …then disengage the straps from the underbody at the rear

b) Accelerator pedal position sensor – informs the ECU of accelerator pedal position, and the rate of throttle opening/closing.

c) Coolant temperature sensor (two) – informs the ECU of engine temperature.

d) Mass airflow sensor (with inlet air temperature sensor) – informs the ECU of the temperature and amount of air passing through the inlet duct.

e) Manifold absolute pressure sensor – informs the ECU of the engine load by monitoring the pressure in the inlet manifold.

f) Charge (boost) pressure sensor (1.4 litre engines) – Informs the ECU of the pressure in the inlet duct.

g) Oxygen sensors (two) – inform the ECU of the oxygen content of the exhaust gases (explained in greater detail in Part C of this Chapter).

h) Crankshaft sensor – informs the ECU of engine speed and crankshaft position.

i) Camshaft sensors (two) – inform the ECU of speed and position of the camshafts.

j) Knock sensor – informs the ECU when pre-ignition ('pinking') is occurring.

k) ABS control unit – informs the ECU of the vehicle speed, based on wheel speed sensor signals (explained in greater detail in Chapter 9).

4 All the above information is analysed by the ECU and, based on this, the ECU determines the appropriate ignition and fuelling requirements for the engine. The ECU controls the fuel injector by varying its pulse width – the length of time the injector is held open – to provide a richer or weaker mixture, as appropriate. The mixture

8.8 Throttle housing retaining bolts

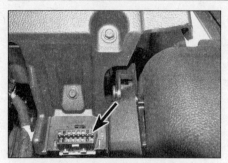

10.2 The vehicle diagnostic socket is located at the base of the facia on the driver's side

is constantly varied by the ECU, to provide the best setting for cranking, starting (with either a hot or cold engine), warm-up, idle, cruising, and acceleration.

5 Idle speed and throttle position is controlled by the throttle valve control motor, which is an integral part of the throttle housing. The motor is controlled by the ECU, in conjunction with signals received from the accelerator pedal position sensor.

6 On 1.6 litre engines, the system incorporates a variable tract inlet manifold to help increase torque output at low engine speeds. Each inlet manifold tract is fitted with a valve. The valve is controlled by the ECU via a solenoid valve and vacuum diaphragm unit.

7 At low engine speeds (below approximately 3600 rpm) the valves remain closed. The air entering the engine is then forced to take the long inlet path through the manifold which leads to an increase in the engine torque output.

8 At higher engine speeds, the ECU switches the solenoid valve which then allows vacuum to act on the diaphragm unit. The diaphragm unit is linked to the valve assemblies and opens up each of the four valves allowing the air passing through the manifold to take the shorter inlet path which is more suited to higher engine speeds.

9 The ECU also controls the evaporative emission control system, which is described in detail in Part C of this Chapter.

10 On 1.4 litre engines, the turbocharger boosts engine power by increasing the mass of oxygen (and therefore the fuel) entering the engine. It does this by raising the pressure in the inlet manifold above atmospheric pressure. Instead of the air simply being sucked into the cylinders, it is forced in.

11 Between the turbocharger and the inlet manifold, the compressed air passes through an intercooler. This is an air-to-air heat exchanger, mounted in front of the radiator, and supplied with cooling air from the front of the vehicle. The purpose of the intercooler is to remove some of the heat gained in being compressed from the inlet air. Because cooler air is denser, removal of this heat further increases engine efficiency.

12 If certain sensors fail, and send abnormal signals to the ECU, the ECU has a back-up programme. In this event, the abnormal signals are ignored, and a pre-programmed value is substituted for the sensor signal, allowing the engine to continue running, albeit at reduced efficiency. If the ECU enters its back-up mode, a warning light on the instrument panel will illuminate, and a fault code will be stored in

the ECU memory. This fault code can be read using suitable specialist test equipment.

10 Fuel injection system components – testing

1 If a fault appears in the engine management system, first ensure that all the system wiring connectors are securely connected and free of corrosion. Ensure that the fault is not due to poor maintenance; ie, check that the air cleaner filter element is clean, the spark plugs are in good condition and correctly gapped, the cylinder compression pressures are correct and that the engine breather hoses are clear and undamaged, referring to Chapters 1A, 2A, 2B and 4C for further information.

2 If these checks fail to reveal the cause of the problem, the vehicle should be taken to a suitably-equipped Vauxhall/Opel dealer or engine management diagnostic specialist for testing. A diagnostic socket is located at the base of the facia, on the driver's side, to which a fault code reader or other suitable test equipment can be connected (**see illustration**). By using the code reader or test equipment, the engine management ECU can be interrogated, and any stored fault codes can be retrieved. Live data can also be captured from the various system sensors and actuators, indicating their operating parameters. This will allow the fault to be quickly and simply traced, alleviating the need to test all the system components individually, which is a time-consuming operation that carries a risk of damaging the ECU.

11 Fuel injection system components (1.4 litre engines) – removal and refitting

Mass airflow sensor

1 Disconnect the wiring connector from the mass airflow sensor (**see illustration**).
2 Undo the two retaining screws and remove the mass airflow sensor from the air cleaner lid.
3 Check the condition of the airflow sensor seal and renew if necessary.
4 Refitting is a reversal of removal.

Fuel rail and injectors

Note: *Refer to the precautions given in Section 1 before proceeding. The seals at both ends of the fuel injectors must be renewed on refitting.*

5 Disconnect the battery negative terminal (refer to battery disconnection and reconnection in Chapter 5A, Section 4).
6 Remove the plastic cover over the top of the engine (**see illustrations**).
7 Depressurise the fuel system as described in Section 5.
8 Open the two crankcase ventilation hose retaining clips at the rear of the camshaft cover (**see illustration**). Pull out the retaining

11.1 Disconnect the wiring connector from the airflow sensor

11.6a Using a plastic spatula or similar tool, release the retaining tabs...

11.6b ...and remove the plastic cover over the top of the engine

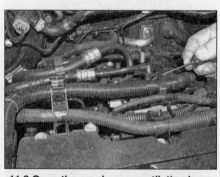

11.8 Open the crankcase ventilation hose retaining clips at the rear of the cover

11.9 Disconnect the fuel return hose quick-release connector

11.10 Disconnect the fuel feed hose quick-release connector

11.13 Undo the retaining nut and disconnect the earth lead from the fuel rail

11.15 Unscrew the two mounting bolts, then lift the fuel rail complete with injectors off of the inlet manifold

11.16a Remove the metal securing clip...

11.16b ...and pull the injector from the fuel rail

spring clip and disconnect the crankcase ventilation hose from the inlet manifold.

9 Disconnect the fuel return hose quick release connector at the fuel rail **(see illustration)**.

10 Disconnect the fuel feed hose quick-release connector at the fuel rail **(see illustration)**. Be prepared for some loss of fuel. Clamp or plug the open end of the hose, to prevent dirt ingress and further fuel spillage.

11 Release the wiring harness from the retaining clips on the camshaft cover and fuel rail.

12 Disconnect the wiring connectors from the fuel injectors and move the wiring harness clear of the fuel rail.

13 Undo the retaining nut and disconnect the earth lead from the fuel rail **(see illustration)**.

14 On engines with a fuel pressure regulator mounted on the fuel rail, disconnect the

pressure regulator vacuum hose from the inlet manifold.

15 Unscrew the two mounting bolts, then lift the fuel rail complete with the injectors off of the inlet manifold **(see illustration)**.

16 To remove an injector from the fuel rail, prise out the metal securing clip using a screwdriver or a pair of pliers, and pull the injector from the fuel rail **(see illustrations)**. Remove and discard the injector sealing rings; new ones must be fitted on refitting.

17 Overhaul of the fuel injectors is not possible, as no spares are available. If faulty, an injector must be renewed.

18 Commence refitting by fitting new O-ring seals to both ends of the fuel injectors **(see illustrations)**. Coat the seals with a thin layer of petroleum jelly before fitting.

19 Refitting is a reversal of removal, bearing in mind the following points:

a) When refitting the injectors to the fuel rail, note that the groove in the metal securing clip must engage with the lug on the injector body.

b) Make sure that the fuel and return hose quick-release connectors audibly engage on the fuel rail.

c) Ensure that all wiring connectors are securely reconnected, and that the wiring is secured in the relevant clips and brackets.

Crankshaft position sensor

Note: *A new O-ring seal must be used on refitting.*

20 The crankshaft position sensor is located at the rear left-hand end of the cylinder block, below the starter motor **(see illustration)**.

21 Apply the handbrake, then jack up the

11.18a Fit a new O-ring seal to the upper end...

11.18b ...and lower end of each injector

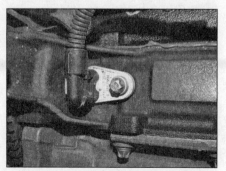

11.20 The crankshaft position sensor is located at the rear left-hand end of the cylinder block

11.25 Inlet camshaft position sensor location

11.55 The turbocharger wastegate solenoid is located on the front of the turbocharger

front of the vehicle and support it on axle stands (see *Jacking and vehicle support*).

22 On 4WD models, remove the transfer gearbox as described in Chapter 7C, Section 3.

23 Disconnect the sensor wiring connector, then undo the retaining bolt and withdraw the sensor from the cylinder block.

24 Refitting is a reversal of removal, but ensure that the mating surfaces of the sensor and baseplate are clean and fit a new O-ring seal to the sensor before refitting. Tighten the bolt to the specified torque. On 4WD models, refit the transfer gearbox as described in Chapter 7C, Section 3.

Camshaft sensor

25 Two sensors are fitted, one for each camshaft. Both sensors are located at the right-hand end of the cylinder head (see illustration).

26 For improved access to the exhaust camshaft sensor, remove the air cleaner outlet air duct as described in Section 2.

27 Disconnect the wiring connector from the relevant sensor, then undo the retaining bolt and remove the sensor from the cylinder head.

28 Refitting is a reversal of removal, tightening the sensor retaining bolt to the specified torque.

Coolant temperature sensor

29 Refer to Chapter 3, Section 7 for removal and refitting details.

Manifold absolute pressure sensor

30 The manifold absolute pressure sensor is located on the top, right-hand side of the inlet manifold.

31 Disconnect the wiring connector, then undo the retaining bolt and remove the sensor from the inlet manifold.

32 Refit the sensor to the manifold, then refit the retaining bolt and tighten it securely.

33 Reconnect the sensor wiring connector.

Charge (boost) pressure sensor

34 The charge (boost) pressure sensor is located in the intercooler outlet air duct.

35 Apply the handbrake, then jack up the

front of the vehicle and support it on axle stands (see *Jacking and vehicle support*).

36 Remove the engine undertray as described in Chapter 11, Section 21.

37 Disconnect the pressure sensor wiring connector.

38 Undo the retaining bolt and remove the sensor from the air duct.

39 Refit the sensor to the air duct, then refit the retaining bolt and tighten it securely.

40 Reconnect the sensor wiring connector.

41 Refit the engine undertray as described in Chapter 11, Section 21.

Knock sensor

42 The knock sensor is located on the rear of the cylinder block, above the starter motor.

43 Remove the inlet manifold as described in Section 13.

44 Disconnect the wiring connector from the knock sensor.

45 Note its position, then unscrew the retaining bolt and remove the knock sensor from the block.

46 Clean the contact surfaces of the sensor and block. Also clean the threads of the sensor mounting bolt.

47 Locate the sensor on the block and insert the mounting bolt. Position the sensor as previously noted, then tighten the bolt to the specified torque. Note that the torque setting is critical for the sensor to function correctly.

48 Reconnect the knock sensor wiring connector.

49 Refit the inlet manifold as described in Section 13.

12.1 Disconnect the wiring connector from the airflow sensor

Electronic control unit (ECU)

Note: *If a new ECU is to be fitted, this work must be entrusted to a Vauxhall/Opel dealer or suitably-equipped specialist as it is necessary to program the new ECU after installation. This work requires the use of dedicated Vauxhall/Opel diagnostic equipment or a compatible alternative.*

50 The ECU is located on the left-hand side of the engine compartment, attached to the side of the battery tray.

51 Disconnect the battery negative terminal (refer to battery disconnection and reconnection in Chapter 5A, Section 4).

52 Lift up the locking bars and disconnect the three ECU wiring connectors.

53 Undo the four retaining nuts and remove the ECU from the mounting bracket.

54 Refitting is a reversal of removal.

Turbocharger wastegate solenoid

55 The turbocharger wastegate solenoid is located on the front of the turbocharger (see illustration).

56 Disconnect the wiring connector from the base of the solenoid.

57 Disconnect the three vacuum hoses, then undo the retaining bolt and remove the solenoid from the turbocharger.

58 Refitting is a reversal of removal.

Oxygen sensors

59 Refer to Chapter 4C, Section 2 for removal and refitting details.

---

12 Fuel injection system components (1.6 litre engines) – removal and refitting

---

Mass airflow sensor

1 Disconnect the wiring connector from the airflow sensor (see illustration).

2 Undo the two retaining screws and remove the airflow sensor from the air cleaner lid.

3 Check the condition of the airflow sensor seal and renew if necessary.

4 Refitting is a reversal of removal.

Fuel rail and injectors

Note: *Refer to the precautions given in Section 1 before proceeding. The seals at both ends of the fuel injectors must be renewed on refitting.*

5 Disconnect the battery negative lead as described in Chapter 5A Section 4.

6 Depressurise the fuel system as described in Section 5.

7 Disconnect the fuel feed hose quick-release connector at the fuel rail (see illustration). Be prepared for some loss of fuel. Clamp or plug the open end of the hose, to prevent dirt ingress and further fuel spillage.

8 Release the retaining clips and move the fuel feed hose assembly to one side.

9 Unclip the wiring harness trough from the rear of the camshaft cover **(see illustration)**.

10 Pull out the retaining wire clip and disconnect the crankcase ventilation hose from the camshaft cover **(see illustration)**.

11 Disconnect the wiring connectors from the following components, labelling each connector to avoid confusion when refitting:

a) *Evaporative emission control system purge valve.*

b) *Fuel injectors.*

c) *Manifold absolute pressure sensor.*

12 Unclip the wiring harness from the support brackets and move the harness to one side.

13 Undo the two bolts securing the evaporative emission control system purge valve mounting bracket to the inlet manifold. Move the valve and bracket to one side.

14 Unscrew the two mounting bolts, then lift the fuel rail complete with the injectors off of the inlet manifold.

15 To remove an injector from the fuel rail, prise out the metal securing clip using a screwdriver or a pair of pliers, and pull the injector from the fuel rail **(see illustrations 11.16a and 11.16b)**. Remove and discard the injector sealing rings; new ones must be fitted on refitting.

16 Overhaul of the fuel injectors is not possible, as no spares are available. If faulty, an injector must be renewed.

17 Commence refitting by fitting new O-ring seals to both ends of the fuel injectors **(see illustrations 11.18a and 11.18b)**. Coat the seals with a thin layer of petroleum jelly before fitting.

18 Refitting is a reversal of removal, bearing in mind the following points:

a) *When refitting the injectors to the fuel rail, note that the groove in the metal securing clip must engage with the lug on the injector body.*

b) *Make sure that the fuel hose quick-release connector audibly engages on the fuel rail.*

c) *Ensure that all wiring connectors are securely reconnected, and that the wiring is secured in the relevant clips and brackets.*

Crankshaft position sensor

Note: *A new O-ring seal must be used on refitting.*

12.10 Pull out the retaining wire clip and disconnect the crankcase ventilation hose from the camshaft cover

12.7 Disconnect the fuel feed hose quick-release connector

19 The crankshaft sensor is located at the rear left-hand end of the cylinder block, below the starter motor

20 Apply the handbrake, then jack up the front of the vehicle and support it on axle stands (see *Jacking and vehicle support*).

21 Remove the starter motor as described in Chapter 5A, Section 11.

22 Disconnect the sensor wiring connector, then undo the retaining bolt and withdraw the sensor from the cylinder block.

23 Refitting is a reversal of removal, but ensure that the mating surfaces of the sensor and baseplate are clean and fit a new O-ring seal to the sensor before refitting. Tighten the retaining bolt securely.

Camshaft sensor

24 Two sensors are fitted, one for each camshaft. Both sensors are located at the left-hand end of the cylinder head.

25 Disconnect the wiring connector from the relevant sensor **(see illustration)**.

26 Undo the retaining bolt and remove the sensor from the cylinder head.

27 Refitting is a reversal of removal, tightening the sensor retaining bolt securely.

Coolant temperature sensor

28 Refer to Chapter 3, Section 7 for removal and refitting details.

Manifold absolute pressure sensor

29 The manifold absolute pressure sensor is located on the top of the inlet manifold at the rear **(see illustration)**.

12.25 Disconnect the wiring connector from the camshaft sensor (exhaust sensor shown)

12.9 Unclip the wiring harness trough from the rear of the camshaft cover

30 Disconnect the wiring connector, then undo the retaining bolt and remove the sensor from the inlet manifold.

31 Refit the sensor to the manifold, then refit the retaining bolt and tighten it securely.

32 Reconnect the sensor wiring connector.

Knock sensor

33 The knock sensor is located on the rear of the cylinder block, adjacent to the starter motor.

34 Remove the inlet manifold as described in Section 14.

35 Disconnect the knock sensor wiring connector, then unscrew the retaining bolt and remove the sensor from the cylinder block.

36 Clean the contact surfaces of the sensor and block. Also clean the threads of the sensor mounting bolt.

37 Locate the sensor on the block and insert the mounting bolt. Tighten the bolt to the specified torque. Note that the torque setting is critical for the sensor to function correctly.

38 Refit the inlet manifold as described in Section 14.

Electronic control unit (ECU)

Note: *If a new ECU is to be fitted, this work must be entrusted to a Vauxhall/Opel dealer or suitably-equipped specialist as it is necessary to program the new ECU after installation. This work requires the use of dedicated Vauxhall/Opel diagnostic equipment or a compatible alternative.*

39 The ECU is located on the left-hand side of the engine compartment, attached to the side of the battery tray.

12.29 The manifold absolute pressure sensor is located on the top of the inlet manifold at the rear

13.11 Disconnect the wiring connector from the evaporative emission control system purge valve

13.15 Disconnect the wiring connector from the charge air bypass regulator solenoid on the underside of the inlet manifold

13.19 Disconnect the ventilation pipe from the evaporative emission control system purge valve

13.20 Disconnect the quick-release fitting and detach the brake servo vacuum hose from the inlet manifold

13.21a Unscrew the six retaining bolts and manoeuvre the manifold assembly away from the cylinder head

13.21b Remove and discard the manifold seal rings

40 Disconnect the battery negative lead as described in Chapter 5A Section 4.
41 Lift up the locking bars and disconnect the three ECU wiring connectors.
42 Undo the four retaining nuts and remove the ECU from the mounting bracket.
43 Refitting is a reversal of removal.

Oxygen sensors

44 Refer to Chapter 4C, Section 2 for removal and refitting details.

13 Inlet manifold (1.4 litre engines) – removal and refitting

Removal

1 Disconnect the battery negative lead as described in Chapter 5A Section 4.
2 Remove the plastic cover over the top of the engine (see illustrations 11.6a and 11.6b).
3 Depressurise the fuel system as described in Section 5.
4 Slacken the retaining clip and disconnect the intercooler outlet air duct from the throttle housing, then disconnect the throttle housing wiring connector.
5 Firmly apply the handbrake, then jack up the front of the car and support it securely on axle stands (see Jacking and vehicle support).
6 From under the car, unclip the coolant hose and the wiring harness from the inlet manifold.

7 Open the two crankcase ventilation hose retaining clips at the rear of the camshaft cover (see illustration 11.8).
8 Pull out the retaining spring clip and disconnect the crankcase ventilation hose from the inlet manifold.
9 Release the retaining clip and disconnect the vacuum hose from the charge air bypass regulator solenoid on the underside of the inlet manifold.
10 Lift the crankcase ventilation hose, together with the vacuum hose out of the support clips and move them to one side.
11 Disconnect the wiring connector from the evaporative emission control system purge valve (see illustration).
12 Disconnect the wiring connectors from the manifold absolute pressure sensor on the top of the manifold.
13 Release the wiring harness from the retaining clips on the camshaft cover and fuel rail.
14 Disconnect the wiring connectors from the fuel injectors and move the wiring harness to one side.
15 Disconnect the wiring connector from the charge air bypass regulator solenoid on the underside of the inlet manifold (see illustration).
16 Release the wiring harness from the clips on the right-hand side of the manifold.
17 Disconnect the fuel return hose quick release connector at the fuel rail (see illustration 11.9).

18 Disconnect the fuel feed hose quick-release connector at the fuel rail (see illustration 11.10). Be prepared for some loss of fuel. Clamp or plug the open end of the hose, to prevent dirt ingress and further fuel spillage.
19 Disconnect the quick-release fitting and detach the ventilation pipe from the evaporative emission control system purge valve (see illustration). Release the pipe from the retainer clip.
20 Disconnect the quick-release fitting and detach the brake servo vacuum hose from the inlet manifold (see illustration).
21 Unscrew the six retaining bolts and manoeuvre the manifold assembly away from the cylinder head. Note that the bolts are captive in the manifold and cannot be completely removed. Remove and discard the seal rings (see illustrations).

Refitting

22 Refitting is the reverse of removal noting the following.
a) Ensure the manifold and cylinder head mating surfaces are clean and dry and fit the new seal rings. Refit the manifold and tighten the retaining bolts evenly and progressively to the specified torque.
b) Ensure that all relevant hoses are reconnected to their original positions, and are securely held (where necessary) by their retaining clips.

14 Inlet manifold (1.6 litre engines) – removal and refitting

Removal

1 Disconnect the battery negative lead as described in Chapter 5A Section 4.
2 Depressurise the fuel system as described in Section 5.
3 Remove the throttle housing as described in Section 8.
4 Remove the evaporative emission control system purge valve as described in Chapter 4C, Section 2.
5 Remove the fuel rail and injectors as described in Section 12.
6 Disconnect the wiring connector from the manifold absolute pressure sensor.
7 Firmly apply the handbrake, then jack up the front of the car and support it securely on axle stands (see *Jacking and vehicle support*).
8 Undo the two retaining bolts and remove the inlet manifold support bracket.
9 Unclip the wiring harness from the base of the inlet manifold.
10 Remove the camshaft cover as described in Chapter 2B, Section 4.
11 Unscrew the seven retaining bolts and manoeuvre the manifold assembly away from the cylinder head. Remove and discard the seal rings.

Refitting

12 Refitting is the reverse of removal noting the following.
a) *Ensure the manifold and cylinder head mating surfaces are clean and dry and fit the new seal rings. Refit the manifold and tighten the retaining bolts evenly and progressively to the specified torque.*
b) *Ensure that all relevant hoses are reconnected to their original positions, and are securely held (where necessary) by their retaining clips.*
c) *Refit the camshaft cover as described in Chapter 2B, Section 4.*
d) *Refit the fuel rail and injectors as described in Section 12.*
e) *Refit the evaporative emission control system purge valve as described in Chapter 4C, Section 2.*
f) *Refit the throttle housing as described in Section 8.*

15 Intercooler – removal and refitting

Removal

1 Disconnect the battery negative lead as described in Chapter 5A, Section 4.
2 Apply the handbrake, then jack up the front of the vehicle and support it on axle stands (see *Jacking and vehicle support*).

3 Remove the front bumper as described in Chapter 11, Section 6.
4 Remove the foam energy absorber from the bumper lower impact bar.
5 Undo the three bolts each side securing the bumper lower impact bar to the subframe and withdraw the bar from under the car **(see illustrations)**.

6 Pull out the centre pins and extract the four plastic expanding rivets, then remove the radiator lower air baffle **(see illustrations)**.
7 Extract the wire spring clip using a small screwdriver and disconnect the inlet and outlet air ducts from the intercooler **(see illustrations)**.
8 Undo the two retaining bolts and lift the intercooler off the car **(see illustrations)**.

15.5a Undo the three bolts each side…

15.5b …and remove the bumper lower impact bar

15.6a Extract the four expanding rivets…

15.6b …and remove the radiator lower air baffle

15.7a Extract the wire spring clip and disconnect the inlet air duct…

15.7b …and outlet air ducts from the intercooler

15.8a Undo the upper retaining bolt each side…

15.8b …and lift the intercooler off the car

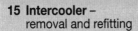

18.7 Open the quick-release connector and disconnect the crankcase ventilation hose from the turbocharger air inlet

18.8 Release the retaining clip and disconnect the vacuum hose from the turbocharger

Refitting

9 Refitting is a reversal of removal, bearing in mind the following points:
a) *Tighten all retaining bolts to the specified torque (where given).*
b) *Refit the front bumper as described in Chapter 11, Section 6.*

16 Turbocharger – description and precautions

Description

1 The turbocharger increases engine efficiency by raising the pressure in the inlet manifold above atmospheric pressure. Instead of the air simply being sucked into the cylinders, it is forced in.

2 Energy for the operation of the turbocharger comes from the exhaust gas. The gas flows through a specially-shaped housing (the turbine housing) and in so doing, spins the turbine wheel. The turbine wheel is attached to a shaft, at the end of which is another vaned wheel known as the compressor wheel. The compressor wheel spins in its own housing, and compresses the inlet air on the way to the inlet manifold.

3 Boost pressure (the pressure in the inlet manifold) is limited by a wastegate, which diverts the exhaust gas away from the turbine

wheel in response to a pressure-sensitive actuator.

4 The turbo shaft is pressure-lubricated by an oil feed pipe from the main oil gallery. The shaft 'floats' on a cushion of oil. A drain pipe returns the oil to the sump.

Precautions

5 The turbocharger operates at extremely high speeds and temperatures. Certain precautions must be observed, to avoid premature failure of the turbo, or injury to the operator.

6 Do not operate the turbo with any of its parts exposed, or with any of its hoses removed. Foreign objects falling onto the rotating vanes could cause excessive damage, and (if ejected) personal injury.

7 Do not race the engine immediately after start-up, especially if it is cold. Give the oil a few seconds to circulate.

8 Always allow the engine to return to idle speed before switching it off – do not blip the throttle and switch off, as this will leave the turbo spinning without lubrication.

9 Allow the engine to idle for several minutes before switching off after a high-speed run.

10 Observe the recommended intervals for oil and filter changing, and use a reputable oil of the specified quality. Neglect of oil changing, or use of inferior oil, can cause carbon formation on the turbo shaft, leading to subsequent failure.

17 Turbocharger – removal and refitting

1 The turbocharger is an integral part of the exhaust manifold and the two components cannot be separated. Removal and refitting procedures for the exhaust manifold are contained in Section.

18 Exhaust manifold (1.4 litre engines) – removal and refitting

Removal

Note: *New manifold and exhaust front pipe retaining nuts, together with new gaskets and seals for all disturbed components must be used on refitting.*

Note: *The turbocharger is an integral part of the exhaust manifold and the two components cannot be separated.*

Note: *Vauxhall/Opel special tool EN-49942 or a suitable alternative will be required to retain the turbocharger coolant supply pipe banjo union as the union bolt is slackened and tightened.*

1 Disconnect the battery negative lead as described in Chapter 5A Section 4.

2 Apply the handbrake, then jack up the front of the vehicle and support it on axle stands (see *Jacking and vehicle support*).

3 Drain the cooling system as described in Chapter 1A, Section 25.

4 Remove the air cleaner outlet air duct as described in Section 2.

5 Open the two crankcase ventilation hose retaining clips at the rear of the camshaft cover.

6 Pull out the retaining spring clip and disconnect the crankcase ventilation hose from the inlet manifold.

7 Open the quick-release connector and disconnect the other end of the crankcase ventilation hose from the turbocharger air inlet **(see illustration)**. Suitably cover the turbocharger inlet to prevent the possible entry of foreign material.

8 Release the retaining clips and disconnect the vacuum hose from the turbocharger and from the charge air bypass regulator solenoid on the underside of the inlet manifold **(see illustration)**.

9 Lift the crankcase ventilation hose, together with the vacuum hose out of the support clips and remove the two hoses from the engine.

10 Undo the three bolts and remove the exhaust manifold heat shield **(see illustrations)**.

11 Disconnect the wiring connector from the turbocharger wastegate solenoid, then release the wiring harness retaining clip from the support bracket **(see illustration)**.

12 Unscrew the turbocharger oil supply pipe banjo union bolt from the top of the

18.10a Undo the three bolts...

18.10b ...and remove the exhaust manifold heat shield

18.11 Disconnect the wiring connector from the turbocharger wastegate solenoid

18.12 Unscrew the turbocharger oil supply pipe banjo union bolt and collect the two sealing washers

18.13a Undo the bolt securing the turbocharger oil supply pipe to the cylinder block...

18.13b ...and remove the pipe from the engine

18.14 Release the retaining clip and disconnect the turbocharger coolant return hose from the oil cooler inlet pipe

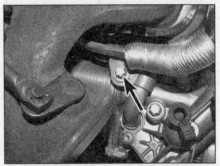

18.15 Undo the bolt securing the turbocharger coolant return pipe to the oil filter housing

turbocharger and collect the two sealing washers (see illustration).

13 Undo the bolt securing the turbocharger oil supply pipe to the cylinder block and remove the pipe from the engine (see illustrations). Collect the two O-ring seals from the end of the pipe.

14 Release the retaining clip and disconnect the turbocharger coolant return hose from the oil cooler inlet pipe (see illustration).

15 Undo the bolt securing the turbocharger coolant return pipe to the oil filter housing (see illustration).

16 Slacken the retaining clip and disconnect the intercooler inlet air duct from the turbocharger (see illustration).

17 Unscrew the turbocharger coolant supply pipe banjo union bolt from the cylinder block and collect the two sealing washers (see illustration). To prevent the pipe banjo union from turning as the bolt is unscrewed, obtain Vauxhall/Opel special tool EN-49942 or a suitable alternative.

18 Remove the warm-up catalytic converter as described in Section 20.

19 Undo the two bolts securing the oil return pipe flange to the base of the turbocharger (see illustration). Separate the flange joint and collect the gasket.

20 Extract the circlip securing the turbocharger wastegate actuator rod to the wastegate bellcrank and lift off the rod (see illustration).

18.16 Slacken the retaining clip and disconnect the intercooler inlet air duct from the turbocharger

18.17 Unscrew the turbocharger coolant supply pipe banjo union bolt and collect the two sealing washers

18.19 Undo the two bolts securing the oil return pipe flange to the base of the turbocharger

18.20 Extract the circlip and lift the turbocharger wastegate actuator rod off the wastegate bellcrank

18.21 Undo the two retaining nuts, lift the turbocharger wastegate actuator from its location and place it to one side

18.22a Undo the eight retaining nuts and withdraw the exhaust manifold from the cylinder head studs

18.22b Recover the inner gasket...

18.22c ...and the outer gasket

21 Undo the two nuts securing the wastegate actuator to the turbocharger. Lift the actuator from its location and place it to one side **(see illustration)**.

22 Undo the eight retaining nuts and withdraw the exhaust manifold from the cylinder head studs. Recover the inner and outer gaskets **(see illustrations)**.

18.27a Locate a new gasket on the turbocharger oil return pipe flange and bend over the tabs to retain the gasket

18.27b Refit the oil return pipe to the turbocharger and secure with the two bolts tightened to the specified torque

Refitting

23 Examine all the exhaust manifold studs for signs of damage and corrosion; remove all traces of corrosion, and repair or renew any damaged studs.

24 Ensure that the manifold and cylinder head sealing faces are clean and flat, and fit the new inner gasket.

25 Refit the manifold then fit the new outer gasket and new retaining nuts. Tighten the nuts progressively, in a diagonal sequence, to the specified torque.

26 Place the wastegate actuator on the turbocharger and engage the actuator rod with the wastegate bellcrank. Refit and securely tighten the actuator retaining nuts then secure the actuator rod with the circlip.

27 Locate a new gasket on the turbocharger oil return pipe flange and bend over the tabs to retain the gasket on the flange. Refit the oil return pipe to the turbocharger and secure with the two bolts tightened to the specified torque **(see illustrations)**.

28 Refit the warm-up catalytic converter as described in Section 20.

29 Fit new sealing washers to the turbocharger coolant supply pipe banjo union bolt, then screw the bolt into the cylinder block **(see illustration)**. Tighten the banjo union bolt to the specified torque while holding the pipe with the tool used during removal.

30 Attach the intercooler inlet air duct to the turbocharger and tighten the retaining clip securely.

31 Refit the bolt securing the turbocharger coolant return pipe to the oil filter housing and tighten the bolt securely.

32 Connect the turbocharger coolant return hose to the oil cooler inlet pipe and tighten the retaining clip securely.

33 Fit two new O-ring seals to one end of the turbocharger oil supply pipe and fit two new sealing washers to the banjo union at the other end **(see illustrations)**.

34 Engage the turbocharger oil supply pipe with the oil filter housing and screw the banjo union bolt into the turbocharger **(see illustrations)**. Tighten the banjo union bolt to the specified torque.

35 Refit and securely tighten the bolt securing the turbocharger oil supply pipe to the cylinder block **(see illustration)**.

18.29 Fit new sealing washers to the coolant supply pipe banjo union bolt, then screw the bolt into the cylinder block

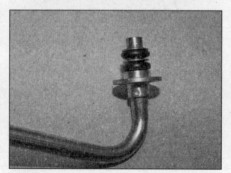

18.33a Fit two new O-ring seals to one end of the turbocharger oil supply pipe...

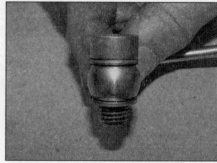

18.33b ...and fit two new sealing washers to the banjo union at the other end

18.34a Engage the turbocharger oil supply pipe with the oil filter housing...

18.34b ...and screw the banjo union bolt into the turbocharger

18.35 Refit and securely tighten the bolt securing the turbocharger oil supply pipe to the cylinder block

36 Reconnect the turbocharger wastegate solenoid wiring connector and attach the wiring harness retaining clip.

37 Refit the exhaust manifold heat shield and tighten the three retaining bolts securely.

38 Refit the crankcase ventilation hose and vacuum hose and secure with the retaining clips, then refit the air cleaner outlet air duct.

39 Lower the car to the ground, then check and if necessary, top-up the engine oil as described in *Weekly checks*.

40 Reconnect the battery negative terminal, then refill the cooling system as described in Chapter 1A, Section 25.

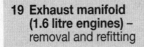

19 Exhaust manifold (1.6 litre engines) – removal and refitting

Removal

Note: *New manifold and exhaust front pipe retaining nuts, a new manifold gasket, exhaust front pipe gasket and oil dipstick guide tube O-rings must be used on refitting.*

1 Disconnect the battery negative lead as described in Chapter 5A Section 4.

2 Unbolt and remove the oil dipstick guide tube, and withdraw it from the cylinder block. Remove and discard the O-ring seals.

3 Trace the wiring back from the manifold oxygen sensor and disconnect its wiring connector. Free the wiring from the support bracket so the sensor is free to be removed with the manifold **(see illustration)**.

4 Apply the handbrake, then jack up the front of the vehicle and support it on axle stands (see *Jacking and vehicle support*).

5 Undo the three nuts securing the exhaust front pipe to the manifold **(see illustration)**. Separate the joint taking care to support the flexible section. Note that angular movement in excess of 10° can cause permanent damage to the flexible section. Recover the gasket.

6 Undo the two lower bolts securing the heat shield to the exhaust manifold.

7 Undo the two bolts securing the manifold to the lower support bracket, and the two bolts securing the support bracket to the cylinder block. Remove the bracket.

8 Undo the bolt securing the wiring harness support bracket, then remove the bracket and the exhaust manifold heat shield.

9 Slacken and remove the nine retaining nuts, and manoeuvre the manifold out of the engine compartment. Recover the gasket.

Refitting

10 Examine all the exhaust manifold studs for signs of damage and corrosion; remove all traces of corrosion, and repair or renew any damaged studs.

11 Ensure that the manifold and cylinder head sealing faces are clean and flat, and fit the new gasket.

12 Refit the manifold then fit the new retaining nuts and tighten them progressively, in a diagonal sequence, to the specified torque.

13 Align the heat shield with the manifold, then refit the wiring harness support bracket and tighten the retaining bolt securely.

14 Refit the lower support bracket to the cylinder block and manifold and tighten the retaining bolts securely.

15 Refit and tighten the two lower bolts securing the heat shield to the exhaust manifold.

16 Reconnect the exhaust front pipe, using a new gasket. Tighten the new nuts to the specified torque.

17 Reconnect the oxygen sensor wiring connector making sure the wiring is correctly routed and retained by the support bracket.

18 Fit the new O-ring seals to the oil dipstick

19.3 Disconnect the oxygen sensor wiring connector and free the connector from the support bracket

guide tube, then insert the tube in the cylinder block. Insert and tighten the retaining bolt.

19 Lower the vehicle to the ground, then reconnect the battery negative terminal.

20 Exhaust system – general information, removal and refitting

Caution: Any work on the exhaust system should only be attempted once the system is completely cool – this may take several hours, especially in the case of the forward sections, such as the manifold and catalytic converter.

General information

1 On 1.4 litre engines, the exhaust system consists of the warm-up catalytic converter, the front pipe which incorporates the main catalytic converter, and the main section which incorporates the front and rear silencers.

2 On 1.6 litre engines, the exhaust system consists of the exhaust manifold with integral catalytic converter, the front pipe which incorporates the oxygen sensor (catalytic converter control), and the main section which incorporates the front and rear silencers.

3 A flexible ('mesh') section is fitted to the front pipe to allow for engine movement.

4 The front pipe is attached to the exhaust manifold (catalytic converter) by a flange joint secured by three nuts. The main section is

19.5 Undo the three nuts securing the exhaust front pipe to the manifold

20.12a Undo the four bolts...

20.12b ...and remove the catalytic converter heat shield – 1.4 litre engines

similarly attached to the front pipe by a flange joint secured by two nuts.

5 When fitted in the factory, the exhaust system from the front pipe flange joint to the end of the system is one piece. If the rear silencer is to be renewed, it will be necessary to cut through the main section using a chain-type pipe cutter.

6 The system is suspended throughout its entire length by rubber mountings.

Removal

7 To remove a part of the system, first jack up the front or rear of the car, and support it on axle stands (see *Jacking and vehicle support*). Alternatively, position the car over an inspection pit, or on car ramps.

8 Where fitted, remove the engine undertray as described in Chapter 11, Section 21.

Manifold and catalytic converter – 1.6 litre engines

9 Refer to Section 19.

Warm-up catalytic converter – 1.4 litre engines

10 Disconnect the battery negative lead as described in Chapter 5A Section 4.

11 Undo the three bolts and remove the exhaust manifold heat shield **(see illustrations 18.10a and 18.10b)**.

12 Undo the four bolts and remove the catalytic converter heat shield **(see illustrations)**.

13 Trace the wiring back from the oxygen sensor, noting its correct routing, and disconnect its wiring connector. Lift the wiring connector out of the support bracket.

14 Undo the three nuts securing the exhaust

front pipe to the catalytic converter **(see illustration)**. Separate the joint taking care to support the flexible section. Note that angular movement in excess of 10° can cause permanent damage to the flexible section. Recover the gasket. Note that new nuts will be required for refitting.

15 Undo the two nuts securing the catalytic converter to the support bracket **(see illustration)**.

16 Undo the catalytic converter clamp bolt nut and remove the bolt. Expand the clamp and disconnect the catalytic converter from the exhaust manifold. Lower the catalytic converter and remove it from under the car. Recover the catalytic converter sealing ring. Note that a new catalytic converter clamp and a new sealing ring will be required for refitting **(see illustrations)**.

Front pipe

17 Trace the wiring back from the oxygen sensor, noting its correct routing, and disconnect its wiring connector. Free the wiring from any clips so the sensor is free to be removed with the front pipe.

18 Undo the three nuts securing the exhaust front pipe to the manifold or catalytic converter **(see illustration 20.14)**. Separate the joint taking care to support the flexible section. Note that angular movement in excess of 10° can cause permanent damage to the flexible section. Recover the gasket.

19 Undo the nuts securing the front pipe to the main section **(see illustration)**. Separate

20.14 Undo the three nuts securing the exhaust front pipe to the catalytic converter – 1.4 litre engines

20.15 Undo the nut each side (one arrowed) securing the catalytic converter to the bracket – 1.4 litre engines

20.16a Undo the catalytic converter clamp bolt nut and remove the bolt – 1.4 litre engines

20.16b Lower the catalytic converter and remove it from under the car...

20.16c ...then recover the catalytic converter sealing ring – 1.4 litre engines

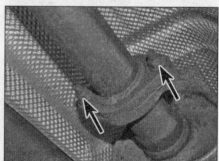

20.19 Undo the nuts securing the front pipe to the main section

the joint and recover the gasket, then remove the front pipe from under the car.

Main section

20 Undo the nuts securing the main section to the front pipe. Separate the joint and recover the gasket.

21 Unhook the main section's rubber mountings, and remove it from under the car.

Rear silencer

Note: *If a new rear silencer is to be fitted, it will be necessary to obtain a connecting sleeve from your Vauxhall/Opel parts supplier.*

22 If the original one-piece exhaust system is still fitted, it will be necessary to cut off the old rear silencer to enable fitment of the new unit. Using the new silencer as a pattern, mark the exhaust main section to determine the cut point.

23 Unhook the rubber mountings and allow the exhaust system to rest on an axle stand or similar support.

24 Using a chain-type pipe cutter, cut through the main section at the marked cut point and remove the silencer.

25 Fit the new silencer and connect it to the main section using the connecting sleeve. Tighten the connecting sleeve retaining bolt nuts securely.

Heat shields

26 The heat shields are secured to the underside of the body by special nuts. Each shield can be removed separately, but note that they may overlap, making it necessary to loosen another section first. If a shield is being removed to gain access to a component located behind it, it may prove sufficient in some cases to remove the retaining nuts and/ or bolts, and simply lower the shield, without disturbing the exhaust system. Otherwise, unhook the exhaust system rubber mountings and rest the system on the rear axle.

Refitting

27 In all cases, refitting is a reversal of removal, but note the following points:

a) *Always use new gaskets, nuts and clamps (as applicable), and coat all threads with copper grease.*

b) *If any of the exhaust mounting rubbers are in poor condition, fit new ones.*

c) *Make sure that the exhaust is suspended properly on its mountings, and will not come into contact with the floor or any suspension parts.*

d) *Tighten all nuts/bolts to the specified torque, where given.*

Chapter 4 Part B
Fuel and exhaust systems – diesel engines

Contents

Degrees of difficulty

| **Easy,** suitable for novice with little experience | | **Fairly easy,** suitable for beginner with some experience | | **Fairly difficult,** suitable for competent DIY mechanic | | **Difficult,** suitable for experienced DIY mechanic | | **Very difficult,** suitable for expert DIY or professional | |

Specifications

Fuel system data

System type:
1.6 litre engines	E98 D1P high-pressure direct injection 'common-rail' system, electronically controlled
1.7 litre engines	Denso Gen II D1 high-pressure direct injection 'common-rail' system, electronically controlled
Firing order..	1–3–4–2 (No 1 at timing belt end of engine)
Fuel system operating pressure..........................	1600 bar (approx)
Idle speed...	Controlled by ECU
Maximum speed..	Controlled by ECU

Fuel supply pump:
Type	Electric, mounted in fuel tank
Delivery pressure	3.3 bar (maximum)

Torque wrench settings

	Nm	lbf ft
1.6 litre engines		
Camshaft sensor retaining bolt	10	7
Crankshaft speed/position sensor retaining bolt	10	7
Diesel particulate filter heat shield bolts.	10	7
Diesel particulate filter retaining nuts	22	16
Diesel particulate filter-to-turbocharger clamp bolt nut	25	18
Exhaust front pipe-to-diesel particulate filter nuts*	22	16
Exhaust front pipe-to-main section nuts	17	13
Exhaust manifold bolts*		
Stage 1	27	20
Stage 2	27	20
Stage 3	27	20
Fuel injector clamp bracket bolts:		
Stage 1	10	7
Stage 2	Angle-tighten a further 60°	
Fuel rail retaining bolts	25	18
High-pressure fuel pipe unions	30	22
High-pressure fuel pump mounting bolts.	25	18
Inlet manifold bolts.	12	9
Inlet manifold brace bolts.	10	7
Inlet manifold EGR pipe bolts.	10	7
Throttle housing retaining bolts	10	7
Turbocharger coolant feed pipe banjo union bolt	25	18
Turbocharger oil return pipe bolts:		
Upper bolt.	10	7
Lower bolt.	25	18
Turbocharger oil supply pipe banjo union bolts	25	18
Turbocharger-to-exhaust manifold.	30	22
Turbocharger upper heat shield bolts/nut	10	7
1.7 litre engines		
Camshaft sensor retaining bolt	10	7
Catalytic converter heat shield bolts	10	7
Catalytic converter-to-support bracket bolts.	25	18
Catalytic converter-to-turbocharger nuts*:		
Stage 1	25	18
Stage 2	48	35
Crankshaft speed/position sensor retaining bolt	10	7
Diesel particulate filter-to-exhaust main section	25	18
Exhaust front pipe-to-catalytic converter nuts.	25	18
Exhaust front pipe-to-diesel particulate filter.	21	16
Exhaust manifold nuts/bolts*	70	52
Exhaust temperature sensor	45	33
Fuel injector clamp bracket bolt:		
Stage 1	40	30
Stage 2	Slacken bolt	
Stage 3	32	24
Fuel rail retaining bolts	25	18
Fuel return pipe pipe banjo union bolt	20	15
High-pressure fuel pipe unions	25	18
High-pressure fuel pump mounting nuts	18	13
Inlet manifold nuts/bolts.	25	18
Inlet manifold support bracket bolts	25	18
Intercooler inlet air duct-to-turbocharger.	10	7
Throttle housing retaining bolts	10	7
Turbocharger oil supply pipe banjo union bolts	21	16
Turbocharger-to-exhaust manifold*:		
Stage 1	25	18
Stage 2	48	35

*Use new fasteners

1 General information and precautions

General information

1 The engines are equipped with a high-pressure direct injection system which incorporates the very latest in diesel injection technology. On this system, a high-pressure fuel pump is used purely to provide the pressure required for the injection system and has no control over the injection timing (unlike conventional diesel injection systems). The injection timing is controlled by the electronic control unit (ECU) via the electrically-operated injectors. The system operates as follows.

2 The fuel system consists of a fuel tank (which is mounted under the rear of the car, with an electric fuel supply pump immersed in it), a fuel filter with integral water separator, a high-pressure fuel pump, injectors and associated components.

3 Fuel is supplied to the fuel filter housing which is attached to the side of the fuel tank. The fuel filter removes all foreign matter and water and ensures that the fuel supplied to the pump is clean.

4 The fuel is heated to ensure no problems occur when the ambient temperature is very low. This is achieved by an electrically-operated fuel heater incorporated in the filter housing, the heater is controlled by the ECU.

5 The high-pressure fuel pump is driven at half-crankshaft speed by the timing chain (1.6 litre engines) or timing belt (1.7 litre engines). The high pressure required in the system (up to 1600 bar) is produced by the three pistons in the pump. The high-pressure pump supplies high pressure fuel to the fuel rail, which acts as a reservoir for the four injectors.

6 The electrical control system consists of the ECU, along with the following sensors:

a) *Accelerator pedal position sensor – informs the ECU of the accelerator pedal position, and the rate of throttle opening/closing.*

b) *Coolant temperature sensor – informs the ECU of engine coolant temperature.*

c) *Mass airflow sensor – informs the ECU of the amount of air passing through the inlet duct.*

d) *Crankshaft speed/position sensor – informs the ECU of the crankshaft position and speed of rotation.*

e) *Camshaft position sensor – informs the ECU of the positions of the pistons.*

f) *Inlet air pressure and temperature sensor (1.6 litre engines) – informs the ECU of the pressure and temperature of the air in the inlet manifold.*

g) *Charge (boost) pressure sensor (1.7 litre engines) – informs ECU of the pressure in the inlet manifold.*

h) *Fuel pressure sensor – informs the ECU of the fuel pressure present in the fuel rail.*

i) *ABS control unit – informs the ECU of the vehicle speed.*

7 All the above signals are analysed by the ECU which selects the fuelling response appropriate to those values. The ECU controls the fuel injectors (varying the pulse width – the length of time the injectors are held open – to provide a richer or weaker mixture, as appropriate). The mixture is constantly varied by the ECU, to provide the best setting for cranking, starting (with either a hot or cold engine), warm-up, idle, cruising and acceleration.

8 The ECU also has full control over the fuel pressure present in the fuel rail via the high-pressure fuel regulator and third piston deactivator solenoid valve which are fitted to the high-pressure pump. To reduce the pressure, the ECU opens the high-pressure fuel regulator which allows the excess fuel to return direct to the tank from the pump. The third piston deactivator is used mainly to reduce the load on the engine, but can also be used to lower the fuel pressure. The deactivator solenoid valve relieves the fuel pressure from the third piston of the pump which results in only two of the pistons pressurising the fuel system.

9 The ECU also controls the exhaust gas recirculation (EGR) system, described in detail in Part C of this Chapter, the pre/post heating system (see Chapter 5A), and the engine cooling fan.

10 On certain 1.7 litre engines, the inlet manifold is fitted with a butterfly valve arrangement to improve efficiency at low engine speeds. Each cylinder has two intake tracts in the manifold, one of which is fitted with a valve; the operation of the valve is controlled by the ECU via a vacuum solenoid. At low engine speeds (below approximately 1500 rpm) the valves remain closed, meaning that air entering each cylinder is passing through only one of the two manifold tracts. At higher engine speeds, the ECU opens up each of the four valves allowing the air passing through the manifold to pass through both inlet tracts.

11 A turbocharger is fitted to increases engine efficiency. It does this by raising the pressure in the inlet manifold above atmospheric pressure. Instead of the air simply being sucked into the cylinders, it is forced in.

12 Between the turbocharger and the inlet manifold, the compressed air passes through an intercooler. This is an air-to-air heat exchanger, mounted in front of the radiator, and supplied with cooling air from the front of the vehicle. The purpose of the intercooler is to remove some of the heat gained in being compressed from the inlet air. Because cooler air is denser, removal of this heat further increases engine efficiency.

13 If certain sensors fail, and send abnormal signals to the ECU, the ECU has a back-up programme. In this event, the abnormal signals are ignored, and a pre-programmed value is substituted for the sensor signal, allowing the engine to continue running, albeit at reduced efficiency. If the ECU enters its back-up mode, a warning light on the instrument panel will illuminate, and a fault code will be stored in the ECU memory. This fault code can be read using suitable

1.13 The vehicle diagnostic socket is located at the base of the facia on the driver's side

specialist test equipment plugged into the system's diagnostic socket. The diagnostic socket is located at the base of the facia, on the driver's side **(see illustration)**.

Precautions

⚠ *Warning: It is necessary to take certain precautions when working on the fuel system components, particularly the high-pressure side of the system. Before carrying out any operations on the fuel system, refer to the precautions given in 'Safety first!', and to any additional warning notes at the start of the relevant Sections. Also refer to the additional information contained in Section 2.*

Caution: Do not operate the engine if any of air intake ducts are disconnected or the filter element is removed. Any debris entering the engine will cause severe damage to the turbocharger.

Caution: To prevent damage to the turbocharger, do not race the engine immediately after start-up, especially if it is cold. Allow it to idle smoothly to give the oil a few seconds to circulate around the turbocharger bearings. Always allow the engine to return to idle speed before switching it off – do not blip the throttle and switch off, as this will leave the turbo spinning without lubrication.

Caution: Observe the recommended intervals for oil and filter changing, and use a reputable oil of the specified quality. Neglect of oil changing, or use of inferior oil, can cause carbon formation on the turbo shaft, leading to subsequent failure.

2 High pressure diesel injection system – special information

Warnings and precautions

1 It is essential to observe strict precautions when working on the fuel system components, particularly the high pressure side of the system. Before carrying out any operations on the fuel system, refer to the precautions given in *Safety first!*, and to the following additional information.

2.4 Typical plastic plug and cap set for sealing disconnected fuel pipes and components

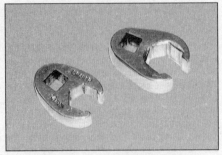

2.6 Two crow-foot adaptors will be necessary for tightening the fuel pipe unions

● *Do not carry out any repair work on the high pressure fuel system unless you are competent to do so, have all the necessary tools and equipment required, and are aware of the safety implications involved.*

● *Before starting any repair work on the fuel system, wait at least 30 seconds after switching off the engine to allow the fuel circuit to return to atmospheric pressure.*

● *Never work on the high pressure fuel system with the engine running.*

● *Keep well clear of any possible source of fuel leakage, particularly when starting the engine after carrying out repair work. A leak in the system could cause an extremely high pressure jet of fuel to escape, which could result in severe personal injury.*

● *Never place your hands or any part of your body near to a leak in the high pressure fuel system.*

● *Do not use steam cleaning equipment or compressed air to clean the engine or any of the fuel system components.*

Repair procedures and general information

2 Strict cleanliness must be observed at all times when working on any part of the fuel system. This applies to the working area in general, the person doing the work, and the components being worked on.

3 Before working on the fuel system components, they must be thoroughly cleaned with a suitable degreasing fluid. Cleanliness is particularly important when working on the fuel system connections at the following components:

a) *Fuel filter.*
b) *High-pressure fuel pump.*
c) *Fuel rail.*
d) *Fuel injectors.*
e) *High pressure fuel pipes.*

4 After disconnecting any fuel pipes or components, the open union or orifice must be immediately sealed to prevent the entry of dirt or foreign material. Plastic plugs and caps in various sizes are available in packs from motor factors and accessory outlets, and are particularly suitable for this application **(see illustration)**. Fingers cut from disposable rubber gloves should be used to protect components such as fuel pipes, fuel injectors and wiring connectors, and can be secured in place using elastic bands. Suitable gloves of this type are available at no cost from most petrol station forecourts.

5 Whenever any of the high pressure fuel pipes are disconnected or removed, a new pipe(s) must be obtained for refitting.

6 The torque wrench settings given in the Specifications must be strictly observed when tightening component mountings and connections. This is particularly important when tightening the high pressure fuel pipe unions. To enable a torque wrench to be used on the fuel pipe unions, two crow-foot adaptors are required. Suitable types are available from motor factors and accessory outlets **(see illustration)**.

3 Air cleaner assembly and air ducts – removal and refitting

Removal

Air cleaner assembly

1 Remove the air cleaner inlet air duct by pulling upward to release the duct from the rubber grommets and the air cleaner assembly **(see illustrations)**.

2 On 1.6 litre engines, release the clip securing the mass airflow sensor wiring harness to the rear of the air cleaner **(see illustration)**.

3 Disconnect the wiring connector at the mass airflow sensor **(see illustration)**.

4 Slacken the retaining clip securing the outlet air duct to the turbocharger or turbocharger inlet air duct **(see illustration)**.

5 Lift up the air cleaner housing to disengage it from the mounting rubbers, then lift the air

3.1a Pull the inlet air duct upward to release it from the grommets...

3.1b ...then disengage the duct from the air cleaner elbow and remove it

3.2 Release the wiring harness retaining clip from the rear of the air cleaner – 1.6 litre engines

3.3 Disconnect the mass airflow sensor wiring connector

3.4 Slacken the retaining clip and detach the outlet air duct from the turbocharger inlet air duct – 1.7 litre model shown

3.5a Lift up the air cleaner housing...

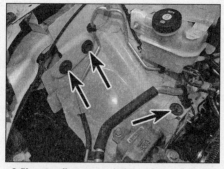

3.5b ...to disengage it from the mounting rubbers

3.9a Remove the engine oil filler cap...

3.9b ...then lift off the plastic cover over the top of the engine – 1.7 litre engines

3.10 Squeeze together the retaining clip legs and disconnect the crankcase ventilation hose – 1.7 litre engines

3.12 Undo the bolt securing the inlet air duct to the camshaft cover – 1.7 litre engines

cleaner housing, complete with air duct, from the engine compartment **(see illustrations)**.

Outlet air duct – 1.6 litre engines

6 Remove the air cleaner inlet air duct by pulling upward to release the duct from the rubber grommets and the air cleaner assembly **(see illustrations 3.1a and 3.1b)**.

7 Slacken the retaining clip securing the outlet air duct to the mass airflow sensor and detach the duct.

8 Slacken the retaining clip securing the outlet air duct to the turbocharger, then detach the duct and remove it from the engine. Suitably cover the turbocharger air inlet to prevent the possible entry of foreign material.

Turbocharger inlet air duct – 1.7 litre engines

9 Remove the engine oil filler cap, then lift off the plastic cover over the top of the engine **(see illustrations)**. Refit the oil filler cap.

10 Squeeze together the legs of the retaining clip and disconnect the crankcase ventilation hose from the inlet air duct **(see illustration)**.

11 Slacken the retaining clip securing the turbocharger inlet air duct to the air cleaner outlet air duct and detach the duct **(see illustration 3.4)**.

12 Undo the bolt securing the inlet air duct to the camshaft cover **(see illustration)**.

13 Disconnect the vacuum hose from the EGR vacuum valve, release it from the retaining clip and move the hose to one side **(see illustration)**.

14 Slacken the clip securing the lower end

of the inlet air duct to the turbocharger. Pull the duct off the turbocharger, disengage the support clip and withdraw the duct up and off the engine **(see illustrations)**.

3.13 Disconnect the vacuum hose from the EGR vacuum valve – 1.7 litre engines

3.14b Pull the duct off the turbocharger...

Inlet air duct

15 Remove the air cleaner inlet air duct by pulling upward to release the duct from

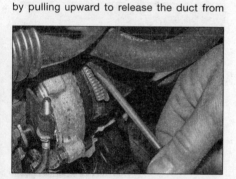

3.14a Slacken the clip securing the lower end of the inlet air duct to the turbocharger – 1.7 litre engines

3.14c ...and withdraw the duct up and off the engine – 1.7 litre engines

3.19 Undo the two bolts and separate the air duct flange from the turbocharger

3.20 Undo the bolt securing the intercooler inlet air duct to the subframe – 1.6 litre engines

3.21 Extract the wire spring clip and disconnect the inlet air duct from the intercooler

the rubber grommets and the air cleaner assembly **(see illustrations 3.1a and 3.1b)**.

Intercooler inlet air duct

16 Apply the handbrake, then jack up the front of the vehicle and support it on axle stands (see *Jacking and vehicle support*).

17 Remove the engine undertray as described in Chapter 11, Section 21.

18 Remove the front bumper as described in Chapter 11, Section 6.

19 On 1.6 litre engines, extract the wire spring clip using a small screwdriver and disconnect the intercooler inlet air duct from the turbocharger. On 1.7 litre engines, undo the two bolts and separate the inlet air duct flange from the turbocharger **(see illustration)**. Collect the gasket.

20 On 1.6 litre engines, undo the bolt

securing the air duct to the front subframe **(see illustration)**.

21 Extract the wire spring clip using a small screwdriver and disconnect the inlet air duct from the intercooler, then manipulate the air duct out from the engine compartment **(see illustration)**.

Intercooler outlet air duct – 1.6 litre engines

22 Remove the battery and battery tray as described in Chapter 5A, Section 4.

23 Apply the handbrake, then jack up the front of the vehicle and support it on axle stands (see *Jacking and vehicle support*).

24 Remove the engine undertray as described in Chapter 11, Section 21.

25 Remove the front bumper as described in Chapter 11, Section 6.

26 Undo the two retaining bolts and move the coolant expansion tank to one side.

27 Remove the plastic cover over the top of the engine.

28 Undo the two bolts securing the air duct to the front subframe **(see illustration)**.

29 Extract the wire spring clip using a small screwdriver and disconnect the outlet air duct from the intercooler **(see illustration)**.

30 Undo the bolt securing the upper section of the outlet air duct to the EGR cooler.

31 Rotate the locking collar and disconnect the outlet air duct from the throttle housing, then manipulate the air duct out from the engine compartment **(see illustration)**.

Intercooler outlet air duct – 1.7 litre engines

32 Spread the sides of the wire retaining clip and disconnect the intercooler outlet air duct from the throttle housing **(see illustration)**.

33 Apply the handbrake, then jack up the front of the vehicle and support it on axle stands (see *Jacking and vehicle support*).

34 Remove the engine undertray as described in Chapter 11, Section 21.

35 Remove the front bumper as described in Chapter 11, Section 6.

36 Undo the two bolts securing the air duct to the front subframe.

37 Extract the wire spring clip using a small screwdriver and disconnect the outlet air duct from the intercooler **(see illustration 3.29)**. Manipulate the duct up and out of the engine compartment.

Refitting

38 Refitting is the reverse of removal, bearing in mind the following points:

a) *Make sure all the air ducts/hoses are securely reconnected.*

b) *Renew all disturbed O-ring seals and gaskets.*

c) *Where applicable, refit the front bumper as described in Chapter 11, Section 6.*

d) *Where applicable, refit the engine undertray as described in Chapter 11, Section 21.*

e) *Where applicable, refit the battery tray and battery as described in Chapter 5A, Section 4.*

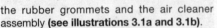

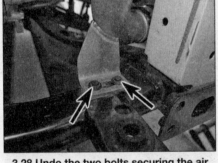

3.28 Undo the two bolts securing the air duct to the subframe – 1.6 litre engines

3.29 Extract the wire spring clip and disconnect the outlet air duct from the intercooler – 1.6 litre engines

3.31 Rotate the locking collar and disconnect the outlet air duct from the throttle housing – 1.6 litre engines

3.32 Spread the sides of the wire clip and disconnect the intercooler outlet air duct – 1.7 litre engines

4 Accelerator pedal/ position sensor – removal and refitting

1 Refer to Chapter 4A, Section 3.

5 Fuel system – priming and bleeding

1 After disconnecting part of the fuel supply system or running out of fuel, it is necessary to prime the fuel system and bleed off any air which may have entered the system components, as follows.

2 Prime the system by switching on the ignition three times for approximately 15 seconds each time. Operate the starter for a maximum of 30 seconds. If the engine does not start within this time, wait 5 seconds and repeat the procedure.

3 When the engine starts, run it at a fast idle speed for a minute or so to purge any trapped air from the fuel lines. After this time the engine should idle smoothly at a constant speed.

4 If the engine idles roughly, then there is still some air trapped in the fuel system. Increase the engine speed again for another minute or so then allow it to idle. Repeat this procedure as necessary until the engine is idling smoothly.

6 Fuel pump/ fuel gauge sender unit – removal and refitting

Note: *Refer to the warnings and precautions in Section 2 before proceeding.*

1 Removal and refitting of the combined fuel pump and fuel gauge sender unit is essentially the same as described in Chapter 4A, Section 6, for petrol engines, bearing in mind the following points:

a) *Ignore all references to evaporative emission control system components.*

b) *When removing the unit from the fuel tank, raise it sufficiently to gain access to the fuel pump/gauge sender unit fuel pickup and disconnect the fuel tank sender from the fuel pickup. Carefully lift out the fuel tank sender unit from the tank. Take care that the sender unit float and arm are not damaged as the unit is removed.*

c) *On completion, prime the fuel system as described in Section 5.*

7 Fuel tank – removal and refitting

Note: *Refer to the warnings and precautions in Section 2 before proceeding.*

Removal

1 Before removing the fuel tank, all fuel must be drained from the tank. Since a fuel

tank drain plug is not provided, it is therefore preferable to carry out the removal operation when the tank is nearly empty. The remaining fuel can then be syphoned or hand-pumped from the tank.

2 Disconnect the battery negative lead as described in Chapter 5A Section 4.

3 Chock the front wheels, then jack up the rear of the vehicle, and support it securely on axle stands (see *Jacking and vehicle support*), Section. Remove the right-hand rear roadwheel.

4 Remove the rear wheel arch liner on the right-hand side as described in Chapter 11, Section 21.

5 Remove the fuel tank filler cap to release any residual pressure in the tank.

6 Remove the exhaust system main section and rear silencer as described in Section 26.

7 On 4WD models, remove the propeller shaft as described in Chapter 8, Section 6.

8 Undo the retaining nut and release the handbrake cable from the fuel tank guard on the right-hand side **(see illustration)**.

9 Undo the three nuts and remove the fuel tank guard on the right-hand side **(see illustrations)**.

10 Disconnect the fuel supply and return line at the underbody quick-release connectors on the right-hand side of the tank **(see illustration)**. Be prepared for some loss of fuel. A Vauxhall/Opel special tool is available to release the fuel line connectors, but provided care is taken, the connectors can be released using a pair of long-nosed pliers, or a similar tool, to depress the retaining tangs.

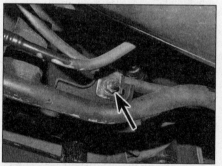

7.8 Undo the retaining nut and release the handbrake cable from the fuel tank guard

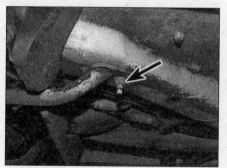

7.9a Undo the front retaining nut...

7.9b ...centre retaining nut...

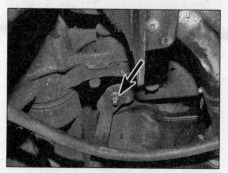

7.9c ...and rear retaining nut...

7.9d ...then remove the fuel tank guard on the right-hand side

7.10 Fuel supply and return line quick-release connectors

7.11 Slacken the clip and disconnect the fuel filler hose from the fuel tank

7.12 Disconnect the filler vent pipe at the quick-release connector

7.14a Undo the bolt securing each fuel tank retaining strap to the underbody...

7.14b ...then disengage the straps from the underbody at the rear

Suitably plug the disconnected fuel and vent hoses to prevent entry of dust and dirt.

11 Place a suitable container under the tank, then slacken the clip and disconnect the fuel filler hose from the fuel tank **(see illustration)**. Collect the escaping fuel in the container.

12 Disconnect the fuel tank filler vent pipe at the quick-release connector **(see illustration)**.

13 Support the weight of the fuel tank on a jack with interposed block of wood.

14 Undo the bolts securing the front of each fuel tank retaining strap to the underbody. Disengage the straps from the underbody at the rear and remove the straps **(see illustrations)**.

15 Taking care not to damage the fuel filter, slowly lower the tank until sufficient clearance exists to enable the fuel pump/fuel gauge sender unit wiring connector to be disconnected **(see illustration)**.

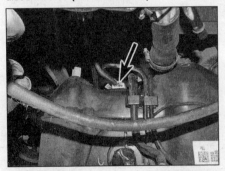

7.15 Lower the fuel tank slightly, then disconnect the fuel pump/fuel gauge sender unit wiring connector

16 Continue to lower the tank until it can be removed from under the vehicle.

17 If necessary, remove the fuel filter, fuel lines and hoses and heat shield from the tank for transfer to the new tank.

18 If the tank contains sediment or water, it may cleaned out with two or three rinses of clean fuel. Remove the fuel pump/fuel gauge sender unit as described in Section 6. Shake the tank vigorously, and change the fuel as necessary to remove all contamination from the tank.

Caution: This procedure should be carried out in a well-ventilated area, and it is vital to take adequate fire precautions.

19 Any repairs to the fuel tank should be carried out by a professional. Do not under any circumstances attempt any form of DIY repair to a fuel tank.

Refitting

20 Refitting is the reverse of the removal procedure, noting the following points:

a) *When lifting the tank back into position, take care to ensure that none of the hoses become trapped between the tank and vehicle body. Refit the retaining straps and tighten the bolts securely.*

b) *Ensure all pipes and hoses are correctly routed and all hose unions are securely joined.*

c) *On 4WD models, refit the propeller shaft as described in Chapter 8, Section 6.*

d) *Refit the exhaust system main section and rear silencer as described in Section 26.*

e) *Refit the rear wheel arch liner as described in Chapter 11, Section 21.*

f) *On completion, prime the fuel system as described in Section 5.*

8 Injection system electrical components – testing

1 If a fault appears in the engine management system, first ensure that all the system wiring connectors are securely connected and free of corrosion. Ensure that the fault is not due to poor maintenance; ie, check that the air cleaner filter element is clean, the cylinder compression pressures are correct and that the engine breather hoses are clear and undamaged, referring to Chapters 1B, 2C, 2D, and 4C for further information.

2 If these checks fail to reveal the cause of the problem, the vehicle should be taken to a suitably-equipped Vauxhall/Opel dealer or engine management diagnostic specialist for testing. A diagnostic socket is located at the base of the facia, on the driver's side, to which a fault code reader or other suitable test equipment can be connected **(see illustration 1.13)**. By using the code reader or test equipment, the engine management ECU can be interrogated, and any stored fault codes can be retrieved. Live data can also be captured from the various system sensors and actuators, indicating their operating parameters. This will allow the fault to be quickly and simply traced, alleviating the need to test all the system components individually, which is a time-consuming operation that carries a risk of damaging the ECU.

9 Injection system electrical components (1.6 litre engines) – removal and refitting

Mass airflow sensor

1 Disconnect the mass airflow sensor wiring connector **(see illustration 3.3)**.

2 Undo the two retaining screws and remove the mass airflow sensor from the air cleaner lid.

3 Check the condition of the airflow sensor seal and renew if necessary.

4 Refitting is a reversal of removal.

Throttle housing

5 Lift off the plastic cover over the top of the engine.

6 Remove the battery and battery tray as described in Chapter 5A, Section 4.

7 Remove the front bumper upper cover by prising out the centre pins and extracting the eight expanding plastic rivets **(see illustrations)**.

8 Undo the two retaining bolts and move the coolant expansion tank to one side **(see illustration)**.

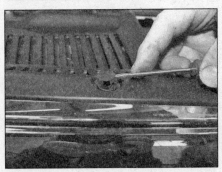

9.7a Prise out the centre pin and extract the eight expanding plastic rivets...

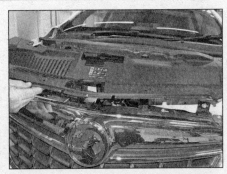

9.7b ...and remove the bumper upper cover

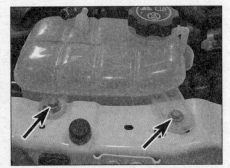

9.8 Undo the two bolts and move the expansion tank to one side

9.11a Undo the four retaining bolts...

9.11b ...and withdraw the throttle housing from the manifold

9.12 Collect the throttle housing O-ring seal

9 Undo the bolt securing the upper section of the intercooler outlet air duct to the EGR cooler.

10 Rotate the locking collar and disconnect the intercooler outlet air duct from the throttle housing.

11 Undo the four retaining bolts and withdraw the throttle housing from the inlet manifold **(see illustrations)**. Twist the unit to gain access to the wiring connector and disconnect the connector.

12 Collect the O-ring seal from the manifold **(see illustration)**.

13 Refitting is a reversal of removal, but thoroughly clean the mating faces of the throttle housing and inlet manifold. Use a new O-ring seal and tighten the retaining bolts to the specified torque. On completion, refit the battery tray and battery as described in Chapter 5A, Section 4.

Crankshaft speed/ position sensor

14 Remove the intercooler outlet air duct as described in Section 3.

15 Remove the diesel particulate filter as described in Section 26.

16 Working at the EGR cooler housing at the left-hand end of the engine, use a forked tool to release the oxygen sensor and exhaust temperature sensor wiring harness retaining clips from the exhaust pressure pipe support bracket.

17 Undo the retaining bolt and remove the exhaust pressure pipe support bracket from the EGR cooler housing.

18 Disconnect the crankshaft speed/position sensor wiring connector.

19 Firmly apply the handbrake, then jack up the front of the car and support it securely on axle stands (see *Jacking and vehicle support*).

20 From under the front of the car, reach up and undo the crankshaft speed/position sensor retaining bolt then slide out the forked retaining plate **(see illustration)**.

21 Working from above, withdraw the crankshaft speed/position sensor upwards from its location and remove it from the engine **(see illustration)**.

22 Refitting is a reversal of removal, bearing in mind the following points:

a) Refit the diesel particulate filter as described in Section 26.

b) Refit the intercooler outlet air duct as described in Section 3.

Camshaft position sensor

23 The camshaft position sensor is located on the rear facing side of the cylinder head, at the right-hand end **(see illustration)**.

24 Lift off the plastic cover over the top of the engine.

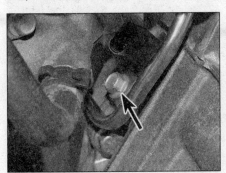

9.20 Undo the crankshaft speed/position sensor retaining bolt

9.21 Withdraw the crankshaft speed/ position sensor upwards from its location

9.23 Camshaft position sensor location

9.29 Inlet air pressure and temperature sensor location

25 Remove the air cleaner assembly as described in Section 3.
26 Disconnect the sensor wiring connector, then undo the retaining bolt and remove the sensor from the cylinder head.
27 Refitting is a reversal of removal, tightening the retaining bolt securely. Refit the air cleaner assembly as described in Section 3.

Coolant temperature sensor

28 Refer to the procedures contained in Chapter 3, Section 7.

Inlet air pressure and temperature sensor

29 The inlet air pressure and temperature sensor is located on the upper face of the inlet manifold, near the centre **(see illustration)**.
30 Lift off the plastic cover over the top of the engine.
31 Using a small screwdriver, lift up the

9.40 Turbocharger wastegate solenoid location

10.7 Disconnect the throttle housing wiring connector

9.37 Release the retaining tab and lift the ECU and mounting bracket off the battery tray

locking catch and disconnect the sensor wiring connector.
32 Undo the retaining bolt and withdraw the sensor from the manifold.
33 Refitting is a reversal of removal.

Electronic control unit (ECU)

Note: *If a new ECU is to be fitted, this work must be entrusted to a Vauxhall/Opel dealer or suitably-equipped specialist as it is necessary to program the new ECU after installation. This work requires the use of dedicated Vauxhall/Opel diagnostic equipment or a compatible alternative.*

34 The ECU is located on the left-hand side of the engine compartment, attached to the side of the battery tray.
35 Remove the battery as described in Chapter 5A, Section 4.
36 Lift up the locking bars and disconnect the ECU wiring connectors.
37 Release the retaining tab and carefully lift the ECU mounting bracket off the battery tray **(see illustration)**.
38 If required, undo the four retaining nuts and remove the ECU from the mounting bracket.
39 Refitting is a reversal of removal, referring to Chapter 5A, Section 4 when refitting the battery.

Turbocharger wastegate solenoid

40 The wastegate (charge pressure) solenoid valve is located at the rear left-hand side of the engine above the starter motor **(see illustration)**.
41 Remove the battery as described in Chapter 5A, Section 4.

10.8 Undo the four retaining bolts and remove the throttle housing from the manifold

42 Disconnect the wiring connector and the two vacuum hoses from the valve.
43 Undo the two retaining bolts and remove the valve from its location.
44 Refitting is a reversal of removal, referring to Chapter 5A, Section 4 when refitting the battery.

10 Injection system electrical components (1.7 litre engines) – removal and refitting

Mass airflow sensor

1 Disconnect the airflow sensor wiring connector **(see illustration 3.3)**.
2 Undo the two retaining screws and remove the mass airflow sensor from the air cleaner lid.
3 Check the condition of the airflow sensor seal and renew if necessary.
4 Refitting is a reversal of removal.

Throttle housing

5 Remove the engine oil filler cap, then lift off the plastic cover over the top of the engine **(see illustrations 3.9a and 3.9b)**. Refit the oil filler cap.
6 Spread the sides of the wire retaining clip and disconnect the intercooler outlet duct from the throttle housing **(see illustration 3.33)**.
7 Disconnect the throttle housing wiring connector **(see illustration)**.
8 Undo the four bolts securing the throttle housing to the inlet manifold and remove the housing from the engine **(see illustration)**. Recover the gasket.
9 Refitting is a reversal of removal, but thoroughly clean the mating faces and use a new gasket. Tighten the retaining bolts to the specified torque.

Crankshaft speed/ position sensor

10 The sensor is located at the rear of the cylinder block, below the starter motor. To gain access, firmly apply the handbrake, then jack up the front of the car and support it securely on axle stands (see *Jacking and vehicle support*).
11 Remove the engine undertray as described in Chapter 11, Section 21.
12 On 4WD models, remove the transfer gearbox as described in Chapter 7C, Section 3.
13 Wipe clean the area around the crankshaft sensor then disconnect the wiring connector.
14 Slacken and remove the retaining bolt and remove the sensor from the cylinder block **(see illustration)**. Recover the sealing ring.
15 Refitting is the reverse of removal, using a new sealing ring. Tighten the sensor retaining bolt to the specified torque. On 4WD models, refit the transfer gearbox as described in Chapter 7C, Section 3.

Camshaft position sensor

16 Remove the engine oil filler cap, then lift off the plastic cover over the top of the engine

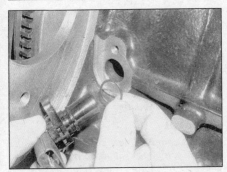

10.14 Remove the crankshaft speed/position sensor and recover the sealing ring

10.18 Disconnect the camshaft position sensor wiring connector

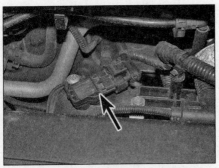

10.22 Charge (boost) pressure sensor location

(see illustrations 3.9a and 3.9b). Refit the oil filler cap.

17 The camshaft position sensor is located on the front right-hand side of the camshaft housing, adjacent to the exhaust camshaft sprocket.

18 Disconnect the camshaft position sensor wiring connector, then undo the sensor retaining bolt and remove the sensor from the mounting bracket **(see illustration)**.

19 Refitting is a reversal of removal, tightening the retaining bolt securely.

Coolant temperature sensor

20 Refer to the procedures contained in Chapter 3, Section 7.

Charge (boost) pressure sensor

21 Remove the engine oil filler cap, then lift off the plastic cover over the top of the engine **(see illustrations 3.9a and 3.9b)**. Refit the oil filler cap.

22 The charge (boost) pressure sensor is located on the upper right-hand side of the inlet manifold **(see illustration)**.

23 Using a small screwdriver, lift up the locking catch and disconnect the sensor wiring connector.

24 Undo the retaining bolt and withdraw the sensor from the manifold. Recover the sealing O-ring.

25 Renew the sealing O-ring, then refit the sensor and tighten the retaining bolts securely. Reconnect the wiring connector and secure with the locking catch.

26 On completion, refit the engine cover.

Electronic control unit (ECU)

Note: *If a new ECU is to be fitted, this work must be entrusted to a Vauxhall/Opel dealer or suitably-equipped specialist as it is necessary to program the new ECU after installation. This work requires the use of dedicated Vauxhall/Opel diagnostic equipment or a compatible alternative.*

27 The ECU is located on the left-hand side of the engine compartment, attached to the side of the battery tray.

28 Remove the battery as described in Chapter 5A, Section 4.

29 Lift up the locking bars and disconnect the ECU wiring connectors.

30 Release the retaining tab and carefully lift the ECU mounting bracket off the battery tray **(see illustration 9.37)**.

31 If required, undo the four retaining nuts and remove the ECU from the mounting bracket.

32 Refitting is a reversal of removal, referring to Chapter 5A, Section 4 when refitting the battery.

Turbocharger wastegate solenoid

33 The wastegate (charge pressure) solenoid valve is located at the rear of the engine adjacent to the high-pressure fuel pump..

34 Firmly apply the handbrake, then jack up the front of the car and support it securely on axle stands (see *Jacking and vehicle support*).

35 Remove the engine undertray as described in Chapter 11, Section 21.

36 Disconnect the wiring connector and the two vacuum hoses from the valve then undo the two retaining bolts and remove the valve from its location.

37 Refitting is the reverse of removal.

11 High-pressure fuel pump (1.6 litre engines) – removal and refitting

> ⚠ **Warning: Refer to the information contained in Section 2 before proceeding.**

Removal

Note: *A new fuel pump-to-fuel rail high-pressure fuel pipe will be required for refitting.*

1 Remove the high-pressure fuel pump sprocket as described in Chapter 2C, Section 8.

2 Remove the inlet manifold as described in Section 17.

3 Disconnect the following attachments at the high-pressure fuel pump **(see illustration)**:
a) Fuel supply pipe.
b) Fuel return pipe.
c) Fuel injector return hose.
d) Wiring connector.

4 Release the fuel pipes and hoses from any support clips and move them to one side. Plug or cover the pump unions, hoses and pipes to prevent dirt entry.

5 Unscrew the two mounting bolts and withdraw the pump from its location. Recover the sealing O-ring.

Caution: *The high-pressure fuel pump is manufactured to extremely close tolerances and must not be dismantled in any way. No parts for the pump are available separately and if the unit is in any way suspect, it must be renewed.*

Refitting

6 Thoroughly clean the fuel pump and engine mating faces.

7 Locate a new sealing O-ring on the pump flange and lubricate the O-ring with clean diesel fuel.

8 Place the pump in position, refit the two mounting bolts and tighten them to the specified torque.

9 Reconnect the pipe, hose and wiring connections removed from the rear of the pump in paragraph three.

10 Refit the inlet manifold as described in Section 17.

11 Remove the high-pressure fuel pump sprocket as described in Chapter 2C, Section 8.

12 High-pressure fuel pump (1.7 litre engines) – removal and refitting

> ⚠ **Warning: Refer to the information contained in Section 2 before proceeding.**

Removal

Note: *A new fuel pump-to-fuel rail high-pressure fuel pipe and new fuel return pipe banjo union sealing washers will be required for refitting.*

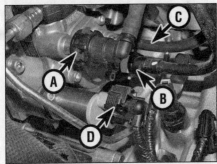

11.3 Wiring, hose and pipe attachments at the high-pressure fuel pump

13.3 Lift the retaining catch and disconnect the fuel pressure sensor wiring connector

1 Disconnect the battery negative lead as described in Chapter 5A Section 4.

2 Remove the engine oil filler cap, then lift off the plastic cover over the top of the engine (see illustrations 3.9a and 3.9b). Refit the oil filler cap.

3 Remove the timing belt as described in Chapter 2D, Section 7.

4 Remove the high-pressure fuel pump sprocket as described in Chapter 2D, Section 8.

5 Disconnect the following attachments at the high-pressure fuel pump:

a) *Fuel supply pipe.*
b) *Fuel return pipe.*
c) *Fuel injector return hose.*
d) *Wiring connector.*

6 Release the fuel pipes and hoses from any support clips and move them to one side. Plug or cover the pump unions, hoses and pipes to prevent dirt entry.

7 Thoroughly clean the fuel pipe unions on the fuel pump and fuel rail. Using an open-ended spanner, unscrew the union nuts securing the high-pressure fuel pipe to the fuel pump and fuel rail. Counterhold the union on the pump with a second spanner, while unscrewing the union nut. Withdraw the pipe and plug or cover the open unions to prevent dirt entry.

8 Unscrew the two retaining nuts and remove the pump from the engine bracket.

Caution: The high-pressure fuel pump is manufactured to extremely close tolerances and must not be dismantled in any way. No parts for the pump are available separately and if the unit is in any way suspect, it must be renewed.

13.7 Open the retaining clip and remove the leak-off hose assembly

13.5 Prise off the locking clip and disconnect the leak-off hose connection at each injector

Refitting

9 Refit the pump to the engine bracket and tighten the two retaining nuts to the specified torque.

10 Remove the blanking plugs from the fuel pipe unions on the pump and fuel rail. Locate a new high-pressure fuel pipe over the unions and screw on the union nuts finger tight at this stage.

11 Using a torque wrench and crow-foot adaptor, tighten the fuel pipe union nuts to the specified torque. Counterhold the union on the pump with an open-ended spanner, while tightening the union nut.

12 Reconnect the pipe, hose and wiring connections removed from the rear of the pump in paragraph five.

13 Refit the high-pressure fuel pump sprocket as described in Chapter 2D, Section 8.

14 Refit the timing belt as described in Chapter 2D, Section 7.

15 Reconnect the battery negative terminal.

16 Observing the precautions listed in Section 2, prime the fuel system as described in Section 5, then start the engine and allow it to idle. Check for leaks at the high-pressure fuel pipe unions with the engine idling. If satisfactory, increase the engine speed to 4000 rpm and check again for leaks. Take the car for a short road test and check for leaks once again on return. If any leaks are detected, obtain and fit a new high-pressure fuel pipe.

17 Refit the engine cover on completion.

13.8 Remove the clamps then release the fuel feed hose from the retaining clips and remove

13 Fuel rail (1.6 litre engines) – removal and refitting

⚠️ *Warning: Refer to the information contained in Section 2 before proceeding.*

Removal

Note: *A complete new set of high-pressure fuel pipes, together with various new retaining clips, O-rings and hose clamps will be required for refitting.*

1 Disconnect the battery negative lead as described in Chapter 5A Section 4.

2 Lift off the plastic cover over the top of the engine.

3 Disconnect the wiring connector at the fuel pressure sensor on the fuel rail (see illustration).

4 Disconnect all the remaining engine wiring harness connectors at the front, top and rear of the engine to enable the harness to be moved clear of the fuel injectors. To avoid confusion when refitting, label each connector as it is disconnected, or alternatively, take a series of photos. Release the wiring harness from all the retaining clips and cable ties, then move the harness to one side.

5 Disconnect the fuel leak-off hose connection at each fuel injector by prising off the locking clip and lifting out the hose fitting (see illustration). Note that new locking clips and new O-rings will be required for refitting. Slip a plastic bag over the disconnected leak-off hose to prevent dirt entry and suitably cap the fuel injectors.

6 Remove the hose clamp securing the fuel leak-off hose to the fuel rail and disconnect the hose. Note that a new clamp will be required for refitting.

7 Release the leak-off hose from the retaining clip and remove the hose assembly from the engine (see illustration).

8 Remove the hose clamps securing the fuel feed hose to the fuel rail and high-pressure fuel pump. Note that new clamps will be required for refitting. Release the hose from the retaining clips on the inlet manifold and support brackets and remove the hose (see illustration).

9 Remove the fuel injector insulator from the fuel injectors (see illustration).

13.9 Remove the insulator from the fuel injectors

13.10a Unscrew the union nuts securing the high pressure fuel pipes to the fuel injectors and fuel rail...

13.10b ...then remove the pipes

13.11 Undo the bolt securing the high-pressure fuel pipe to the manifold bracket

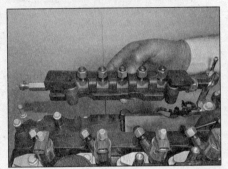

13.12 Undo the two retaining bolts and remove the fuel rail

13.14a Tighten the fuel pipe union nuts on the injectors...

13.14b ...and fuel rail with a torque wrench and crow-foot adaptor

10 Thoroughly clean all the high-pressure fuel pipe unions on the fuel rail, fuel pump and injectors. Unscrew the union nuts securing the high-pressure fuel pipes to the fuel injectors. Unscrew the union nuts securing the high-pressure fuel pipes to the fuel rail, withdraw the pipes and plug or cover the open unions to prevent dirt entry **(see illustrations)**. It is advisable to label each fuel pipe (1 to 4) to avoid confusion when refitting.

11 Using an open-ended spanner, unscrew the union nuts securing the high-pressure fuel pipe to the fuel pump and fuel rail. Counterhold the union on the pump with a second spanner, while unscrewing the union nut. Undo the bolt securing the high-pressure fuel pipe to the manifold bracket **(see illustration)**, then withdraw the pipe and plug or cover the open unions to prevent dirt entry.

12 Undo the two fuel rail retaining bolts, then lift the fuel rail from its location **(see illustration)**.

Refitting

13 Locate the fuel rail in position and refit the two retaining bolts. Only tighten the bolts finger tight at this stage.

14 Working on one fuel injector at a time, remove the blanking plugs from the fuel pipe unions on the fuel rail and the relevant injector. Locate the new high-pressure fuel pipe over the unions and screw on the union nuts finger tight. Tighten the union nuts to the specified torque using a torque wrench and crow-foot adaptor **(see illustrations)**.

Repeat this operation for the remaining three injectors.

15 Similarly, fit the new high-pressure fuel pipe to the fuel pump and fuel rail, and tighten the union nuts to the specified torque. Counterhold the union on the pump with an open-ended spanner, while tightening the union nut.

16 With all the high-pressure fuel pipes in place, tighten the two fuel rail retaining bolts to the specified torque.

17 Locate the new high-pressure fuel pipe in place and screw on the union nuts at the fuel rail and fuel pump. Tighten the union nuts to the specified torque.

18 Refit and tighten the bolt securing the high-pressure fuel pipe to the manifold bracket.

19 Refit the fuel injector insulator to the fuel injectors.

20 Refit the fuel feed hose to the fuel rail and high-pressure fuel pump and secure with new clamps at each end. Secure the hose with the clips on the inlet manifold and support brackets.

21 Fit a new O-ring to each injector connection, then connect the fuel leak-off hose to each fuel injector and secure with new locking clips. Secure the hose in the retaining clip.

22 Attach the leak-off hose to the fuel rail and secure with a new clamp.

23 Reconnect the wiring connector to the fuel pressure sensor.

24 Reconnect all the remaining engine wiring

harness connectors at the front, top and rear of the engine and secure with the relevant retaining clips and cable ties.

25 Observing the precautions listed in Section 2, prime the fuel system as described in Section 5, then start the engine and allow it to idle. Check for leaks at the high-pressure fuel pipe unions with the engine idling. If satisfactory, increase the engine speed to 4000 rpm and check again for leaks. Take the car for a short road test and check for leaks once again on return. If any leaks are detected, obtain and fit a new high-pressure fuel pipe(s).

26 Refit the engine cover on completion.

14 Fuel rail (1.7 litre engines) – removal and refitting

⚠ *Warning: Refer to the information contained in Section 2 before proceeding.*

Removal

Note: *A complete new set of high-pressure fuel pipes will be required for refitting.*

1 Disconnect the battery negative lead as described in Chapter 5A Section 4.

2 Remove the engine oil filler cap, then lift off the plastic cover over the top of the engine **(see illustrations 3.9a and 3.9b)**. Refit the oil filler cap.

3 Disconnect the vacuum hose from the EGR vacuum valve. Release the vacuum hose from

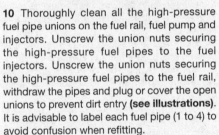

14.3a Disconnect the vacuum hose from the EGR vacuum valve

14.3b Release the vacuum hose from the retaining clips and move the hose to one side

14.5 Undo the three bolts securing the wiring harness trough to the camshaft cover

14.8a Open the retaining clip securing the fuel leak-off hose to the fuel rail banjo union...

14.8b ...then disconnect the hose from the banjo union

11 Thoroughly clean the fuel pipe unions on the fuel pump and fuel rail. Using an open-ended spanner, unscrew the union nuts securing the high-pressure fuel pipe to the fuel pump and fuel rail. Counterhold the union on the pump with a second spanner, while unscrewing the union nut. Remove the pipe and plug or cover the open unions to prevent dirt entry.

12 Unscrew and remove the two fuel rail retaining bolts and the two fuel rail support bracket retaining nuts. Lift away the fuel rail and remove it from the engine.

Refitting

13 Locate the fuel rail in position and refit the two retaining bolts and two retaining nuts. Tighten the bolts/nuts finger tight only at this stage.

14 Remove the blanking plugs from the fuel pipe unions on the fuel rail and fuel injectors. Locate the new high-pressure fuel pipes over the unions and screw on the union nuts finger tight.

15 Similarly, fit a new high-pressure fuel pipe to the fuel pump and fuel rail, and tighten the union nuts finger tight.

16 Tighten the two fuel rail retaining bolts and two retaining nuts to the specified torque.

17 Using a torque wrench and crow-foot adaptor, tighten the fuel pipe unions on the injectors and fuel rail to the specified torque **(see illustration)**. Similarly tighten the fuel pipe unions on the fuel pump and fuel rail while counterholding the unions with an open-ended spanner.

18 Refit the bolt securing the leak-off hose support clip to the inlet manifold and tighten it securely.

19 Reconnect the leak-off hose to the banjo union on the fuel rail and secure the hose with a new retaining clip.

20 Reconnect all the disconnected wiring connectors and secure the wiring harness troughs with the retaining bolts.

21 Reconnect the vacuum hose to the EGR vacuum valve and secure the hose with the retaining clips.

22 Observing the precautions listed in Section 2, prime the fuel system as described in Section 5, then start the engine and allow it to idle. Check for leaks at the high-pressure

the retaining clips and move the hose to one side **(see illustrations)**.

4 Undo the bolt securing the glow plug wiring harness to the inlet manifold.

5 Undo the three bolts securing the wiring harness trough to the top of the camshaft cover **(see illustration)**.

6 Disconnect the wiring connectors from the following components:

a) Coolant temperature sensor.
b) Throttle housing.
c) Fuel injectors.
d) Glow plugs.
e) Camshaft sensor.
f) Mass airflow sensor.
g) Fuel rail pressure sensor.
h) Inlet manifold vacuum solenoid valves.
i) Charge (boost) pressure sensor.
j) Engine management electronic control unit.
k) Wiring harness connection plug.

7 Release the wiring harnesses from the cable ties and retaining clips and move the harnesses to one side.

8 Open the retaining clip securing the fuel leak-off hose to the banjo union on the fuel rail, then disconnect the hose from the banjo union **(see illustrations)**. Plug or cover the banjo union and hose to prevent dirt entry.

9 Undo the bolt securing the fuel leak-off hose support clip to the inlet manifold and move the hose to one side.

10 Thoroughly clean the fuel pipe unions on the fuel injectors and fuel rail. Using an open-ended spanner, unscrew the union nuts securing the high-pressure fuel pipes to the injectors and fuel rail. Counterhold the union on the fuel rail with a second spanner, while unscrewing the union nut. Withdraw the high-pressure fuel pipes and plug or cover the open unions to prevent dirt entry **(see illustration)**.

14.10 Unscrew the union nuts and remove the high-pressure fuel pipes from the injectors and fuel rail

14.17 Tighten the fuel pipe union nuts using a torque wrench and crow-foot adaptor

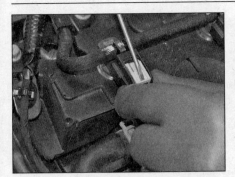

15.3 Disconnect the wiring plug at each fuel injector

15.7 Unscrew the retaining bolt then remove the injector clamp bracket

15.8 Use a twisting motion to free the injectors then lift them from the cylinder head

fuel pipe unions with the engine idling. If satisfactory, increase the engine speed to 4000 rpm and check again for leaks. Take the car for a short road test and check for leaks once again on return. If any leaks are detected, obtain and fit a new high-pressure fuel pipe(s).

23 Refit the engine cover on completion.

15 Fuel injectors (1.6 litre engines) – removal and refitting

⚠️ *Warning: Refer to the information contained in Section 2 before proceeding.*

Removal

Note: A new copper sealing washer, high-pressure fuel pipe and fuel leak-off pipe O-ring will be required for each disturbed injector when refitting.

1 Disconnect the battery negative lead as described in Chapter 5A Section 4.

2 Lift off the plastic cover over the top of the engine.

3 Pull out the yellow locking catch, depress the retainer and disconnect the wiring plug at each fuel injector **(see illustration)**.

4 Disconnect the fuel leak-off hose connection at each fuel injector by prising off the locking clip and lifting out the hose fitting **(see illustration 13.5)**. Note that new locking clips and O-rings will be required for refitting. Slip a plastic bag over the disconnected

leak-off hose to prevent dirt entry and suitably cap the fuel injectors.

5 Remove the fuel injector insulator from the fuel injectors **(see illustration 13.9)**.

6 Thoroughly clean all the high-pressure fuel pipe unions on the fuel rail, fuel pump and injectors. Unscrew the union nuts securing the high-pressure fuel pipes to the fuel injectors. Unscrew the union nuts securing the high-pressure fuel pipes to the fuel rail, withdraw the pipes and plug or cover the open unions to prevent dirt entry **(see illustrations 13.10a and 13.10b)**. It is advisable to label each fuel pipe (1 to 4) to avoid confusion when refitting.

7 Starting with injector No 1, unscrew the retaining bolt then remove the injector clamp bracket **(see illustration)**.

8 Withdraw the injector from the cylinder head using a twisting motion while pulling upward **(see illustration)**. If difficulty is experienced removing the injector, liberally apply penetrating oil to the base of the injector and allow time for the oil to penetrate. If the injector is still reluctant to free, it will be necessary to use a small slide hammer engaged under the flange of the injector body casting, and gently tap it free. If available, use Vauxhall/Opel special tools for this purpose.

9 Examine each injector visually for any signs of obvious damage or deterioration. If any defects are apparent, renew the injector(s).

Caution: The injectors are manufactured to extremely close tolerances and must not be dismantled in any way. Do not attempt to clean carbon deposits from the injector

nozzle or carry out any form of ultra-sonic or pressure testing.

10 If the injectors are in a satisfactory condition, plug the fuel pipe union (if not already done) and suitably cover the electrical element and the injector nozzle.

11 If the original injectors are to be refitted, it is absolutely essential that they are refitted in their original locations.It is advisable to mount the injectors in a suitable stand, the right way up.This will help to prevent difficulties when priming and bleeding after refitting **(see illustration)**.

12 Prior to refitting, obtain a new copper washer and high-pressure fuel pipe for each removed injector.

Refitting

13 If new injectors are being fitted, take of note of the identification numbers. These need to be uploaded into the ECU on completion of the work.

14 Thoroughly clean the injector seat in the cylinder head, ensuring all traces of carbon and other deposits are removed.

15 Starting with injector No 4, locate a new copper washer on the base of the injector **(see illustration)**.

16 Refit the injector to the cylinder head, then engage the injector clamp bracket with the slot in the injector body. Refit the clamp bracket retaining bolt but only tighten it finger tight at this stage **(see illustrations)**.

17 Repeat this procedure for the remaining injectors.

18 Working on one fuel injector at a time,

15.11 It is advisable to mount the injectors in a suitable stand after removal

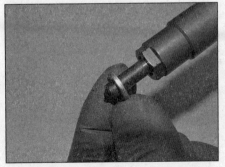

15.15 Fit a new copper washer to each injector

15.16a Refit the injector to the cylinder head...

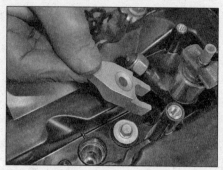

15.16b ...engage the injector clamp bracket with the slot in the injector body...

15.16c ...then fit the retaining bolt and tighten it finger tight only at this stage

remove the blanking plugs from the fuel pipe unions on the fuel rail and the relevant injector. Locate the new high-pressure fuel pipe over the unions and screw on the union nuts. Take care not to cross-thread the nuts or strain the fuel pipes as they are fitted. Once the union nut threads have started, tighten the nuts moderately tight only at this stage.

19 When all the fuel pipes are in place, tighten the injector clamp bracket retaining bolts to the specified torque, then through the specified angle.

20 Tighten the fuel pipe union nuts to the specified torque using a torque wrench and crow-foot adaptor **(see illustrations 13.14a and 13.14b)**. Repeat this operation for the remaining three injectors.

21 Refit the fuel injector insulator to the fuel injectors.

22 Fit a new O-ring to each injector connection, then connect the fuel leak-off hose to each fuel injector and secure with new locking clips.

23 Reconnect the wiring connectors to the fuel injectors.

24 Reconnect the battery negative terminal.

25 If new injectors have been fitted, their classification numbers must be programmed into the engine management ECU using dedicated diagnostic equipment. If this equipment is not available, entrust this task to a Vauxhall/Opel dealer or suitably-equipped repairer. Note that it should be possible to drive the vehicle, albeit with reduced performance/increased emissions, to a repairer for the numbers to be programmed.

26 Observing the precautions listed in Section 2, prime the fuel system as described in Section 5, then start the engine and allow it to idle. Check for leaks at the high-pressure fuel pipe unions with the engine idling. If satisfactory, increase the engine speed to 4000 rpm and check again for leaks. Take the car for a short road test and check for leaks once again on return. If any leaks are detected, obtain and fit a new high-pressure fuel pipe(s).

27 Refit the engine cover on completion.

16 Fuel injectors (1.7 litre engines) – removal and refitting

> **Warning: Refer to the information contained in Section 2 before proceeding.**

Removal

Note: *A new copper washer, O-ring seal and high-pressure fuel pipe will be required for each removed injector when refitting.*

1 Remove the camshaft cover as described in Chapter 2D, Section 4.

2 Before removing the injectors it is advisable to mark their fitted position in relation to the camshaft housing. This will ensure that the injector extension will fit centrally through the seal rings in the camshaft cover and will also prevent any difficulties when screwing on the fuel pipe union nuts. Place a square or similar tool on the camshaft housing and in contact with the injector extension. Using a permanent ink marker pen, mark the position of the edge of the square on the camshaft housing surface **(see illustration)**.

3 Starting with injector No 1, unscrew the retaining bolt then remove the injector clamp bracket **(see illustration)**.

4 Using pointed-nose pliers, lift out the clamp bracket locating dowel from the camshaft housing **(see illustration)**.

5 Withdraw the injector from the cylinder head, twisting it from side-to-side if it is initially tight.

6 Once the injector has been withdrawn, remove the copper washer from the injector base. The copper washer may have remained in place at the base of the injector orifice in the cylinder head. If so, hook it out with a length of wire.

7 Remove the remaining injectors in the same way.

8 If the original injectors are to be refitted, it is absolutely essential that they are refitted in their original locations.It is advisable to mount the injectors in a suitable stand, the right way up.This will help to prevent difficulties when priming and bleeding after refitting **(see illustration 15.11)**.

9 Examine the injector visually for any signs of obvious damage or deterioration. If any defects are apparent, renew the injector.

Caution: The injectors are manufactured to extremely close tolerances and must not be dismantled in any way. Do not unscrew the fuel pipe union on the side of the injector, or separate any parts of the injector body. Do not attempt to clean carbon deposits from the injector nozzle or carry out any form of ultra-sonic or pressure testing.

10 If the injectors are in a satisfactory condition, plug the fuel pipe union (if not already done) and suitably cover the electrical element and the injector nozzle.

11 Prior to refitting, obtain a new copper washer and high-pressure fuel pipe for each injector.

Refitting

12 Thoroughly clean the injector seat in the cylinder head, ensuring all traces of carbon and other deposits are removed.

16.2 Mark the position of the injectors on the camshaft housing surface

16.3 Unscrew the retaining bolt then remove the injector clamp bracket

16.4 Lift out the clamp bracket locating dowel from the camshaft housing

17.6 Move the disconnected wiring harness clear of the inlet manifold

17.7a Disconnect the wiring connector at the exhaust pressure sensor...

17.7b ...EGR vacuum solenoid...

17.7c ...EGR valve...

17.7d ...oxygen sensor...

17.7e ...and throttle housing

13 Locate a new copper washer on the base of the injector.

14 Place the injector in position in the cylinder head.

15 Refit the clamp bracket locating dowel to the camshaft housing, then engage the clamp bracket over the injector. Refit the clamp bracket retaining bolt finger tight only at this stage.

16 Using the square, align the injector with the marks made on the camshaft housing during removal. To be absolutely sure that the injector is correctly aligned, temporarily refit the high-pressure fuel pipe to the injector and fuel rail and lightly tighten the union nuts.

17 Tighten the clamp bracket retaining bolt to the specified torque in the three stages given in the specifications. Remove the high-pressure fuel pipe.

18 Repeat this procedure for the remaining injectors.

19 Once all the injectors are installed, refit the camshaft cover as described in Chapter 2D, Section 4.

20 Observing the precautions listed in Section 2, prime the fuel system as described in Section 5, then start the engine and allow it to idle. Check for leaks at the high-pressure fuel pipe unions with the engine idling. If satisfactory, increase the engine speed to 4000 rpm and check again for leaks. Take the car for a short road test and check for leaks once again on return. If any leaks are detected, obtain and fit a new high-pressure fuel pipe(s).

21 Refit the engine cover on completion.

17 Inlet manifold (1.6 litre engines) – removal and refitting

Removal

1 Remove the battery and battery tray as described in Chapter 5A, Section 4.

2 Lift off the plastic cover over the top of the engine.

3 Remove the air cleaner assembly as described in Section 3.

4 Undo the three bolts securing the wiring harness trough to the inlet manifold.

5 Disconnect the wiring connectors at the following components:
a) Inlet manifold actuator.
b) Camshaft position sensor.
c) Inlet air pressure and temperature sensor.

17.9 Disconnect the fuel return hose from the fuel rail...

d) Fuel rail pressure sensor.
e) Exhaust temperature sensor.
f) Fuel injectors.
g) Glow plugs.

6 Undo the remaining wiring trough retaining bolts and support bracket retaining bolts, release the retaining clips and move the wiring harness clear of the inlet manifold (see illustration).

7 Working at the left-hand end of the engine, disconnect the wiring connectors at the following components (see illustrations):
a) Exhaust pressure sensor.
b) EGR vacuum solenoid.
c) EGR valve.
d) Oxygen sensor.
e) Throttle housing.

8 Release the wiring harness from the retaining clips and move it to one side.

9 Release the retaining clip and disconnect the fuel return hose from the fuel rail (see illustration), then release the fuel return hose from the clips on the inlet manifold support bracket

10 Undo the bolt securing the upper section of the intercooler outlet air duct to the EGR cooler.

11 Rotate the locking collar and disconnect the intercooler outlet air duct from the throttle housing (see illustration 3.32).

12 Disconnect the brake servo vacuum hose at the quick-release connector above the inlet manifold.

13 Undo the four bolts securing the EGR pipe to the inlet manifold and EGR cooler. Lift away the pipe and collect the gasket and rubber seal (see illustrations).

17.13a Undo the EGR pipe retaining bolts...

17.13b ...then remove the pipe and collect the gasket and seal

17.14a Release the wiring harness retaining clip...

17.14b ...then undo the four bolts and remove the inlet manifold brace

14 Release the wiring harness retaining clip from the inlet manifold brace, then undo the four bolts and remove the brace **(see illustrations)**.

15 Using an open-ended spanner, unscrew the union nuts securing the high-pressure fuel pipe to the fuel pump and fuel rail. Counterhold the union on the pump with a second spanner, while unscrewing the union nut. Undo the bolt securing the high-pressure fuel pipe to the inlet

manifold bracket **(see illustration 3.11)**, then withdraw the pipe and plug or cover the open unions to prevent dirt entry.

16 Remove the engine wiring harness bracket from the engine lifting bracket.

17 Undo the four retaining bolts and remove the wiring harness bracket and fuel return hose bracket from the inlet manifold **(see illustrations)**.

18 Remove the throttle housing as described in Section 9.

19 Work around the manifold and release any remaining wiring harness retaining clips and mounting brackets likely to impede removal of the manifold.

20 Working in the reverse of the tightening sequence **(see illustration 17.22)** slacken, then remove the nine inlet manifold retaining bolts. Lift the manifold off the cylinder head and collect the gasket **(see illustrations)**.

Refitting

21 Thoroughly clean the inlet manifold and cylinder head mating faces, then locate a new rubber gasket on the inlet manifold flange.

22 Locate the manifold in position and refit the retaining bolts. Tighten the bolts in the sequence shown to the specified torque **(see illustration)**.

23 The remainder of refitting is the reverse of removal, bearing in mind the following points:

a) *Use new gaskets and seals at all disturbed connections.*

b) *Refit the throttle housing as described in Section 9.*

c) *Refit the air cleaner assembly as described in Section 3.*

17.17a Undo the retaining bolts...

17.17b ...then remove the wiring harness bracket...

17.17c ...and fuel return hose bracket from the manifold

17.20a Lift the inlet manifold off the cylinder head...

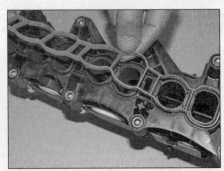

17.20b ...and collect the gasket

17.22 Inlet manifold retaining bolt tightening sequence

d) Refit the battery tray and battery as described in Chapter 5A, Section 4.
e) Observing the precautions listed in Section 2, prime the fuel system as described in Section 5, then start the engine and allow it to idle. Check for leaks at the high-pressure fuel pipe unions with the engine idling. If satisfactory, increase the engine speed to 4000 rpm and check again for leaks. If any leaks are detected, obtain and fit a new high-pressure fuel pipe.
f) Refit the engine cover on completion.

18 Inlet manifold (1.7 litre engines) – removal and refitting

Removal

1 Remove the engine oil filler cap, then lift off the plastic cover over the top of the engine (see illustrations 3.9a and 3.9b). Refit the oil filler cap.
2 Remove the battery and battery tray as described in Chapter 5A, Section 4.
3 Drain the cooling system as described in Chapter 1B, Section 27.
4 Remove the front bumper upper cover by prising out the centre pins and extracting the eight expanding plastic rivets (see illustrations 9.7a and 9.7b).
5 Slacken the retaining clips and disconnect the coolant hoses at the expansion tank. Undo the two retaining bolts and remove the expansion tank (see illustration 9.8).
6 Remove the plenum chamber cover as described in Chapter 11, Section 21.
7 Undo the three bolts securing the wiring harness trough to the top of the camshaft cover (see illustration 14.5).
8 Disconnect the wiring connectors from the following components:
a) Charge (boost) pressure sensor.
b) Throttle housing.
c) Fuel injectors.
d) Glow plugs.
e) Fuel pressure sensor.
f) EGR valve.
g) Crankshaft speed/position sensor.
h) Turbocharger wastegate solenoid.
i) EGR vacuum solenoid.
j) Oxygen sensor.

k) Starter motor.
l) High-pressure fuel pump.
m) Oil pressure warning light switch.
n) Camshaft sensor.
o) Mass airflow sensor.
9 Release the wiring harness from the cable ties and retaining clips and move the harness to one side.
10 Remove the fuel rail as described in Section 14.
11 Disconnect the brake servo unit vacuum hose.
12 Remove the EGR valve as described in Chapter 4C, Section 3.
13 Remove the throttle housing as described in Section 10.
14 On 4WD models, remove the transfer gearbox as described in Chapter 7C, Section 3.
15 Remove the starter motor as described in Chapter 5A, Section 11.
16 Remove the oil filter housing as described in Chapter 2D, Section 17.
17 Undo the retaining nut and three retaining bolts and remove the inlet manifold rear support bracket.
18 Undo the retaining nut and retaining bolt and remove the inlet manifold front support bracket.
19 Disconnect the vacuum hose from the vacuum pump.
20 Disconnect the two vacuum hoses at the EGR vacuum solenoid.
21 Unclip the vacuum hoses and fuel return hose from the inlet manifold.
22 Undo the eight bolts and two nuts, then remove the inlet manifold from the cylinder head. Recover the gasket.

Refitting

23 Thoroughly clean the inlet manifold and cylinder head mating faces.
24 Place a new gasket on the inlet manifold flange, then position the manifold over the cylinder head studs.
25 Refit the manifold retaining bolts and nuts and tighten them finger tight only at this stage.
26 Refit the manifold front and rear support brackets and tighten the retaining nuts and bolts finger tight only at this stage.
27 Working in a diagonal sequence, progressively, tighten the manifold retaining bolts and nuts to the specified torque.
28 Tighten the manifold support bracket nuts and bolts to the specified torque.

29 The remainder of refitting is the reverse of removal, bearing in mind the following points:
a) Refit the oil filter housing as described in Chapter 2D, Section 17.
b) Refit the starter motor as described in Chapter 5A, Section 11.
c) On 4WD models, refit the transfer gearbox as described in Chapter 7C, Section 3.
d) Refit the throttle housing as described in Section 10.
e) Refit the EGR valve as described in Chapter 4C, Section 3.
f) Refit the fuel rail as described in Section 14.
g) Refit the plenum chamber cover as described in Chapter 11, Section 21.
h) Refit the battery tray and battery as described in Chapter 5A, Section 4.
i) Refill the cooling system as described in Chapter 1B, Section 27.
j) Observing the precautions listed in Section 1, prime the fuel system as described in Section 5, then start the engine and allow it to idle. Check for leaks at the high-pressure fuel pipe unions with the engine idling. If satisfactory, increase the engine speed to 4000 rpm and check again for leaks. If any leaks are detected, obtain and fit a new high-pressure fuel pipe.

19 Intercooler – removal and refitting

Removal

1 Disconnect the battery negative lead as described in Chapter 5A Section 4.
2 Apply the handbrake, then jack up the front of the vehicle and support it on axle stands (see Jacking and vehicle support).
3 Remove the front bumper as described in Chapter 11, Section 6.
4 Remove the foam energy absorber from the bumper lower impact bar.
5 Undo the three bolts each side securing the bumper lower impact bar to the subframe and withdraw the bar from under the car (see illustrations).
6 Pull out the centre pins and extract the four plastic expanding rivets, then remove the radiator lower air baffle (see illustrations).
7 Extract the wire spring clip using a small

19.5a Undo the three bolts each side...

19.5b ...and remove the bumper lower impact bar

19.6a Extract the four expanding rivets...

19.6b ...and remove the radiator lower air baffle

19.7a Extract the wire spring clip and disconnect the inlet air duct...

19.7b ...and outlet air ducts from the intercooler

19.8a Undo the upper retaining bolt each side...

19.8b ...and lift the intercooler off the car

screwdriver and disconnect the inlet and outlet air ducts from the intercooler **(see illustrations)**.

8 Undo the two retaining bolts and lift the intercooler off the car **(see illustrations)**.

Refitting

9 Refitting is a reversal of removal, bearing in mind the following points:
a) *Tighten all retaining bolts to the specified torque (where given).*
b) *Refit the front bumper as described in Chapter 11, Section 6.*

20 Turbocharger – description and precautions

Description

1 The turbocharger increases engine efficiency by raising the pressure in the inlet manifold above atmospheric pressure. Instead of the air simply being sucked into the cylinders, it is forced in.

2 Energy for the operation of the turbocharger comes from the exhaust gas. The gas flows through a specially-shaped housing (the turbine housing) and, in so doing, spins the turbine wheel. The turbine wheel is attached to a shaft, at the end of which is another vaned wheel known as the compressor wheel. The compressor wheel spins in its own housing, and compresses the inlet air on the way to the intercooler and inlet manifold.

3 Bboost pressure (the pressure in the inlet manifold) is limited by a wastegate, which diverts the exhaust gas away from the turbine wheel in response to a pressure-sensitive actuator.

4 The turbo shaft is pressure-lubricated by an oil feed pipe from the main oil gallery. The shaft 'floats' on a cushion of oil. A drain pipe returns the oil to the sump.

Precautions

5 The turbocharger operates at extremely high speeds and temperatures. Certain precautions must be observed, to avoid premature failure of the turbo, or injury to the operator.

6 Do not operate the turbo with any of its parts exposed, or with any of its hoses removed. Foreign objects falling onto the

21.6 Undo the retaining bolt and withdraw the turbocharger oil return pipe end fitting from the block

rotating vanes could cause excessive damage, and (if ejected) personal injury.

7 Do not race the engine immediately after start-up, especially if it is cold. Give the oil a few seconds to circulate.

8 Always allow the engine to return to idle speed before switching it off – do not blip the throttle and switch off, as this will leave the turbo spinning without lubrication.

9 Allow the engine to idle for several minutes before switching off after a high-speed run.

10 Observe the recommended intervals for oil and filter changing, and use a reputable oil of the specified quality. Neglect of oil changing, or use of inferior oil, can cause carbon formation on the turbo shaft, leading to subsequent failure.

21 Turbocharger (1.6 litre engines) – removal and refitting

Removal

1 Disconnect the battery negative lead as described in Chapter 5A Section 4.

2 Lift off the plastic cover over the top of the engine.

3 Drain the cooling system as described in Chapter 1B, Section 27.

4 Remove the air cleaner outlet air duct and intercooler inlet air duct as described in Section 3.

5 Remove the diesel particulate filter as described in Section 26.

6 Undo the bolt securing the turbocharger oil return pipe to the cylinder block and withdraw the end fitting from the block **(see illustration)**.

7 Undo the bolt securing the oil return pipe to the turbocharger, then remove the pipe and collect the two O-ring seals **(see illustrations)**.

8 Release the two clips securing the positive crankcase ventilation hose to the camshaft cover and turbocharger and remove the hose **(see illustration)**.

9 Undo the bolts securing the turbocharger wastegate actuator pipe and turbocharger coolant pipe support brackets to their attachments. Disconnect the pipe hose

21.7a Undo the bolt and withdraw the oil return pipe from the turbocharger...

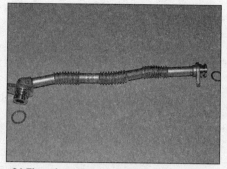

21.7b ...then remove the pipe and collect the O-ring seals

21.8 Release the retaining clips and remove the crankcase ventilation hose

21.9a Undo the turbocharger coolant pipe bracket bolt...

21.9b ...and wastegate actuator pipe bracket bolt at the left-hand end of the engine...

21.9c ...at the centre of the engine...

21.9d ...and at the turbocharger...

21.9e ...then disconnect the hose ends and remove the wastegate actuator pipe

21.10 Disconnect the wastegate actuator wiring connector

ends from the vacuum pipe and wastegate actuator and remove the pipe (see illustrations).

10 Disconnect the turbocharger wastegate actuator wiring connector (see illustration).

11 Unscrew the turbocharger oil supply pipe banjo union bolt from the top of the turbocharger and collect the two sealing washers (see illustration).

12 Similarly, unscrew the turbocharger oil

supply pipe banjo union bolt from the cylinder block and collect the two sealing washers. Undo the bolt securing the oil supply pipe to the turbocharger and remove the pipe (see illustrations).

21.11 Unscrew the turbocharger oil supply pipe banjo union bolt and collect the two sealing washers

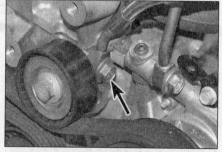

21.12a Unscrew the turbocharger oil supply pipe banjo union bolt from the cylinder block...

21.12b Unbolt the oil supply pipe from the turbocharger and remove it from the engine

21.13 Unbolt and remove the turbocharger inner heat shield

21.14 Disconnect the turbocharger coolant return pipe from the thermostat bypass pipe hose

21.16 Unscrew the turbocharger coolant feed pipe banjo union bolt from the coolant pump

13 Undo the two bolts and remove the turbocharger inner heat shield **(see illustration)**.

14 Slacken the retaining clip and disconnect

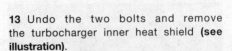

21.17a Undo the upper retaining nut...

the turbocharger coolant return pipe from the thermostat bypass pipe hose connection **(see illustration)**.

15 Remove the wiring harness bracket from the coolant pump.

16 Unscrew the turbocharger coolant feed pipe banjo union bolt from the coolant pump and collect the two sealing washers **(see illustration)**.

17 Undo the three nuts (one from above and two from below) securing the turbocharger to the exhaust manifold **(see illustrations)**. Lift off the turbocharger and recover the gasket.

18 With the turbocharger on the bench, undo the two retaining bolts, remove the coolant feed and return pipe and collect the gasket **(see illustrations)**.

Refitting

19 Refitting is the reverse of removal, noting the following points.

a) Ensure all mating surfaces are clean and dry and renew all gaskets and sealing washers.

b) Tighten all retaining nuts and bolts to the specified torque (where given).

c) Refit the diesel particulate filter as described in Section 26.

d) Refit the air cleaner outlet air duct and intercooler inlet air duct as described in Section 3.

e) Refill the cooling system as described in Chapter 1B, Section 27.

21.17b ...and two lower retaining nuts...

21.17c ...then lift the turbocharger off the exhaust manifold...

21.17d ...and recover the gasket

21.18a Undo the two retaining bolts...

21.18b ...remove the coolant feed and return pipe from the turbocharger...

21.18c ...and collect the gasket

22 Turbocharger (1.7 litre engines) – removal and refitting

Removal

1 Remove the exhaust manifold as described in Section 25.

2 With the manifold on the bench, undo the three bolts and remove the catalytic converter heat shield (see illustration).

3 Undo the three bolts, remove the catalytic converter from the turbocharger, and recover the gasket (see illustrations). Note that new bolts will be required for refitting.

4 Undo the three nuts, remove the turbocharger from the manifold, and recover the gasket (see illustration). Note that new nuts will be required for refitting.

Refitting

5 Refitting is the reverse of removal, noting the following points.

a) Ensure all mating surfaces are clean and dry and renew all gaskets.

b) Tighten all retaining nuts and bolts to the specified torque (where given).

c) Refit the exhaust manifold as described in Section 25.

23 Turbocharger – examination and overhaul

1 With the turbocharger removed, inspect the housing for cracks or other visible damage.

2 Spin the turbine or the compressor wheel to verify that the shaft is intact and to feel for excessive shake or roughness. Some play is normal since in use the shaft is 'floating' on a film of oil. Check that the wheel vanes are undamaged.

3 If the exhaust or induction passages are oil-contaminated, the turbocharger shaft oil seals have probably failed. (On the induction side, this will also have contaminated the intercooler, which if necessary should be flushed with a suitable solvent.)

4 No DIY repair of the turbocharger is

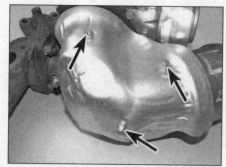

22.2 Undo the three bolts and remove the catalytic converter heat shield

22.3a Undo the three bolts...

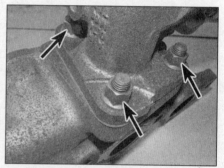
22.3b ...then remove the catalytic converter from the turbocharger and recover the gasket

22.4 Turbocharger retaining nuts

possible. A new unit may be available on an exchange basis.

24 Exhaust manifold (1.6 litre engines) – removal and refitting

Removal

Note: New exhaust manifold retaining bolts will be required for refitting.

1 Remove the turbocharger as described in Section 21.

2 Undo the two bolts securing the manifold EGR pipe to the EGR cooler (see illustration).

3 Undo the eight retaining bolts and remove the manifold, together with the gasket, from the cylinder head (see illustration).

Refitting

4 Using a suitable tap, recut the exhaust manifold retaining bolt threads in the cylinder head (see illustration).

5 Thoroughly clean the mating surfaces of the cylinder head, exhaust manifold, EGR pipe and EGR cooler.

6 Locate a new gasket on the EGR pipe and bend over the tabs to retain the gasket in position.

7 Place a new gasket on the manifold, fit the manifold to the cylinder head and screw in the new retaining bolts.

8 Working in the sequence shown, tighten the manifold retaining bolts to the specified torque in the three stages given in the Specifications (see illustration).

9 Attach the EGR pipe to the EGR cooler and

24.2 Undo the bolts securing the EGR pipe to the EGR cooler

24.3 Undo the retaining bolts and remove the exhaust manifold from the cylinder head

24.4 Recut the manifold retaining bolt threads in the cylinder head using a suitable tap

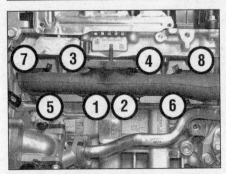

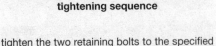

24.8 Exhaust manifold retaining bolt tightening sequence

25.3 Release the retaining clip and disconnect the coolant hose from the right-hand side coolant pipe

25.4 Undo the two bolts and remove the exhaust manifold heat shield on the left-hand side

tighten the two retaining bolts to the specified torque.

10 Refit the turbocharger as described in Section 21.

25 Exhaust manifold (1.7 litre engines) – removal and refitting

Removal

Note: *The exhaust manifold is removed complete with the catalytic converter and turbocharger. The three components are then separated on the bench.*

Note: *New exhaust manifold retaining nuts/bolts will be required for refitting.*

1 Disconnect the battery negative lead as described in Chapter 5A Section 4.

2 Remove the EGR valve cooler as described in Chapter 4C, Section 3.

3 Release the retaining clip and disconnect the coolant hose from the right-hand side coolant pipe **(see illustration)**.

4 Undo the two bolts and remove the exhaust manifold heat shield on the left-hand side **(see illustration)**.

5 Release the retaining clip and disconnect the coolant hose from the left-hand side coolant pipe **(see illustration)**.

6 Unscrew the turbocharger oil supply pipe banjo union from the top of the turbocharger and collect the two sealing washers.

7 Disconnect the vacuum hose from the

turbocharger wastegate actuator **(see illustration)**.

8 Release the retaining clip and disconnect the exhaust pressure sensor hose from the pressure sensor pipe **(see illustration)**.

9 Disconnect the oxygen sensor wiring connector and release the wiring harness retaining clip from the support bracket.

10 Undo the three nuts securing the exhaust system front pipe to the catalytic converter. Separate the joint taking care to support the flexible section. Note that angular movement in excess of 10° can cause permanent damage to the flexible section. Recover the gasket **(see illustrations)**.

11 Undo the two bolts securing the catalytic converter to the support bracket **(see illustration)**.

25.5 Release the retaining clip and disconnect the coolant hose from the left-hand side coolant pipe

25.7 Disconnect the vacuum hose from the turbocharger wastegate actuator

25.8 Release the retaining clip and disconnect the exhaust pressure sensor hose from the pressure sensor pipe

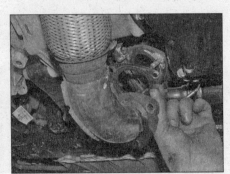

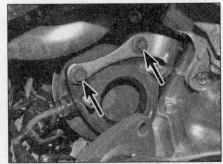

25.10a Undo the three nuts securing the exhaust system front pipe to the catalytic converter...

25.10b ...then separate the joint and recover the gasket

25.11 Undo the two bolts securing the catalytic converter to the support bracket

25.12 Undo the two bolts and separate the intercooler inlet air duct flange from the turbocharger

25.13 Release the two retaining clips then slide the oil return hose down and release it from the turbocharger pipe

25.14 Disconnect the exhaust temperature sensor wiring connector

25.15 Undo the exhaust manifold retaining nuts and bolts and collect the washers

25.16 Withdraw the exhaust manifold complete with turbocharger and catalytic converter

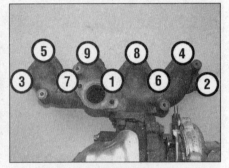

25.18 Exhaust manifold retaining nut/bolt tightening sequence

12 Undo the two bolts and separate the intercooler inlet air duct flange from the turbocharger **(see illustration)**. Collect the gasket.

13 Release the two turbocharger oil return hose retaining clips. Slide the hose down and release it from the turbocharger pipe **(see illustration)**.

14 Disconnect the exhaust temperature sensor wiring connector and release the wiring harness from the support clip above the alternator **(see illustration)**.

15 Undo the seven nuts and two bolts securing the exhaust manifold to the cylinder head and collect the washers **(see illustration)**.

16 Withdraw the exhaust manifold complete with turbocharger and catalytic converter and lift it up and out of the engine compartment **(see illustration)**.

17 If required, the catalytic converter and turbocharger can now be removed from the manifold as described in Section 22.

Refitting

18 Refitting is the reverse of removal, noting the following points.

d) *If removed, refit the turbocharger and catalytic converter to the manifold as described in Section 22.*

e) *Ensure all mating surfaces are clean and dry and renew all gaskets and sealing washers.*

f) *Tighten the new manifold nuts and bolts evenly and progressively to the specified torque, working in the sequence shown (see illustration).*

g) *Tighten all other retaining nuts and bolts to the specified torque (where given).*

h) *Refit the EGR valve cooler as described in Chapter 4C, Section 3.*

26 Exhaust system – general information, removal and refitting

Caution: Any work on the exhaust system should only be attempted once the system is completely cool – this may take several hours, especially in the case of the forward sections, such as the manifold and catalytic converter.

General information

1 On 1.6 litre engines, the exhaust system consists of the catalytic converter/diesel particulate filter, the front pipe and the main section which incorporates the silencer. The diesel particulate filter is integral with the catalytic converter. On 1.7 litre engines, the system consists of a catalytic converter, front pipe, diesel particulate filter and main section which incorporates the silencer.

2 A flexible ('mesh') section is fitted to the front pipe to allow for engine movement.

3 The front pipe is attached to the catalytic converter/diesel particulate filter by a flange joint secured by three nuts. The main section is similarly attached at the front by a flange joint secured by two nuts.

4 When fitted in the factory, the exhaust system from the front pipe flange joint to the end of the system is one piece. If the silencer is to be renewed, it will be necessary to cut through the main section using a chain-type pipe cutter.

5 The system is suspended throughout its entire length by rubber mountings.

Removal

6 To remove a part of the system, first jack up the front or rear of the car, and support it on axle stands (see *Jacking and vehicle support*). Alternatively, position the car over an inspection pit, or on car ramps.

Catalytic converter/diesel particulate filter – 1.6 litre engines

7 Disconnect the battery negative lead as described in Chapter 5A Section 4.

8 Remove the plastic cover from the top of the engine.

9 Remove the engine undertray as described in Chapter 11, Section 21.

10 Remove the oxygen sensor (position 1) from the top of the diesel particulate filter as described in Chapter 4C, Section 3.

11 Undo the nut and four bolts securing the upper heat shield to the turbocharger. Remove the washer from the particulate

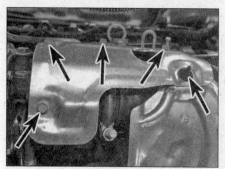

26.11a Undo the nut and four bolts...

26.11b ...and lift off the turbocharger upper heat shield

26.14a Undo the four bolts...

26.14b ...and remove the particulate filter heat shield

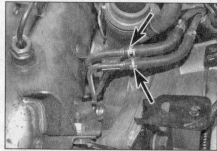

26.15 Release the retaining clips and disconnect the two exhaust pressure sensor hoses

14 Undo the four bolts and remove the particulate filter heat shield **(see illustrations)**.
15 Release the retaining clips and disconnect the two exhaust pressure sensor hoses from the pressure sensor pipes **(see illustration)**.
16 Undo the three nuts and separate the exhaust front pipe from the particulate filter, taking care to support the flexible section. Collect the flange joint gasket noting that a new gasket and three new nuts will be required for refitting **(see illustrations)**.
17 Undo the two nuts securing the particulate filter to the cylinder block studs **(see illustrations)**.
18 Slacken the particulate filter clamp bolt nut. Expand the clamp and disconnect the particulate filter from the exhaust manifold. Lift the filter up and out of the engine compartment and collect the particulate filter-to-manifold gasket **(see illustrations)**. Note

filter stud, then lift off the heat shield **(see illustrations)**.
12 Remove the two exhaust temperature sensors (positions 1 and 2) from the left-hand side of the diesel particulate filter as described

in Chapter 4C, Section 3.
13 Where fitted, remove the oxygen sensor (position 2) from the left-hand side of the diesel particulate filter as described in Chapter 4C, Section 3.

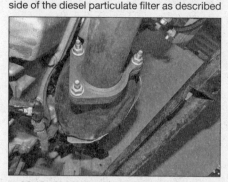

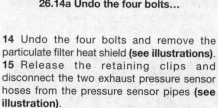

26.16a Undo the three retaining nuts...

26.16b ...then separate the front pipe flange and collect the gasket

26.17a Undo the right-hand retaining nut...

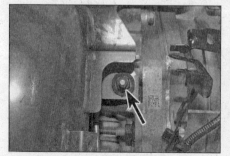

26.17b ...and left-hand retaining nut securing the particulate filter to the cylinder block studs

26.18a Expand the clamp and disconnect the particulate filter from the exhaust manifold...

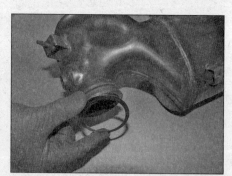

26.18b ...then remove the particulate filter and collect the gasket

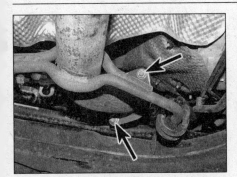

26.33 Undo the nuts securing the front pipe to the main section

26.41 Release the retaining clip and disconnect the exhaust pressure sensor hose

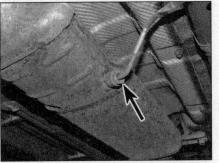

26.42 Unscrew the exhaust temperature sensor from the diesel particulate filter

that a new clamp and new gasket will be required for refitting.

Catalytic converter – 1.7 litre engines

19 Disconnect the battery negative lead as described in Chapter 5A Section 4.

20 Remove the engine oil filler cap, then lift off the plastic cover over the top of the engine (**see illustrations 3.9a and 3.9b**). Refit the oil filler cap.

21 Remove the alternator as described in Chapter 5A, Section 8.

22 Drain the engine oil as described in Chapter 1B, Section 5.

23 Undo the oil level dipstick guide tube upper retaining bolt. Undo the two lower bolts securing the dipstick tube flange to the sump and remove the dipstick tube. Collect the O-ring seal.

24 Undo the three bolts and remove the catalytic converter heat shield (**see illustration 22.2**).

25 Remove the exhaust system front pipe as described later in this Section.

26 Undo the two bolts securing the base of the catalytic converter to the support bracket (**see illustration 25.11**).

27 Undo the two bolts and separate the intercooler inlet air duct flange from the turbocharger (**see illustration 25.12**). Collect the gasket.

28 Disconnect the exhaust temperature sensor wiring connector and release the wiring harness from the support clip above the alternator (**see illustration 25.14**).

29 Undo the three bolts securing the catalytic converter to the turbocharger (**see illustration 22.3a**). Lower the catalytic converter out from under the car and collect the gasket.

Front pipe – 1.6 litre engines

30 Disconnect the battery negative lead as described in Chapter 5A Section 4.

31 Remove the engine undertray as described in Chapter 11, Section 21.

32 Undo the three nuts securing the exhaust front pipe to the diesel particulate filter (**see illustrations 26.16a and 26.16b**). Separate the joint taking care to support the flexible section. Note that angular movement in excess of 10° can cause permanent damage to the flexible section. Recover the gasket.

33 Undo the nuts securing the front pipe to the main section (**see illustration**).

34 Separate the main section joint, recover the gasket, then remove the front pipe from under the car.

Front pipe – 1.7 litre engines

35 Disconnect the battery negative lead as described in Chapter 5A Section 4.

36 Remove the engine undertray as described in Chapter 11, Section 21.

37 Undo the three nuts securing the exhaust front pipe to the catalytic converter (**see illustration 25.10a**). Separate the joint taking care to support the flexible section. Note that angular movement in excess of 10° can cause permanent damage to the flexible section. Recover the gasket.

38 Undo the two nuts securing the front pipe to the diesel particulate filter.

39 Separate the particulate filter joint, recover the gasket, then remove the front pipe from under the car.

Diesel particulate filter – 1.7 litre engines

40 Disconnect the battery negative lead as described in Chapter 5A Section 4.

41 Release the retaining clip and disconnect the exhaust pressure sensor hose from the pressure sensor pipe (**see illustration**).

42 Unscrew the exhaust temperature sensor from the diesel particulate filter (**see illustration**). Withdraw the temperature sensor and hang it to one side.

43 Suitably support the particulate filter and exhaust main section, then undo the two retaining nuts at the flange joint. Unhook the main section's adjacent rubber mounting and separate the joint. Recover the gasket.

44 Undo the two nuts securing the exhaust system front pipe to the particulate filter. Unhook the particulate filter from the two rubber mountings, then separate the joint taking care to support the front pipe flexible section. Recover the gasket and remove the particulate filter from under the car.

Main section

45 Undo the nuts securing the main section to the front pipe or diesel particulate filter. Separate the joint and recover the gasket.

46 Unhook the main section's rubber mountings, and remove it from under the car.

Rear silencer

Note: *If a new rear silencer is to be fitted, it will be necessary to obtain a connecting sleeve from your Vauxhall/Opel parts supplier.*

47 If the original one-piece exhaust system is still fitted, it will be necessary to cut off the old rear silencer to enable fitment of the new unit. Using the new silencer as a pattern, mark the exhaust main section to determine the cut point.

48 Unhook the rubber mountings and allow the exhaust system to rest on an axle stand or similar support.

49 Using a chain-type pipe cutter, cut through the main section at the marked cut point and remove the silencer.

50 Fit the new silencer and connect it to the main section using the connecting sleeve. Tighten the connecting sleeve retaining bolt nuts securely.

Heat shields

51 The heat shields are secured to the underside of the body by special nuts. Each shield can be removed separately, but note that they may overlap, making it necessary to loosen another section first. If a shield is being removed to gain access to a component located behind it, it may prove sufficient in some cases to remove the retaining nuts and/or bolts, and simply lower the shield, without disturbing the exhaust system. Otherwise, unhook the exhaust system rubber mountings and rest the system on suitable stands.

Refitting

52 In all cases, refitting is a reversal of removal, but note the following points:

i) *Always use new gaskets, nuts and clamps (as applicable), and coat all threads with copper grease.*

j) *If any of the exhaust mounting rubbers are in poor condition, fit new ones.*

k) *Make sure that the exhaust is suspended properly on its mountings, and will not come into contact with the floor or any suspension parts.*

l) *Tighten all nuts/bolts to the specified torque, where given.*

Chapter 4 Part C
Emission control systems

Contents

Degrees of difficulty

| Easy, suitable for novice with little experience | 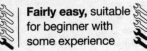 | Fairly easy, suitable for beginner with some experience | | Fairly difficult, suitable for competent DIY mechanic | | Difficult, suitable for experienced DIY mechanic | | Very difficult, suitable for expert DIY or professional | |

Specifications

Torque wrench settings	Nm	lbf ft
Petrol engines		
Oxygen sensor	42	31
Diesel engines		
EGR valve bolts:		
1.6 litre engines	10	7
1.7 litre engines	25	18
EGR valve cooler mounting bolts/nuts:		
1.6 litre engines	25	18
1.7 litre engines	29	21
EGR valve pipe bolts:		
1.6 litre engines:		
Exhaust manifold EGR pipe	10	7
EGR valve cooler outlet pipe	10	7
1.7 litre engines	29	21
Inlet manifold support bracket (1.6 litre engines)	10	7
Oxygen sensor:		
1.6 litre engines	42	31
1.7 litre engines	45	33
Temperature sensor	45	33

1 General Information

1 All petrol engine models use unleaded petrol and also have various other features built into the fuel system to help minimise harmful emissions. These include a crankcase emission control system, a catalytic converter and an evaporative emission control system to keep fuel vapour/exhaust gas emissions down to a minimum.

2 All diesel engine models are also designed to meet strict emission requirements. The engines are fitted with a crankcase emission control system, and a catalytic converter/diesel particulate filter to keep exhaust emissions down to a minimum. An exhaust gas recirculation (EGR) system is also fitted to further decrease exhaust emissions.

3 The emission control systems function as follows.

Petrol engines

Crankcase emission control

4 To reduce the emission of unburned hydrocarbons from the crankcase into the atmosphere, the engine is sealed and the blow-by gases and oil vapour are drawn from inside the crankcase, through an oil separator, into the inlet tract to be burned by the engine during normal combustion.

5 Under all conditions the gases are forced out of the crankcase by the (relatively) higher crankcase pressure; if the engine is worn, the raised crankcase pressure (due to increased blow-by) will cause some of the flow to return under all manifold conditions.

Exhaust emission control

6 To minimise the amount of pollutants which escape into the atmosphere, all engines are fitted with a catalytic converter which is attached to (1.4 litre engines), or integral with (1.6 litre engines), the exhaust manifold. The system is of the closed-loop type, in which the oxygen sensors in the exhaust system provides the engine management system ECU with constant feedback, enabling the ECU to adjust the mixture to provide the best possible conditions for the converter to operate.

2.3 Charcoal canister location on the side of the fuel tank

7 Two heated oxygen sensors are fitted to the exhaust system. The sensor in the exhaust manifold (before the catalytic converter) determines the residual oxygen content of the exhaust gasses for mixture correction. The sensor in the exhaust front pipe (after the catalytic converter) monitors the function of the catalytic converter to give the driver a warning signal if there is a fault.

8 The oxygen sensor's tip is sensitive to oxygen and sends the ECU a varying voltage signal depending on the amount of oxygen in the exhaust gases. Peak conversion efficiency of all major pollutants occurs if the intake air/fuel mixture is maintained at the chemically-correct ratio for the complete combustion of petrol of 14.7 parts (by weight) of air to 1 part of fuel (the 'stoichiometric' ratio). The sensor output voltage alters in a large step at this point, the ECU using the signal change as a reference point and correcting the intake air/fuel mixture accordingly by altering the fuel injector pulse width.

Evaporative emission control

9 To minimise the escape into the atmosphere of unburned hydrocarbons, an evaporative emissions control system is also fitted to all models. The fuel tank filler cap is sealed and a charcoal canister is mounted on the fuel tank. The canister collects the petrol vapours generated in the tank when the car is parked and stores them until they can be cleared from the canister (under the control of the engine management system ECU) via the purge valve into the inlet tract to be burned by the engine during normal combustion.

10 To ensure that the engine runs correctly when it is cold and/or idling and to protect the catalytic converter from the effects of an over-rich mixture, the purge control valve is not opened by the ECU until the engine has warmed-up, and the engine is under load; the valve solenoid is then modulated on and off to allow the stored vapour to pass into the inlet tract.

Diesel engines

Crankcase emission control

11 Refer to paragraphs 4 and 5.

Exhaust emission control

12 To minimise the level of exhaust pollutants released into the atmosphere, a catalytic converter and a diesel particulate filter are fitted in the exhaust system.

13 The catalytic converter consists of a canister containing a fine mesh impregnated with a catalyst material, over which the hot exhaust gases pass. The catalyst speeds up the oxidation of harmful carbon monoxide and unburned hydrocarbons, effectively reducing the quantity of harmful products released into the atmosphere via the exhaust gases.

14 The diesel particle filter (integral with the catalytic converter on 1.6 litre engines, and fitted after the exhaust system front pipe on 1.7 litre engines) contains a silicon carbide

honeycomb block containing microscopic channels in which the exhaust gasses flow. As the gasses flow through the honeycomb channels, soot particles are deposited on the channel walls. To prevent clogging of the honeycomb channels, the soot particles are burned off at regular intervals during what is known as a 'regeneration phase'. Under the control of the engine management system ECU, the injection characteristics are altered to raise the temperature of the exhaust gasses to approximately 600°C. At this temperature, the soot particles are effectively burned off the honeycomb walls as the exhaust gasses pass through. A differential pressure sensor and a temperature sensor are used to inform the ECU of the condition of the particulate filter, and the temperature of the exhaust gasses during the regeneration phase. When the ECU detects that soot build-up is reducing the efficiency of the particulate filter, it will instigate the regeneration process. This occurs at regular intervals under certain driving conditions and will normally not be detected by the driver.

Exhaust gas recirculation (EGR) system

15 This system is designed to recirculate small quantities of exhaust gas into the inlet tract, and therefore into the combustion process. This process reduces the level of unburnt hydrocarbons present in the exhaust gas before it reaches the catalytic converter. The system is controlled by the engine management system ECU, using the information from its various sensors, via the electrically-operated EGR valve.

2 Petrol engine emission control systems – testing and component renewal

Crankcase emission control

1 The components of this system require no attention other than to check that the hose(s) are clear and undamaged at regular intervals.

Evaporative emission control system

Testing

2 If the system is thought to be faulty, disconnect the hoses from the charcoal canister and purge control valve and check that they are clear by blowing through them. Full testing of the system can only be carried out using specialist electronic equipment which is connected to the engine management system diagnostic connector. If the purge control valve or charcoal canister are thought to be faulty, they must be renewed.

Charcoal canister renewal

3 The charcoal canister is located on the side of the fuel tank (see illustration). To gain access to the canister, remove the fuel tank as described in Chapter 4A, Section 7.

2.7 Purge valve location on the inlet manifold

2.14a Undo the three bolts…

2.14b …and remove the exhaust manifold heat shield

4 With the fuel tank removed, disconnect the vapour hose quick-release fittings at the charcoal canister.

5 Detach the locking clamp and release the canister from its mounting bracket on the fuel tank. Remove the canister from the tank.

6 Refitting is a reverse of the removal procedure, ensuring the hoses are correctly and securely reconnected.

Purge valve renewal

7 The purge valve is mounted on the inlet manifold **(see illustration)**.

8 To renew the valve, ensure the ignition is switched off then depress the retaining clip and disconnect the wiring connector from the valve.

9 Disconnect the hoses from the valve, noting their correct fitted locations then unclip and remove the valve from the engine.

10 Refitting is a reversal of the removal procedure, ensuring the valve is fitted the correct way around and the hoses are securely connected.

Exhaust emission control

Testing

11 The performance of the catalytic converter can be checked only by measuring the exhaust gases using a good-quality, carefully-calibrated exhaust gas analyser.

12 If the CO level at the tailpipe is too high, the vehicle should be taken to a Vauxhall/ Opel dealer or engine diagnostic specialist so that the complete fuel injection and ignition

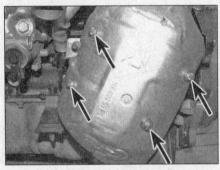

2.15a Undo the four bolts…

2.15b …and remove the catalytic converter heat shield

systems, including the oxygen sensors, can be thoroughly checked using diagnostic equipment. This equipment will give an indication as to where the fault lies and the necessary components can then be renewed.

Catalytic converter renewal – 1.4 litre engines

13 Disconnect the battery negative lead as described in Chapter 5A Section 4.

14 Undo the three bolts and remove the exhaust manifold heat shield **(see illustrations)**.

15 Undo the four bolts and remove the catalytic converter heat shield **(see illustrations)**.

16 Trace the wiring back from the oxygen sensor, noting its correct routing, and disconnect its wiring connector. Lift the wiring connector out of the support bracket.

17 Undo the three nuts securing the exhaust front pipe to the catalytic converter **(see illustration)**. Separate the joint taking care to support the flexible section. Note that angular movement in excess of 10° can cause permanent damage to the flexible section. Recover the gasket. Note that new nuts will be required for refitting.

18 Undo the two nuts securing the catalytic converter to the support bracket **(see illustration)**.

19 Undo the catalytic converter clamp bolt nut and remove the bolt **(see illustration)**. Expand the clamp and disconnect the catalytic converter from the exhaust manifold. Lower the catalytic converter and remove it from under the car. Recover the catalytic converter sealing ring. Note that a new catalytic converter clamp and a new

2.17 Undo the three nuts securing the exhaust front pipe to the catalytic converter

2.18 Undo the nut each side (one arrowed) securing the catalytic converter to the bracket

2.19a Undo the catalytic converter clamp bolt nut and remove the bolt

2.19b Lower the catalytic converter and remove it from under the car...

2.19c ...then recover the catalytic converter sealing ring

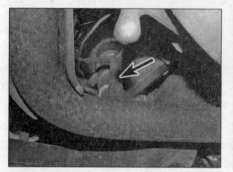

2.24a Oxygen sensors are located in the exhaust system front pipe...

apply a smear of high temperature grease to the sensor threads (Vauxhall/Opel recommend the use of special grease available from your dealer). Tighten the sensor to the specified torque and ensure that the wiring is correctly routed and in no danger of contacting either the exhaust system or engine.

3 Diesel engine emission control systems – testing and component renewal

Crankcase emission control

1 The components of this system require no attention other than to check that the hose(s) are clear and undamaged at regular intervals.

Exhaust emission control

2 The performance of the catalytic converter and diesel particulate filter can only be checked using special diagnostic equipment. If a system fault is suspected, the vehicle should be taken to a Vauxhall/Opel dealer so that the complete fuel injection system can be thoroughly checked.

Catalytic converter/ diesel particulate filter renewal

3 Refer to Chapter 4B, Section 26 for removal and refitting details.

Oxygen sensor renewal – 1.6 litre engines

2.24b ...and in the exhaust manifold

sealing ring will be required for refitting **(see illustrations)**.

Catalytic converter renewal – 1.6 litre engines

20 On 1.6 litre engines, the catalytic converter is an integral part of the exhaust manifold. Exhaust manifold removal and refitting procedures are contained in Chapter 4A, Section 19.

Oxygen sensor renewal

Note: *There are two heated oxygen sensors fitted to the exhaust system. The sensor in the exhaust manifold is for mixture regulation and the sensor in the exhaust front pipe is to check the operation of the catalytic converter (see Section 1).*

Note: *The oxygen sensor is delicate and will not work if it is dropped or knocked, if its power supply is disrupted, or if any cleaning materials are used on it.*

21 Warm the engine up to normal operating temperature then stop the engine and

disconnect the battery negative terminal as described in Chapter 5A Section 4.

22 For the sensor fitted in the exhaust front pipe, firmly apply the handbrake, then jack up the front of the car and support it securely on axle stands (see *Jacking and vehicle support*).

23 Trace the wiring back from the oxygen sensor which is to be replaced, and disconnect its wiring connector, freeing the wiring from any relevant retaining clips or ties and noting its correct routing.

Caution: Take great care not burn yourself on the hot manifold/sensor.

24 Unscrew the sensor and remove it from the exhaust system front pipe/manifold **(see illustrations)**. Where applicable recover the sealing washer and discard it; a new one should be used on refitting.

25 Refitting is a reverse of the removal procedure, using a new sealing washer (where applicable). Prior to installing the sensor,

Note: *On certain engines, a second oxygen sensor, located near the base of the diesel particulate filter, may be fitted. Removal and refitting procedures are similar to those for the upper sensor described below, but access is from under the car.*

4 Warm the engine up to normal operating temperature then stop the engine and disconnect the battery negative terminal as described in Chapter 5A Section 4). Remove the plastic cover from the top of the engine.

5 Extract the centre pins, then remove the eight plastic expanding rivets securing the bumper upper panel to the bumper and bonnet lock platform. Lift up the upper panel and remove it from the engine compartment **(see illustrations)**.

6 Undo the two retaining bolts and move the coolant expansion tank to one side **(see illustration)**.

3.5a Extract the centre pins and remove the eight expanding rivets...

3.5b ...then lift away the bumper upper panel

3.6 Undo the two bolts and move the expansion tank to one side

3.7a Disconnect the oxygen sensor wiring connector...

3.7b ...and release the wiring retaining clips from the support brackets

3.8 Unscrew the oxygen sensor and remove it from the particulate filter

7 Trace the sensor wiring back to the wiring connector and support bracket. Using a screwdriver, ease the wiring connector out of the support bracket. Open the locking catch and disconnect the wiring connector. Release the wiring retaining clips from the support brackets around the top of the engine **(see illustrations)**.

Caution: Take great care not to burn yourself on the hot manifold/sensor.

8 Unscrew the sensor and remove it from the diesel particulate filter **(see illustration)**.

9 Refitting is a reverse of the removal procedure. Prior to installing the sensor, apply a smear of high temperature grease to the sensor threads (Vauxhall/Opel recommend the use of a special grease available from your dealer). Tighten the sensor to the specified torque and ensure that the wiring is correctly routed and in no danger of contacting either the exhaust system or engine.

Oxygen sensor renewal – 1.7 litre engines

10 Warm the engine up to normal operating temperature then stop the engine and disconnect the battery negative terminal as described in Chapter 5A Section 4). Remove the plastic cover from the top of the engine.

11 Trace the sensor wiring back to the wiring connector and support bracket. Using a screwdriver, ease the wiring connector out of the support bracket. Open the locking catch and disconnect the wiring connector. Release the wiring retaining clip from the support bracket.

12 Unscrew the oxygen sensor and remove it from the top of the diesel particulate filter.

13 Refitting is a reverse of the removal procedure. Prior to installing the sensor, apply a smear of high temperature grease to the sensor threads (Vauxhall/Opel recommend the use of a special grease available from your dealer). Tighten the sensor to the specified torque and ensure that the wiring is correctly routed and in no danger of contacting either the exhaust system or engine.

Particulate filter temperature sensor renewal – 1.6 litre engines

Note: *Two temperature sensors are fitted, both located on the left-hand side of the particulate filter. The removal and refitting procedures are basically the same for each sensor.*

14 Warm the engine up to normal operating temperature then stop the engine and disconnect the battery negative terminal as described in Chapter 5A Section 4. Remove the plastic cover from the top of the engine.

15 If working on the upper sensor (position 1), undo the two retaining bolts and move the coolant expansion tank to one side **(see illustration 3.6)**. If working on the lower sensor (position 2), firmly apply the handbrake, then jack up the front of the vehicle and support it securely on axle stands (see *Jacking and vehicle support*). Remove the engine undertray as described in Chapter 11, Section 21.

16 Trace the sensor wiring back to the wiring connector and support bracket. Using a screwdriver, ease the wiring connector out of the support bracket. Open the locking catch

and disconnect the wiring connector. Release the wiring retaining clips from the support brackets.

Caution: Take great care not burn yourself on the hot particulate filter/sensor.

17 Unscrew the sensor and remove it from the particulate filter **(see illustration)**.

18 Refitting is a reverse of the removal procedure. Prior to installing the sensor, apply a smear of high temperature grease to the sensor threads (Vauxhall/Opel recommend the use of a special grease available from your dealer). Tighten the sensor to the specified torque and ensure that the wiring is correctly routed and in no danger of contacting either the exhaust system or engine.

Particulate filter temperature sensor renewal – 1.7 litre engines

19 On 1.7 litre engines, two temperature sensors are fitted; one located in the catalytic converter, and one located in the diesel particulate filter. To renew the sensor located in the catalytic converter, proceed as described previously in paragraphs 14 to 18 **(see illustrations)**.

20 To remove the sensor located in the diesel particulate filter, warm the engine up to normal operating temperature then stop the engine and disconnect the battery negative terminal as described in Chapter 5A Section 4.

21 Firmly apply the handbrake, then jack up the front of the vehicle and support it securely on axle stands (see *Jacking and vehicle support*).

22 Remove the engine undertray as described in Chapter 11, Section 21.

3.17 Removing the position 1 particulate filter temperature sensor

3.19a Particulate filter temperature sensor wiring connector

3.19b One of the particulate filter temperature sensors is located in the catalytic converter

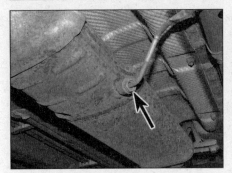

3.25 Unscrew the sensor and remove it from the particulate filter

3.30 Disconnect the pressure sensor wiring connector

3.37 Disconnect the EGR valve wiring connector

23 Using a screwdriver, ease the wiring connector out of the support bracket below the starter motor. Open the locking catch and disconnect the wiring connector.

24 Trace the wiring harness back to the temperature sensor, releasing it from all the retaining clips on the subframe and underbody.

25 Unscrew the sensor and remove it from the particulate filter **(see illustration)**.
Caution: Take great care not burn yourself on the hot filter/sensor.

26 Refitting is a reverse of the removal procedure. Prior to installing the sensor, apply a smear of high temperature grease to the sensor threads (Vauxhall/Opel recommend the use of a special grease available from your dealer). Tighten the sensor to the specified torque and ensure that the wiring is correctly routed and in no danger of contacting either the exhaust system or engine.

Exhaust pressure sensor renewal

27 The exhaust pressure sensor is attached to a bracket located on the EGR valve cooler (1.6 litre engines), or at the rear right-hand side of the engine compartment (1.7 litre engines).

28 Remove the plastic cover from the top of the engine.

29 On 1.6 litre engines, remove the battery and battery tray as described in Chapter 5A, Section 4.

30 Disconnect the pressure hose(s) and the sensor wiring connector **(see illustration)**.

31 Undo the retaining bolt/nut and remove the sensor from the mounting bracket.

32 Refitting is the reverse of removal.

Exhaust gas recirculation (EGR) system

Testing

33 Comprehensive testing of the system can

only be carried out using specialist electronic equipment which is connected to the injection system diagnostic wiring connector.

EGR valve renewal – 1.6 litre engines

34 Remove the battery and battery tray as described in Chapter 5A, Section 4.

35 Remove the bumper upper panel as described in paragraph 5, then undo the two retaining bolts and move the coolant expansion tank to one side **(see illustration 3.6)**.

36 Remove the throttle housing as described in Chapter 4B, Section 9.

37 Disconnect the EGR valve wiring connector **(see illustration)**.

38 Unscrew the two bolts and remove the valve from the EGR cooler. Collect the two O-ring seals.

39 Refitting is the reverse of removal using new O-rings and tightening the retaining bolts to the specified torque.

EGR valve renewal – 1.7 litre engines

40 Disconnect the battery negative lead as described in Chapter 5A Section 4.

41 Remove the plastic cover from the top of the engine.

42 Undo the two bolts securing the EGR valve pipe to the EGR valve. Withdraw the pipe slightly and collect the gasket **(see illustrations)**.

43 Disconnect the EGR valve wiring connector, then undo the three retaining bolts (noting the location of the coolant hose support bracket) and remove the valve. Collect the EGR valve gasket.

44 Refitting is the reverse of removal using a new gasket and tightening the retaining bolts to the specified torque.

EGR valve cooler renewal – 1.6 litre engines

45 Remove the battery and battery tray as described in Chapter 5A, Section 4.

46 Drain the cooling system as described in Chapter 1B, Section 27.

47 Remove the intercooler outlet air duct as described in Chapter 4B, Section 3.

48 Undo the four bolts and remove the EGR valve cooler outlet pipe. Recover the sealing ring and gasket noting that new components will be required for refitting **(see illustrations)**.

3.42a Undo the two bolts securing the EGR valve pipe to the EGR valve

3.42b Withdraw the pipe slightly and collect the gasket

3.48a Undo the EGR valve cooler outlet pipe retaining bolts...

3.48b ...then remove the pipe and recover the sealing ring and gasket

3.49 Unclip the wiring harness retainers from the inlet manifold support bracket

3.50 Undo the four bolts and remove the inlet manifold support bracket

3.51a Disconnect the wiring connectors at the EGR vacuum solenoid…

3.51b …EGR valve…

3.51c …temperature sensor…

3.51d …and throttle housing

49 Using a suitable forked tool, unclip the wiring harness retainers from the inlet manifold support bracket **(see illustration)**.
50 Disconnect the exhaust pressure sensor wiring connector, then undo the four retaining bolts and remove the inlet manifold support bracket **(see illustration)**.
51 Disconnect the wiring connectors at the EGR vacuum solenoid, EGR valve, exhaust temperature sensor, and throttle housing **(see illustrations)**.
52 Disconnect the vacuum hose from the base of the EGR vacuum solenoid **(see illustration)**.
53 Undo the two bolts securing the coolant expansion tank hose support bracket to the EGR valve cooler.
54 Release the retaining clip and disconnect the coolant outlet hose from the coolant expansion tank.
55 Undo the two bolts and release the coolant outlet hose support bracket from the EGR valve cooler.
56 Disconnect the coolant air bleed hose from the top of the EGR valve cooler, then release the hose support clip from the wiring harness bracket.
57 Undo the bolt securing the exhaust pressure sensor to the EGR valve cooler. Unclip the exhaust temperature sensor wiring harness plug from the EGR cooler bracket, then remove the exhaust pressure sensor and temperature sensor wiring harness from the cooler **(see illustration)**. Release the wiring and vacuum pipe retaining clips as necessary to allow removal.

58 Undo and remove the turbocharger wastegate actuator pipe bracket bolt from the EGR valve cooler **(see illustration)**.
59 Undo the two bolts securing the exhaust

3.52 Disconnect the EGR solenoid vacuum hose

3.58 Remove the wastegate actuator pipe bracket bolt

manifold EGR pipe to the EGR valve cooler **(see illustration)**. Lift up the tabs on the gasket and slide the gasket out of position.
60 Release the thermostat bypass hose

3.57 Remove the exhaust pressure sensor and temperature sensor wiring harness plug from the EGR valve cooler

3.59 Undo the two exhaust manifold EGR pipe retaining bolts

3.60 Release the retaining clip and disconnect the thermostat bypass hose from the EGR valve cooler

3.61 Extract the retaining spring clip and disconnect the heater outlet hose

3.62a Undo the rear retaining bolt...

3.62b ...and the two front retaining bolts...

3.62c ...then remove the EGR valve cooler

62 Undo the three retaining bolts and remove the EGR valve cooler **(see illustrations)**.
63 Refitting is the reverse of removal, bearing in mind the following points:
a) *Renew all disturbed gaskets and seals.*
b) *Tighten the retaining bolts/nuts to the specified torque (where given).*
c) *Refit the intercooler outlet air duct as described in Chapter 4B, Section 3.*
d) *Refill the cooling system as described in Chapter 1B, Section 27.*
e) *Refit the battery tray and battery as described in Chapter 5A, Section 4.*

EGR valve cooler renewal – 1.7 litre engines

retaining clip and disconnect the bypass hose from the EGR valve cooler **(see illustration)**.
61 Extract the retaining spring clip and disconnect the heater outlet hose from the EGR valve cooler **(see illustration)**.

64 Disconnect the battery negative lead as described in Chapter 5A Section 4.
65 Firmly apply the handbrake, then jack up the front of the vehicle and support it securely on axle stands (see *Jacking and vehicle support*).
66 Remove the engine undertray as described in Chapter 11, Section 21.
67 Drain the cooling system as described in Chapter 1B, Section 27.
68 Drain the engine oil as described in Chapter 1B, Section 5.
69 Remove the auxiliary drivebelt as described in Chapter 1B, Section 30.
70 Remove the air cleaner assembly and air ducts as described in Chapter 4B, Section 3.
71 Undo the oil level dipstick guide tube upper retaining bolt. Undo the two lower bolts securing the dipstick tube flange to the sump and remove the dipstick tube. Collect the O-ring seal.
72 Undo the three bolts securing the air conditioning compressor to the compressor mounting bracket. Lift the compressor off the mounting bracket and secure it to the front body panel using cable ties or similar **(see illustration)**. Do not disconnect the refrigerant lines from the compressor.
73 Undo the three upper bolts and two lower bolts securing the compressor mounting bracket to the cylinder head and cylinder block. Withdraw the mounting bracket from the engine, release the wiring harness retaining clip and remove the mounting bracket **(see illustration)**.
74 Undo the three bolts and remove the exhaust manifold heat shield on the right-hand side **(see illustrations)**.

3.72 Unbolt the compressor and secure it to the front body panel using cable ties or similar

3.73 Undo the five retaining bolts and remove the compressor mounting bracket

3.74a Undo the three bolts...

3.74b ...and remove the exhaust manifold heat shield on the right-hand side

3.76 Undo the bolt securing the EGR valve cooler bracket

3.77 Undo the bolt securing the EGR valve cooler coolant pipe

3.78 Release the retaining clip securing the EGR valve cooler pipe to the bypass hose

3.79 Undo the two EGR valve cooler flange bolts

3.80a Undo the EGR valve cooler upper mounting bolt...

3.80b ...and remove the cooler assembly from the engine

75 Undo the two bolts securing the EGR valve pipe to the EGR valve. Withdraw the pipe slightly and collect the gasket (see illustrations 3.42a and 3.42b).

76 Undo the bolt securing the EGR valve cooler mounting bracket to the cylinder head on the left-hand side (see illustration).

77 Undo the bolt securing the EGR valve cooler coolant pipe to the thermostat housing (see illustration).

78 Release the retaining clip securing the EGR valve cooler pipe to the bypass hose (see illustration).

79 Undo the two EGR valve cooler flange bolts (see illustration). Withdraw the EGR cooler pipe from the bypass hose and collect the cooler flange gasket.

80 Undo the EGR valve cooler upper mounting bolt and remove the cooler assembly from the engine (see illustrations).

81 Refitting is the reverse of removal, bearing in mind the following points:
a) Renew all disturbed gaskets and seals.
b) Tighten the retaining bolts/nuts to the specified torque (where given).
c) Refit the auxiliary drivebelt as described in Chapter 1B, Section 30.
d) Refit the air cleaner assembly and outlet air duct as described in Chapter 4B, Section 3.
e) Fill the engine with fresh oil as described in Chapter 1B, Section 5.
f) Refit the engine undertray as described in Chapter 11, Section 21.
g) Refill the cooling system as described in Chapter 1B, Section 27.

4 Catalytic converter – general information and precautions

General information

Note: *The information contained in this Section is, in the main, equally applicable to the particulate filter fitted to diesel engines.*

1 The catalytic converter reduces harmful exhaust emissions by chemically converting the more poisonous gases to ones that (in theory at least) are less harmful. The chemical reaction is known as an 'oxidising' reaction, or one where oxygen is 'added'.

2 Inside the converter is a honeycomb structure, made of ceramic material and coated with the precious metals palladium, platinum and rhodium (the 'catalyst' which promotes the chemical reaction). The chemical reaction generates heat, which itself promotes the reaction – therefore, once the car has been driven several miles, the body of the converter will be very hot.

3 The ceramic structure contained within the converter is understandably fragile, and will not withstand rough treatment. Since the converter runs at a high temperature, driving through deep standing water (in flood conditions, for example) is to be avoided, since the thermal stresses imposed when plunging the hot converter into cold water may well cause the ceramic internals to fracture, resulting in a 'blocked' converter – a common cause of failure. A catalytic converter that has

been damaged in this way can be checked by shaking it, do not strike it – if a rattling noise is heard, this indicates probable failure.

Precautions

4 The catalytic converter is a reliable and simple device which needs no maintenance in itself, but there are some facts of which an owner should be aware if the converter is to function properly for its full service life.

Petrol engines

a) DO NOT use leaded petrol (or lead-replacement petrol, LRP) in a car equipped with a catalytic converter – the lead (or other additives) will coat the precious metals, reducing their converting efficiency and will eventually destroy the converter.
b) Always keep the ignition and fuel systems well maintained in accordance with the manufacturer's schedule.
c) If the engine develops a misfire, do not drive the car at all (or at least as little as possible) until the fault is cured.
d) DO NOT push- or tow-start the car – this will soak the catalytic converter in unburned fuel, causing it to overheat when the engine does start.
e) DO NOT switch off the ignition at high engine speeds.
f) DO NOT use fuel or engine oil additives – these may contain substances harmful to the catalytic converter.
g) DO NOT continue to use the car if the engine burns oil to the extent of leaving a visible trail of blue smoke.

h) Remember that the catalytic converter operates at very high temperatures. DO NOT, therefore, park the car in dry undergrowth, over long grass or piles of dead leaves after a long run.

i) Remember that the catalytic converter is FRAGILE – do not strike it with tools during servicing work.

j) In some cases a sulphurous smell (like that of rotten eggs) may be noticed from the exhaust. This is common to many catalytic converter-equipped cars and once the car has covered a few thousand miles the problem should disappear.

k) The catalytic converter, used on a well-maintained and well-driven car, should last at least 100 000 miles – if the converter is no longer effective it must be renewed.

l) If a substantial loss of power is experienced, remember that this could be due to the converter being blocked. This can occur simply as a result of high mileage, but may be due to the ceramic element having fractured and collapsed internally (see paragraph 3). A new converter is the only cure in this instance.

m) As mentioned above, driving through deep water should be avoided if possible. The sudden cooling effect may fracture the ceramic honeycomb, damaging it beyond repair.

Diesel engines

5 Refer to the information given in parts F, G, H, I, K, L, and M of the petrol engines information given above.

Chapter 5 Part A
Starting and charging systems

Contents

Degrees of difficulty

Easy, suitable for novice with little experience	**Fairly easy,** suitable for beginner with some experience	**Fairly difficult,** suitable for competent DIY mechanic	**Difficult,** suitable for experienced DIY mechanic	**Very difficult,** suitable for expert DIY or professional

Specifications

General

Electrical system type . . . . 12 volt negative earth

Battery

Type . . . . Lead-acid, 'maintenance-free' (sealed for life)
Charge condition:
　Poor . . . . 12.5 volts
　Normal . . . . 12.6 volts
　Good . . . . 12.7 volts

Torque wrench settings

	Nm	lbf ft
Alternator:		
Petrol engine models	35	26
Diesel engine models:		
1.6 litre engines.	22	16
1.7 litre engines:		
M8 bolt	22	16
M10 bolt	35	26
Auxiliary drivebelt tensioner:		
Petrol engine models:		
1.4 litre engines:		
M8 bolt	22	16
M10 bolt	55	41
1.6 litre engines.	55	41
Diesel engine models:		
1.6 litre engines.	55	41
1.7 litre engines.	49	36
Battery retaining bracket securing bolt	22	16
Battery tray retaining bolts	14	10
Driveshaft intermediate shaft support bracket bolts	58	43
Glow plugs:		
1.6 litre engines	10	7
1.7 litre engines	14	10
Oil pressure warning light switch:		
Petrol engine models:		
1.4 litre engines.	20	15
1.6 litre engines.	30	22
Diesel engine models:		
1.6 litre engines.	35	26
1.7 litre engines.	20	15
Roadwheel nuts	140	103
Starter motor:		
Petrol engine models	25	18
Diesel engine models:		
1.6 litre engines.	22	16
1.7 litre engines:		
Upper bolt.	60	44
Lower bolt.	38	28

1 General information, precautions and battery disconnection

General information

1 The engine electrical system consists mainly of the charging and starting systems, and the diesel engine pre/post-heating system. Because of their engine-related functions, these components are covered separately from the body electrical devices such as the lights, instruments, etc (which are covered in Chapter 12). On petrol engine models refer to Chapter 5B for information on the ignition system.

2 The electrical system is of 12-volt negative earth type.

3 The battery is of the maintenance-free (sealed for life) type, and is charged by the alternator, which is belt-driven from the crankshaft pulley.

4 The starter motor is of pre-engaged type incorporating an integral solenoid. On starting, the solenoid moves the drive pinion into engagement with the flywheel/driveplate ring gear before the starter motor is energised. Once the engine has started, a one-way clutch prevents the motor armature being driven by the engine until the pinion disengages.

5 Further details of the various systems are given in the relevant Sections of this Chapter. While some repair procedures are given, the usual course of action is to renew the component concerned.

Precautions

6 It is necessary to take extra care when working on the electrical system to avoid damage to semi-conductor devices (diodes and transistors), and to avoid the risk of personal injury. In addition to the precautions given in *Safety first!*, observe the following when working on the system:

● *Always remove rings, watches, etc before working on the electrical system. Even with the battery disconnected, capacitive discharge could occur if a component's live terminal is earthed through a metal object. This could cause a shock or nasty burn.*

● *Do not reverse the battery connections. Components such as the alternator, electronic control units, or any other components having semi-conductor circuitry could be irreparably damaged.*

● *If the engine is being started using jump leads and a slave battery, connect the batteries positive-to-positive and negative-to-negative (see Jump starting). This also applies when connecting a battery charger but in this case both of the battery terminals should first be disconnected.*

● *Never disconnect the battery terminals, the alternator, any electrical wiring or any test instruments when the engine is running.*

● *Do not allow the engine to turn the alternator when the alternator is not connected.*

● *Never test for alternator output by flashing the output lead to earth.*

● *Never use an ohmmeter of the type incorporating a hand-cranked generator for circuit or continuity testing.*

● *Always ensure that the battery negative lead is disconnected when working on the electrical system.*

● *Before using electric-arc welding equipment on the car, disconnect the battery, alternator and components such as the fuel injection/ignition electronic control unit to protect them from the risk of damage.*

Battery disconnection

7 Refer to the precautions listed in Section 4.

2 Electrical fault finding – general information

1 Refer to Chapter 12, Section 2.

3 Battery – testing and charging

Testing

Traditional and low maintenance battery

1 If the vehicle covers a small annual mileage, it is worthwhile checking the specific gravity of the electrolyte every three months to determine the state of charge of the battery. Use a hydrometer to make the check and compare the results with the following table. Note that the specific gravity readings assume an electrolyte temperature of 15°C (60°F); for every 10°C (18°F) below 15°C (60°F) subtract 0.007. For every 10°C (18°F) above 15°C (60°F) add 0.007.

Ambient temperature:	above 25°C (77°F)	below 25°C (77°F)
Fully-charged	1.210 to 1.230	1.270 to 1.290
70% charged	1.170 to 1.190	1.230 to 1.250
Discharged	1.050 to 1.070	1.110 to 1.130

2 If the battery condition is suspect, first check the specific gravity of electrolyte in each cell. A variation of 0.040 or more between any cells indicates loss of electrolyte or deterioration of the internal plates.
3 If the specific gravity variation is 0.040 or more, the battery should be renewed. If the cell variation is satisfactory but the battery is discharged, it should be charged as described later in this Section.

Maintenance-free battery

4 Where a 'sealed for life' maintenance-free battery is fitted, topping-up and testing of the electrolyte is not possible. The condition of the battery can therefore only be tested using a battery condition indicator or a voltmeter.
5 Some models may be fitted with a maintenance-free battery with a built-in 'magic-eye' charge condition indicator. The indicator is located in the top of the battery casing, and indicates the condition of the battery from its colour. If the indicator shows green, then the battery is in a good state of charge. If the indicator turns darker, eventually to black, then the battery requires charging, as described later in this Section. If the indicator shows clear/yellow, then the electrolyte level in the battery is too low to allow further use, and the battery should be renewed. Do not attempt to charge, load or jump start a battery when the indicator shows clear/yellow.

All battery types

6 If testing the battery using a voltmeter, connect the voltmeter across the battery and compare the result with those given in the Specifications under 'charge condition'. The test is only accurate if the battery has not been subjected to any kind of charge for the previous six hours. If this is not the case, switch on the headlights for 30 seconds, then wait four to five minutes before testing the battery after switching off the headlights. All other electrical circuits must be switched off, so check that the doors and tailgate are fully shut when making the test.
7 If the voltage reading is less than 12.2 volts, then the battery is discharged, whilst a reading of 12.2 to 12.4 volts indicates a partially-discharged condition.
8 If the battery is to be charged, remove it from the vehicle (Section 4) and charge it as described later in this Section.

Charging

Note: *The following is intended as a guide only. Always refer to the manufacturer's recommendations (often printed on a label attached to the battery) before charging a battery.*

Traditional and low maintenance battery

9 Charge the battery at a rate of 3.5 to 4 amps and continue to charge the battery at this rate until no further rise in specific gravity is noted over a four hour period.
10 Alternatively, a trickle charger charging at the rate of 1.5 amps can safely be used overnight.
11 Specially rapid 'boost' charges which are claimed to restore the power of the battery in 1 to 2 hours are not recommended, as they can cause serious damage to the battery plates through overheating.
12 While charging the battery, note that the temperature of the electrolyte should never exceed 37.8°C (100°F).

Maintenance-free battery

13 This battery type takes considerably longer to fully recharge than the standard type, the time taken being dependent on the extent of discharge, but it will take anything up to three days.
14 A constant voltage type charger is required, to be set, when connected, to 13.9 to 14.9 volts with a charger current below 25 amps. Using this method, the battery should be usable within three hours, giving a voltage reading of 12.5 volts, but this is for a partially-discharged battery and, as mentioned, full charging can take considerably longer.
15 If the battery is to be charged from a fully-discharged state (condition reading less than 12.2 volts), have it recharged by your Vauxhall dealer or local automotive electrician, as the charge rate is higher and constant supervision during charging is necessary.

4 Battery – disconnection, reconnection, removal and refitting

Battery

Disconnection and reconnection

1 Numerous systems fitted to the vehicle require battery power to be available at all times, either to ensure their continued operation (such as the clock) or to maintain control unit memories which would be erased if the battery were to be disconnected. Whenever the battery is to be disconnected therefore, first note the following, to ensure that there are no unforeseen consequences of this action:

a) *First, on any vehicle with central locking, it is a wise precaution to remove the key from the ignition, and to keep it with you, so that it does not get locked in, if the central locking should engage accidentally when the battery is reconnected.*

b) *The engine management electronic control unit is of the 'self-learning' type, meaning that as it operates, it also monitors and stores the settings which give optimum engine performance under all operating conditions. When the battery is disconnected, these settings are lost and the ECU reverts to the base settings programmed into its memory at the factory. On restarting, this may lead to the engine running/idling roughly for a short while, until the ECU has re-learned the optimum settings. This process is best accomplished by taking the vehicle on a road test (for approximately 15 minutes), covering all engine speeds and loads, concentrating mainly in the 2500 to 3500 rpm region.*

c) *It will be necessary to reprogramme the electric window motors to restore the one-touch function of the buttons, after reconnection of the battery. To do this, open then fully close both front windows. With the windows closed, depress the up button of the driver's side window for approximately 5 seconds, then release it and depress the passenger side window up button for approximately 5 seconds. Repeat this procedure for the rear windows.*

d) *On models equipped with automatic transmission, the transmission selector lever assembly incorporates an electrically-operated selector lever lock mechanism that prevents the lever being moved out of the P position unless the ignition is switched on and the brake pedal is depressed. If the selector lever is in the P position and the battery is disconnected, it will not be possible to move the selector lever out of position P by the normal means. Although it*

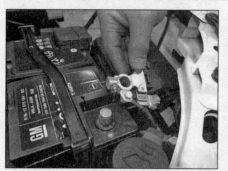

4.3 Slacken the clamp nut and disconnect the lead at the battery negative (–) terminal

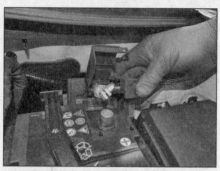

4.4 Slacken the clamp nut and disconnect the lead at the battery positive (+) terminal

4.7a Undo the battery retaining bracket securing bolt...

4.7b ...then disengage the retaining bracket from the battery stay and remove it

4.8 Carefully lift the battery from its location

8 Carefully lift the battery from its location and remove it from the car **(see illustration)**. Make sure the battery is kept upright at all times.

Refitting

9 Refitting is a reversal of removal, tightening the retaining bracket bolt to the specified torque. Reconnect the battery terminals as described previously in this Section.

Battery tray

Removal

10 Remove the battery as described previously in this Section.
11 Unclip the engine management ECU mounting bracket from the battery tray and move the ECU and bracket to one side **(see illustration)**.
12 Using a suitable forked tool, release the retaining clips securing the wiring harness to the battery tray **(see illustration)**.
13 Extract the plastic fastener securing the wiring harness trough to the side of the battery tray **(see illustration)**.
14 Where fitted, unclip the battery current sensor from the battery tray and move it to one side.
15 Undo the three retaining bolts and lift out the battery tray **(see illustrations)**.

Refitting

16 Refitting is a reversal of removal.

---

is possible to manually override the system (see Chapter 7B, Section 3), it is sensible to move the selector lever to the N position before disconnecting the battery.
e) On models with an electric sliding sunroof, it will be necessary to fully open and fully close the sunroof after battery reconnection, to recalibrate the sensors.
f) On all models, when reconnecting the battery after disconnection, switch on the ignition and wait 10 seconds to allow the electronic vehicle systems to stabilise and re-initialise.
2 The battery is located at the front, left-hand side of the engine compartment.
3 Disconnect the lead at the negative (–) terminal by slackening the retaining nut and removing the terminal clamp **(see illustration)**.

Note that the battery negative (–) and positive (+) terminal connections are stamped on the battery case.
4 Disconnect the lead at the battery positive (+) terminal by lifting up the terminal cover, then slackening the retaining nut and removing the terminal clamp **(see illustration)**.
5 Reconnection is a reversal of disconnection, ensuring that the battery positive (+) lead is reconnected first.

Removal

6 Disconnect the battery terminals as described previously in this Section.
7 Undo the bolt securing the battery retaining bracket to the front of the battery tray. Disengage the retaining bracket from the battery tray stay and remove the retaining bracket **(see illustrations)**.

4.11 Unclip the ECU mounting bracket from the battery tray

4.12 Release the wiring harness retaining clips from the battery tray

4.13 Extract the plastic fastener and release the wiring harness trough from the battery tray

5 Charging system – testing

Note: *Refer to the precautions given in 'Safety first!' and in Section 1 of this Chapter before starting work.*

1 If the ignition no-charge warning light fails to illuminate when the ignition is switched on, first check the alternator wiring connections for security. If satisfactory, check the condition of all related fuses, fusible links, wiring connections and earthing points. If this fails to reveal the fault, the vehicle should be taken to a Vauxhall/Opel dealer or auto-electrician, as further testing entails the use of specialist diagnostic equipment.

2 If the ignition warning light illuminates when the engine is running, stop the engine and check the condition of the auxiliary drivebelt (see Chapter 1A, Section 20 or Chapter 1B, Section 22) and the security of the alternator wiring connections. If satisfactory, have the alternator checked by a Vauxhall/Opel dealer or auto-electrician.

3 If the alternator output is suspect even though the warning light functions correctly, the regulated voltage may be checked as follows.

4 Connect a voltmeter across the battery terminals, and start the engine.

5 Increase the engine speed until the voltmeter reading remains steady; the reading should be approximately 12 to 13 volts, and no more than 14 volts.

6 Switch on as many electrical accessories (eg, the headlights, heated rear window and heater blower) as possible, and check that the alternator maintains the regulated voltage at around 13.5 to 14.5 volts.

7 If the regulated voltage is not as stated, the fault may be due to worn brushes, weak brush springs, a faulty voltage regulator, a faulty diode, a severed phase winding, or worn or damaged slip rings. The alternator should be renewed or taken to a Vauxhall/Opel dealer or auto-electrician for testing and repair.

6 Auxiliary drivebelt – removal and refitting

1 Refer to Chapter 1A, Section 29 or Chapter 1B, Section 30 as applicable.

7 Auxiliary drivebelt tensioner – removal and refitting

Removal

1 Firmly apply the handbrake, then jack up the front of the car and support it securely on axle stands (see *Jacking and vehicle support*). Remove the right-hand front roadwheel.

4.15a Undo the three retaining bolts…

4.15b …and lift out the battery tray

7.3a On 1.4 litre engines, undo the two bolts and remove the tensioner assembly

7.3b On 1.6 litre engines, undo the central mounting bolt and remove the tensioner assembly

2 Remove the auxiliary drivebelt as described in Chapter 1A, Section 29, or Chapter 1B, Section 30, as applicable.

3 On petrol engine models, undo the two retaining bolts (1.4 litre engines), or the central mounting bolt (1.6 litre engines) and remove the tensioner assembly from the engine **(see illustrations)**.

4 On diesel engine models, undo the central mounting bolt and remove the tensioner assembly from the engine.

Refitting

5 Place the tensioner assembly in position ensuring that the locating pegs (where fitted) on the tensioner mounting surface engage correctly with the corresponding holes in the mounting bracket. Tighten the tensioner retaining bolts, or the central mounting bolt to the specified torque.

8.6 Alternator mounting bolts – petrol engine models

6 Refit the auxiliary drivebelt as described in Chapter 1A, Section 29 or Chapter 1B, Section 30 as applicable.

7 Refit the roadwheel then lower the car to the ground. Tighten the roadwheel nuts to the specified torque.

8 Alternator – removal and refitting

Removal

Petrol engine models

1 Disconnect the battery negative lead as described in Section 4.

2 Firmly apply the handbrake, then jack up the front of the car and support it securely on axle stands (see *Jacking and vehicle support*). Remove the right-hand front roadwheel.

3 Remove the auxiliary drivebelt as described in Chapter 1A, Section 29.

4 On 1.4 litre engines, remove the auxiliary drivebelt tensioner as described in Section 7.

5 On 1.4 litre engines, unscrew the retaining nut and disconnect the wiring terminal, then unplug the wiring block connector from the rear of the alternator. On 1.6 litre engines, unscrew the two retaining nuts and disconnect the wiring from the rear of the alternator.

6 Unscrew the upper and lower mounting bolts and withdraw the alternator upwards from the block **(see illustration)**.

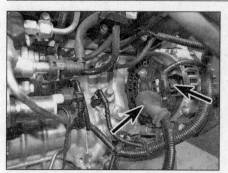

8.14 Wiring connections at the rear of the alternator

8.17 Alternator mounting bolts – 1.6 litre diesel engine models

8.25 Alternator mounting bolts – 1.7 litre diesel engine models

1.6 litre diesel engine models

7 Disconnect the battery negative lead as described in Section 4.

8 Firmly apply the handbrake, then jack up the front of the car and support it securely on axle stands (see *Jacking and vehicle support*). Remove the right-hand front roadwheel.

9 Remove the auxiliary drivebelt as described in Chapter 1B, Section 30.

10 Remove the front suspension lower arm on the right-hand side as described in Chapter 10, Section 5.

11 Remove the driveshaft intermediate shaft as described in Chapter 8, Section 3.

12 On 4WD models, remove the transfer gearbox as described in Chapter 7C, Section 3.

13 Disconnect the wiring connector from the oil pressure warning light switch to the rear of the alternator.

14 Remove the protective boot, then unscrew the retaining nut and disconnect the battery positive cable wiring from the alternator stud **(see illustration)**.

15 Unplug the wiring block connector from the rear of the alternator.

16 Release the retaining clips and free the wiring harness from the support bracket on the alternator.

17 Undo the three mounting bolts, then lower the alternator and remove it from under the car **(see illustration)**.

1.7 litre diesel engine models

18 Disconnect the battery negative lead as described in Section 4.

19 Firmly apply the handbrake, then jack up the front of the car and support it securely on axle stands (see *Jacking and vehicle support*). Remove the right-hand front roadwheel.

20 Remove the auxiliary drivebelt as described in Chapter 1B, Section 30.

21 Remove the intercooler inlet air duct as described in Chapter 4B, Section 3.

22 Remove the protective boot, then unscrew the retaining nut and disconnect the battery positive cable wiring from the alternator stud.

23 Unplug the wiring block connector from the rear of the alternator.

24 Release the retaining clip and free the oxygen sensor wiring harness from the alternator.

25 Undo the two mounting bolts, then lower the alternator and remove it from under the car **(see illustration)**.

Refitting

Petrol engine models

26 Manoeuvre the alternator into position and refit the upper and lower mounting bolts. Tighten the bolts to the specified torque.

27 Reconnect the wiring to the alternator terminals and tighten the retaining nuts securely. On 1.4 litre engines, reconnect the wiring block connector.

28 On 1.4 litre engines, refit the auxiliary drivebelt tensioner as described in Section 7.

29 Refit the auxiliary drivebelt as described in Chapter 1A, Section 29.

30 Refit the roadwheel then lower the car to the ground. Tighten the roadwheel nuts to the specified torque.

31 On completion, reconnect the battery negative terminal.

1.6 litre diesel engine models

32 Manoeuvre the alternator into position and refit the three mounting bolts. Tighten the bolts to the specified torque.

33 Reconnect the alternator wiring block connector and the terminal wiring, tightening the retaining nut securely.

34 Reconnect the wiring connector to the oil pressure warning light switch.

35 Attach the retaining clip and secure the wiring harness to the support bracket on the alternator.

36 On 4WD models, refit the transfer gearbox as described in Chapter 7C, Section 3.

37 Refit the driveshaft intermediate shaft as described in Chapter 8, Section 3.

38 Refit the front suspension lower arm as described in Chapter 10, Section 5.

39 Refit the auxiliary drivebelt as described in Chapter 1B, Section 30.

40 Refit the roadwheel then lower the car to the ground. Tighten the roadwheel nuts to the specified torque.

41 On completion, reconnect the battery negative terminal.

1.7 litre diesel engine models

42 Manoeuvre the alternator into position and refit the upper and lower mounting bolts. Tighten the bolts to the specified torque.

43 Attach the retaining clip and secure the oxygen sensor wiring harness to the alternator.

44 Reconnect the wiring to the alternator terminal and tighten the retaining nut securely. Reconnect the wiring plug.

45 Refit the intercooler inlet air duct as described in Chapter 4B, Section 3.

46 Refit the auxiliary drivebelt as described in Chapter 1B, Section 30.

47 Refit the roadwheel then lower the car to the ground. Tighten the roadwheel nuts to the specified torque.

48 On completion, reconnect the battery negative terminal.

9 Alternator – testing and overhaul

1 If the alternator is thought to be suspect, it should be removed from the vehicle and taken to an auto-electrician for testing. Most auto-electricians will be able to supply and fit brushes at a reasonable cost. However, check on the cost of repairs before proceeding as it may prove more economical to obtain a new or exchange alternator.

10 Starting system – testing

Note: *Refer to the precautions given in 'Safety first!' and in Section 1 of this Chapter before starting work.*

1 If the starter motor fails to operate when the ignition key is turned to the appropriate position, the possible causes are as follows:

a) *The engine immobiliser is faulty.*

b) *The battery is faulty.*

c) *The electrical connections between the switch, solenoid, battery and starter motor are somewhere failing to pass the necessary current from the battery through the starter to earth.*

d) *The solenoid is faulty.*

e) *The starter motor is mechanically or electrically defective.*

2 To check the battery, switch on the headlights. If they dim after a few seconds, this indicates that the battery is discharged – recharge (see Section 3) or renew the battery. If the headlights glow brightly, operate the starter switch while watching the headlights. If they dim, then this indicates that current is reaching the starter motor, therefore the fault must lie in the starter motor. If the lights continue to glow brightly (and no clicking sound can be heard from the starter motor solenoid), this indicates that there is a fault in the circuit or solenoid – see the following paragraphs. If the starter motor turns slowly when operated, but the battery is in good condition, then this indicates either that the starter motor is faulty, or there is considerable resistance somewhere in the circuit.

3 If a fault in the circuit is suspected, disconnect the battery leads (including the earth connection to the body), the starter/solenoid wiring and the engine/transmission earth strap. Thoroughly clean the connections, and reconnect the leads and wiring. Use a voltmeter or test light to check that full battery voltage is available at the battery positive lead connection to the solenoid. Smear petroleum jelly around the battery terminals to prevent corrosion – corroded connections are among the most frequent causes of electrical system faults.

4 If the battery and all connections are in good condition, check the circuit by disconnecting the ignition switch supply wire from the solenoid terminal. Connect a voltmeter or test lamp between the wire end and a good earth (such as the battery negative terminal), and check that the wire is live when the ignition switch is turned to the 'start' position. If it is, then the circuit is sound – if not the circuit wiring can be checked as described in Chapter 12, Section 2.

5 The solenoid contacts can be checked by connecting a voltmeter or test light between the battery positive feed connection on the starter side of the solenoid and earth. When the ignition switch is turned to the 'start' position, there should be a reading or lighted bulb, as applicable. If there is no reading or lighted bulb, the solenoid is faulty and should be renewed.

6 If the circuit and solenoid are proved sound, the fault must lie in the starter motor. In this

11.7 Starter motor retaining bolts

event, it may be possible to have the starter motor overhauled by a specialist, but check on the cost of spares before proceeding, as it may prove more economical to obtain a new or exchange motor.

11 Starter motor –
removal and refitting

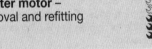

Removal

Petrol engine models

1 Disconnect the battery negative lead as described in Section 4.

2 Firmly apply the handbrake, then jack up the front of the car and support it securely on axle stands (see *Jacking and vehicle support*).

3 On 4WD models, remove the transfer gearbox as described in Chapter 7C, Section 3.

4 On 1.4 litre engines, undo the retaining bolt and move the oxygen sensor wiring harness support bracket to one side. On 1.6 litre engines, undo the two bolts and remove the inlet manifold support bracket.

5 Slacken and remove the two retaining nuts and disconnect the wiring from the starter motor solenoid. Recover the washers under the nuts.

6 Unscrew the retaining nut and disconnect the earth lead from the starter motor lower stud bolt.

7 Slacken and remove the retaining bolts then manoeuvre the starter motor out from underneath the engine **(see illustration)**.

1.6 litre diesel engine models

8 Disconnect the battery negative lead as described in Section 4.

9 Firmly apply the handbrake, then jack up the front of the car and support it securely on axle stands (see *Jacking and vehicle support*).

10 Remove the engine undertray as described in Chapter 11, Section 21.

11 On 4WD models, remove the transfer gearbox as described in Chapter 7C, Section 3.

12 Undo the three bolts securing the driveshaft intermediate shaft support bracket to the cylinder block **(see illustration)**. Rotate the support bracket away from the starter motor.

13 Disconnect the starter motor solenoid wiring plug, then undo the retaining nut and disconnect the battery positive cable from the solenoid stud **(see illustration)**.

14 Slacken and remove the two retaining bolts then manoeuvre the starter motor out from underneath the engine **(see illustration)**.

1.7 litre diesel engine models

15 Remove the battery as described in Section 4.

16 Unscrew and remove the starter motor upper mounting bolt.

17 Apply the handbrake, then jack up the front of the vehicle and support it on axle stands (see *Jacking and vehicle support*).

18 Remove the engine undertray as described in Chapter 11, Section 21.

19 On 4WD models, remove the transfer gearbox as described in Chapter 7C, Section 3.

20 Slacken and remove the two retaining nuts and disconnect the wiring from the starter motor solenoid. Recover the washers under the nuts.

21 Unscrew the retaining nut and disconnect the earth lead from the starter motor lower stud bolt.

22 Unscrew the lower stud bolt and withdraw the starter motor from the engine.

Refitting

23 Refitting is a reversal of removal, bearing in mind the following points:

a) *Tighten the starter retaining bolts to the specified torque.*

11.12 Driveshaft intermediate shaft support bracket retaining bolts

11.13 Disconnect the starter solenoid wiring connectors

11.14 Starter motor removal

14.1 Oil pressure warning light switch location – 1.4 litre petrol engine

14.8 Oil pressure warning light switch location – 1.6 litre diesel engine

b) On 4WD models, refit the transfer gearbox as described in Chapter 7C, Section 3.

c) Where applicable, refit the engine undertray as described in Chapter 11, Section 21.

d) On 1.7 litre diesel engine models, refit the battery as described in Section 4.

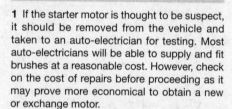

12 Starter motor – testing and overhaul

1 If the starter motor is thought to be suspect, it should be removed from the vehicle and taken to an auto-electrician for testing. Most auto-electricians will be able to supply and fit brushes at a reasonable cost. However, check on the cost of repairs before proceeding as it may prove more economical to obtain a new or exchange motor.

13 Ignition switch – removal and refitting

1 The switch is integral with the steering column lock, and removal and refitting is described in Chapter 10, Section 14.

14 Oil pressure warning light switch – removal and refitting

Removal

1.4 litre petrol engine models

1 The oil pressure warning light switch is screwed into the oil filter housing at the front of the cylinder block **(see illustration)**.

2 Pull back the cover and disconnect the wiring from the switch.

3 Place some cloth rags beneath the switch, then unscrew the switch and recover the sealing washer. Be prepared for oil spillage, and if the switch is to be left removed from the engine for any length of time, plug the switch aperture

1.6 litre petrol engine models

4 The oil pressure warning light switch is

screwed into the front of the cylinder block.

5 Firmly apply the handbrake, then jack up the front of the car and support it securely on axle stands (see *Jacking and vehicle support*).

6 Refer to Chapter 3, Section 13 and unbolt the air conditioning compressor from the cylinder block without disconnecting the refrigerant lines. Support the compressor to one side for access to the oil pressure warning light switch.

7 Disconnect the wiring connector then unscrew the switch and recover the sealing washer. Be prepared for oil spillage, and if the switch is to be left removed from the engine for any length of time, plug the switch aperture.

1.6 litre diesel engine models

8 The oil pressure warning light switch is screwed into the rear of the cylinder block, above the oil cooler **(see illustration)**.

9 Firmly apply the handbrake, then jack up the front of the car and support it securely on axle stands (see *Jacking and vehicle support*).

10 Remove the engine undertray as described in Chapter 11, Section 21.

11 Disconnect the wiring connector from the switch.

12 Place some cloth rags beneath the switch, then unscrew the switch from the cylinder block. Be prepared for oil spillage, and if the switch is to be left removed from the engine for any length of time, plug the switch aperture

1.7 litre diesel engine models

13 The oil pressure warning light switch is

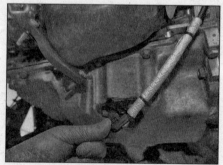

15.4 Disconnect the wiring connector from the oil level sensor

screwed into the oil filter housing at the rear of the engine.

14 Remove the plastic cover over the top of the engine.

15 Remove the starter motor as described in Section 11.

16 Disconnect the switch wiring connector, then unscrew the switch and collect the O-ring seal. Be prepared for oil spillage, and if the switch is to be left removed from the engine for any length of time, plug the switch aperture.

Refitting

17 Refitting is a reversal of removal, bearing in mind the following points:

e) Examine the sealing washer for signs of damage or deterioration and if necessary renew.

f) Tighten the switch to the specified torque.

g) On 1.6 litre petrol engine models, refit the air conditioning compressor with reference to Chapter 3, Section 13.

h) Where applicable, refit the engine undertray as described in Chapter 11, Section 21.

i) Check, and if necessary top-up the engine oil as described in 'Weekly checks'.

15 Oil level sensor – removal and refitting

Note: *An oil level sensor is only fitted to diesel engine models.*

Removal

1.6 litre engines

1 Disconnect the battery negative lead as described in Section 4.

2 Firmly apply the handbrake, then jack up the front of the car and support it securely on axle stands (see *Jacking and vehicle support*).

3 Drain the engine oil as described in Chapter 1B, Section 5.

4 Disconnect the wiring connector from the oil level sensor located on the underside of the sump **(see illustration)**.

5 Undo the three retaining bolts, withdraw the sensor from the sump and collect the seal.

6 Check the condition of the seal and renew it if there is any doubt about its condition.

1.7 litre engines

7 Remove the sump lower pan from the main casting as described in Chapter 2D, Section 14.

8 Disconnect the wiring plug, then extract the retaining clip securing the level sensor wiring connector to the sump main casting.

9 Undo the three bolts securing the oil level sensor mounting bracket to the sump main casting. Pull the wiring connector out of the sump casting, then remove the level sensor and mounting bracket from under the car.

10 Undo the two nuts and remove the level sensor from the mounting bracket.

11 Remove the O-ring seal from the level sensor wiring connector and obtain a new seal for refitting.

Refitting

12 Refitting is a reversal of removal, bearing in mind the following points:
j) Thoroughly clean all mating surfaces and obtain new seals as necessary.
k) On 1.7 litre engines, refit the sump as described in Chapter 2D, Section 14.
l) Refill the engine with fresh oil as described in Chapter 1B, Section 5.

---

16 Pre/post-heating system (diesel engine models) – description and testing

Description

1 Each cylinder of the engine is fitted with a heater plug (commonly called a glow plug) screwed into it. The plugs are electrically-operated before and during start-up when the engine is cold. Electrical feed to the glow plugs is controlled by the engine management system ECU.
2 A warning light in the instrument panel tells the driver that pre/post-heating is taking place. When the light goes out, the engine is ready to be started. The voltage supply to the glow plugs continues for several seconds after the light goes out. If no attempt is made to start, the timer then cuts off the supply, in order to avoid draining the battery and overheating the glow plugs.
3 The glow plugs also provide a 'post-heating' function, whereby the glow plugs remain switched on after the engine has started. The length of time 'post-heating' takes place is also determined by the control unit, and is dependant on engine temperature.
4 The fuel filter is fitted with a heating element to prevent the fuel 'waxing' in extreme cold temperature conditions and to improve combustion. The heating element is an integral part of the fuel filter housing and is controlled by the ECU.

Testing

5 If the system malfunctions, testing is ultimately by substitution of known good units, but some preliminary checks may be made as follows.
6 Connect a voltmeter or 12 volt test lamp between the glow plug supply cable and earth (engine or vehicle metal). Make sure that the live connection is kept clear of the engine and bodywork.
7 Have an assistant switch on the ignition, and check that voltage is applied to the glow plugs. Note the time for which the warning light is lit, and the total time for which voltage is applied before the system cuts out. Switch off the ignition.
8 At an underbonnet temperature of 20°C (68°F), typical times noted should be approximately 3 seconds for warning light operation. Warning light time will increase with lower temperatures and decrease with higher temperatures.
9 If there is no supply at all, the control unit or associated wiring is at fault.
10 To locate a defective glow plug, disconnect the wiring connector from each plug.
11 Use a continuity tester, or a 12 volt test lamp connected to the battery positive terminal, to check for continuity between each glow plug terminal and earth. The resistance of a glow plug in good condition is very low (less than 1 ohm), so if the test lamp does not light or the continuity tester shows a high resistance, the glow plug is certainly defective.
12 If an ammeter is available, the current draw of each glow plug can be checked. After an initial surge of 15 to 20 amps, each plug should draw 12 amps. Any plug which draws much more or less than this is probably defective.
13 As a final check, the glow plugs can be removed and inspected as described in the following Section.

---

17 Glow plugs (diesel engine models) – removal, inspection and refitting

Caution: If the pre/post-heating system has just been energised, or if the engine has been running, the glow plugs will be very hot.

Removal

1 The glow plugs are located at the rear of the cylinder head above the inlet manifold.
2 Lift off the plastic cover over the top of the engine by pulling it upwards off the mounting studs.
3 Disconnect the wiring from the glow plugs by squeezing the connectors with thumb and forefinger, and pulling them from the plugs.
4 Unscrew the glow plugs and remove them from the cylinder head.

Inspection

5 Inspect each glow plug for physical damage. Burnt or eroded glow plug tips can be caused by a bad injector spray pattern. Have the injectors checked if this type of damage is found.
6 If the glow plugs are in good physical condition, check them electrically using a 12 volt test lamp or continuity tester as described in the previous Section.
7 The glow plugs can be energised by applying 12 volts to them to verify that they heat up evenly and in the required time. Observe the following precautions.
m) Support the glow plug by clamping it carefully in a vice or self-locking pliers. Remember it will become red-hot.
n) Make sure that the power supply or test lead incorporates a fuse or overload trip to protect against damage from a short-circuit.
o) After testing, allow the glow plug to cool for several minutes before attempting to handle it.
8 A glow plug in good condition will start to glow red at the tip after drawing current for 5 seconds or so. Any plug which takes much longer to start glowing, or which starts glowing in the middle instead of at the tip, is defective.

Refitting

9 Carefully refit the plugs and tighten to the specified torque. Do not overtighten, as this can damage the glow plug element. Push the electrical connectors firmly onto the glow plugs.
10 The remainder of refitting is a reversal of removal, checking the operation of the glow plugs on completion.

Chapter 5 Part B
Ignition system – petrol engines

Contents

Degrees of difficulty

Easy, suitable for novice with little experience	**Fairly easy,** suitable for beginner with some experience	**Fairly difficult,** suitable for competent DIY mechanic	**Difficult,** suitable for experienced DIY mechanic	**Very difficult,** suitable for expert DIY or professional

Specifications

General

System type	Distributorless ignition system
System application:	
1.4 litre (1364cc) engines	Bosch E78
1.4 litre (1398cc) and 1.6 litre engines	Hitachi E83
Location of No 1 cylinder	Timing chain/timing belt end of engine
Firing order	1-3-4-2

Torque wrench setting	Nm	lbf ft
Ignition module retaining bolts	8	6

1 Ignition system – general information

1 The ignition system is integrated with the fuel injection system to form a combined engine management system under the control of one electronic control unit (ECU) – see Chapter 4A for further information. The ignition side of the system is of the distributorless type, and consists of the ignition module and the knock sensor.

2 The ignition module consists of four ignition coils, one per cylinder, in one casing mounted directly above the spark plugs. This module eliminates the need for any HT leads as the coils locate directly onto the relevant spark plug. The ECU uses its inputs from the various sensors to calculate the required ignition advance setting and coil charging time.

3 The knock sensor is mounted onto the cylinder block and informs the ECU when the engine is 'pinking' under load. The sensor is sensitive to vibration and detects the knocking which occurs when the engine starts to 'pink' (pre-ignite). The knock sensor sends an electrical signal to the ECU which in turn retards the ignition advance setting until the 'pinking' ceases.

4 The ignition systems fitted to the engines covered by this manual all operate in a similar fashion under the overall control of the engine management ECU. The systems comprise various sensors (whose inputs also provide data to control the fuel injection system), and the ECU, in addition to the ignition module and spark plugs. Further details of the system sensors and the ECU are given in Chapter 4A.

5 The ECU selects the optimum ignition advance setting based on the information received from the various sensors, and fires the relevant ignition coil accordingly. The degree of advance can thus be constantly varied to suit the prevailing engine operating conditions.

⚠ **Warning: Due to the high voltages produced by the electronic ignition system, extreme care must be taken when working on the system with the ignition switched on.**

Persons with surgically-implanted cardiac pacemaker devices should keep well clear of the ignition circuits, components and test equipment.

2 Ignition system – testing

1 If a fault appears in the engine management system, first ensure that all the system wiring connectors are securely connected and free of corrosion. Ensure that the fault is not due to poor maintenance; ie, check that the air cleaner filter element is clean, the spark plugs are in good condition and correctly gapped, the cylinder compression pressures are correct and that the engine breather hoses are clear and undamaged, referring to Chapters 1A, 2A and 2B for further information.

2 If these checks fail to reveal the cause of the problem, the vehicle should be taken to a suitably-equipped Vauxhall/Opel dealer or engine management diagnostic specialist for testing. A diagnostic socket is located at

2.2 The vehicle diagnostic socket is located at the base of the facia on the driver's side

3.1a Using a plastic spatula or similar tool, release the retaining tabs...

3.1b...and remove the plastic cover over the top of the engine – 1.4 litre engines

3.2 Remove the ignition module cover by sliding it towards the transmission and lifting off – 1.6 litre engines

3.3 Disconnect the ignition module wiring connector

3.4 Undo the two ignition module retaining bolts

the base of the facia, on the driver's side, to which a fault code reader or other suitable test equipment can be connected (**see illustration**). By using the code reader or test equipment, the engine management ECU can be interrogated, and any stored fault codes can be retrieved. This will allow the fault to be quickly and simply traced, alleviating the need to test all the system components individually, which is a time-consuming operation that carries a risk of damaging the ECU.

3 The only ignition system checks which can be carried out by the home mechanic are those described in Chapter 1A relating to the spark plugs. If necessary, the system wiring and wiring connectors can be checked as described in Chapter 12, ensuring that the ECU wiring connector(s) have first been disconnected.

3 Ignition module – removal and refitting

Removal

1 On 1.4 litre engines, lift off the plastic cover over the top of the engine (**see illustrations**).
2 On 1.6 litre engines, unclip the wiring trough from the left-hand end of the cylinder head, then remove the cover from the ignition module by sliding it towards the transmission and lifting off (**see illustration**).
3 Disconnect the wiring connector at the left-hand end of the ignition module (**see illustration**).
4 Undo the two ignition module retaining bolts (**see illustration**).

5 Insert two long 8 mm bolts into the threaded holes in the top of the module, and pull up on the bolts to free the module from the spark plugs (**see illustrations**).

Refitting

6 Refitting is the reversal of removal, tightening the retaining bolts to the specified torque.

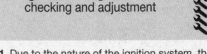

4 Ignition timing – checking and adjustment

1 Due to the nature of the ignition system, the ignition timing is constantly being monitored and adjusted by the engine management ECU.
2 The only way in which the ignition timing can be checked is by using specialist diagnostic test equipment, connected to the engine management system diagnostic socket. No adjustment of the ignition timing is possible. Should the ignition timing be incorrect, then a fault is likely to be present in the engine management system.

5 Knock sensor – removal and refitting

1 Refer to Chapter 4A, Section 11 (1.4 litre engines) or Chapter 4A, Section 12 (1.6 litre engines).

3.5a Insert two long 8 mm bolts into the threaded holes in the top of the module, and pull up on the bolts...

3.5b...to free the module from the spark plugs

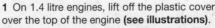

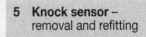

Chapter 6
Clutch

Contents

Degrees of difficulty

Easy, suitable for novice with little experience		Fairly easy, suitable for beginner with some experience		Fairly difficult, suitable for competent DIY mechanic		Difficult, suitable for experienced DIY mechanic		Very difficult, suitable for expert DIY or professional	

Specifications

Type . . . . Single dry plate with diaphragm spring, hydraulically-operated

Friction disc
Diameter:

1.4 litre petrol engine models	228 mm
1.6 litre petrol engine models	225 mm
1.6 litre diesel engine models	Information not available
1.7 litre diesel engine models	239 mm

Torque wrench settings
Pressure plate retaining bolts:

	Nm	lbf ft
1.4 litre petrol engine models	28	21
1.6 litre petrol engine models	15	11
1.6 litre diesel engine models	28	21
1.7 litre diesel engine models	28	21

1 General Information

1 The clutch consists of a friction disc, a pressure plate assembly, and the hydraulic release cylinder (which incorporates the release bearing); all of these components are contained in the large cast-aluminium alloy bellhousing, sandwiched between the engine and the transmission.

2 The friction disc is fitted between the engine flywheel and the clutch pressure plate, and is allowed to slide on the transmission input shaft splines.

3 The pressure plate assembly is bolted to the engine flywheel. When the engine is running, drive is transmitted from the crankshaft, via the flywheel, to the friction disc (these components being clamped securely together by the pressure plate assembly) and from the friction disc to the transmission input shaft.

4 To interrupt the drive, the spring pressure must be relaxed. This is achieved using a hydraulic release mechanism which consists of the master cylinder, the release cylinder and the pipe/hose linking the two components. Depressing the pedal pushes on the master cylinder pushrod which hydraulically forces the release cylinder piston against the pressure plate spring fingers. This causes the springs to deform and releases the clamping force on the friction disc.

5 The clutch is self-adjusting and requires no manual adjustment.

2 Clutch hydraulic system – bleeding

⚠️ *Warning: Hydraulic fluid is poisonous; wash off immediately and thoroughly in the case of skin contact, and seek immediate medical advice if any fluid is swallowed or gets into the eyes. Certain types of hydraulic fluid are flammable, and may ignite when allowed into contact with hot components; when servicing any hydraulic system, it is safest to assume that the fluid is flammable, and to take precautions against the risk of fire as though it is petrol that is being handled. Hydraulic fluid is also an effective paint stripper, and will attack plastics; if any is spilt, it should be washed off immediately, using copious quantities of fresh water. Finally, it is hygroscopic (it absorbs moisture from the air) – old fluid may be contaminated and unfit for further use. When topping-up or renewing the fluid, always use the recommended type, and ensure that it comes from a freshly-opened sealed container.*

General information

1 The correct operation of any hydraulic system is only possible after removing all air from the components and circuit; this is achieved by bleeding the system.

2 The manufacturer's stipulate that the system must be initially bled by the 'back-bleeding' method using Vauxhall/Opel special bleeding equipment. This entails connecting a pressure bleeding unit containing fresh brake fluid, to the release cylinder bleed screw, with a collecting vessel connected to the brake fluid master cylinder reservoir. The pressure bleeding unit is then switched on, the bleed screw is opened and hydraulic fluid is delivered under pressure, backwards, to be expelled from the reservoir into the collecting vessel. Final bleeding is then carried out in the conventional way.

3 In practice, this method would normally only be required if new hydraulic components have been fitted, or if the system has been

2.8 Clutch bleed screw

completely drained of hydraulic fluid. If the system has only been disconnected to allow component removal and refitting procedures to be carried out, such as removal and refitting of the transmission (for example for clutch replacement) or engine removal and refitting, then it is quite likely that normal bleeding will be sufficient.

4 Our advice would therefore be as follows:

a) If the hydraulic system has only been partially disconnected, try bleeding by the conventional methods described in paragraphs 10 to 15, or 16 to 19.

b) If the hydraulic system has been completely drained and new components have been fitted, try bleeding by using the pressure bleeding method described in paragraphs 20 to 22.

c) If the above methods fail to produce a firm pedal on completion, it will be necessary to 'back-bleed' the system using Vauxhall/Opel bleeding equipment, or suitable alternative equipment as described in paragraphs 23 to 28.

5 During the bleeding procedure, add only clean, unused hydraulic fluid of the recommended type; never re-use fluid that has already been bled from the system. Ensure that sufficient fluid is available before starting work.

6 If there is any possibility of incorrect fluid being already in the system, the hydraulic circuit must be flushed completely with uncontaminated, correct fluid.

7 If hydraulic fluid has been lost from the system, or air has entered because of a leak, ensure that the fault is cured before continuing further.

8 The bleed screw is located in the hose end fitting which is situated on the top of the transmission housing (see illustration).

9 Check that all pipes and hoses are secure, unions tight and the bleed screw is closed. Clean any dirt from around the bleed screw.

Bleeding procedure

Note: *Two different types of bleed screw may be encountered. If the bleed screw has a hexagonal section, a spanner will be needed to unscrew it. If the bleed screw has a serrated section, it can be unscrewed by hand. The following procedure assumes the bleed screw has a hexagonal section.*

Conventional method

10 Collect a clean glass jar, a suitable length of plastic or rubber tubing which is a tight fit over the bleed screw, and a ring spanner to fit the screw. The help of an assistant will also be required.

11 Unscrew the master cylinder fluid reservoir cap (the clutch shares the same fluid reservoir as the braking system), and top the master cylinder reservoir up to the upper (MAX) level line. Ensure that the fluid level is maintained at least above the lower level line in the reservoir throughout the procedure.

12 Remove the dust cap from the bleed

screw. Fit the spanner and tube to the screw, place the other end of the tube in the jar, and pour in sufficient fluid to cover the end of the tube.

13 Have the assistant fully depress the clutch pedal several times to build-up pressure, then maintain it on the final down stroke.

14 While pedal pressure is maintained, unscrew the bleed screw (approximately one turn) and allow the compressed fluid and air to flow into the jar. The assistant should maintain pedal pressure and should not release it until instructed to do so. When the flow stops, tighten the bleed screw again, have the assistant release the pedal slowly, and recheck the reservoir fluid level.

15 Repeat the steps given in paragraphs 13 and 14 until the fluid emerging from the bleed screw is free from air bubbles. If the master cylinder has been drained and refilled allow approximately five seconds between cycles for the master cylinder passages to refill.

Conventional method using a one-way valve kit

16 As their name implies, these kits consist of a length of tubing with a one-way valve fitted, to prevent expelled air and fluid being drawn back into the system; some kits include a translucent container, which can be positioned so that the air bubbles can be more easily seen flowing from the end of the tube.

17 The kit is connected to the bleed screw, which is then opened.

18 The user returns to the driver's seat, depresses the clutch pedal with a smooth, steady stroke, and slowly releases it; this is repeated until the expelled fluid is clear of air bubbles.

19 Note that these kits simplify work so much that it is easy to forget the clutch fluid reservoir level; ensure that this is maintained at least above the lower level line at all times.

Pressure-bleeding method

20 These kits are usually operated by the reservoir of pressurised air contained in the spare tyre. However, note that it will probably be necessary to reduce the pressure to a lower level than normal; refer to the instructions supplied with the kit.

21 By connecting a pressurised, fluid-filled container to the clutch fluid reservoir, bleeding can be carried out simply by opening the bleed screw and allowing the fluid to flow out until no more air bubbles can be seen in the expelled fluid.

22 This method has the advantage that the large reservoir of fluid provides an additional safeguard against air being drawn into the system during bleeding.

'Back-bleeding' method

23 The following procedure describes the bleeding method using Vauxhall/Opel equipment. Alternative equipment is available and should be used in accordance with the maker's instructions.

24 Connect the pressure hose of the bleeding kit (DT-6174-A) to the bleed screw located in the hose end fitting situated on the top of the transmission housing (**see illustration 2.8**). Connect the other end of the hose to a suitable pressure bleeding device set to operate at approximately 2.0 bar.

25 Attach the cap of the bleeding kit to the master cylinder reservoir, and place the hose in a collecting vessel.

26 Switch on the pressure bleeding equipment, open the bleed screw, and allow fresh hydraulic fluid to flow from the pressure bleeding unit, through the system and out through the top of the reservoir and into the collecting vessel. When fluid, free from air bubbles appears in the reservoir, close the bleed screw and switch off the bleeding equipment.

27 Disconnect the bleeding equipment from the bleed screw and reservoir.

28 Carry out a final conventional bleeding procedure as described in paragraphs 10 to 15, or 16 to 19.

All methods

29 When bleeding is complete, no more bubbles appear and correct pedal feel is restored, tighten the bleed screw securely (do not overtighten). Remove the tube and spanner, and wash off any spilt fluid. Refit the dust cap to the bleed screw.

30 Check the hydraulic fluid level in the master cylinder reservoir, and top-up if necessary.

31 Discard any hydraulic fluid that has been bled from the system; it will not be fit for re-use.

32 Check the operation of the clutch pedal. If the clutch is still not operating correctly, air must still be present in the system, and further bleeding is required. Failure to bleed satisfactorily after a reasonable repetition of the bleeding procedure may be due to worn master cylinder/release cylinder seals.

3 Master cylinder – removal and refitting

Removal

1 Where fitted, remove the plastic cover over the top of the engine.

2 Release the clutch hydraulic pipe from the support clip(s) on the bulkhead.

3 Unscrew the brake/clutch hydraulic fluid reservoir filler cap, and top-up the reservoir to the MAX mark (see *Weekly checks*). Place a piece of polythene over the filler neck, and secure the polythene with the filler cap. This will minimise brake fluid loss during subsequent operations.

4 Remove all traces of dirt from the outside of the master cylinder and the brake/clutch hydraulic fluid reservoir, then position some

cloth beneath the cylinder to catch any spilt fluid.

5 Extract the retaining clip and disconnect the hydraulic pipe from the connector on the end of the master cylinder (**see illustration**). Plug the pipe end and master cylinder port to minimise fluid loss and prevent the entry of dirt.

6 Disconnect the hydraulic fluid supply hose from the master cylinder.

7 Remove the brake and clutch pedal assembly from the car as described in Chapter 9, Section 10.

8 Separate the clutch pedal from the master cylinder piston rod by releasing the retaining clip at the pedal. Vauxhall technicians use a special tool to do this, however, the clip may be released by pressing the retaining tabs together using screwdrivers, while at the same time pulling the clutch pedal rearwards (**see illustration**). Do not remove the clip from the master cylinder piston rod, just release it from the pedal.

9 Unscrew the two bolts securing the master cylinder to the brake pedal mounting bracket and remove the master cylinder from the bracket (**see illustration**). If the master cylinder is faulty it must be renewed; overhaul of the unit is not possible.

Refitting

10 Locate the master cylinder on the pedal mounting bracket whilst ensuring that the piston rod and its retaining clip align correctly with the pedal. Fit the two master cylinder retaining bolts and tighten them securely.

11 Push the master cylinder piston rod retaining clip into the clutch pedal, ensuring that the two lugs on the clip fully engage.

12 Refit the brake and clutch pedal assembly as described in Chapter 9, Section 10.

13 Connect the fluid supply hose to the master cylinder and, where applicable, secure with the retaining clip.

14 Press the hydraulic pipe back into the connector on the end of the master cylinder and refit the retaining clip. Ensure that the retaining clip engages fully and the pipe is securely retained. Secure the pipe with the support clip(s).

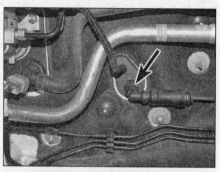

3.5 Extract the retaining clip and disconnect the hydraulic pipe from the master cylinder connector

15 Bleed the clutch hydraulic system as described in Section 2.

16 Where applicable, refit the plastic cover over the top of the engine.

4 Release cylinder – removal and refitting

Note: *Due to the amount of work necessary to remove and refit clutch components, it is usually considered good practice to renew the clutch friction disc, pressure plate assembly and release cylinder as a matched set, even if only one of these is actually worn enough to require renewal. It is also worth considering the renewal of the clutch components on a preventative basis if the engine and/or transmission have been removed for some other reason.*

Note: *Refer to the warning concerning the dangers of asbestos dust at the beginning of Section 6.*

5-speed transmission

Removal

1 Unless the complete engine/transmission unit is to be removed from the car and separated for major overhaul (see Chapter 2E, Section 4), the clutch release cylinder can be reached by removing the transmission only, as described in Chapter 7A, Section 8.

2 Wipe clean the outside of the release

3.8 Separate the clutch pedal from the master cylinder piston rod by releasing the retaining clip at the pedal

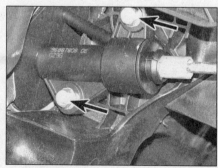

3.9 Unscrew the two bolts and remove the master cylinder from the brake pedal mounting bracket

4.2 Clutch release cylinder hydraulic pipe union nut – 5-speed transmissions

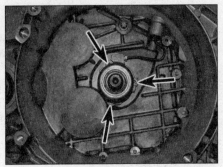

4.3 Clutch release cylinder retaining bolts – 5-speed transmissions

cylinder then slacken the union nut and disconnect the hydraulic pipe **(see illustration)**. Wipe up any spilt fluid with a clean cloth.

3 Unscrew the three retaining bolts and slide the release cylinder off from the transmission input shaft **(see illustration)**. Remove the sealing ring which is fitted between the cylinder and transmission housing and discard it; a new one must be used on refitting. Whilst the cylinder is removed, take care not to allow any debris to enter the transmission unit.

4 The release cylinder is a sealed unit and cannot be overhauled. If the cylinder seals are leaking or the release bearing is noisy or rough in operation, then the complete unit must be renewed.

5 To remove the hydraulic pipe, unclip and remove the hydraulic hose end fitting from the fastening sleeve on top of the transmission housing.

6 Using a small screwdriver, carefully spread the retaining lugs of the fastening sleeve to release the hydraulic pipe connection, and remove the pipe from inside the transmission housing. Check the condition of the sealing ring on the hydraulic pipe and renew if necessary.

7 If required, the fastening sleeve can be removed by squeezing the lower retaining lugs together with pointed-nose pliers, then withdrawing the sleeve upwards and out of the transmission. Note that if the fastening sleeve is removed, a new one must be obtained for refitting.

Refitting

8 Ensure the release cylinder and transmission mating surfaces are clean and dry and fit the new sealing ring to the transmission recess.

9 Carefully ease the cylinder along the input shaft and into position. Ensure the sealing ring is still correctly seated in its groove then refit the release cylinder retaining bolts and tighten them securely.

10 If removed, fit the new fastening sleeve, engaging the lug on the sleeve with the cut-out in the housing. Ensure that the sleeve can be felt to positively lock in position.

11 Insert the hydraulic pipe into the fastening sleeve until the end fitting can be felt to positively lock in position.

12 Reconnect the hydraulic pipe to the release cylinder, tightening its union nut securely.

13 Refit the hydraulic hose end fitting to the fastening sleeve ensuring that it is positively retained.

14 Refit the transmission unit as described in Chapter 7A, Section 8.

15 Bleed the clutch hydraulic system as described in Section 2.

6-speed transmission

Removal

16 Unless the complete engine/transmission unit is to be removed from the car and separated for major overhaul (see Chapter 2E, Section 4), the clutch release cylinder can be

reached by removing the transmission only, as described in Chapter 7A, Section 8.

17 Extract the retaining clip and remove the hydraulic hose end fitting from the fastening sleeve on top of the transmission housing **(see illustration)**.

18 Squeeze the retaining lugs on the fastening sleeve together with pointed-nose pliers, then withdraw the sleeve upwards and out of the transmission.

19 Unscrew the three retaining bolts and slide the release cylinder off from the transmission input shaft **(see illustration)**. Remove the sealing ring which is fitted between the cylinder and transmission housing and discard it; a new one must be used on refitting. Also check the condition of the sealing ring on the hydraulic pipe and renew if necessary. Whilst the cylinder is removed, take care not to allow any debris to enter the transmission unit.

20 The release cylinder is a sealed unit and cannot be overhauled. If the cylinder seals are leaking or the release bearing is noisy or rough in operation, then the complete unit must be renewed.

Refitting

21 Ensure the release cylinder and transmission mating surfaces are clean and dry and fit the new sealing ring to the transmission recess.

22 Carefully ease the cylinder along the input shaft and into position.

23 Refit the fastening sleeve and hydraulic hose end fitting to the top of the transmission.

24 Refit the release cylinder retaining bolts and tighten them securely.

25 Refit the transmission unit as described in Chapter 7A, Section 8.

26 Bleed the clutch hydraulic system as described in Section 2.

---

5 Clutch pedal – removal and refitting

1 The clutch pedal is an integral part of the brake pedal and mounting bracket assembly and cannot be individually removed. Removal and refitting details for the brake pedal and mounting bracket assembly are contained in Chapter 9, Section 10.

---

6 Clutch assembly – removal, inspection and refitting

⚠️ *Warning: Dust created by clutch wear and deposited on the clutch components may contain asbestos, which is a health hazard. DO NOT blow it out with compressed air, or inhale any of it. DO NOT use petrol or petroleum-based solvents to clean off the dust. Brake system cleaner or methylated spirit should be used to flush the dust into a suitable receptacle. After the clutch components*

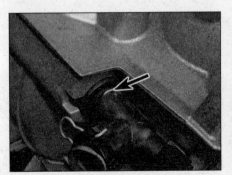

4.17 Extract the clip and remove the hose end fitting from the sleeve – 6-speed transmissions

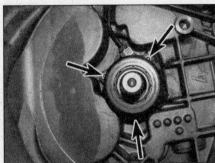

4.19 Clutch release cylinder retaining bolts – 6-speed transmissions

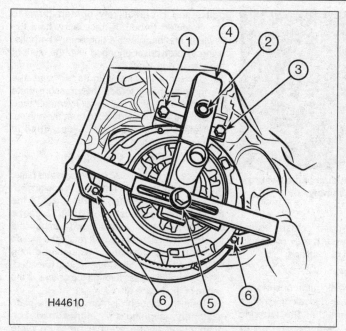

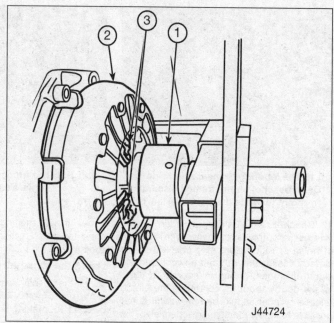

6.3a Vauxhall special jig DT-6263 for removing the clutch pressure plate and friction disc

1, 3 and 6 Bolts securing the jig to the engine
2 and 5 Bolts for adjusting the jig to the centre of the crankshaft

6.3b Thrust piece (1) in contact with the diaphragm spring fingers (3) of the pressure plate (2)

are wiped clean with rags, dispose of the contaminated rags and cleaner in a sealed, marked container.

Note: *To prevent possible damage to the ends of the pressure plate diaphragm spring fingers, Vauxhall/Opel recommend the use of a special jig (DT-6263) to remove the clutch assembly, however, with care it is possible to carry out the work without the jig. If the Vauxhall/Opel jig is not being used, a commercially-available clutch aligning tool of the type that clamps the friction disc to the pressure plate, will be required.*

Removal

1 Unless the complete engine/transmission unit is to be removed from the car and separated for major overhaul (see Chapter 2E, Section 4), the clutch can be reached by removing the transmission as described in Chapter 7A, Section 8.

2 Before disturbing the clutch, use chalk or a marker pen to mark the relationship of the pressure plate assembly to the flywheel.

3 At this stage, Vauxhall technicians fit the special jig DT-6263 to the rear of the engine and compress the diaphragm spring fingers until the friction disc is released **(see illustrations)**. The pressure plate mounting bolts are then unscrewed, and the jig spindle backed off.

4 If the jig is not available, progressively unscrew the pressure plate retaining bolts in diagonal sequence by half a turn at a time, until spring pressure is released and the bolts can be unscrewed by hand.

5 Remove the pressure plate assembly and collect the friction disc, noting which way

round the disc is fitted **(see illustration)**. It is recommended that new pressure plate retaining bolts are obtained.

Inspection

Note: *Due to the amount of work necessary to remove and refit clutch components, it is usually considered good practice to renew the clutch friction disc, pressure plate assembly and release cylinder as a matched set, even if only one of these is actually worn enough to require renewal. It is also worth considering the renewal of the clutch components on a preventative basis if the engine and/or transmission have been removed for some other reason.*

6 When cleaning clutch components, read first the warning at the beginning of this Section; remove dust using a clean, dry cloth, and working in a well-ventilated atmosphere.

7 Check the friction disc facings for signs of wear, damage or oil contamination. If the

6.5 Remove the pressure plate assembly and collect the friction disc, noting which way round the disc is fitted

friction material is cracked, burnt, scored or damaged, or if it is contaminated with oil or grease (shown by shiny black patches), the friction disc must be renewed.

8 If the friction material is still serviceable, check that the centre boss splines are unworn, that the torsion springs are in good condition and securely fastened, and that all the rivets are tight. If any wear or damage is found, the friction disc must be renewed.

9 If the friction material is fouled with oil, this must be due to an oil leak from the crankshaft oil seal, from the sump-to-cylinder block joint, or from the release cylinder assembly (either the main seal or the sealing ring). Renew the crankshaft oil seal or repair the sump joint as described in Chapter 2A, 2B, 2C, or 2D, before installing the new friction disc. The clutch release cylinder is covered in Section 4.

10 Check the pressure plate assembly for obvious signs of wear or damage; shake it to check for loose rivets, or worn or damaged fulcrum rings, and check that the drive straps securing the pressure plate to the cover do not show signs of overheating (such as a deep yellow or blue discoloration). If the diaphragm spring is worn or damaged, or if its pressure is in any way suspect, the pressure plate assembly should be renewed.

11 Examine the machined bearing surfaces of the pressure plate and of the flywheel; they should be clean, completely flat, and free from scratches or scoring. If either is discoloured from excessive heat, or shows signs of cracks, it should be renewed – although minor damage of this nature can sometimes be polished away using emery paper.

6.15 The lettering 'transmission side' or 'Getriebeseite' on the friction disc must point towards the transmission

6.21 Refit the friction disc and pressure plate assembly

12 Check that the release cylinder bearing rotates smoothly and easily, with no sign of noise or roughness. Also check that the surface itself is smooth and unworn, with no signs of cracks, pitting or scoring. If there is any doubt about its condition, the clutch release cylinder should be renewed (it is not possible to renew the bearing separately).

Refitting

13 On reassembly, ensure that the friction surfaces of the flywheel and pressure plate are completely clean, smooth, and free from oil or grease. Use solvent to remove any protective grease from new components.

14 Lightly grease the teeth of the friction disc hub with high melting-point grease. Do not apply too much, otherwise it may eventually contaminate the friction disc linings.

Using the Vauxhall jig

15 Fit the special Vauxhall guide bush to the centre of the crankshaft, and locate the friction disc on it, making sure that the lettering 'transmission side' or 'Getriebeseite' points towards the transmission **(see illustration)**.

16 Locate the pressure plate on the special centring pins on the flywheel, then compress the diaphragm spring fingers with the jig, until the friction disc is in full contact with the flywheel.

17 Insert new pressure plate retaining bolts, and progressively tighten them to the specified torque. If necessary, hold the flywheel stationary while tightening the bolts, using a screwdriver engaged with the teeth of the starter ring gear.

18 Back off the jig spindle so that the diaphragm spring forces the pressure plate against the friction disc and flywheel, then remove the jig and guide bush from the engine.

19 Refit the transmission as described in Chapter 7A, Section 8.

Without using the Vauxhall jig

20 Place the friction disc on the pressure plate, making sure that the lettering 'transmission side' or 'Getriebeseite' points towards the transmission **(see illustration 6.15)**. Using the clutch aligning tool, clamp the two components together. The aligning tool will retain the friction disc and pressure plate in alignment during refitting so that when the transmission is refitted, its input shaft will pass through the splines at the centre of the friction disc.

21 Refit the friction disc and pressure plate assembly, aligning the marks made on dismantling (if the original pressure plate is re-used) **(see illustration)**. Fit the new pressure plate bolts and tighten them evenly and progressively in a diagonal sequence to the specified torque. When the bolts have all been tightened, remove the aligning tool.

22 Refit the transmission as described in Chapter 7A, Section 8.

Chapter 7 Part A
Manual transmission

Contents

Degrees of difficulty

Easy, suitable for novice with little experience 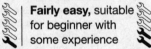	**Fairly easy,** suitable for beginner with some experience	**Fairly difficult,** suitable for competent DIY mechanic	**Difficult,** suitable for experienced DIY mechanic	**Very difficult,** suitable for expert DIY or professional

Specifications

General

Type	5 or 6 forward speeds and reverse. Synchromesh on all forward speeds and reverse

Manufacturer's designation:

1.4 litre petrol engine models and all diesel engine models	M32 (6-speed)
1.6 litre petrol engine models	D16 (5-speed)

Lubrication

Lubricant type	See *Lubricants and fluids* on page 0•18
Lubricant capacity	See Specifications in Chapter 1A (petrol) or Chapter 1B (diesel)

Torque wrench settings

	Nm	lbf ft
Differential lower cover plate bolts (D16 transmissions)	30	22
Engine/transmission mountings:		
Petrol engine models	See Chapter 2A (1.4 litre) or Chapter 2B (1.6 litre)	
Diesel engine models	See Chapter 2C (1.6 litre) or Chapter 2D (1.7 litre)	
Engine-to-transmission bolts:		
Petrol engine models	See Chapter 2A (1.4 litre) or Chapter 2B (1.6 litre)	
Diesel engine models	See Chapter 2C (1.6 litre) or Chapter 2D (1.7 litre)	
Oil drain plug (M32 transmissions)	20	15
Oil filler plug (M32 transmissions)	30	22
Oil seal carrier-to-differential (M32 transmissions)*:		
Stage 1	20	15
Stage 2	Angle-tighten a further 45°	
Reversing light switch:		
D16 transmissions	20	15
M32 transmissions	25	18
Roadwheel nuts	140	103
Transmission breather (D16 transmissions)	35	26

Use new fasteners

**2.4 Transmission oil level plug –
D16 transmissions**

1 General Information

1 The transmission is contained in a cast-aluminium alloy casing bolted to the engine's left-hand end, and consists of the gearbox and final drive differential – often called a transaxle.

2 Drive is transmitted from the crankshaft via the clutch to the input shaft, which has a splined extension to accept the clutch friction disc, and rotates in tapered roller bearings. From the input shaft, drive is transmitted to the main shaft(s), which also rotate in tapered roller bearings. From the main shaft, the drive is transmitted to the differential crown-wheel, which rotates with the differential case and planetary gears, thus driving the sun gears and driveshafts. The rotation of the planetary gears on their shaft allows the inner roadwheel to rotate at a slower speed than the outer roadwheel when the car is cornering.

3 The input shaft and main shaft(s) are arranged side-by-side, so that their gear pinion teeth are in constant mesh. In the neutral position, the main shaft gear pinions rotate freely, so that drive cannot be transmitted to the crownwheel.

4 Gear selection is via a floor-mounted lever and cable-operated selector linkage mechanism. The selector linkage causes the appropriate selector fork to move its

**2.11 Differential cover plate retaining bolts
– D16 transmissions**

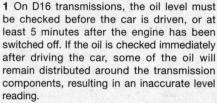

**2.6 Top-up/fill the transmission through the transmission breather aperture –
D16 transmissions**

respective synchro-sleeve along the shaft, to lock the gear pinion to the synchro-hub. Since the synchro-hubs are splined to the main shafts, this locks the pinion to the shaft, so that drive can be transmitted. To ensure that gearchanging can be made quickly and quietly, a synchromesh system is fitted to all gears, consisting of baulk rings and spring-loaded fingers, as well as the gear pinions and synchro-hubs. The synchro-mesh cones are formed on the mating faces of the baulk rings and gear pinions.

2 Transmission oil – level check, draining and refilling

1 On D16 transmissions, the oil level must be checked before the car is driven, or at least 5 minutes after the engine has been switched off. If the oil is checked immediately after driving the car, some of the oil will remain distributed around the transmission components, resulting in an inaccurate level reading.

2 Draining the oil is much more efficient if the car is first taken on a journey of sufficient length to warm the engine/transmission up to normal operating temperature.

Caution: If the procedure is to be carried out on a hot transmission unit, take care not to burn yourself on the hot exhaust or the transmission/engine unit.

3 Position the vehicle over an inspection pit, on vehicle ramps, or jack it up and support it securely on axle stands (see *Jacking and vehicle support*), but make sure that it is level. Where fitted, remove the engine undertray as described in Chapter 11, Section 21.

D16 transmissions

Level check

Note: *A new transmission oil level plug will be needed for this operation.*

4 Wipe clean the area around the level plug. The level plug is located behind the driveshaft inner joint on the left-hand side of the transmission **(see illustration)**. Unscrew the plug and remove it.

5 The oil level should reach the lower edge of the level plug aperture.

6 The transmission is topped-up via the transmission breather aperture adjacent to the selector mechanism on the top of the transmission **(see illustration)**. Wipe clean the area around the transmission breather and remove the cap. Unscrew the breather and remove it from the transmission. Refill the transmission with the specified grade of oil (see *Lubricants and fluids*), until it reaches the bottom of the level plug aperture. Allow any excess oil to drain, then fit and securely tighten the new level plug.

7 Refit the transmission breather and tighten it to the specified torque.

8 Where applicable, refit the engine undertray as described in Chapter 11, Section 21 then lower the vehicle to the ground.

Draining

Note: *A new differential lower cover plate gasket will be required for this operation.*

9 Since the transmission oil is not renewed as part of the manufacturer's maintenance schedule, no drain plug is fitted to the transmission. If for any reason the transmission needs to be drained, the only way of doing so is to remove the differential lower cover plate.

10 Wipe clean the area around the differential cover plate and position a suitable container underneath the cover.

11 Evenly and progressively slacken and remove the retaining bolts then withdraw the cover plate and allow the transmission oil to drain into the container **(see illustration)**. Remove the gasket and discard it; a new one should be used on refitting.

12 Allow the oil to drain completely into the container. If the oil is hot, take precautions against scalding. Remove all traces of dirt and oil from the cover and transmission mating surfaces and wipe clean the inside of the cover plate.

13 Once the oil has finished draining, ensure the mating surfaces are clean and dry then refit the cover plate to the transmission unit, complete with a new gasket. Refit the retaining bolts and evenly and progressively tighten them to the specified torque.

Refilling

14 Refer to paragraphs 4 to 8.

M32 transmissions

Level check

15 On the M32 transmissions, there is no provision for oil level checking once the transmission has been initially filled. If for any reason it is thought that the oil level may be low, the transmission oil must be completely drained, then refilled with an exact specified quantity of oil as described below.

Draining

Note: *A new transmission oil drain plug will be required for this operation.*

16 Wipe clean the area around the drain

2.16 Transmission oil drain plug –
M32 transmissions

2.19 The transmission is refilled via the oil
filler plug on the top of the casing –
M32 transmissions

2.20 Using a funnel and graduated
container, refill the transmission with the
specified grade and quantity of oil

plug, located on the lower left-hand side
of the differential housing, and position
a suitable container under the plug **(see
illustration)**.

17 Undo the drain plug and allow the oil to
drain.

18 Once the oil has finished draining, fit the
new drain plug and tighten the plug to the
specified torque.

Refilling

19 The transmission is refilled via the oil filler
plug on the top of the casing **(see illustration)**.
To gain access to the plug, remove the battery
and battery tray as described in Chapter 5A,
Section 4.

20 Wipe clean the area around the plug and
unscrew it. Using a funnel and graduated
container, refill the transmission with the
specified grade of oil as given in *Lubricants
and fluids* and the specified quantity of oil as
given in Chapter 1A, Section (petrol engine
models) or Chapter 1B, Section (diesel engine
models) **(see illustration)**. On completion,
refit and tighten the oil filler plug to the
specified torque.

21 Refit the engine undertray (where
applicable), then lower the vehicle to the
ground.

22 Refit the battery tray and battery as
described in Chapter 5A, Section 4.

3 Selector cables –
adjustment

Note: *A 5 mm drill bit or dowel rod will be
required to carry out this procedure.*

1 Remove the centre console as described in
Chapter 11, Section 26.

2 Position the gear lever in the 'neutral'
position.

3 Working inside the car at the gear lever
housing, lift the adjustment locks up while
pushing the adjustment retainers rearward,
one for each cable, to release the shift lever
and selector lever cables.

4 Depress the retaining tab and lift up the
reverse detent block at the base of the gear
lever. Rotate the detent block clockwise
through 90º so that the two lugs on the detent

3.4 Rotate the reverse detent block so
that the lugs on the block engage with the
holes in the housing

block are aligned with the two holes in the
gear lever housing **(see illustration)**. Push
the detent block down so that the two lugs
engage with the holes. The gear lever is now
locked in the adjustment position.

5 Set the gearchange selector on the
transmission to the 'neutral' position. Press
the selector down (D16 transmissions) or pull
it up (M32 transmission) and lock it in the
adjustment position by inserting a 5 mm drill
bit or dowel rod through the hole in the side
of the housing **(see illustration)**. Ensure that
the drill bit or dowel rod fully engages with the
selector.

6 Working back inside the car, push down
on the cable adjustment locks. There
should be an audible click as the locks
engage, indicating that the cables are now
locked.

4.4 Prise the inner cable end fittings from
the transmission selector lever ballpins

3.5 Using a 5 mm drill bit to lock the
selector mechanism in the adjustment
position

7 Depress the retaining tab and lift up the
reverse detent block at the base of the gear
lever. Rotate the detent block anti-clockwise
through 90º to return it to its original position.
Push the detent block down until the tab
engages with the detent.

8 Remove the drill bit or dowel rod used
to lock the gearchange selector on the
transmission.

9 Check the gearchange mechanism for
correct operation and if necessary, repeat the
adjustment procedure.

10 When all is satisfactory, refit the centre
console as described in Chapter 11,
Section 26.

4 Selector cables –
removal and refitting

Removal

1 Remove the battery and battery tray as
described in Chapter 5A, Section 4.

2 Remove the centre console as described in
Chapter 11, Section 26.

3 Working in the engine compartment, note
the fitted locations of the cables at their
transmission attachments.

4 Using a large screwdriver, prise the inner
cable end fittings from the transmission
selector lever ballpins **(see illustration)**.

5 Push the retaining sleeves toward the rear
of the car and detach the outer cables from

4.5 Push the retaining sleeves rearward and detach the outer cables from the transmission bracket

4.8 Depress the lugs on the sides of the outer cable ends and lift the outer cables from the gear lever housing

the mounting bracket on the transmission **(see illustration)**.

6 From inside the car, move the floor carpet in order to gain access to the selector cable grommet. Undo the two retaining nuts and release the grommet from the floor.

7 Working at the gear lever housing, lift the adjustment locks up while pushing the adjustment retainers rearward, one for each cable, to release the shift lever and selector lever cables from the end fittings.

8 Note the fitted locations of the two cables, then depress the lugs on the sides of the outer cable ends and lift the outer cables from the gear lever housing **(see illustration)**.

9 Pull the two cables (complete with grommet) out of the bulkhead and remove them from inside the car.

Refitting

10 Push the selector cables through the bulkhead and locate the rubber grommet back into position. Secure the grommet with the two retaining nuts.

11 Engage the two inner cables with their end fittings at the gear lever housing, then refit the outer cables to the housing.

12 Refit the outer cables to the mounting bracket on the transmission.

13 Engage the inner cable end fittings with the transmission selector lever ballpins, squeezing them together with pliers if necessary.

14 Carry out the selector cable adjustment procedure, as described in Section 3.

6.14 Undo the four bolts and remove the oil seal carrier from the side of the differential housing

15 Refit the centre console as described in Chapter 11, Section 26.

16 Refit the battery tray and battery as described in Chapter 5A, Section 4.

5 Gear lever housing – removal and refitting

Removal

1 Remove the centre console as described in Chapter 11, Section 26.

2 Working at the gear lever housing, lift the adjustment locks up while pushing the adjustment retainers rearward, one for each cable, to release the shift lever and selector lever cables from the end fittings.

3 Note the fitted locations of the two cables, then depress the lugs on the sides of the outer cable ends and lift the outer cables from the gear lever housing **(see illustration 4.8)**.

4 Undo the four retaining bolts and lift the gear lever housing from the floor.

Refitting

5 Position the gear lever housing on the floor, engage the selector inner cables with the end fittings, then refit the outer cables to the gear lever housing.

6 Insert the gear lever housing retaining bolts and tighten them securely.

7 Carry out the selector cable adjustment procedure, as described in Section 3.

8 On completion, refit the centre console as described in Chapter 11, Section 26.

6 Oil seals – renewal

Driveshaft oil seals

D16 transmissions

1 Firmly apply the handbrake, then jack up the front of the car and support it securely on axle stands (see *Jacking and vehicle support*).

2 Drain the transmission oil as described in Section 2.

3 Remove the driveshaft/intermediate shaft as described in Chapter 8, Section 2.

4 Note the correct fitted depth of the seal in its housing then carefully prise it out of position using a large flat-bladed screwdriver.

5 Remove all traces of dirt from the area around the oil seal aperture, then lubricate the outer lip of the new oil seal with transmission oil. Ensure the seal is correctly positioned, with its sealing lip facing inwards, and tap it squarely into position, using a suitable tubular drift (such as a socket) which bears only on the hard outer edge of the seal. Ensure the seal is fitted at the same depth in its housing that the original was.

6 Refit the driveshaft/intermediate shaft as described in Chapter 8, Section 2.

7 Refill the transmission with the specified type and amount of oil, as described in Section 2.

M32 transmissions – right-hand oil seal

8 On 4WD models, remove the transfer gearbox as described in Chapter 7C, Section 3.

9 Proceed as described above in paragraphs 1 to 7.

10 On 4WD models, refit the transfer gearbox as described in Chapter 7C, Section 3.

M32 transmissions – left-hand oil seal

Note: *A hydraulic press, together with tubes and mandrels of suitable diameters will be needed for this operation.*

11 Firmly apply the handbrake, then jack up the front of the car and support it securely on axle stands (see *Jacking and vehicle support*).

12 Drain the transmission oil as described in Section 2.

13 Remove the driveshaft as described in Chapter 8, Section 2.

14 Undo the four bolts and remove the oil seal carrier from the side of the differential housing **(see illustration)**. Note that new bolts will be required for refitting.

15 Place the oil seal carrier on the press bed with its outer surface facing down. Using a suitable mandrel, press the oil seal out of the carrier.

16 Using a small screwdriver, remove the sealing O-ring from the oil seal carrier.

17 Support the inner surface of the oil seal carrier on the press bed. Lubricate the new oil seal with transmission oil, then press it fully into position in the carrier using a suitable tube which bears only on the hard outer edge of the seal.

18 Fit a new O-ring to the oil seal carrier, then refit the carrier to the differential housing, tightening the new retaining bolts to the specified torque, then through the specified angle.

19 Refit the driveshaft as described in Chapter 8, Section 2.

20 Refill the transmission with the specified type and amount of oil, as described in Section 2.

Input shaft oil seal

21 The input shaft oil seal is an integral part of the clutch release cylinder; if the seal is leaking the complete release cylinder assembly must be renewed. Before condemning the release cylinder, check that the leak is not coming from the sealing ring which is fitted between the cylinder and the transmission housing; the sealing ring can be renewed once the release cylinder assembly has been removed. Refer to Chapter 6, Section 4 for removal and refitting details.

---

7 Reversing light switch – testing, removal and refitting

1 The reversing light circuit is controlled by a plunger-type switch screwed into the front of the transmission casing (see illustration).

Testing

2 If a fault develops in the circuit, first ensure that the circuit fuse has not blown and that the reversing light bulbs are sound.
3 To test the switch, disconnect the wiring connector. Use a multimeter (set to the resistance function) or a battery-and-bulb test circuit to check that there is continuity between the switch terminals only when reverse gear is selected. If this is not the case, and there are no obvious breaks or other damage to the wires, the switch is faulty, and must be renewed.

Removal

4 Disconnect the wiring connector, then

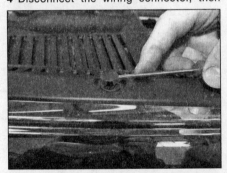

8.4a Extract the centre pins and remove the eight expanding rivets...

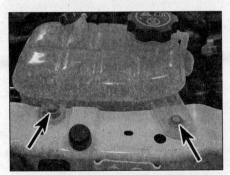

8.5 Undo the two bolts and move the expansion tank to one side

---

unscrew the switch and remove it from the transmission casing along with its sealing washer.

Refitting

5 Fit a new sealing washer to the switch, then screw it back into position in the transmission casing and tighten it to the specified torque. Reconnect the wiring connector, then test the operation of the circuit.

---

8 Transmission – removal and refitting

Removal

1 Apply the handbrake, then jack up the front of the vehicle and support it on axle stands (see *Jacking and vehicle support*). Allow sufficient working room to remove the transmission from under the left-hand side of the engine compartment. Remove both front roadwheels. Also remove the engine top cover where fitted.
2 On diesel engine models, remove the engine undertray as described in Chapter 11, Section 21.
3 Remove the battery and battery tray as described in Chapter 5A, Section 4, and the starter motor as described in Chapter 5A, Section 11.
4 Extract the centre pins, then remove the eight plastic expanding rivets securing the bumper upper panel to the bumper and bonnet lock platform. Lift up the upper panel and remove it from the engine compartment (see illustrations).

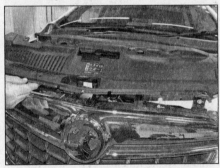

8.4b ...then lift away the bumper upper panel

8.8 Prise out the retaining clip then detach the clutch hydraulic hose from the end fitting on the transmission bellhousing

---

7.1 The reversing light switch is screwed into the front of the transmission casing

5 Undo the two retaining bolts and move the coolant expansion tank to one side (see illustration).
6 Drain the transmission oil as described in Section 2.
7 Remove the filler cap from the brake/clutch fluid reservoir on the bulkhead, then tighten it onto a piece of polythene. This will reduce the loss of fluid when the clutch hydraulic hose is disconnected. Alternatively, fit a hose clamp to the flexible hose next to the clutch hydraulic connection on the transmission housing.
8 Place some cloth rags beneath the hose, then partially prise out the retaining clip securing the clutch hydraulic hose to the end fitting on top of the transmission bellhousing. Detach the hose from the end fitting (see illustration). Discard the sealing ring from the hose end; a new sealing ring must be used on refitting. Plug/cover both the end fitting and hose end to minimise fluid loss and prevent the entry of dirt into the hydraulic system. Whilst the hose is disconnected, do not depress the clutch pedal.
9 Note the fitted locations of the gearchange selector cables at their transmission attachments. Using a large screwdriver, prise the inner cable end fittings from the transmission selector lever ballpins (see illustration 4.4).
10 Push the retaining sleeves toward the rear of the car and detach the outer cables from the mounting bracket on the transmission (see illustration 4.5). Undo the retaining bolts and remove the cable mounting bracket(s) from the transmission (see illustrations).

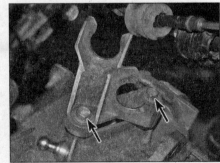

8.10a Undo the two bolts...

8.10b ...and remove the selector cable mounting bracket

8.12 Where fitted, disconnect the wiring connector from the neutral position switch

8.18 Connect a hoist or support bar to the engine left-hand lifting bracket

11 Disconnect the wiring connector from the reversing light switch and speedometer driven gear (D16 transmissions), then free the wiring from the transmission unit and retaining brackets.

12 Where fitted, disconnect the wiring connector from the neutral position switch on the selector housing **(see illustration)**.

13 Remove both driveshafts as described in Chapter 8, Section 2.

14 Remove the exhaust system front pipe as described in Chapter 4A, Section 20 (petrol engines) or Chapter 4B, Section 26 (diesel engines).

15 On 4WD models, remove the propeller shaft as described in Chapter 8, Section 6, then remove the transfer gearbox as described in Chapter 7C, Section 3.

16 Unscrew and remove the upper bolts

securing the transmission to the rear of the engine.

17 Remove the front subframe assembly as described in Chapter 10, Section 7.

18 Connect a hoist or support bar to the engine left-hand lifting bracket. The type of support bar which locates in the engine compartment side channels is to be preferred. This can be adapted to rest on wooden blocks in the engine compartment **(see illustration)**.

19 Place a jack with a block of wood beneath the transmission, and raise the jack to take the weight of the transmission.

20 Unbolt and remove the left-hand engine/transmission mounting bracket from the transmission with reference to Chapter 2A, Chapter 2B, Chapter 2C or Chapter 2D, as applicable.

21 Lower the engine and transmission by

approximately 5 cm making sure that the coolant hoses and wiring harnesses are not stretched.

22 Slacken and remove the remaining bolts securing the transmission to the engine and sump flange. Note the correct fitted positions of each bolt, and the relevant brackets, as they are removed to use as a reference on refitting **(see illustration)**. Make a final check that all components have been disconnected, and are positioned clear of the transmission so that they will not hinder the removal procedure.

23 With the bolts removed, move the trolley jack and transmission, to free it from its locating dowels. Once the transmission is free, lower the jack and manoeuvre the unit out from under the car **(see illustration)**. Remove the transmission spacer plate (where fitted), then remove the locating dowels from the transmission or engine if they are loose, and keep them in a safe place.

Refitting

24 Prior to refitting, use a clean flat brush to apply very light coating of multi-purpose grease to input shaft, then clean the entire lead surface of the input shaft with a new, lint-free cloth **(see illustrations)**.

25 Ensure that the locating dowels are correctly positioned then, where applicable, place the transmission spacer plate in position over the dowels.

26 Raise the transmission with the trolley jack and position the transmission on the engine. Refit the retaining bolts together with the relevant brackets as noted during removal and tighten the bolts to the specified torque.

27 Refit the left-hand engine/transmission mounting to the transmission with reference to Chapter 2A, Chapter 2B, Chapter 2C or Chapter 2D, as applicable.

28 The remainder of refitting is a reversal of the removal procedure, bearing in mind the following points:

a) *Renew the driveshaft oil seals (see Section 6) before refitting the driveshafts/ intermediate shaft.*

b) *Refit the front subframe assembly as described in Chapter 10, Section 7.*

c) *On 4WD models, refit the transfer gearbox as described in Chapter 7C, Section 3,*

8.22 Note the correct fitted position of the transmission retaining bolts and any relevant brackets

8.23 Once the transmission is free, lower the jack and manoeuvre the unit out from under the car

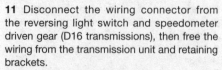

8.24a Apply a very light coating of multi-purpose grease to the input shaft...

8.24b ...then clean the entire lead surface of the shaft with a new, lint-free cloth

then refit the propeller shaft as described in Chapter 8, Section 6.

d) Fit a new sealing ring to the clutch hydraulic hose before clipping the hose into the end fitting. Ensure the hose is securely retained by its clip then bleed the hydraulic system as described in Chapter 6, Section 2.

e) Refill the transmission with the specified type and quantity of oil, as described in Section 2.

f) Refit the battery tray and battery as described in Chapter 5A, Section 4.

9 Transmission overhaul – general information

1 Overhauling a manual transmission unit is a difficult and involved job for the DIY home mechanic. In addition to dismantling and reassembling many small parts, clearances must be precisely measured and, if necessary, changed by selecting shims and spacers. Internal transmission components are also often difficult to obtain, and in many instances, extremely expensive. Because of this, if the transmission develops a fault or becomes noisy, the best course of action is to have the unit overhauled by a specialist repairer, or to obtain an exchange reconditioned unit.

Chapter 7 Part B
Automatic transmission

Contents

Degrees of difficulty

Easy, suitable for novice with little experience	**Fairly easy,** suitable for beginner with some experience	**Fairly difficult,** suitable for competent DIY mechanic	**Difficult,** suitable for experienced DIY mechanic	**Very difficult,** suitable for expert DIY or professional

Specifications

General

Type . . . . Electronically-controlled adaptive automatic, six forward speeds and reverse, with sequential manual gear selection capability

Manufacturer's designation 6T40/45/50

Lubrication

Lubricant type . . . . See *Lubricants and fluids* on page 0•18
Lubricant capacity . . . . See Chapter 1A (petrol) or Chapter 1B (diesel)

Torque wrench settings

	Nm	lbf ft
Automatic transmission fluid drain/level check plugs:		
Drain plug	12	9
Level checking plug	12	9
Engine/transmission mountings:		
Petrol engine models	See Chapter 2A (1.4 litre) or Chapter 2B (1.6 litre)	
Diesel engine models	See Chapter 2C (1.6 litre) or Chapter 2D (1.7 litre)	
Engine-to-transmission bolts:		
Petrol engine models	See Chapter 2A (1.4 litre) or Chapter 2B (1.6 litre)	
Diesel engine models	See Chapter 2C (1.6 litre) or Chapter 2D (1.7 litre)	
Fluid cooler pipes-to-transmission	22	16
Roadwheel nuts	140	103
Torque converter-to-driveplate bolts*	60	44

*Use new fasteners

1 General Information

1 All except 1.6 litre petrol engine models are optionally available with a six-speed, electronically-controlled automatic transmission. The transmission consists of a torque converter, an epicyclic geartrain, and hydraulically-operated clutches and brakes. The unit is controlled by the transmission electronic control unit (ECU) via electrically-operated solenoid valves. In addition to the fully automatic operation, the transmission can also be operated manually with a six-speed sequential gear selection.

2 The torque converter provides a fluid coupling between engine and transmission, which acts as an automatic clutch, and also provides a degree of torque multiplication when accelerating. The torque converter incorporates a lock-up function whereby the engine and transmission can be directly coupled by means of a clutch unit inside the torque converter. The lock-up function is controlled by the ECU according to operating conditions.

3 The epicyclic geartrain provides either of the six forward or one reverse gear ratios, according to which of its component parts are held stationary or allowed to turn. The components of the geartrain are held or released by hydraulically actuated brakes and

clutches. A fluid pump within the transmission provides the necessary hydraulic pressure to operate the brakes and clutches.

4 In automatic mode, the transmission is fully adaptive, whereby the shift points are dependant on driver input, road speed, engine speed and vehicle operating conditions. The ECU receives inputs from various engine and drive train related sensors, and determines the appropriate shift point for each gear.

5 Driver control of the transmission is by a five-position selector lever. The drive D position, allows automatic changing throughout the range of forward gear ratios. An automatic kickdown facility shifts the transmission down a gear if the accelerator pedal is fully depressed. If the selector lever is moved to

the M position, the transmission enters manual mode. In manual mode the gear selector lever can be used to shift the transmission up or down each gear sequentially.

6 Due to the complexity of the automatic transmission, any repair or overhaul work must be left to a Vauxhall/Opel dealer or transmission specialist with the necessary special equipment for fault diagnosis and repair. The contents of the following Sections are therefore confined to supplying general information, and any service information and instructions that can be used by the owner.

2 Automatic transmission fluid – draining and refilling

Draining

1 Position the vehicle over an inspection pit, on vehicle ramps, or jack it up and support it securely on axle stands (see *Jacking and vehicle support*), but make sure that it is level.
2 Position a container under the drain plug at the base of the transmission. Unscrew the drain plug and remove it from the transmission. Allow the fluid to drain completely into the container.
3 When the fluid has finished draining, clean the drain plug threads and those of the transmission casing. Refit the drain plug, tightening it to the specified torque.

Refilling

4 Wipe clean the area around the transmission fluid filler cap, located on the top of the transmission housing. Remove the transmission vent hose (where fitted) from the filler cap, then unscrew and remove the cap.
5 Refill the transmission with the specified grade of oil as given in *Lubricants and fluids* and the specified quantity of oil as given in Chapter 1A, Section or Chapter 1B, Section. Use a funnel with a fine mesh gauze, to avoid spillage, and to ensure that no foreign matter enters the transmission. On completion, refit the fluid filler cap and, where applicable, reconnect the vent hose.
6 Start the engine, and with the footbrake firmly applied, slowly move the selector lever from position P to position D and back to position P, stopping at each position for at least three seconds. Allow the engine to idle for a further three minutes to allow any fluid frothing to dissipate and the fluid level to stabilize.
7 Using the driver information centre, observe the transmission fluid temperature. Before checking the fluid level, the fluid temperature must be between 85° C and 95° C.
8 With the engine still idling and the selector lever in the P position, wipe clean the area around the level plug and position a container under it. The level plug is located behind the driveshaft inner joint on the left-hand side of the transmission. Unscrew the plug and clean it.

9 Allow the excess fluid to run from the level plug aperture until it is only dripping out. If no fluid runs from the level plug aperture, remove the filler cap and add further fluid until it does.
10 Once the level is correct, switch off the engine and refit the filler cap, then refit the level plug and tighten it securely.
11 Lower the vehicle to the ground.

3 Selector cable – adjustment

Note: *If the battery is disconnected with the selector lever in the P (park) position, the lever will be locked in position. To manually release the lever, carefully release the cap (located in front of the selector lever on the right-hand side) from the centre console using a small screwdriver. Insert a screwdriver into the opening as far as it will go to release the selector lever lock, then move the selector lever to the N (neutral) position.*

1 Operate the selector lever throughout its entire range and check that the transmission engages the correct gear indicated on the selector lever position indicator. If adjustment is necessary, continue as follows.
2 Position the selector lever in the P (park) position.
3 Remove the battery and battery tray as described in Chapter 5A, Section 4.
4 Working in the engine compartment, release the selector cable adjuster clip on the cable end fitting to unlock the cable.
5 Verify that the transmission selector lever is also positioned in the P position. With both the selector lever and transmission correctly positioned, lock the adjuster clip on the selector cable end fitting by pushing it down until it clicks in position.
6 Check the operation of the selector lever and, if necessary, repeat the adjustment procedure.
7 When all is satisfactory, refit the battery tray and battery as described in Chapter 5A, Section 4.

4 Selector cable – removal and refitting

Removal

1 Remove the battery and battery tray as described in Chapter 5A, Section 4.
2 Remove the centre console as described in Chapter 11, Section 26.
3 Using a large screwdriver or similar tool, prise the selector inner cable end fitting off the balljoint on the transmission selector lever.
4 Press the locking tabs inwards and lift the outer cable up and out of the support bracket on the transmission. Release the outer cable from the retaining clip on the transmission.
5 From inside the car, using a large

screwdriver or similar tool, prise the selector inner cable end fitting off the balljoint on the selector lever.
6 Release the selector cable retainer and remove the cable from the selector lever base.
7 Move the floor carpet in order to gain access to the selector cable grommet. Undo the two retaining nuts and release the grommet from the floor.
8 Pull the cable (complete with grommet) out of the bulkhead and remove it from inside the car.

Refitting

9 Refitting is a reversal of removal bearing in mind the following points:
a) Refit the centre console as described in Chapter 11, Section 26.
b) Adjust the selector cable as described in Section 3.
c) Refit the battery tray and battery as described in Chapter 5A, Section 4.

5 Selector lever assembly – removal and refitting

Removal

1 Remove the centre console as described in Chapter 11, Section 26.
2 Using a large screwdriver or similar tool, prise the selector inner cable end fitting off the balljoint on the selector lever.
3 Release the selector cable retainer and remove the cable from the selector lever base.
4 Disconnect the selector lever assembly wiring connector and release the wiring harness from the retaining clips and ties.
5 Undo the four mounting bolts and remove the selector lever assembly from the car

Refitting

6 Manoeuvre the selector lever assembly into place, engaging it with the selector cable. Refit the mounting bolts and tighten them securely.
7 Engage the selector outer cable with the selector lever housing.
8 Reconnect the selector lever assembly wiring connector and secure the wiring harness with the retaining clips and ties.
9 Adjust the selector cable as described in Section 3.
10 Refit the centre console as described in Chapter 11, Section 26.

6 Oil seals – renewal

Driveshaft oil seals

1 Remove the driveshaft/intermediate shaft as described in Chapter 8, Section 2.
2 Note the correct fitted depth of the seal in its housing then carefully prise it out of position using a large flat-bladed screwdriver.

3 Remove all traces of dirt from the area around the oil seal aperture, then lubricate the new oil seal with automatic transmission fluid. Ensure the seal is correctly positioned, with its sealing lip facing inwards, and tap it squarely into position, using a suitable tubular drift (such as a socket) which bears only on the hard outer edge of the seal. Ensure the seal is fitted at the same depth in its housing that the original was.

4 Refit the driveshaft/intermediate shaft as described in Chapter 8, Section 2.

5 Refill the transmission with the specified type and amount of oil, as described in Section 2.

Torque converter oil seal

6 Remove the transmission as described in Section 9.

7 Carefully slide the torque converter off of the transmission shaft whilst being prepared for fluid spillage.

8 Note the correct fitted position of the seal in the oil pump housing then carefully lever the seal out of position taking care not to mark the housing or input shaft.

9 Remove all traces of dirt from the area around the oil seal aperture then press the new seal into position, ensuring its sealing lip is facing inwards.

10 Lubricate the seal with clean transmission fluid then carefully ease the torque converter into position. Slide the torque converter onto the transmission shaft by turning it until it fully engages with the oil pump.

11 Refit the transmission as described in Section 9.

7 Fluid cooler – removal and refitting

Removal

1 The transmission main fluid cooler is an integral part of the radiator assembly. Refer to Chapter 3, Section 3 for removal and refitting details.

2 An additional auxiliary fluid cooler is also used and is located in front of the radiator. Removal and refitting is as follows.

3 Remove the front bumper as described in Chapter 11, Section 6.

4 Pull the plastic cap on the fluid cooler upper hydraulic pipe quick-release connector back by approximately 5 cm.

5 Using a bent-tip screwdriver or similar tool, pull on one of the open ends of the quick-release connector retaining ring to rotate the ring out of position. Remove the retaining ring from the connector, then pull the hydraulic pipe out of the union. Note that a new retaining ring will be required for refitting.

6 Repeat paragraphs 4 and 5 to remove the lower hydraulic pipe.

7 Undo the bolt securing the fluid cooler hydraulic pipe connector block to the side of the radiator.

8 Unclip and remove the air seal plate over the top of the radiator and condenser.

9 Undo the two retaining bolts and remove the fluid cooler from the radiator.

Refitting

10 Refitting is a reversal of removal bearing in mind the following points:

a) *Use new retaining rings to secure the hydraulic pipe quick-release connectors.*

b) *When fitting the retaining rings, hook one of the open ends of the ring into one of the slots in the quick-release connector. Rotate the retaining ring around the connector until it is positioned with all three ears through the three slots on the connector.*

c) *Refit the hydraulic pipes to the quick-release connectors by pushing the pipe into the union until a click is either heard or felt.*

d) *On completion, check the transmission fluid level as described in Section 2.*

e) *Refit the front bumper as described in Chapter 11, Section 6.*

8 Electronic control unit – removal and refitting

The ECU is an integral part of the solenoid valve assembly and the transmission must be partially dismantled to gain access. This work is considered beyond the scope of this manual and should therefore be entrusted to a Vauxhall/Opel dealer or suitably-equipped specialist.

9 Automatic transmission – removal and refitting

Note: *New torque converter-to-driveplate bolts will be required for refitting.*

Removal

1 Apply the handbrake, then jack up the front of the vehicle and support it on axle stands (see *Jacking and vehicle support*). Allow sufficient working room to remove the transmission from under the left-hand side of the engine compartment. Remove both front roadwheels. Where fitted, remove the engine top cover.

2 Drain the transmission fluid as described in Section 2, then refit the drain plug and tighten it to the specified torque.

3 Remove the battery and battery tray as described in Chapter 5A, Section 4.

4 Remove the front bumper as described in Chapter 11, Section 6.

5 Remove the exhaust system front pipe as described in Chapter 4A, Section 20 or Chapter 4B, Section 26.

6 Undo the two retaining bolts and move the cooling system expansion tank to one side.

7 Using a large screwdriver or similar tool, prise the selector inner cable end fitting off the balljoint on the transmission selector lever.

8 Press the locking tabs inwards and lift the outer cable up and out of the support bracket on the transmission. Release the outer cable from the retaining clip on the transmission.

9 Disconnect the transmission ECU wiring harness connector and release the harness from the clips or ties securing it to the engine/transmission unit.

10 Unscrew the retaining nuts and detach the fluid cooler inlet and outlet pipes from the transmission. Collect the sealing washer from each pipe noting that new washers will be required for refitting Suitably cover the pipe ends and plug the transmission orifices to prevent dirt entry.

11 Unclip the wiring harness retainers from the battery cable bracket and transmission wiring harness bracket.

12 Remove both driveshafts and the intermediate shaft as described in Chapter 8, Section 2.

13 Unscrew and remove the upper bolts securing the transmission to the rear of the engine and collect the wiring harness brackets.

14 Remove the front subframe assembly as described in Chapter 10, Section 7.

15 Connect a hoist or support bar to the engine left-hand lifting bracket. The type of support bar which locates in the engine compartment side channels is to be preferred. This can be adapted to rest on wooden blocks in the engine compartment **(see illustration)**.

16 Remove the engine/transmission rear mounting/torque link as described in Chapter 2A, Section 16, Chapter 2B, Section 19, Chapter 2C, Section 18 or Chapter 2D, Section 20 as applicable.

17 Remove the starter motor as described in Chapter 5A, Section 11.

18 Using a socket and extension bar, turn the crankshaft pulley until one of the bolts securing the torque converter to the driveplate becomes accessible through the starter motor aperture (1.6 and 1.7 litre diesel engines) or through the aperture just below the starter motor location (1.4 and 1.6 litre petrol engines). Slacken and remove the bolt, then turn the crankshaft pulley as necessary, and undo the two remaining bolts as they become

9.15 Connect a hoist or support bar to the engine left-hand lifting bracket

accessible. Discard the bolts, new ones must be used on refitting.

19 Disconnect the breather hose from the top of the transmission.

20 Place a jack with a block of wood beneath the transmission, and raise the jack to take the weight of the transmission.

21 Remove the engine/transmission left-hand mounting as described in Chapter 2A, Section 16, Chapter 2B, Section 19, Chapter 2C, Section 18 or Chapter 2D, Section 20 as applicable.

22 By manipulating the engine hoist and the jack under the transmission, lower the engine/transmission approximately 50mm. Ensure that the various coolant hoses and wiring harnesses are not stretched.

23 Slacken and remove the remaining bolts securing the transmission to the engine and sump flange. Note the correct fitted positions of each bolt, and the relevant brackets, as they are removed to use as a reference on refitting. Make a final check that all components have been disconnected, and are positioned clear of the transmission so that they will not hinder the removal procedure.

24 With all the bolts removed, move the trolley jack and transmission, to free it from its locating dowels. Once the transmission is free, lower the jack and manoeuvre the unit out from under the car, taking care to ensure that the torque converter does not fall off. Remove

the locating dowels from the transmission or engine if they are loose, and keep them in a safe place. Retain the torque converter while the transmission is removed by bolting a strip of metal across the transmission bellhousing end face.

Refitting

25 The transmission is refitted by a reversal of the removal procedure, bearing in mind the following points.

a) *Prior to refitting, remove all traces of old locking compound from the torque converter threads by running a tap of the correct thread diameter and pitch down the holes. In the absence of a suitable tap, use one of the old bolts with slots cut in its threads.*

b) *Ensure the engine/transmission locating dowels are correctly positioned and apply a smear of molybdenum disulphide grease to the torque converter locating pin and its centering bush in the crankshaft end.*

c) *Once the transmission and engine are correctly joined, refit the securing bolts, tightening them to the specified torque setting.*

d) *Fit the new torque converter-to-driveplate bolts and tighten them lightly only to start, then go around and tighten them to the specified torque setting in a diagonal sequence.*

e) *Tighten all nuts and bolts to the specified torque (where given).*

f) *Renew the driveshaft oil seals (see Section 6) and refit the driveshafts/intermediate shaft to the transmission as described in Chapter 8, Section 2.*

g) *Refit the front subframe assembly as described in Chapter 10, Section 7.*

h) *On completion, refill the transmission with the specified type and quantity of fluid as described in Section 2 and adjust the selector cable as described in Section 3.*

10 Automatic transmission overhaul – general information

1 In the event of a fault occurring with the transmission, it is first necessary to determine whether it is of a mechanical, electrical or hydraulic nature, and to do this, special test equipment is required. It is therefore essential to have the work carried out by a Vauxhall/Opel dealer or suitably equipped specialist if a transmission fault is suspected.

2 Do not remove the transmission from the car for possible repair before professional fault diagnosis has been carried out, since most tests require the transmission to be in the vehicle.

Chapter 7 Part C
Transfer gearbox and final drive – 4WD models

Contents

Degrees of difficulty

Easy, suitable for novice with little experience	**Fairly easy,** suitable for beginner with some experience	**Fairly difficult,** suitable for competent DIY mechanic	**Difficult,** suitable for experienced DIY mechanic	**Very difficult,** suitable for expert DIY or professional

Specifications

Lubrication

Lubricant type	See *Lubricants and fluids* on page 0•18
Lubricant capacity:	
Transfer gearbox	0.35 litre
Final drive diferential	0.5 litre

Torque wrench settings

	Nm	lbf ft
Electronic clutch housing retaining bolts	40	30
Final drive front mounting bolt	100	74
Final drive oil filler/level plug and drain plug	40	30
Final drive rear mounting bolts	108	80
Intermediate shaft flange-to-transfer gearbox	58	43
Propeller shaft-to-final drive flange bolts	36	27
Rear driveshaft constant velocity joint bolts	58	43
Rear mounting torque link-to-transfer gearbox adapter through bolt: *		
Stage 1	80	59
Stage 2	Angle-tighten a further 60°	
Roadwheel nuts	140	103
Transfer gearbox adapter bolts: *		
Stage 1	100	74
Stage 2	Angle-tighten a further 40°	
Transfer gearbox oil level/filler plug and drain plug	40	30
Transfer gearbox support bracket bolts	22	16
Transfer gearbox-to-transmission: *		
Stage 1	100	74
Stage 2	Angle-tighten a further 45°	

*Use new fasteners

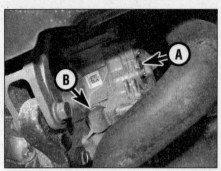

2.3 Transfer gearbox oil filler/level plug (A)
and drain plug (B)

1 General information

1 The four-wheel drive (4WD) system controls the distribution of drive between the front and rear wheels according to road conditions. During normal driving, only the front wheels are driven. The 4WD control unit, which is located on the final drive housing, receives signals from an array of sensors and uses this information to determine when it is time to deploy up to 50% of available torque to the rear wheels for better traction.

2 The 4WD system uses an electronically-controlled, multi-plate clutch which is flange-mounted on the final drive housing. The clutch plates operate in an oil bath and their locking effect is seamlessly adjusted in milliseconds by the 4WD control unit.

3 The advantages of having an intelligent on-demand control of power distribution are immense in comparison with purely mechanical systems still sometimes used today. The inherent disadvantages of such simple 4WD systems – for example, twisting in the drivetrain, substantially higher fuel consumption and a reduction in driving comfort due to stronger vibrations – are almost completely eliminated by the intelligent 4WD system. The system provides a precise, on-demand power distribution which reacts in milliseconds to changing driving situations. This makes it clearly superior to other on-demand 4WD systems which rely

3.13 Rear mounting torque link-to-transfer
gearbox adapter through bolt

on the purely mechanical control of power distribution when wheel slip occurs. This usually happens after a noticeable time delay, which has a corresponding negative impact on driving dynamics and comfort levels.

2 Transfer gearbox oil – draining and refilling

Note: *A new filler/level plug and drain plug will be required for this operation.*

1 Draining the oil is much more efficient if the car is first taken on a journey of sufficient length to warm the engine/transmission up to normal operating temperature.

Note: *If the procedure is to be carried out on a hot transmission unit, take care not to burn yourself on the hot exhaust or the transmission/engine unit.*

2 Position the vehicle over an inspection pit, on vehicle ramps, or jack it up and support it securely on axle stands (see *Jacking and vehicle support*), but make sure that it is level. Where fitted, remove the engine undertray as described in Chapter 11, Section 21.

3 Wipe clean the area around the filler/level plug which is situated on the right-hand side of the transfer gearbox **(see illustration)**.

4 Unscrew the filler/level plug and be prepared for some oil spillage as the plug is removed.

5 Position a suitable container under the drain plug which is located under the filler/level plug at the rear of the transfer gearbox **(see illustration 2.3)**.

6 Wipe clean the area around the drain plug, then unscrew the plug and allow the oil to drain completely into the container.

7 When the oil had finished draining, clean the threads in the transfer gearbox casing and fit the new drain plug. Tighten the plug to the specified torque.

8 Refill the transfer gearbox through the filler/level plug aperture with the specified grade of oil as given in *Lubricants and fluids*. Continue adding oil until the level is up to the bottom of the filler/level plug aperture.

9 When the level is correct, clean the threads in the transfer gearbox casing and fit the new filler/level plug. Tighten the plug to the specified torque.

10 Where applicable, refit the engine undertray as described in Chapter 11, Section 21, then lower the vehicle to the ground.

3 Transfer gearbox – removal and refitting

Removal

1 Apply the handbrake, then jack up the front of the vehicle and support it on axle stands (see *Jacking and vehicle support*).

2 Disconnect the breather hose from the top of the transfer gearbox.

3 Drain the transfer gearbox oil as described in Section 2.

4 Remove the exhaust system front pipe as described in Chapter 4A, Section 20 or Chapter 4B, Section 26.

5 Remove the right-hand driveshaft as described in Chapter 8, Section 2.

6 Undo the two bolts securing the intermediate shaft flange to the right-hand side of the transfer gearbox. Remove the intermediate shaft from the transfer gearbox, then note the fitted depth and position of the intermediate shaft oil seal. Using a suitable hooked tool, remove the intermediate shaft oil seal. Note that a new oil seal will be required for refitting.

7 Undo the three bolts securing the transfer gearbox support bracket to the cylinder block and remove the support bracket.

8 Remove the propeller shaft as described in Chapter 8, Section 6.

9 Place a jack with a block of wood beneath the transfer gearbox, and raise the jack to take the weight of the assembly. Ensure that the gearbox is well supported and stable on the jack.

10 Engage the help of an assistant to support the transfer gearbox, then undo the four bolts securing the gearbox to the transmission. Note that new bolts will be required for refitting.

11 With the bolts removed, lower the trolley jack, withdraw the transfer gearbox from the transmission and manoeuvre the unit out from under the car.

12 Remove the transfer gearbox-to-transmission oil seal from the transfer gearbox adapter, noting that a new seal will be required for refitting.

13 If necessary, the transfer gearbox adapter can be removed after undoing the five retaining bolts and the through bolt securing it to the rear mounting torque link **(see illustration)**. Note that new bolts will be required for refitting.

Refitting

Note: *Vauxhall/Opel special tool DT-48877 or a suitable alternative will be required to protect the oil seal in the transfer gearbox as the intermediate shaft is refitted.*

14 If removed, refit the transfer gearbox adapter and fit the five new retaining bolts and rear mounting torque link through bolt. Tighten the bolts to the specified torque, then through the specified angle.

15 Lightly lubricate the transfer gearbox oil seal with transmission oil, then fit the seal to the transfer gearbox adapter.

16 Position the transfer gearbox securely on the trolley jack and with the help of an assistant, raise it into position on the transmission.

17 Insert the four new transfer case retaining bolts and tighten them to the specified torque, then through the specified angle.

18 Locate the transfer gearbox support bracket in position on the cylinder block. Refit the three bolts and tighten to the specified torque.

19 Lightly lubricate the intermediate shaft oil

seal with transmission oil, then carefully ease the seal into position in the transfer gearbox. Using a drift or suitable mandrel, drive the seal into the transfer gearbox in the position noted on removal. Take care not to damage the seal lips during refitting.

20 Locate the special tool (DT-48877) on the transfer gearbox to protect the oil seal, then slide the intermediate shaft back into position. Refit the two retaining bolts and tighten them to the specified torque. Remove the special tool.

21 Refit the right-hand driveshaft as described in Chapter 8, Section 2.

22 Refit the propeller shaft as described in Chapter 8, Section 6.

23 Refit the exhaust system front pipe as described in Chapter 4A, Section 20 or Chapter 4B, Section 26.

24 Refill the transfer gearbox with oil as described in Section 2.

25 Reconnect the breather hose to the top of the transfer gearbox, then lower the vehicle to the ground.

4 Transfer gearbox – overhaul

1 Overhauling a transfer gearbox unit is a difficult and involved job for the DIY home mechanic. In addition to dismantling and reassembling many small parts, clearances must be precisely measured and, if necessary, changed by selecting shims and spacers. Internal components are also often difficult to obtain, and in many instances, extremely expensive. Because of this, if the transfer gearbox develops a fault or becomes noisy, the best course of action is to have the unit overhauled by a specialist repairer, or to obtain an exchange reconditioned unit.

5 Final drive – draining and refilling

1 Draining the oil is much more efficient if the car is first taken on a journey of sufficient length to warm the transmission up to normal operating temperature.

2 Position the vehicle over an inspection pit, on vehicle ramps, or jack it up and support it securely on axle stands (see *Jacking and vehicle support*), but make sure that it is level.

3 Wipe clean the area around the filler/level plug which is situated at the rear of the final drive housing **(see illustration)**.

4 Unscrew the filler/level plug and be prepared for some oil spillage as the plug is removed.

5 Position a suitable container under the drain plug which is located on the underside of the final drive housing **(see illustration 5.3)**.

6 Wipe clean the area around the drain plug, then unscrew the plug and allow the oil to drain completely into the container.

7 When the oil had finished draining, clean the threads of the drain plug and final drive housing and refit the drain plug. Tighten the plug to the specified torque.

8 Refill the final drive housing through the filler/level plug aperture with the specified grade of oil as given in *Lubricants and fluids*. Continue adding oil until the level is up to the bottom of the filler/level plug aperture.

9 When the level is correct, clean the threads of the filler/level plug and final drive housing and refit the filler/level plug. Tighten the plug to the specified torque.

10 On completion, lower the vehicle to the ground.

6 Final drive – removal and refitting

Removal

1 Chock the front wheels, then jack up the rear of the vehicle, and support it securely on axle stands (see *Jacking and vehicle support*.

2 Drain the final drive unit as described in Section 5.

3 Undo the six retaining bolts and collect the three washer plates securing the propeller shaft to the final drive flange **(see illustration)**. Make alignment marks on the propeller shaft flange and final drive flange to ensure correct refitting, then separate the flanges and move the rear of the propeller shaft to one side.

4 Disconnect the wiring connector from the

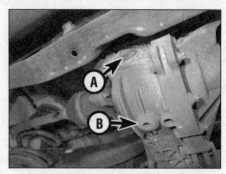

5.3 Final drive oil filler/level plug (A) and drain plug (B)

4WD control unit and remove the cable ties securing the wiring harness to the final drive casing **(see illustration)**.

5 Undo the three bolts each side and separate the rear driveshaft constant velocity joints from the final drive flanges **(see illustration)**. Suitably support the driveshafts clear of the final drive unit.

6 Place a trolley jack with interposed block of wood beneath the final drive unit, and raise the jack to take the weight of the assembly. Ensure that the unit is well supported and stable on the jack.

7 Undo the two rear mounting bolts securing the final drive unit to the support beam.

8 Using paint or a suitable marker, mark the outline of the front final drive mounting on the body to ensure correct alignment when refitting **(see illustration)**, then undo and remove the mounting bolt.

6.3 Undo the six bolts and separate the propeller shaft and final drive flanges

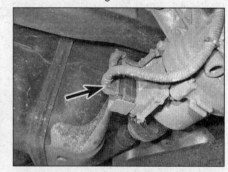

6.4 Disconnect the wiring connector from the 4WD control unit

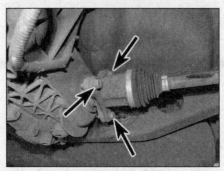

6.5 Undo the retaining bolts and separate the driveshaft constant velocity joints from the final drive flanges

6.8 Mark the outline of the front final drive mounting on the body to ensure correct alignment when refitting

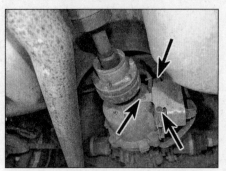

8.4 Undo the nut and two screws and remove the protective shield

9 Engage the help of an assistant to support the final drive unit, then lower the trolley jack and manoeuvre the unit out from under the car.

Refitting

10 Position the final drive unit securely on the trolley jack and with the help of an assistant, raise it into position.
11 Refit the three final drive mounting bolts but only tighten them finger tight at this stage.
12 Align the final drive front mounting with the markings on the underbody made on removal. With the unit aligned, tighten the two rear mounting bolts to the specified torque, then tighten the front mounting bolt to the specified torque.
13 Remove the trolley jack from under the final drive unit.
14 Reconnect the wiring connector to the 4WD control unit, then secure the wiring harness to the final drive unit using new cable ties.
15 Locate the rear driveshaft constant velocity joints back in position on the final drive flanges, refit the retaining bolts and tighten the bolts to the specified torque.
16 Thoroughly clean the threads of the propeller shaft retaining bolts and the bolt hole threads in the final drive flange.
17 Locate the propeller shaft back in position on the final drive flange, aligning the marks on the flanges made on removal. Apply a suitable thread locking compound to the bolt threads, then insert the six retaining bolts together with the three washer plates and tighten the bolts to the specified torque.
18 Refill the final drive with oil as described in Section 5, then lower the car to the ground.

7 Final drive – overhaul

1 Overhauling a final drive unit is a difficult and involved job for the DIY home mechanic. In addition to dismantling and reassembling many small parts, clearances must be precisely measured and, if necessary, changed by selecting shims and spacers.

Internal components are also often difficult to obtain, and in many instances, extremely expensive. Because of this, if the final drive develops a fault or becomes noisy, the best course of action is to have the unit overhauled by a specialist repairer, or to obtain an exchange reconditioned unit.

8 Four wheel drive (4WD) control unit – removal and refitting

Note: *If a new control unit is to be fitted, this work must be entrusted to a Vauxhall/Opel dealer or suitably-equipped specialist as it is necessary to program the new control unit after installation. This work requires the use of dedicated Vauxhall/Opel diagnostic equipment or a compatible alternative.*

Removal

1 The 4WD control unit is located under the car, attached to the left-hand side of the final drive housing.
2 Chock the front wheels, then jack up the rear of the vehicle, and support it securely on axle stands (see *Jacking and vehicle support*).
3 Disconnect the wiring connector from the 4WD control unit **(see illustration 6.4)**.
4 Undo the nut and two screws and remove the control unit protective shield **(see illustration)**.
5 Disconnect the control unit wiring connector at the electronic clutch housing, then free the wiring harness from the cable tie.
6 Undo the two retaining bolts and remove the control unit from the side of the electronic clutch housing.

Refitting

7 Refitting is a reversal of removal, ensuring that the wiring harness is properly secured with a new cable tie.

9 Electronic clutch housing – removal and refitting

Removal

1 Chock the front wheels, then jack up the rear of the vehicle, and support it securely on axle stands (see *Jacking and vehicle support*).
2 Undo the six retaining bolts and collect the three washer plates securing the propeller shaft to the final drive flange **(see illustration 6.3)**. Make alignment marks on the propeller shaft flange and final drive flange to ensure correct refitting, then separate the flanges and move the rear of the propeller shaft to one side.
3 Remove the 4WD control unit as described in Section 8.
4 Undo the four bolts securing the electronic clutch housing to the final drive housing.

5 Suitably support the electronic clutch housing. Using a large screwdriver, carefully prise the clutch housing free of the final drive housing. Note that there are three projecting lugs around the housing flange to allow the unit to be prised free.
6 Once the unit is released from the final drive housing, remove it from under the car and collect the sealing O-ring. Note that a new O-ring will be required for refitting.

Refitting

7 Refitting is a reversal of removal, bearing in mind the following points:
a) Use a new O-ring when refitting the clutch housing.
b) Position the unit on the final drive housing so that the control unit location is on the left-hand side.
c) Tighten the retaining bolts to the specified torque, working in a diagonal sequence.
d) Refit the 4WD control unit as described in Section 8.
e) Locate the propeller shaft back in position on the final drive flange, aligning the marks on the flanges made on removal. Insert the six retaining bolts together with the three washer plates and tighten the bolts to the specified torque.

10 Rear axle shaft oil seal – renewal

1 Remove the rear driveshaft on the side concerned as described in Chapter 8, Section 2.
2 Attach a suitable slide hammer or similar tool to the legs of the axle shaft and draw the shaft out of the final drive housing.
3 Remove the circlip from the end of the axle shaft splines and discard it. When refitting, fit a new circlip, making sure it is correctly located in the groove.
4 Measure and record the fitted depth of the oil seal in the final drive housing, then prise it out using a stout screwdriver or hooked tool. Take care not to damage the final drive housing as the seal is removed.
5 Remove all traces of dirt from the area around the oil seal aperture, then lubricate the outer lip of the new oil seal with oil. Ensure the seal is correctly positioned, with its sealing lip facing inwards, and tap it squarely into position, using a suitable tubular drift (such as a socket) which bears only on the hard outer edge of the seal. Ensure the seal is fitted at the same depth in its housing that the original was.
6 With a new circlip in position, insert the axle shaft into the final drive housing. Using a suitable hammer, lightly tap the axle shaft until it locks into place.
7 Refit the rear driveshaft as described in Chapter 8, Section 2.

Chapter 8
Driveshafts and propeller shafts

Contents

Degrees of difficulty

Easy, suitable for novice with little experience	Fairly easy, suitable for beginner with some experience	Fairly difficult, suitable for competent DIY mechanic	Difficult, suitable for experienced DIY mechanic	Very difficult, suitable for expert DIY or professional

Specifications

General

Driveshaft type	Solid steel shafts with inner and outer constant velocity (CV) joints. Front right-hand driveshaft incorporating intermediate shaft
Constant velocity joint type:	
Front and rear driveshafts:	
Inner joints	Tripod
Outer joints	Ball-and-cage
Propeller shaft type	Two-piece variable length steel tube with centre joint

Lubrication (overhaul only – see text)

Lubricant type/specification	Use only special grease supplied in sachets with gaiter kits – joints are otherwise pre-packed with grease and sealed

Torque wrench settings

	Nm	lbf ft
Anti-roll bar link rod nut	65	48
Driveshaft retaining nut*:		
Stage 1	50	37
Stage 2	Angle-tighten through a further 60°	
Intermediate shaft (FWD models):		
Bearing flange-to-support bracket bolts	22	16
Support bracket-to-cylinder block bolts	58	43
Intermediate shaft (4WD models):		
Intermediate shaft flange-to-transfer gearbox bolts	58	43
Lower arm balljoint clamp bolt nut*:		
Stage 1	42	31
Stage 2	Slacken by 120°	
Stage 3	35	26
Stage 4	Angle-tighten through a further 45°	
Propeller shaft centre bearing bracket bolts	42	31
Propeller shaft flange bolts	36	27
Roadwheel nuts	140	103
Track rod end-to-swivel hub*:		
Stage 1	30	22
Stage 2	Angle-tighten through a further 135°	

*Use new fasteners

2.4 Knock back the staking securing the retaining nut to the driveshaft groove

2.5 Attach the holding tool to the wheel hub using two wheel nuts

A tool to hold the wheel hub stationary whilst the driveshaft retaining nut is slackened can be fabricated from two lengths of steel strip (one long, one short) and a nut and bolt; the nut and bolt forming the pivot of a forked tool.

1 General Information

FWD models

1 Drive is transmitted from the differential to the front wheels by means of two, solid steel driveshafts.

2 Both driveshafts are splined at their outer ends to accept the wheel hubs, and are threaded so that each hub can be fastened by a large nut. The inner end of each driveshaft is splined to accept the intermediate shaft or differential sun gear and is held in place by an internal circlip.

3 Constant velocity (CV) joints are fitted to each end of the driveshafts, to ensure the smooth and efficient transmission of drive at all the angles possible as the roadwheels move up-and-down with the suspension, and as they turn from side-to-side under steering. The outer constant velocity joints are of the ball-and-cage type and the inner joints are of the tripod type.

4 The right-hand driveshaft has an intermediate shaft supported by a bracket attached to the rear of the cylinder block.

4WD models

5 The front wheels are driven the same as they are on FWD models, by means of two solid steel driveshafts.

6 Drive is transmitted from the transfer gearbox to the rear final drive unit through a two-piece propeller shaft with three joints. The joints on each end are a flange type connecting to the transfer gearbox and final drive unit, and there is a universal joint at the centre.

7 The rear final drive unit then transmits the drive through two solid steel driveshafts to the rear wheels. Both driveshafts are splined at their outer ends, to accept the wheel hubs, and are threaded so that each hub can be fastened to the driveshaft by a large nut. The inner end of each driveshaft is bolted to the final drive axle shaft.

8 Constant velocity (CV) joints are fitted to each end of the driveshafts, to ensure the smooth and efficient transmission of drive at all the angles possible as the roadwheels move up-and-down with the suspension. The outer constant velocity joints are of the ball-and-cage type and the inner joints are of the tripod type.

2 Driveshaft – removal and refitting

Front driveshaft

Removal

Note: *The driveshaft outer joint splines may be a tight fit in the hub and it is possible that a puller/extractor will be required to draw the hub assembly off the driveshaft during removal.*

1 Firmly apply the handbrake, then jack up the front of the car and support it securely on axle stands (see *Jacking and vehicle support*). Remove the relevant front roadwheel.

2 Remove the engine undertray as described in Chapter 11, Section 21.

3 Drain the transmission oil/fluid as described in Chapter 7A, Section 2 or Chapter 7B, Section 2.

4 Using a hammer and small chisel, knock back the staking securing the retaining nut to the driveshaft groove **(see illustration)**.

5 To prevent rotation of the wheel hub as the driveshaft retaining nut is slackened, make up a holding tool and attach the tool to the wheel hub using two wheel nuts **(see Tool Tip and illustration)**.

6 With the holding tool in place, slacken and remove the driveshaft retaining nut using a socket and long bar. Where necessary, support the socket on an axle stand to prevent it slipping off the nut. This nut is very tight; make sure that there is no risk of pulling the car off the axle stands as the nut is slackened. Note that a new nut will be required for refitting.

7 Slacken, but do not fully remove, the nut securing the track rod end to the steering arm on the swivel hub then use a two-legged puller to separate the track rod end balljoint taper. Once the taper is released, fully unscrew the nut and remove the track rod end from the steering arm **(see illustrations)**. Note that a new nut will be required for refitting.

2.7a Slacken the nut securing the track rod end to the steering arm...

2.7b ...separate the track rod end balljoint taper using a two-legged puller...

2.7c ...then unscrew the nut and remove the track rod end

2.8a Unscrew the anti-roll bar link rod nut...

2.8b ...then disconnect the link rod from the strut

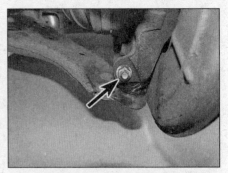

2.9a Unscrew the retaining nut...

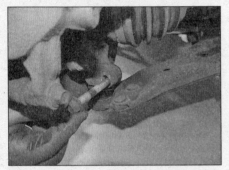

2.9b ...and remove the clamp bolt from the swivel hub

2.10 Tap a chisel into the slot in the swivel hub to spread the slot

2.11 Push down on the suspension lower arm to free the balljoint from the swivel hub

8 Unscrew the nut and disconnect the anti-roll bar link rod from the suspension strut. Use a suitable Torx key inserted into the link to hold the link while the nut is being loosened **(see illustrations)**.

9 Unscrew the nut and remove the clamp bolt securing the front suspension lower arm to the swivel hub **(see illustrations)**. Note that a new nut and bolt will be required for refitting.

10 Tap a suitable chisel into the slot in the swivel hub to slightly spread the slot and aid removal of the ball joint **(see illustration)**.

11 Using a lever, push down on the suspension lower arm to free the balljoint from the swivel hub, then move the swivel hub to one side and release the arm, taking care not to damage the balljoint rubber boot **(see illustration)**. It is advisable to place a protective cover over the rubber boot such as the plastic cap from an aerosol can, suitably cut to fit.

12 The hub must now be freed from the end of the driveshaft **(see illustration)**. It may be possible to pull the hub off the driveshaft, but if the end of the driveshaft is tight in the hub, temporarily refit the driveshaft retaining nut to protect the driveshaft threads, then tap the end of the driveshaft with a soft-faced hammer while pulling outwards on the swivel hub. Alternatively, use a suitable puller to press the driveshaft through the hub.

13 Collect the spacer from the end of the constant velocity joint **(see illustration)**.

14 With the driveshaft detached from the swivel hub, tie the suspension strut to one side and support the driveshaft on an axle stand.

15 If working on the left-hand driveshaft, place a suitable container beneath the differential, to collect escaping transmission oil/fluid when the driveshaft is withdrawn.

16 Using a stout bar, release the inner

end of the driveshaft from the differential or intermediate shaft. Lever between the constant velocity joint and differential housing, or between the constant velocity joint and intermediate shaft bearing housing, to release the driveshaft retaining circlip **(see illustration)**.

17 Withdraw the driveshaft, ensuring that the constant velocity joints are not placed under excessive strain, and remove the driveshaft from beneath the vehicle. Whilst the driveshaft is removed, plug or tape over the differential aperture to prevent dirt entry. Where applicable, remove the circlip and O-ring seal from the intermediate shaft splined end and obtain new components for refitting.

Caution: Do not allow the vehicle to rest on its wheels with one or both driveshafts removed, as damage to the wheel bearings(s) may result. If the vehicle must be moved on its wheels, clamp the

2.12 Pull the swivel hub outwards, then withdraw the driveshaft from the hub splines

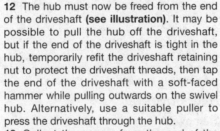

2.13 Collect the spacer from the end of the driveshaft constant velocity joint

2.16 Using a stout bar, release the inner end of the driveshaft from the differential or intermediate shaft

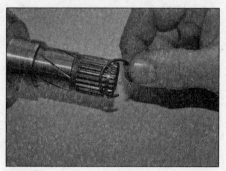

2.19 Remove the circlip from the end of the driveshaft inner joint splines

wheel bearings using spacers and a long threaded rod to take the place of the driveshaft.

Refitting

18 Before refitting the driveshaft, examine the oil seal in the differential housing and renew it if necessary as described in Chapter 7A, Section 6 or Chapter 7B, Section 6.

19 Remove the circlip from the end of the driveshaft inner joint splines or intermediate shaft and discard it **(see illustration)**. Fit a new circlip, making sure it is correctly located in the groove.

20 Where applicable, fit a new O-ring to the splined end of the intermediate shaft.

21 Thoroughly clean the driveshaft splines, intermediate shaft splines (where applicable), and the apertures in the transmission and hub assembly. Apply a thin film of grease to the oil seal lips, and to the driveshaft splines and shoulders. Check that all gaiter clips are securely fastened.

22 Offer up the driveshaft, and engage the inner joint splines with those of the differential sun gear or intermediate shaft, taking care not to damage the oil seal. Push the joint fully into position, then check that the circlip is correctly located and securely holds the joint in position. If necessary, use a soft-faced mallet or drift to drive the driveshaft inner joint fully into position.

23 Ensure that he spacer is in position on the constant velocity joint, then align the outer constant velocity joint splines with those of the hub, and slide the joint back into position in the hub.

24 Using a lever, push down on the lower

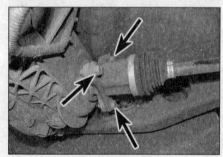

2.38 Undo the retaining bolts and separate the driveshaft constant velocity joint from the final drive flange

suspension arm, then relocate the balljoint and release the arm. Make sure that the balljoint stub is fully entered in the swivel hub.

25 Insert a new lower arm balljoint clamp bolt with its head facing the front of the vehicle, fit a new retaining nut and tighten the nut to the specified torque and through the specified angle in the stages given in the Specifications.

26 Connect the anti-roll bar link rod to the suspension strut and use a Torx key to hold the link while tightening the nut to the specified torque.

27 Engage the track rod end in the swivel hub, then fit the new retaining nut and tighten it to the specified torque and through the specified angle given in the Specifications. Counterhold the balljoint shank using a second spanner to prevent rotation as the nut is tightened.

28 Lubricate the inner face and threads of the new driveshaft retaining nut with clean engine oil, and refit it to the end of the driveshaft. Use the method employed on removal to prevent the hub from rotating, and tighten the driveshaft retaining nut to the specified torque and through the specified angle in the stages given in the Specifications. Check that the hub rotates freely.

29 Using a hammer and small chisel, securely stake the retaining nut into the driveshaft groove.

30 Refill the transmission with oil/fluid as described in Chapter 7A, Section 2 or Chapter 7B, Section 2.

31 Refit the engine undertray as described in Chapter 11, Section 21.

32 Refit the roadwheel, lower the vehicle to the ground and tighten the roadwheel nuts to the specified torque.

Rear driveshaft

Removal

Note: *The driveshaft outer joint splines may be a tight fit in the hub flange and it is possible that a puller/extractor will be required to release the driveshaft during removal.*

33 Chock the front wheels, then jack up the rear of the vehicle, and support it securely on axle stands (see *Jacking and vehicle support*.

34 Using a hammer and small chisel, knock back the staking securing the retaining nut to the driveshaft groove **(see illustration 2.4)**.

35 To prevent rotation of the wheel hub as the driveshaft retaining nut is slackened, make up a holding tool and attach the tool to the wheel hub using two wheel nuts (see Tool Tip and illustration 2.5).

36 With the holding tool in place, slacken and remove the driveshaft retaining nut using a socket and long bar. Where necessary, support the socket on an axle stand to prevent it slipping off the nut. This nut is very tight; make sure that there is no risk of pulling the car off the axle stands as the nut is slackened. Note that a new nut will be required for refitting.

37 The end of the driveshaft must now be freed from the hub flange. It may be possible to push the driveshaft out of the hub, but if

the end of the driveshaft is tight in the hub, temporarily refit the driveshaft retaining nut to protect the driveshaft threads, then tap the end of the driveshaft with a soft-faced hammer. Alternatively, use a suitable puller to press the driveshaft through the hub.

38 Undo the three bolts and separate the driveshaft constant velocity joint from the final drive flange, then remove the driveshaft from under the car **(see illustration)**.

Refitting

39 Offer up the driveshaft, then align the outer constant velocity joint splines with those of the hub flange, and slide the joint back into position in the hub.

40 Locate the driveshaft constant velocity joint back in position on the final drive flange, refit the retaining bolts and tighten the bolts to the specified torque.

41 Lubricate the inner face and threads of the new driveshaft retaining nut with clean engine oil, and refit it to the end of the driveshaft. Use the method employed on removal to prevent the hub flange from rotating, and tighten the driveshaft retaining nut to the specified torque in the stages given in the Specifications. Check that the hub rotates freely.

42 Using a hammer and small chisel, securely stake the retaining nut into the driveshaft groove.

43 Refit the roadwheel, lower the vehicle to the ground and tighten the roadwheel nuts to the specified torque.

---

3 Intermediate shaft – removal and refitting

FWD models

Removal

1 Remove the right-hand driveshaft as described in Section 2.

2 Place a suitable container beneath the differential, to collect escaping transmission oil/fluid when the intermediate shaft is withdrawn.

3 Undo the three bolts (two on some engines) securing the intermediate shaft support bracket to the cylinder block **(see illustration)**.

4 Rotate the support bracket away from the engine, pull the intermediate shaft out of the

3.3 Undo the intermediate shaft support bracket bolts (1.6 litre diesel engine shown)...

transmission and remove it from under the car **(see illustration)**.

5 Clamp the intermediate shaft in a vice, then undo the three bolts securing the intermediate shaft bearing flange to the support bracket and lift off the flange.

6 If not already done, remove the O-ring from the end of the shaft, then using a small screwdriver, remove the two circlips. Note that new circlips and a new O-ring will be required for reassembly.

7 Remove the bearing and support bracket from the intermediate shaft.

Refitting

8 Before refitting the intermediate shaft, examine the oil seal in the transmission housing and renew it if necessary as described in Chapter 7A, Section 6 or Chapter 7B, Section 6.

9 Refit the bearing and support bracket to the intermediate shaft, then fit the two new circlips and new O-ring.

10 Place the bearing flange on the support bracket and secure with the three retaining bolts tightened to the specified torque.

11 Thoroughly clean the intermediate shaft splines, and the aperture in the transmission. Apply a thin film of grease to the oil seal lips, and to the intermediate shaft splines and shoulders.

12 Insert the intermediate shaft into the transmission, engaging the splines with the differential sun gear, taking care not to damage the oil seal.

13 Position the support bracket on the cylinder block, refit the retaining bolts and tighten them to the specified torque.

3.4 ...then pull the intermediate shaft out of the transmission and remove it

14 Refit the right-hand driveshaft as described in Section 2.

4WD models

Removal

15 Remove the right-hand driveshaft as described in Section 2.

16 Drain the transfer gearbox oil as described in Chapter 7C, Section 2.

17 Undo the two bolts securing the intermediate shaft flange to the right-hand side of the transfer gearbox. Remove the intermediate shaft from the transfer gearbox, then note the fitted depth and position of the intermediate shaft oil seal. Using a suitable hooked tool, remove the intermediate shaft oil seal. Note that a new oil seal will be required for refitting.

Refitting

Note: *Vauxhall/Opel special tool DT-48877 or a suitable alternative will be required to*

protect the oil seal in the transfer gearbox as the intermediate shaft is refitted.

18 Lightly lubricate the intermediate shaft oil seal with transmission oil, then carefully ease the seal into position in the transfer gearbox. Using a drift or suitable mandrel, drive the seal into the transfer gearbox in the position noted on removal. Take care not to damage the seal lips during refitting.

19 Locate the special tool (DT-48877) on the transfer gearbox to protect the oil seal, then slide the intermediate shaft back into position. Refit the two retaining bolts and tighten them to the specified torque. Remove the special tool.

20 Refit the right-hand driveshaft as described in Section 2.

21 Refill the transfer gearbox with oil as described in Chapter 7C, Section 2.

4 Driveshaft joint gaiters – renewal

Outer joint

1 Remove the driveshaft from the car as described in Section 2.

2 Release the rubber gaiter inner and outer retaining clips by cutting through them using side cutters **(see illustration)**. Spread the clips and remove them from the gaiter.

3 Slide the rubber gaiter down the shaft to expose the CV joint or, alternatively, cut the gaiter open using a suitable knife and remove it from the driveshaft **(see illustration)**.

4 Using old rags, clean away as much of the old grease as possible from the CV joint. It is advisable to wear disposable rubber gloves during this operation.

5 Using a brass drift and hammer, sharply strike the centre part of the outer joint to drive it off the end of the shaft **(see illustrations)**. The joint is retained on the driveshaft by a circlip, and striking the joint in this manner forces the circlip into its groove, so allowing the joint to slide off.

6 Once the joint has been removed, extract the circlip from the groove in the driveshaft splines, noting that a new circlip must be fitted on reassembly **(see illustration)**.

4.2 Release the rubber gaiter retaining clips using side cutters

4.3 Slide the rubber gaiter down the shaft to expose the CV joint

4.5a Using a soft metal drift...

4.5b ...to release the outer joint

4.6 Remove the circlip from the groove in the driveshaft splines...

4.7 ...then withdraw the rubber gaiter from the driveshaft

4.12a Locate the new small retaining clip on the gaiter..

4.12b ...then slide the gaiter and retaining clip onto the driveshaft

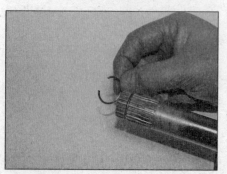

4.13 Fit a new circlip to the groove in the driveshaft

4.14a Engage the joint with the driveshaft splines…

4.14b ...then tap the joint onto the driveshaft until the circlip engages in its groove

7 If still in place, withdraw the rubber gaiter from the driveshaft **(see illustration)**.

8 With the CV joint removed from the driveshaft, wipe away the remaining grease (do not use any solvent) to allow the joint components to be inspected.

9 Move the inner splined driving member from side-to-side, to expose each ball in turn at the top of its track. Examine the balls for cracks, flat spots, or signs of surface pitting.

10 Inspect the ball tracks on the inner and outer members. If the tracks have widened, the balls will no longer be a tight fit. At the same time, check the ball cage windows for wear or cracking between the windows.

11 If on inspection any of the constant velocity joint components are found to be worn or damaged, it will be necessary to renew the complete joint assembly. If the joint is in satisfactory condition, obtain a repair kit consisting of a new gaiter and retaining clips, a constant velocity joint circlip, and the correct type and quantity of grease.

12 Locate the new small retaining clip on the new rubber gaiter, then slide the gaiter and retaining clip onto the driveshaft **(see illustrations)**.

13 Fit a new circlip to the groove in the driveshaft **(see illustration)**.

14 Screw on the driveshaft retaining nut two or three turns to protect the threads, then engage the joint with the driveshaft splines. Tap the joint onto the driveshaft until the circlip engages in its groove **(see illustrations)**. Make sure that the joint is securely retained, by pulling on the joint, not the shaft.

15 Pack the joint with the grease supplied in the repair kit **(see illustration)**. Work the grease well into the bearing tracks whilst twisting the joint, and fill the rubber gaiter with any excess.

16 Ease the gaiter over the joint, and ensure that the gaiter lips are correctly located in the grooves on both the driveshaft and constant velocity joint. Lift the outer sealing lip of the gaiter, to equalise air pressure within the gaiter.

17 Fit the new large retaining clip to the gaiter. Pull the retaining clip as tight as possible, and locate the hooks on the clip in their slots. Remove any slack in the gaiter retaining clip by carefully compressing the raised section of the clip. In the absence of the special tool, a pair of side-cutters may be used. Secure the small retaining clip using the same procedure **(see illustrations)**.

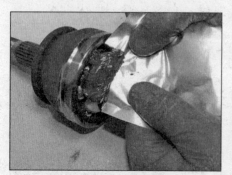

4.15 Pack the joint with the grease supplied in the repair kit

4.17a Fit the new large retaining clip to the gaiter…

4.17b ...then secure the gaiter retaining clip in position by compressing the raised section of the clip

4.17c The small inner retaining clip is secured in the same way

4.22 Slide the rubber gaiter and the gaiter adapter down the driveshaft

4.26 Pack the joint with half the grease supplied in the repair kit

4.27a Locate the gaiter adapter in position…

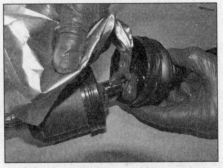

4.27b …fill the rubber gaiter with the remaining grease

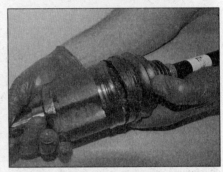

4.27c …then engage the gaiter with the gaiter adapter

18 Check that the constant velocity joint moves freely in all directions, then refit the driveshaft to the car as described in Section.

Inner joint

Note: *The inner tripod type CV joint cannot be detached from the driveshaft. If inner CV joint gaiter renewal is required, it will be necessary to remove the outer CV joint and slide the inner joint gaiter off the outer end of the driveshaft.*

19 Remove the driveshaft from the car as described in Section 2.

20 Remove the outer CV joint and gaiter as described previously in this Section.

21 Release the rubber gaiter inner and outer retaining clips by cutting through them using side cutters. Spread the clips and remove them from the gaiter.

22 Slide the rubber gaiter and the gaiter adapter down the driveshaft and remove them from the outer joint end of the shaft **(see illustration)**.

23 Using old rags, clean away as much of the old grease as possible from the CV joint. It is advisable to wear disposable rubber gloves during this operation.

24 Carry out a visual inspection of the joint. If any of the joint components are worn, a new driveshaft will be required. The tripod assembly, driveshaft and the joint body cannot be separated and are not individually available. If the joint is in satisfactory condition, obtain a repair kit consisting of a new gaiter and retaining clips, and the correct type and quantity of grease.

25 Commence reassembly by sliding the new gaiter adapter and rubber gaiter onto the driveshaft.

26 Pack the joint with half the grease supplied in the repair kit **(see illustration)**. Work the grease well into the joint rollers and tracks whilst twisting the joint.

27 Locate the gaiter adapter in position over the joint. Fill the rubber gaiter with the remaining grease, then engage the gaiter with the gaiter adapter **(see illustrations)**.

28 Ensure that the gaiter lips are correctly located in the grooves on both the driveshaft and gaiter adapter. Lift the outer sealing lip of the gaiter, to equalise air pressure within the gaiter.

29 Fit the new small retaining clip to the

gaiter. Pull the retaining clip as tight as possible, and locate the hooks on the clip in their slots. Remove any slack in the gaiter retaining clip by carefully compressing the raised section of the clip. In the absence of the special tool, a pair of side-cutters may be used **(see illustration)**.

30 Locate the new large retaining clip on the gaiter. Using circlip pliers and a C-clamp as shown, or a similar arrangement, pull the retaining clip very tight until the hooks engage and retain the compressed clip **(see illustration)**.

31 Check that the constant velocity joint moves freely in all directions, then refit the driveshaft to the car as described in Section 2.

4.29 Secure the gaiter retaining clip in position by compressing the raised section of the clip

4.30 Using circlip pliers and a C-clamp to pull the retaining clip tight

6.5 Undo the two centre bearing bracket retaining bolts

6.6 Undo the six bolts securing the propeller shaft and final drive flanges

6.7 Undo the six bolts securing the propeller shaft and transfer gearbox flanges

5 Driveshaft overhaul – general information

1 If any of the checks described in Chapter 1A, Section 10 or Chapter 1B, Section 11 reveal possible wear in any driveshaft joint, carry out the following procedures to identify the source of the problem.

2 Firmly apply the handbrake, then jack up the front of the vehicle and support it securely on axle stands (see *Jacking and vehicle support*).

3 Referring to the information contained in Section 2, make up a tool to hold the wheel hub, and attach the tool to the hub using two wheel nuts. Use a torque wrench to check that the driveshaft retaining nut is securely fastened, then repeat this check on the remaining driveshaft nut.

4 Road test the vehicle, and listen for a metallic clicking from the front as the vehicle is driven slowly in a circle on full-lock. If a clicking noise is heard, this indicates wear in the outer constant velocity joint.

5 If vibration, consistent with road speed, is felt through the car when accelerating, there is a possibility of wear in the inner constant velocity joints.

6 To check the joints for wear, remove the driveshafts, then dismantle them as described in Section 4; if any wear or free play is found, the affected joint must be renewed.

6 Propeller shaft – removal and refitting

Removal

1 Jack up the front and rear of the car and support it securely on axle stands (see *Jacking and vehicle support*).

2 Make sure that the transmission is in neutral and the handbrake is released.

3 Remove the exhaust system main section and the intermediate underbody heat shield as described in Chapter 4A, Section 20 or Chapter 4B, Section 26.

4 Using paint or a suitable marker pen, mark the position of the propeller shaft centre bearing bracket on the underbody.

5 Position a suitable jack under the centre bearing bracket, then undo the two bolts securing the centre bearing bracket to the underbody (see illustration).

6 Undo the six retaining bolts and collect the three washer plates securing the propeller shaft to the final drive flange (see illustration). Make alignment marks on the propeller shaft flange and final drive flange to ensure correct refitting.

7 Suitably support the front section of the propeller shaft then undo the six retaining bolts and collect the three washer plates securing the propeller shaft to the transfer

gearbox flange (see illustration). Make alignment marks on the propeller shaft flange and transfer gearbox flange to ensure correct refitting.

8 Lower the propeller shaft from its location and remove it from under the car.

Refitting

9 Refitting is a reversal of removal, bearing in mind the following points:

a) Thoroughly clean the retaining bolt threads and the threads in the drive flanges and apply thread locking compound to the bolt threads.

b) Align the marks made on the flanges and centre bearing bracket during removal.

c) Tighten all retaining bolts to the specified torque.

d) Refit the exhaust system heat shield and main section as described in Chapter 4A, Section 20 or Chapter 4B, Section 26.

7 Propeller shaft – overhaul

1 The propeller shaft is effectively a complete sealed assembly and individual component parts are not available separately.

2 If there is any damage to the propeller shaft, or if any of the joints are defective, the complete shaft will need to be replaced.

Chapter 9
Braking system

Contents

Degrees of difficulty

Easy, suitable for novice with little experience	**Fairly easy,** suitable for beginner with some experience	**Fairly difficult,** suitable for competent DIY mechanic	**Difficult,** suitable for experienced DIY mechanic	**Very difficult,** suitable for expert DIY or professional

Specifications

Front brakes

Type	Ventilated disc, with single-piston sliding caliper
Disc diameter	300.0 mm
Disc thickness:	
New	26.0 mm
Minimum.	23.0 mm
Maximum disc thickness variation	0.025 mm
Maximum disc run-out	0.6 mm
Minimum brake pad friction material thickness	2.0 mm

Rear Brakes

Type	Solid disc, with single-piston sliding caliper
Disc diameter	268 mm
Disc thickness:	
New	12.0 mm
Minimum.	10.0 mm
Maximum disc thickness variation	0.025 mm
Maximum disc run-out	0.5 mm
Minimum brake pad friction material thickness	2.0 mm

Handbrake

Type	Self-adjusting cable-operated, acting on rear brake calipers

Torque wrench settings

	Nm	lbf ft
ABS hydraulic modulator mounting bracket nuts	17	12
ABS hydraulic modulator-to-mounting bracket	10	7
Brake caliper bleed screws	17	12
Brake caliper guide pin bolts	28	21
Brake caliper mounting bracket bolts: *		
Front caliper mounting bracket bolts	160	118
Rear caliper mounting bracket bolts	100	74
Brake fluid pipe unions.	18	13
Brake hydraulic hose banjo union bolts.	40	30
Brake master cylinder retaining nuts	13	10
Brake pedal mounting bracket nuts/bolts	19	14
Handbrake cable bracket bolts/nuts	22	16
Handbrake lever retaining nuts	22	16
Roadwheel nuts	140	103
Vacuum pump mounting bolts/nuts:		
Petrol engine models:		
Vacuum pump bracket nuts (5-speed manual transmission only) .	58	43
Vacuum pump bracket nuts (all other models).	22	16
1.7 litre diesel engine models.	24	18
Vacuum servo unit nuts	22	16

Use new fasteners

1 General Information

1 The braking system is of servo-assisted, dual-circuit hydraulic type split diagonally. The arrangement of the hydraulic system is such that each circuit operates one front and one rear brake from a tandem master cylinder. Under normal circumstances, both circuits operate in unison. However, in the event of hydraulic failure in one circuit, full braking force will still be available at two wheels.

2 All models are fitted with front and rear disc brakes. The disc brakes are actuated by single-piston sliding type calipers, which ensure that equal pressure is applied to each disc pad.

3 An Anti-lock Braking System (ABS) incorporating traction control and an electronic stability program, is fitted as standard equipment to all vehicles covered in this manual. Refer to Section 17 for further information on ABS operation.

4 The self-adjusting, cable-operated handbrake provides an independent mechanical means of rear brake application.

5 On petrol engine models, to ensure that an adequate source of vacuum to operate the vacuum servo unit under all engine operating conditions is available (particularly during cold start, high engine RPM and high altitude) an auxiliary vacuum pump is fitted. The pump is located in the engine compartment on the lower left-hand side and is driven by an integral electric motor. The pump only operates on demand and is controlled by a vacuum switch located on the vacuum servo unit.

6 On diesel engine models, since there is no throttling as such of the inlet manifold, the manifold is not a suitable source of vacuum to operate the vacuum servo unit. The servo unit

is therefore connected to a separate vacuum pump. On 1.6 litre engines, the vacuum pump is integral with the engine oil pump (refer to Chapter 2C, Section 13 for further information). On 1.7 litre engines, the vacuum pump is bolted to the left-hand end of the cylinder head and driven by the camshaft.

⚠ *Warning: When servicing any part of the system, work carefully and methodically; also observe scrupulous cleanliness when overhauling any part of the hydraulic system. Always renew components (in axle sets, where applicable) if in doubt about their condition, and use only genuine Vauxhall/Opel replacement parts, or at least those of known good quality. Note the warnings given in Safety first! and at relevant points in this Chapter concerning the dangers of asbestos dust and hydraulic fluid.*

2 Hydraulic system – bleeding

⚠ *Warning: Hydraulic fluid is poisonous; wash off immediately and thoroughly in the case of skin contact, and seek immediate medical advice if any fluid is swallowed or gets into the eyes. Certain types of hydraulic fluid are inflammable, and may ignite when allowed into contact with hot components; when servicing any hydraulic system, it is safest to assume that the fluid is inflammable, and to take precautions against the risk of fire as though it is petrol that is being handled. Hydraulic fluid is also an effective paint stripper, and will attack plastics; if any is spilt, it should be washed off immediately, using copious quantities of fresh water. Finally, it is hygroscopic (it absorbs moisture from the air) – old fluid*

may be contaminated and unfit for further use. When topping-up or renewing the fluid, always use the recommended type, and ensure that it comes from a freshly-opened sealed container.

General

1 The correct operation of any hydraulic system is only possible after removing all air from the components and circuit; this is achieved by bleeding the system.

2 During the bleeding procedure, add only clean, unused hydraulic fluid of the recommended type; never re-use fluid that has already been bled from the system. Ensure that sufficient fluid is available before starting work.

3 If there is any possibility of incorrect fluid being already in the system, the brake components and circuit must be flushed completely with uncontaminated, correct fluid, and new seals should be fitted to the various components.

4 If hydraulic fluid has been lost from the system, or air has entered because of a leak, ensure that the fault is cured before proceeding further.

5 Park the vehicle over an inspection pit or on car ramps. Alternatively, apply the handbrake then jack up the front and rear of the vehicle and support it on axle stands (see *Jacking and vehicle support*). For improved access with the vehicle jacked up, remove the roadwheels.

6 Check that all pipes and hoses are secure, unions tight and bleed screws closed. Clean any dirt from around the bleed screws.

7 Unscrew the master cylinder reservoir cap, and top the master cylinder reservoir up to the MAX level line; refit the cap loosely, and remember to maintain the fluid level at least above the MIN level line throughout the procedure, otherwise there is a risk of further air entering the system.

8 There are a number of one-man, do-it-yourself brake bleeding kits currently available from motor accessory shops. It is recommended that one of these kits is used whenever possible, as they greatly simplify the bleeding operation, and also reduce the risk of expelled air and fluid being drawn back into the system. If such a kit is not available, the basic (two-man) method must be used, which is described in detail below.

Caution: Vauxhall/Opel recommend using a pressure bleeding kit for this operation (see paragraphs 24 to 27).

9 If a kit is to be used, prepare the vehicle as described previously, and follow the kit manufacturer's instructions, as the procedure may vary slightly according to the type being used; generally, they are as outlined below in the relevant sub-section.

10 Whichever method is used, the same sequence should be followed (paragraphs 11 and 12) to ensure the removal of all air from the system.

Bleeding sequence

11 If the system has been only partially disconnected, and precautions were taken to minimise fluid loss, it should only be necessary to bleed that part of the system (ie, the primary or secondary circuit). If the master cylinder or main brake lines have been disconnected, then the complete system must be bled.

12 If the complete system is to be bled, then it should be done in the following sequence:
a) *Right-hand rear brake.*
b) *Left-hand rear brake.*
c) *Right-hand front brake.*
d) *Left-hand front brake.*

Bleeding

Basic (two-man) method

13 Collect together a clean glass jar or similar container, a suitable length of plastic or rubber tubing which is a tight fit over the bleed screw, and a ring spanner to fit the screw. The help of an assistant will also be required.

14 Remove the dust cap from the first bleed screw in the sequence **(see illustrations)**. Fit the spanner and tube to the screw, place the other end of the tube in the jar, and pour in sufficient fluid to cover the end of the tube.

15 Ensure that the master cylinder reservoir fluid level is maintained at least above the MIN level line throughout the procedure.

16 Have the assistant fully depress the brake pedal several times to build-up pressure, then maintain it on the final downstroke.

17 While pedal pressure is maintained, unscrew the bleed screw (approximately one turn) and allow the compressed fluid and air to flow into the jar. The assistant should maintain pedal pressure, following it down to the floor if necessary, and should not release it until instructed to do so. When the flow stops, tighten the bleed screw again, have

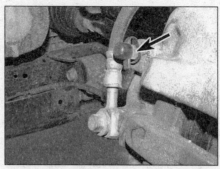

2.14a Front brake caliper bleed screw...

the assistant release the pedal slowly, and recheck the reservoir fluid level.

18 Repeat the steps given in paragraphs 16 and 17 until the fluid emerging from the bleed screw is free from air bubbles. If the master cylinder has been drained and refilled, and air is being bled from the first screw in the sequence, allow approximately five seconds between cycles for the master cylinder passages to refill.

19 When no more air bubbles appear, securely tighten the bleed screw, remove the tube and spanner, and refit the dust cap. Do not overtighten the bleed screw.

20 Repeat the procedure on the remaining screws in the sequence, until all air is removed from the system and the brake pedal feels firm again.

Using a one-way valve kit

21 As the name implies, these kits consist of a length of tubing with a one-way valve fitted, to prevent expelled air and fluid being drawn back into the system; some kits include a translucent container, which can be positioned so that the air bubbles can be more easily seen flowing from the end of the tube.

22 The kit is connected to the bleed screw, which is then opened. The user returns to the driver's seat, depresses the brake pedal with a smooth, steady stroke, and slowly releases it; this is repeated until the expelled fluid is clear of air bubbles.

23 Note that these kits simplify work so much that it is easy to forget the master cylinder reservoir fluid level; ensure that this is maintained at least above the MIN level line at all times.

Using a pressure-bleeding kit

24 These kits are usually operated by a reservoir of pressurised air contained in the spare tyre. However, note that it will probably be necessary to reduce the pressure to a lower level than normal; refer to the instructions supplied with the kit. Note also that if the car is equipped with a puncture repair kit instead of a spare tyre, it will be necessary to borrow a friend's spare wheel and tyre.

25 By connecting a pressurised, fluid-filled container to the master cylinder reservoir, bleeding can be carried out simply by opening each screw in turn (in the specified

2.14b ...and rear brake caliper bleed screw

sequence), and allowing the fluid to flow out until no more air bubbles can be seen in the expelled fluid.

26 This method has the advantage that the large reservoir of fluid provides an additional safeguard against air being drawn into the system during bleeding.

27 Pressure-bleeding is particularly effective when bleeding 'difficult' systems, or when bleeding the complete system at the time of routine fluid renewal.

All methods

28 When bleeding is complete, and firm pedal feel is restored, wash off any spilt fluid, securely tighten the bleed screws, and refit the dust caps.

29 Check the hydraulic fluid level in the master cylinder reservoir, and top-up if necessary (see *Weekly checks*).

30 Discard any hydraulic fluid that has been bled from the system; it will not be fit for re-use.

31 Check the feel of the brake pedal. If it feels at all spongy, air must still be present in the system, and further bleeding is required. Failure to bleed satisfactorily after a reasonable repetition of the bleeding procedure may be due to worn master cylinder seals.

3 Hydraulic pipes and hoses – renewal

Note: *Before starting work, refer to the note at the beginning of Section 2 concerning the dangers of hydraulic fluid.*

1 If any pipe or hose is to be renewed, minimise fluid loss by first removing the master cylinder reservoir cap and screwing it down onto a piece of polythene. Alternatively, flexible hoses can be sealed, if required, using a proprietary brake hose clamp. Metal brake pipe unions can be plugged (if care is taken not to allow dirt into the system) or capped immediately they are disconnected. Place a wad of rag under any union that is to be disconnected, to catch any spilt fluid.

2 If a flexible hose is to be disconnected, unscrew the brake pipe union nut before removing the spring clip which secures the

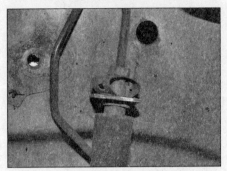

3.2 Front flexible brake hose attachment at the mounting bracket

hose to its mounting bracket **(see illustration)**. Where applicable, unscrew the banjo union bolt securing the hose to the caliper and recover the copper washers. When removing the front flexible hose, withdraw the hose and grommet from the bracket on the suspension strut.

3 To unscrew union nuts, it is preferable to obtain a brake pipe spanner of the correct size; these are available from most motor accessory shops. Failing this, a close-fitting open-ended spanner will be required, though if the nuts are tight or corroded, their flats may be rounded-off if the spanner slips. In such a case, a self-locking wrench is often the only way to unscrew a stubborn union, but it follows that the pipe and the damaged nuts must be renewed on reassembly. Always clean a union and surrounding area before disconnecting it. If disconnecting a component with more than one union, make

a careful note of the connections before disturbing any of them.

4 If a brake pipe is to be renewed, it can be obtained, cut to length and with the union nuts and end flares in place, from Vauxhall/Opel dealers. All that is then necessary is to bend it to shape, following the line of the original, before fitting it to the car. Alternatively, most motor accessory shops can make up brake pipes from kits, but this requires very careful measurement of the original, to ensure that the replacement is of the correct length. The safest answer is usually to take the original to the shop as a pattern.

5 On refitting, do not overtighten the union nuts.

6 When refitting hoses to the calipers, always use new copper washers and tighten the banjo union bolts to the specified torque. Make sure that the hoses are positioned so that they will not touch surrounding bodywork or the roadwheels.

7 Ensure that the pipes and hoses are correctly routed, with no kinks, and that they are secured in the clips or brackets provided. After fitting, remove the polythene from the reservoir, and bleed the hydraulic system as described in Section 2. Wash off any spilt fluid, and check carefully for fluid leaks.

4 Front brake pads – renewal

⚠ **Warning: Renew BOTH sets of front brake pads at the same time – NEVER renew the pads on**

only one wheel, as uneven braking may result.

⚠ **Warning: Note that the dust created by wear of the pads may contain asbestos, which is a health hazard. Never blow it out with compressed air, and don't inhale any of it. An approved filtering mask should be worn when working on the brakes. DO NOT use petrol or petroleum-based solvents to clean brake parts; use brake cleaner or methylated spirit only.**

1 Firmly apply the handbrake, and then jack up the front of the vehicle and support it securely on axle stands (see *Jacking and vehicle support*). Remove the front roadwheels.

2 Working on one side of the vehicle, push the caliper piston into its bore by pulling the caliper outwards.

3 Unscrew the caliper lower guide pin bolt (use a slim open-ended spanner to counterhold the guide pin), then remove the bolt **(see illustrations)**.

4 Pivot the caliper body upwards to expose the brake pads and secure the caliper in place using wire or a cable tie **(see illustration)**. Do not depress the brake pedal until the caliper is refitted. Take care not to strain the brake fluid hose.

5 Lift out the brake pads, followed by the anti-rattle clips **(see illustrations)**.

6 First measure the thickness of each brake pad's friction material **(see illustration)**. If either pad is worn at any point to the specified minimum thickness or less, all four pads must be renewed. Also, the pads should be

4.3a Use an open-ended spanner to hold the guide pin...

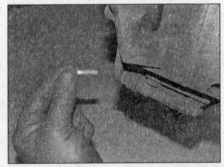

4.3b ...then unscrew and remove the lower guide pin bolt

4.4 Pivot the caliper upwards and away from the brake pads

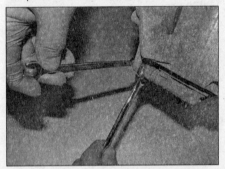

4.5a Withdraw the outer brake pad...

4.5b ...and inner brake pad from the caliper mounting bracket...

4.5c ...then remove the lower...

4.5d ...and upper anti-rattle clips

4.6 Measure the thickness of the pads friction material

4.11 Using a piston retractor tool to push back the caliper piston

renewed if any are fouled with oil or grease; there is no satisfactory way of degreasing friction material, once contaminated. If any of the brake pads are worn unevenly, or are fouled with oil or grease, trace and rectify the cause before reassembly. New brake pads and clip kits are available from Vauxhall/Opel dealers. Do not be tempted to swap brake pads over to compensate for uneven wear.

7 If the brake pads are still serviceable, carefully clean them using a clean, fine wire brush or similar, paying particular attention to the sides and back of the metal backing. Where applicable, clean out the grooves in the friction material, and pick out any large embedded particles of dirt or debris.

8 Clean the anti-rattle clips, and the brake pad locations in the caliper body/mounting bracket.

9 Prior to fitting the pads, check that the guide pins are free to slide easily in the caliper body/mounting bracket, and check that the rubber guide pin gaiters are undamaged.

10 Brush the dust and dirt from the caliper and piston, but do not inhale it, as it is a health hazard. Inspect the dust seal around the piston for damage, and the piston for evidence of fluid leaks, corrosion or damage. If attention to any of these components is necessary, refer to Section 7.

11 If new brake pads are to be fitted, the caliper piston must be pushed back into the cylinder, to make room for them. Either use a piston retractor tool, a G-clamp or a similar tool, or use pieces of wood as levers **(see illustration)**. Provided that the master cylinder reservoir has not been overfilled with hydraulic fluid, there should be no spillage, but keep a careful watch on the fluid level while retracting the piston. If the fluid level rises above the

MAX level line at any time, the surplus fluid should be syphoned off.

⚠️ *Warning: Do not syphon the fluid by mouth, as it is poisonous; use a syringe or an old poultry baster.*

12 Refit the upper and lower anti-rattle clips to the caliper mounting bracket **(see illustrations)**.

13 Apply a little high-temperature, anti-squeal brake lubricant to the contact surfaces of the pad backing plates, but take great care not to allow any grease onto the pad friction linings **(see illustrations)**.

14 Fit the inner and outer brake pads, noting that the pad with the metal wear indicator tab is the inner pad. Ensure that the pad friction material is against the disc **(see illustrations)**.

15 Pivot the caliper back into position, over the pads and mounting bracket.

16 Refit the caliper lower guide pin bolt, and

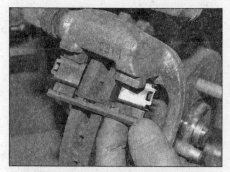

4.12a Refit the upper anti-rattle clip...

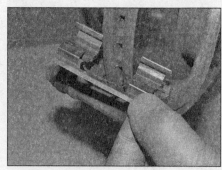

4.12b ...and lower anti-rattle clip to the caliper mounting bracket

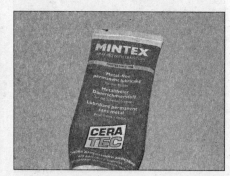

4.13a Apply a little high-temperature brake lubricant...

4.13b ...to the contact surfaces of the pad backing plates

4.14a Fit the inner brake pad...

4.14b ...and outer brake pad to the caliper mounting bracket

4.16a Refit the lower guide pin bolt…

4.16b …and tighten it to the specified torque while counterholding the guide pin

then tighten it to the specified torque **(see illustrations)**.

17 Check that the caliper body slides smoothly on the guide pins.

18 Repeat the procedure on the remaining front caliper.

19 With both sets of front brake pads fitted, depress the brake pedal repeatedly until the

pads are pressed into firm contact with the brake disc, and normal pedal pressure is restored.

20 Refit the roadwheels, and lower the vehicle to the ground, then tighten the roadwheel nuts to the specified torque

21 Finally, check the brake hydraulic fluid level as described in *Weekly checks*.

Caution: New pads will not give full braking efficiency until they have bedded-in. Be prepared for this, and avoid hard braking as far as possible for the first hundred miles or so after pad renewal.

5.2a Use an open-ended spanner to hold the guide pin…

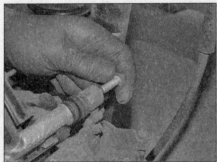

5.2b …then unscrew and remove the upper guide pin bolt

5 Rear brake pads – renewal

⚠️ *Warning: Renew BOTH sets of rear brake pads at the same time – NEVER renew the pads on only one wheel, as uneven braking may result.*

⚠️ *Warning: Note that the dust created by wear of the pads may contain asbestos, which is a health hazard. Never blow it out with compressed air, and don't inhale any of it. An approved filtering mask should be worn when working on the brakes. DO NOT use petrol or petroleum-based solvents to clean brake parts; use brake cleaner or methylated spirit only.*

1 Firmly apply the handbrake, and then jack up the rear of the vehicle and support it securely on axle stands (see *Jacking and vehicle support*). Remove the rear roadwheels.

2 Unscrew the caliper upper guide pin bolt (use an open-ended spanner to counterhold the guide pin), then remove the bolt **(see illustrations)**.

3 Pivot the caliper body down to gain access to the brake pads **(see illustration)**. Do not depress the brake pedal until the caliper is refitted. Take care not to strain the brake fluid hose.

4 Lift out the brake pads, followed by the anti-rattle clips **(see illustrations)**.

5 First measure the thickness of each brake pad's friction material **(see illustration)**. If either pad is worn at any point to the specified

5.3 Pivot the caliper down and away from the brake pads

5.4a Withdraw the inner brake pad…

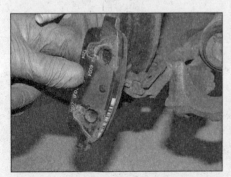

5.4b …and outer brake pad from the caliper mounting bracket…

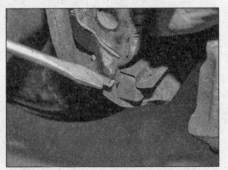

5.4c …noting that it may be necessary to depress the small tab on the anti-rattle clips to free the pads

5.4d Remove the lower…

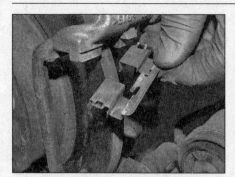

5.4e ...and upper anti-rattle clips

5.5 Measure the thickness of the pads friction material

5.7 Clean the caliper body/mounting bracket with brake cleaner and a fine wire brush

minimum thickness or less, all four pads must be renewed. Also, the pads should be renewed if any are fouled with oil or grease; there is no satisfactory way of degreasing friction material, once contaminated. If any of the brake pads are worn unevenly, or are fouled with oil or grease, trace and rectify the cause before reassembly. New brake pads and clip kits are available from Vauxhall/Opel dealers. Do not be tempted to swap brake pads over to compensate for uneven wear.

6 If the brake pads are still serviceable, carefully clean them using a clean, fine wire brush or similar, paying particular attention to the sides and back of the metal backing. Where applicable, clean out the grooves in the friction material, and pick out any large embedded particles of dirt or debris.

7 Clean the anti-rattle clips, and the brake pad locations in the caliper body/mounting bracket **(see illustration)**.

8 Prior to fitting the pads, check that the guide pins are free to slide easily in the caliper body/mounting bracket, and check that the rubber guide pin gaiters are undamaged.

9 Brush the dust and dirt from the caliper and piston, but do not inhale it, as it is a health hazard. Inspect the dust seal around the piston for damage, and the piston for evidence of fluid leaks, corrosion or damage. If attention to any of these components is necessary, refer to Section 7.

10 If new brake pads are to be fitted, it will be necessary to retract the piston fully into the caliper bore by rotating it in a clockwise direction. Special tools are readily available at moderate cost to achieve this **(see illustration)**. Provided that the master cylinder reservoir has not been overfilled with hydraulic fluid, there should be no spillage, but keep a careful watch on the fluid level while retracting the piston. If the fluid level rises above the MAX level line at any time, the surplus fluid should be syphoned off.

 Warning: Do not syphon the fluid by mouth, as it is poisonous; use a syringe or an old poultry baster.

11 Once the piston is fully retracted, position it so that the notches in the piston align correctly with the pins on the inner brake pad **(see illustration)**.

12 Refit the lower and upper anti-rattle

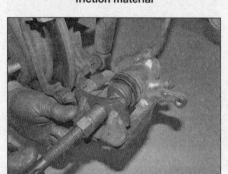

5.10 Using a piston retractor tool to push back the caliper piston

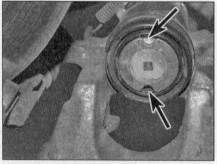

5.11 Position the retracted piston so that the notches align with the pins on the inner brake pad

clips to the caliper mounting bracket **(see illustrations)**.

13 Apply a little high-temperature, anti-squeal brake lubricant to the contact surfaces of the

5.12a Refit the lower anti-rattle clip...

5.13a Apply a high-temperature brake lubricant...

pad backing plates, but take great care not to allow any grease onto the pad friction linings **(see illustrations)**.

14 Fit the inner and outer brake pads, noting

5.12b ...and upper anti-rattle clip to the caliper mounting bracket

5.13b ...to the contact surfaces of the pad backing plates

5.14a Fit the inner brake pad...

5.14b ...and outer brake pad to the caliper mounting bracket

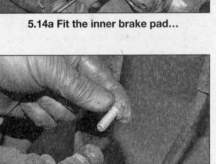

5.16a Refit the upper guide pin bolt...

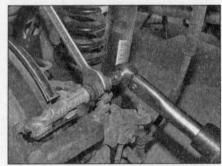

5.16b ...and tighten it to the specified torque while counterholding the guide pin

that the pad with the metal wear indicator tab is the inner pad. Ensure that the pad friction material is against the disc **(see illustrations)**.

15 Pivot the caliper back into position, over the pads and mounting bracket.

16 Refit the caliper upper guide pin bolt, and

6.2 Release the brake hydraulic hose from the suspension strut

6.3a Unscrew the two bolts securing the front brake caliper mounting bracket to the swivel hub...

6.3b ...slide the caliper/bracket assembly from the swivel hub and brake disc

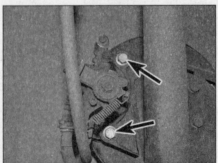

6.5a Unscrew the two bolts securing the rear brake caliper mounting bracket to the trailing arm carrier plate

then tighten it to the specified torque **(see illustrations)**.

17 Check that the caliper body slides smoothly on the guide pins.

18 Repeat the procedure on the remaining rear caliper.

19 With both sets of rear brake pads fitted, depress the brake pedal repeatedly until the pads are pressed into firm contact with the brake disc, and normal pedal pressure is restored.

20 Refit the roadwheels, and lower the vehicle to the ground, then tighten the roadwheel nuts to the specified torque

21 Finally, check the brake hydraulic fluid level as described in *Weekly checks*.

Caution: New pads will not give full braking efficiency until they have bedded-in. Be prepared for this, and avoid hard braking as far as possible for the first hundred miles or so after pad renewal.

6 Front/rear brake disc – inspection, removal and refitting

Note: *Before starting work, refer to the warning at the beginning of Section 4 or 5 concerning the dangers of asbestos dust. If either disc requires renewal, both should be renewed at the same time together with new pads, to ensure even and consistent braking.*

Inspection

Front disc

1 Firmly apply the handbrake, then jack up the front of the vehicle and support it securely on axle stands (see *Jacking and vehicle support*). Remove the roadwheel.

2 Release the brake hydraulic hose from the suspension strut bracket **(see illustration)**.

3 Unscrew the two bolts securing the brake caliper mounting bracket to the swivel hub, noting that new bolts will be required for refitting. Slide the caliper/bracket assembly from the swivel hub and brake disc (there is no need to remove the brake pads). Suspend the caliper/bracket assembly from the strut coil spring using wire or a cable tie – do not allow the caliper to hang on the brake hose **(see illustrations)**.

Rear disc

4 Chock the front wheels, then jack up the rear of the vehicle, and support it securely on axle stands (see *Jacking and vehicle support*). Remove the roadwheel, then ensure that the handbrake is released.

5 Unscrew the two bolts securing the brake caliper mounting bracket to the trailing arm carrier plate noting that new bolts will be required for refitting. Slide the caliper/bracket assembly from the carrier bracket and brake disc (there is no need to remove the brake pads). Suspend the caliper/bracket assembly from a convenient place under the wheel arch using wire or a cable tie – do not allow the caliper to hang on the brake hose **(see illustrations)**.

6.5b Slide the caliper/bracket assembly from the carrier plate and brake disc...

6.5c ...and suspend it from a convenient place under the wheel arch using wire or a cable tie

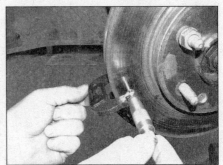

6.7 Using a micrometer to measure the thickness of the brake disc

6.8 Using a dial gauge to measure the brake disc run-out

6.11a Undo the securing screw...

6.11b ...and withdraw the brake disc from the hub

Front and rear discs

6 Temporarily refit two of the wheel nuts to diagonally-opposite studs, with the flat sides of the nuts against the disc. Tighten the nuts progressively, to hold the disc firmly.

7 Scrape any corrosion from the disc. Rotate the disc, and examine it for deep scoring, grooving or cracks. Using a micrometer, measure the thickness of the disc in several places **(see illustration)**. The minimum thickness is given in the Specifications. Light wear and scoring is normal, but if excessive, the disc should be removed, and either reground by a specialist, or renewed. If regrinding is undertaken, the minimum thickness must be maintained. Obviously, if the disc is cracked, it must be renewed.

8 Using a dial gauge, check that the disc run-out, approximately 13 mm from the outer edge, does not exceed the limit given in the Specifications. To do this, fix the measuring equipment, and rotate the disc, noting the variation in measurement as the disc is rotated **(see illustration)**. The difference between the minimum and maximum measurements recorded is the disc run-out.

9 If the run-out is greater than the specified amount, check for variations of the disc thickness as follows. Mark the disc at eight positions 45° apart then, using a micrometer, measure the disc thickness at the eight positions, 15 mm in from the outer edge. If the variation between the minimum and maximum readings is greater than the specified amount, the disc should be renewed.

Removal

10 If not already done, remove the brake caliper and mounting bracket as described in paragraph 3 (front disc) or 5 (rear disc). Where applicable, remove the two wheel nuts used when checking the disc.

11 Undo the securing screw and withdraw the disc from the hub **(see illustrations)**.

Refitting

12 Refit the disc, making sure that the mating faces of the disc and hub are perfectly clean, then refit and tighten the securing screw.

13 Refit the caliper mounting bracket, together with the caliper and brake pads and tighten the retaining bolts to the specified torque and through the specified angle. Ensure that new bolts are used.

14 Where applicable refit the brake hydraulic hose to the suspension strut bracket.

15 Refit the roadwheel, then lower the vehicle

7.4 Front brake caliper guide pin bolts

to the ground and tighten the roadwheel nuts to the specified torque.

7 Front brake caliper – removal, overhaul and refitting

Note: *New brake hose copper washers will be required when refitting. Before starting work, refer to the note at the beginning of Section 2 concerning the dangers of hydraulic fluid, and to the warning at the beginning of Section 4 concerning the dangers of asbestos dust.*

Removal

1 Firmly apply the handbrake, then jack up the front of the car and support it securely on axle stands (see *Jacking and vehicle support*). Remove the appropriate roadwheel.

2 Minimise fluid loss by first removing the master cylinder reservoir cap, then tightening it down onto a piece of polythene to obtain an airtight seal.

3 Clean the area around the caliper brake hose union. Unscrew and remove the union bolt, and recover the copper sealing washer from each side of the hose union. Discard the washers; new ones must be used on refitting. Plug the hose end and caliper hole, to minimise fluid loss and prevent the ingress of dust and dirt into the hydraulic system.

4 Slacken and remove the lower and upper caliper guide pin bolts, while counter-holding the guide pins with a second spanner, and remove the brake caliper from the mounting bracket **(see illustration)**. Note that the brake

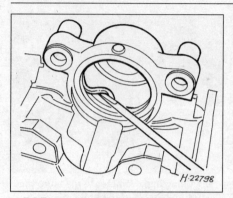

7.8 Removing the piston seal from the caliper body

pads need not be disturbed, and can be left in position in the caliper mounting bracket.

5 If required, the caliper mounting bracket can be unbolted from the swivel hub. Discard the bolts as new ones must be fitted.

Overhaul

Note: *Before starting work, check with your local dealer for the availability of parts to overhaul the caliper.*

6 With the caliper on the bench, wipe away all traces of dust and dirt, but avoid inhaling the dust, as it is a health hazard.

7 Withdraw the partially-ejected piston from the caliper body, and remove the dust seal. The piston can be withdrawn by hand, or if necessary pushed out by applying compressed air to the brake hose union hole. Only low pressure should be required, such as is generated by a foot pump, and as a precaution a block of wood should be positioned to prevent any damage to the piston.

8 Using a small screwdriver, carefully remove the piston seal from the caliper, taking great care not mark the bore **(see illustration)**.

9 Thoroughly clean all components, using only methylated spirit, isopropyl alcohol or clean hydraulic fluid as a cleaning medium. Never use mineral-based solvents such as petrol or paraffin, which will attack the hydraulic system's rubber components. Dry the components immediately, using compressed air or a clean, lint-free cloth. If compressed air is available, use it to blow through the fluid passages to make sure they are clear

> ⚠️ **Warning: Always wear eye protection when using compressed air.**

10 Check all components, and renew any that are worn or damaged. Check particularly the cylinder bore and piston; these should be renewed (note that this means the renewal of the complete body assembly) if they are scratched, worn or corroded in any way. Similarly check the condition of the guide pin rubber gaiters and check that the guide pins are a snug fit in the caliper mounting bracket. If there is any doubt about the condition of any component, renew it.

11 If the assembly is fit for further use, obtain the necessary components from your Vauxhall/Opel dealer. Renew the caliper seals as a matter of course; these should never be re-used.

12 On reassembly, ensure that all components are absolutely clean and dry.

13 Soak the piston and the new piston (fluid) seal in clean hydraulic fluid. Smear clean fluid on the cylinder bore surface.

14 Fit the new piston (fluid) seal, using only the fingers to manipulate it into the cylinder bore groove.

15 Fit the new dust seal to the piston, refit it to the cylinder bore using a twisting motion, and ensure that the piston enters squarely into the bore. Press the dust seal fully into the caliper body, and push the piston fully into the caliper bore.

Refitting

16 If previously removed, refit the caliper mounting bracket to the swivel hub, and tighten the new bolts to the specified torque.

17 Refit the brake pads as described in Section 4, together with the caliper which at this stage will not have the hose attached.

18 Position a new copper sealing washer on each side of the hose union, and connect the brake hose to the caliper. Ensure that the hose is correctly positioned, then install the union bolt and tighten it to the specified torque setting.

19 Remove the brake hose clamp or polythene, and bleed the hydraulic system as described in Section 2. Note that, providing the precautions described were taken to minimise brake fluid loss, it should only be necessary to bleed the relevant front brake circuit.

20 Refit the roadwheel, then lower the vehicle to the ground and tighten the roadwheel nuts to the specified torque.

8 Rear brake caliper – removal, overhaul and refitting

Note: *Before starting work, refer to the note at the beginning of Section 2 concerning the dangers of hydraulic fluid, and to the warning at the beginning of Section 5 concerning the dangers of asbestos dust.*

Removal

1 Chock the front wheels, then jack up the rear of the vehicle and support on axle stands (see *Jacking and vehicle support*). Remove the roadwheel.

2 Minimise fluid loss by first removing the master cylinder reservoir cap and screwing it down onto a piece of polythene. Alternatively, use a brake hose clamp to clamp the flexible hose leading to the brake caliper.

3 Clean the area around the caliper brake hose union **(see illustration)**. Unscrew and remove the union bolt, and recover the copper sealing washer from each side of the hose union. Discard the washers; new ones must be used on refitting. Plug the hose end and caliper hole, to minimise fluid loss and prevent the ingress of dust and dirt into the hydraulic system.

4 Disengage the handbrake inner cable from the caliper lever. Extract the retaining circlip then pull the outer cable out of the caliper bracket **(see illustrations)**.

5 Slacken and remove the lower and upper caliper guide pin bolts, while counter-holding the guide pins with a second spanner, and remove the brake caliper from the mounting bracket **(see illustrations)**. Note that the brake pads need not be disturbed, and can be left in position in the caliper mounting bracket.

Overhaul

6 No overhaul procedures, or parts, were available at the time of writing. Check the

8.3 Clean the area around the caliper brake hose union

8.4a Disengage the handbrake inner cable from the caliper lever...

8.4b ...then extract the circlip and pull the outer cable out of the caliper bracket

availability of spares before dismantling the caliper. Do not attempt to dismantle the handbrake mechanism inside the caliper; if the mechanism is faulty, the complete caliper assembly must be renewed.

Refitting

7 Locate the brake caliper in position over the brake pads and refit the upper and lower guide pin bolts. Tighten the guide pin bolts to the specified torque.

8 Position a new copper sealing washer on each side of the hose union, and connect the brake hose to the caliper. Ensure that the hose is correctly positioned then install the union bolt and tighten it to the specified torque setting.

9 Engage the handbrake outer cable with the caliper bracket and refit the circlip. Pull back the operating lever and reconnect the handbrake inner cable.

10 Remove the brake hose clamp or polythene, and bleed the hydraulic system as described in Section 2. Note that, providing the precautions described were taken to minimise brake fluid loss, it should only be necessary to bleed the relevant front brake circuit.

11 Refit the roadwheel, then lower the vehicle to the ground and tighten the roadwheel nuts to the specified torque.

9 Master cylinder – removal, overhaul and refitting

Note: *Before starting work, refer to the warning at the beginning of Section 2 concerning the dangers of hydraulic fluid.*

Removal

1 Remove the master cylinder reservoir cap, and syphon the hydraulic fluid from the reservoir. Do not syphon the fluid by mouth, as it is poisonous; use a syringe or an old hydrometer. Alternatively, open any convenient bleed screw in the system, and gently pump the brake pedal to expel the fluid through a plastic tube connected to the screw (see Section 2).

2 Remove the air cleaner assembly as described in Chapter 4A, Section 2 or Chapter 4B, Section 3.

3 Disconnect the wiring connector from the brake fluid level sensor on the underside of the reservoir.

4 On models with manual transmission, release the clip (if fitted) and disconnect the clutch hydraulic pipe from the fluid reservoir **(see illustration)**. Tape over or plug the outlet.

5 Place cloth rags beneath the master cylinder to collect escaping brake fluid. Release the quick-release connectors and disconnect the two reservoir fluid hoses from the master cylinder **(see illustration)**.

6 Undo the retaining nut, depress the retaining tab and remove the reservoir from the bulkhead **(see illustration)**.

8.5a Remove the caliper guide pin bolts...

7 Identify the brake pipes at the master cylinder for position, then unscrew the union nuts and move the pipes to one side. Tape over or plug the pipe outlets.

8 Unscrew the mounting nuts and withdraw the master cylinder from the vacuum servo unit **(see illustration)**. Recover the seal. Take care not to spill fluid on the vehicle paintwork.

Overhaul

9 At the time of writing, master cylinder overhaul is not possible as no spares are available.

10 The only parts available individually are the fluid reservoir, its rubber seals, the filler cap and the master cylinder mounting seal.

11 If the master cylinder is worn excessively, it must be renewed.

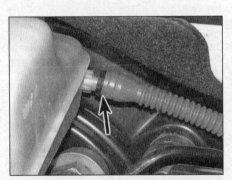

9.4 On manual transmission models disconnect the clutch hydraulic pipe from the fluid reservoir

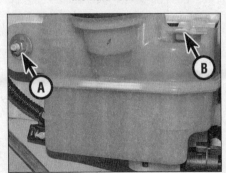

9.6 Undo the retaining nut (A), depress the retaining tab (B) and remove the reservoir from the bulkhead

8.5b ...then remove the brake caliper from the mounting bracket

Refitting

12 Ensure that the mating surfaces are clean and dry then fit the new seal to the rear of the master cylinder.

13 Fit the master cylinder to the servo unit, ensuring that the servo unit pushrod enters the master cylinder piston centrally. Refit the retaining nuts and tighten them to the specified torque setting.

14 Refit the brake pipes and tighten the union nuts securely.

15 Locate the fluid reservoir in position on the bulkhead, ensuring that the retaining tab engages fully. Refit the retaining nut and tighten the nut securely.

16 Reconnect the two fluid reservoir hoses to the pipe fittings on the master cylinder.

9.5 Disconnect the two reservoir fluid hoses from the master cylinder

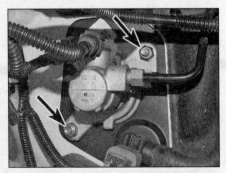

9.8 Master cylinder mounting nuts

10.5 Separate the clutch pedal from the master cylinder piston rod by releasing the retaining clip at the pedal

10.6 Unscrew the two bolts securing the clutch master cylinder to the pedal mounting bracket

10.7 Pull the brake pedal rearward to release the connector from the servo unit pushrod

17 On manual transmission models, reconnect the clutch hydraulic pipe and secure with the clip (where fitted).

18 Reconnect the wiring connector to the brake fluid level sensor.

19 Refit the air cleaner assembly as described in Chapter 4A, Section 2 or Chapter 4B, Section 3.

20 Remove the reservoir filler cap and polythene, then top-up the reservoir with fresh hydraulic fluid to the MAX mark (see *Weekly checks*).

21 Bleed the hydraulic systems as described in Section 2 and Chapter 6, Section 2 then refit the filler cap. Thoroughly check the operation of the brakes and clutch before using the vehicle on the road.

10 Brake pedal – removal and refitting

Note: *The brake pedal mounting bracket, the brake pedal and, on manual transmission models, the clutch pedal are one assembly and must be renewed as a complete unit. In the event of a frontal collision, the brake pedal is released from its bearing in the mounting bracket to prevent injury to the driver's feet and legs (this also applies to the clutch pedal). If an airbag has been deployed, inspect the pedal and mounting bracket assembly and if necessary renew the complete unit.*

Removal

1 Disconnect the battery negative lead as described in Chapter 5A Section 4.

10.8 Pull out the vacuum sensor or check valve from the servo unit grommet

2 Remove the facia crossmember as described in Chapter 11, Section 28.

3 Remove the windscreen cowl panel as described in Chapter 11, Section 21.

4 Remove the air cleaner assembly as described in Chapter 4A, Section 2 or Chapter 4B, Section 3.

5 On manual transmission models, separate the clutch pedal from the master cylinder piston rod by releasing the retaining clip at the pedal. Vauxhall technicians use a special tool to do this, however, the clip may be released by pressing the retaining tabs together using screwdrivers, while at the same time pulling the clutch pedal rearwards **(see illustration)**. **Note:** *Do not remove the clip from the master cylinder piston rod, just release it from the pedal.*

6 On manual transmission models, unscrew the two bolts securing the clutch master cylinder to the brake pedal mounting bracket **(see illustration)**.

7 Pull the brake pedal rearward to release the connector from the vacuum servo unit pushrod **(see illustration)**. Remove the connector from the brake pedal noting that a new connector will be required for refitting.

8 Pull out the vacuum sensor or check valve from the rubber grommet on the servo unit **(see illustration)**.

9 Undo the two nuts securing the brake master cylinder to the servo unit **(see illustration 9.8)**.

10 Disconnect the wiring connectors from the accelerator pedal/position sensor, brake pedal position sensor and, where fitted, the clutch pedal position sensor. Release the

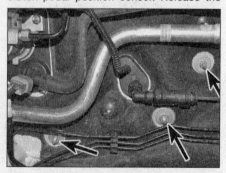

10.11 Pedal mounting bracket-to-bulkhead retaining nuts

wiring harness from the retaining clips on the pedal mounting bracket.

11 Working in the engine compartment, undo the pedal mounting bracket upper retaining bolt located beneath the windscreen wiper linkage location, then undo the nuts securing the mounting bracket to the bulkhead **(see illustration)**.

12 Withdraw the mounting bracket and pedal assembly and remove it from inside the car.

Refitting

13 Manoeuvre the mounting bracket and pedal assembly into position, whilst ensuring that the vacuum servo unit pushrod locates in the brake pedal connector. Fit the pedal mounting bracket retaining nuts and bolt and tighten them securely.

14 Reconnect the wiring connectors to the accelerator pedal/position sensor, brake pedal position sensor and, where fitted, the clutch pedal position sensor. Secure the wiring harness with the retaining clips on the pedal mounting bracket.

15 Manoeuvre the brake master cylinder back into position on the servo unit. Refit the two retaining nuts and tighten them to the specified torque.

16 On manual transmission models, manoeuvre the clutch master cylinder into position whilst ensuring that the piston rod and its retaining clip align correctly with the pedal. Fit the two master cylinder retaining bolts and tighten them securely.

17 Push the master cylinder piston rod retaining clip into the clutch pedal, ensuring that the two lugs on the clip fully engage.

18 Locate a new pushrod connector in the brake pedal, then attach the pedal to the servo unit pushrod, ensuring that is engages fully.

19 Refit the vacuum sensor or check valve to the rubber grommet on the servo unit.

20 Refit the air cleaner assembly as described in Chapter 4A, Section 2 or Chapter 4B, Section 3.

21 Refit the windscreen cowl panel as described in Chapter 11, Section 21.

22 Refit the facia crossmember as described in Chapter 11, Section 28.

23 Reconnect the battery negative terminal on completion.

11 Vacuum servo unit –
testing, removal and refitting

Testing

1 To test the operation of the servo unit, with the engine off, depress the footbrake several times to exhaust the vacuum. Now start the engine, keeping the pedal firmly depressed. As the engine starts, there should be a noticeable 'give' in the brake pedal as the vacuum builds-up. Allow the engine to run for at least two minutes, then switch it off. The brake pedal should now feel normal, but further applications should result in the pedal feeling firmer, the pedal stroke decreasing with each application.

2 If the servo does not operate as described, first inspect the servo unit check valve as described in Section 12.

3 If the servo unit still fails to operate satisfactorily, the fault lies within the unit itself. Repairs to the unit are not possible; if faulty, the servo unit must be renewed.

Removal

4 Remove the brake pedal and mounting bracket assembly as described in Section 10.

5 Unscrew the two retaining nuts and remove the servo unit from the brake pedal mounting bracket. Collect the servo unit seal.

Refitting

6 Inspect the servo unit seal and renew the seal if necessary.

7 Locate the vacuum servo unit in position on the brake pedal mounting bracket. Refit the two retaining nuts and tighten them to the specified torque.

8 Refit the brake pedal and mounting bracket assembly as described in Section 10.

12 Vacuum servo unit
check valve and hose –
removal, testing and refitting

Removal

Petrol engine models

1 Working at the left-hand rear of the engine compartment, disconnect the secondary vacuum hose quick-release fitting from the vacuum pump (**see illustration**).

2 Unclip the secondary vacuum hose from its supports in the engine compartment.

3 Disconnect the primary vacuum hose quick-release fitting from the inlet manifold (**see illustration**).

4 Disconnect the wiring connector from the vacuum sensor on the servo unit (**see illustration**).

5 Pull out the vacuum sensor from the rubber grommet in the servo unit.

6 Unclip the primary vacuum hose from its supports on the bulkhead and engine, then remove the sensor, hoses and check valve assembly from the car.

Diesel engine models

7 Where applicable, disconnect the wiring connector from the vacuum sensor on the servo unit (**see illustration 12.04**).

8 Pull out the vacuum sensor or vacuum hose end fitting from the rubber grommet in the vacuum servo unit (**see illustration**).

9 Remove the plastic cover over the top of the engine.

10 Disconnect the hose quick-release fitting from the main vacuum pipe assembly (1.6 litre engines) or vacuum pump (1.7 litre engines) (**see illustration**).

11 Unclip the vacuum hose from its supports on the bulkhead and engine, then remove the hose and check valve from the car.

Testing

12 Examine the check valve and hose for signs of damage, and renew if necessary. The valve may be tested by blowing through it in both directions. Air should flow through the valve in one direction only – when blown through from the servo unit end. If air flows in both directions, or not at all, renew the valve and hose as an assembly.

13 Examine the servo unit rubber sealing grommet for signs of damage or deterioration, and renew as necessary.

Refitting

14 Refitting is a reversal of removal ensuring that the quick-release connectors audibly lock in position, and that the hose/adaptor is correctly seated in the servo grommet.

15 On completion, start the engine and check that there are no air leaks.

13 Handbrake lever –
removal and refitting

Removal

1 Remove the centre console as described in Chapter 11, Section 26.

2 Ensure that the handbrake lever is released (off).

12.1 Disconnect the secondary vacuum hose quick-release fitting from the vacuum pump

12.3 Disconnect the primary vacuum hose quick-release fitting from the inlet manifold

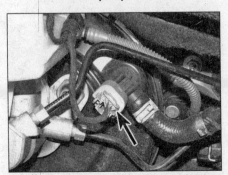

12.4 Disconnect the wiring connector from the vacuum sensor on the servo unit

12.8 Pull out the vacuum hose end fitting from the rubber grommet in the vacuum servo unit

12.10 On 1.7 litre engines, disconnect the hose quick-release fitting from the vacuum pump

13.3 Unscrew and remove the handbrake cable adjuster nut

3 Unscrew and remove the handbrake cable adjuster nut from the primary (front) handbrake cable **(see illustration)**.
4 Disconnect the wiring connector from the handbrake warning light switch on the side of the lever.
5 Undo the four nuts securing the handbrake lever to the floor **(see illustration)**. Lift the lever assembly up, disengage the primary handbrake cable and remove the handbrake lever assembly from inside the car.

Refitting

6 Refitting is a reversal of removal, bearing in mind the following points:
a) Feed the handbrake primary cable through the lever as the lever assembly is fitted.
b) Refit the centre console as described in Chapter 11, Section 26.
c) On completion, adjust the handbrake as described in Chapter 1A, Section 21 or Chapter 1B, Section 23.

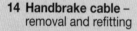

14 Handbrake cable –
removal and refitting

Removal

1 The handbrake cable consists of two sections, a short primary (front) section which connects the lever to the compensator plate, and the secondary (main) cables which link the compensator plate to the rear brake calipers. Each section can be removed individually as follows.

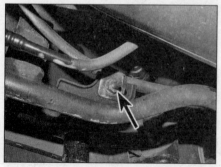

14.12 Undo the retaining nut and release the handbrake cable from the fuel tank guard

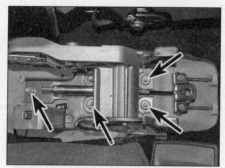

13.5 Undo the four retaining nuts and remove the handbrake lever assembly

Primary (front) cable

2 Remove the centre console as described in Chapter 11, Section 26.
3 Ensure that the handbrake lever is released (off).
4 Unscrew and remove the handbrake cable adjuster nut from the primary (front) handbrake cable **(see illustration 13.03)**.
5 Remove the spacer tube from the primary (front) cable, then disengage the cable from the compensator plate and remove it from the handbrake lever.

Secondary (main) cable

Note: *The secondary cables are supplied as one part, together with the compensator plate.*
6 Remove the primary (front) cable as described previously.
7 Remove the rear seat cushion as described in Chapter 11, Section 22.
8 Lift up the rear floor carpet for access to the handbrake cables.
9 Release the handbrake cable retaining clips from the floor, then release the rubber grommets at the cable entry points.
10 Disengage the handbrake outer cable collars from the floor tunnel bracket.
11 Chock the front wheels, then jack up the rear of the vehicle, and support it securely on axle stands (see *Jacking and vehicle support*).
12 Undo the retaining nut (or bolt) and release the handbrake cables from the underbody or fuel tank guard **(see illustration)**.
13 Undo the retaining bolt and release each handbrake cable from the rear axle training arm **(see illustration)**.

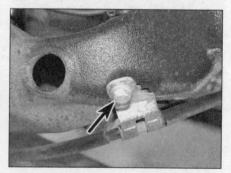

14.13 Undo the retaining bolt and release each cable from the trailing arm

14 Working on one side at a time, disengage the handbrake inner cable from the brake caliper lever. Extract the retaining circlip then pull the outer cable out of the caliper bracket **(see illustrations 8.4a and 8.4b)**.
15 Pull the cables through into the vehicle interior and remove the cables from inside the car.

Refitting

16 Refitting is a reversal of removal, bearing in mind the following points:
a) Refit the rear seat cushion as described in Chapter 11, Section 22.
b) Refit the centre console as described in Chapter 11, Section 26.
c) On completion, adjust the handbrake as described in Chapter 1A, Section 21 or Chapter 1B, Section 23.

15 Stop-light switch –
removal and refitting

Removal

Note: *The stop-light switch is also a brake pedal position sensor. The body control module (BCM) uses the switch/sensor signal to determine brake pedal position and rate of brake application. If a new stop-light switch is to be fitted, this work must be entrusted to a Vauxhall/Opel dealer or suitably-equipped specialist as it is necessary to calibrate the BCM after installation. This work requires the use of dedicated Vauxhall/Opel diagnostic equipment or a compatible alternative.*
1 The stop-light switch is located on the brake pedal mounting bracket in the driver's footwell, behind the facia.
2 Remove the facia right-hand lower trim panel as described in Chapter 11, Section 27.
3 Disconnect the wiring connector from the stop-light switch **(see illustration)**.
4 Undo the retaining bolt and remove the switch from the pedal mounting bracket.

Refitting

5 Refitting is a reversal of removal.

15.3 Disconnect the wiring connector from the stop-light switch

16 Handbrake warning light switch – removal and refitting

Removal

1 Remove the centre console as described in Chapter 11, Section 26.
2 Disconnect the wiring connector from the warning light switch on the side of the handbrake lever.
3 Unscrew the mounting bolt and remove the switch from the handbrake lever bracket.

Refitting

4 Refitting is a reversal of removal.

17 Anti-lock braking and stability control systems – general information

1 ABS is fitted as standard to all models and also incorporates traction control, an electronic stability program and various other additional safety features.
2 The ABS system comprises a hydraulic modulator and electronic control unit together with four wheel speed sensors. The hydraulic modulator contains the electronic control unit (ECU), the hydraulic solenoid valves (one set for each brake) and the electrically-driven pump. The purpose of the system is to prevent the wheel(s) locking during heavy braking. This is achieved by automatic release of the brake on the relevant wheel, followed by re-application of the brake.
3 The solenoid valves are controlled by the ECU, which itself receives signals from the four wheel speed sensors which monitor the speed of rotation of each wheel. By comparing these signals, the ECU can determine the speed at which the vehicle is travelling. It can then use this speed to determine when a wheel is decelerating at an abnormal rate, compared to the speed of the vehicle, and therefore predicts when a wheel is about to lock. During normal operation, the system functions in the same way as a conventional braking system.
4 If the ECU senses that a wheel is about to lock, it operates the relevant solenoid valve(s) in the hydraulic modulator, which then isolates from the master cylinder the relevant brake(s) on the wheel(s) which is/are about to lock, effectively sealing-in the hydraulic pressure.
5 If the speed of rotation of the wheel continues to decrease at an abnormal rate, the ECU operates the electrically-driven pump which pumps the hydraulic fluid back into the master cylinder, releasing the brake. Once the speed of rotation of the wheel returns to an acceptable rate, the pump stops, and the solenoid valves switch again, allowing the hydraulic master cylinder pressure to return to the caliper, which then re-applies the brake. This cycle can be carried out many times a second.

6 The action of the solenoid valves and return pump creates pulses in the hydraulic circuit. When the ABS system is functioning, these pulses can be felt through the brake pedal.
7 The ABS hydraulic modulator incorporates an additional set of solenoid valves which operate the traction control system. The system operates using the signals supplied by the wheel speed sensors. If the ECU senses that a driving wheel is about to lose traction, the ECU communicates with the engine management ECU which will reduce engine power. In severe cases of traction loss the ABS ECU will momentarily apply the relevant front brake to assist with traction recovery.
8 The electronic stability program (ESP) is a further development of ABS and traction control. Using additional sensors to monitor steering wheel position, vehicle yaw rate, acceleration and deceleration, in conjunction with the ABS sensors, the ECU can intervene under conditions of vehicle instability. Using the signals from the various sensors, the ECU can determine driver intent (steering wheel position, throttle position, vehicle speed and engine speed). From the sensor inputs from the wheel speed sensors, yaw rate sensors and acceleration sensors the ECU can calculate whether the vehicle is responding to driver input, or whether an unstable driving situation is occurring. If instability is detected, the ECU will intervene by applying or releasing the relevant front or rear brake, in conjunction with a power reduction, until vehicle stability returns.
9 The operation of the ABS, traction control, and stability programs is entirely dependent on electrical signals. To prevent the system responding to any inaccurate signals, a built-in safety circuit monitors all signals received by the ECU. If an inaccurate signal or low battery voltage is detected, the system is automatically shut down, and the relevant warning light on the instrument panel is illuminated, to inform the driver that the system is not operational. Normal braking is still available, however.
10 If a fault develops in the ABS/traction control/ESP system, the vehicle must be taken to a Vauxhall/Opel dealer for fault diagnosis and repair.

18.3 Disconnect the wiring harness plug from the ECU

18 Anti-lock braking and stability control system components – removal and refitting

Note: *Faults on the ABS system can only be diagnosed using Vauxhall/Opel diagnostic equipment or compatible alternative equipment.*
Note: *Before starting work, refer to the note at the beginning of Section 2 concerning the dangers of hydraulic fluid.*

Hydraulic modulator and ECU

Removal

1 Remove the battery and battery tray as described in Chapter 5A, Section 1).
2 Thoroughly clean the area around the brake pipe unions and wiring harness plug on the hydraulic modulator and ECU assembly.
3 Pull out the locking bar and disconnect the wiring harness plug from the ECU **(see illustration)**.
4 Minimise fluid loss by first removing the master cylinder reservoir cap and screwing it down onto a piece of polythene.
5 Note and record the fitted position of the brake pipes at the modulator, then unscrew the union nuts and release the pipes. As a precaution, place absorbent rags beneath the brake pipe unions when unscrewing them. Suitably plug or cap the disconnected unions to prevent dirt entry and fluid loss.
6 Remove the protective caps from the two hydraulic modulator mounting bracket retaining nuts.
7 Unscrew the two nuts securing the hydraulic modulator mounting bracket, then remove the modulator and mounting bracket from the engine compartment.
8 Undo the three bolts and separate the modulator from the mounting bracket.

Refitting

9 Refitting is a reversal of removal, noting the following points:
a) *Tighten the modulator retaining bolts and modulator mounting bracket retaining nuts to the specified torque.*
b) *Refit the brake pipes to their respective locations, and tighten the union nuts to the specified torque.*
c) *Ensure that the wiring is correctly routed, and that the ECU wiring harness plug is firmly pressed into position and secured with the locking bar.*
d) *Refit the battery tray and battery as described in Chapter 5A, Section 4.*
e) *On completion, bleed the complete hydraulic system as described in Section 2. Ensure that the system is bled in the correct order, to prevent air entering the modulator return pump.*

Electronic control unit (ECU)

Note: *If a new electronic control unit is to be fitted, this work must be entrusted to a Vauxhall/Opel dealer or suitably-equipped*

18.17 Disconnect the front wheel speed sensor wiring connector, then release the wiring harness from the support brackets

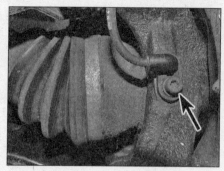

18.18a Undo the retaining bolt...

18.18b ...and withdraw the sensor from the swivel hub

specialist as it is necessary to calibrate the unit after installation. This work requires the use of dedicated Vauxhall/Opel diagnostic equipment or a compatible alternative.

Removal

10 Remove the hydraulic modulator from the car as described previously in this Section.
11 Undo the four retaining bolts and carefully separate the ECU from the hydraulic modulator.

Refitting

12 Prior to refitting, clean the sealing surfaces on the ECU and hydraulic modulator. If the surfaces are in any way deformed, damaged, or rough to the extent that a perfect seal cannot be maintained, the complete modulator and ECU assembly must be renewed.
13 Holding the ECU by the outer edges, carefully lower it over the solenoid valves on the modulator, keeping it square and level.
14 Fit the four retaining bolts and tighten securely.
15 On completion, refit the hydraulic modulator as described previously in this Section.

Front wheel speed sensors

Removal

16 Firmly apply the handbrake, then jack up the front of the vehicle and support it securely on axle stands (see *Jacking and vehicle support*). Remove the relevant front roadwheel.

17 Disconnect the wheel speed sensor wiring at the connector under the wheel arch. Release the wiring harness from the support brackets under the wheel arch and on the suspension strut **(see illustration)**.
18 Undo the retaining bolt and withdraw the sensor from the swivel hub **(see illustrations)**. Remove the sensor and wiring harness from under the wheel arch.

Refitting

19 Refitting is a reversal of removal. Tighten the sensor retaining bolt securely and ensure that the wiring harness is correctly routed and secured by the retaining clips.

Rear wheel speed sensors

Removal

20 Chock the front wheels, then jack up the rear of the vehicle, and support it securely on axle stands (see *Jacking and vehicle support*). Remove the relevant rear roadwheel.
21 Remove the rear wheel arch liner as described in Chapter 11, Section 21.
22 Disconnect the wheel speed sensor wiring at the connector under the wheel arch **(see illustration)**.
23 Free the wiring harness from the clips on the underbody and rear axle trailing arm.
24 Undo the retaining bolt and withdraw the sensor from the hub carrier, then remove the sensor and wiring harness from under the wheel arch **(see illustration)**.

Refitting

25 Refitting is a reversal of removal. Tighten the sensor retaining bolt securely and ensure that the wiring harness is correctly routed and secured by the retaining clips.

Steering angle sensor

Note: *If a new steering angle sensor is to be fitted, this work must be entrusted to a Vauxhall/Opel dealer or suitably-equipped specialist as it is necessary to calibrate the unit after installation. This work requires the use of dedicated Vauxhall/Opel diagnostic equipment or a compatible alternative.*

Removal

26 The steering angle sensor is fitted to the rear of the airbag rotary connector. Remove the airbag rotary connector as described in Chapter 12, Section 22.
27 Press the retaining tabs on the rear of the airbag rotary connector and slide the steering angle sensor off the rotary connector boss **(see illustration)**.

Refitting

28 Refitting is a reversal of removal.

Yaw rate sensor

Note: *If a new yaw rate sensor is to be fitted, this work must be entrusted to a Vauxhall/Opel dealer or suitably-equipped specialist as it is necessary to calibrate the*

18.22 Disconnect the rear wheel speed sensor wiring at the connector under the wheel arch

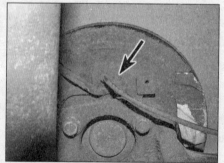

18.24 Undo the retaining bolt then remove the sensor and wiring harness from under the wheel arch

18.27 Press the retaining tabs and slide the steering angle sensor off the airbag rotary connector boss

19.9 Disconnect the two vacuum hoses (arrowed) from the pump – 1.7 litre diesel engines

19.10 Undo the three mounting bolts and remove the pump from the camshaft housing – 1.7 litre diesel engines

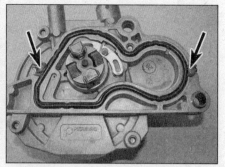

19.11 Apply a bead of silicone sealant over the sealing ring protrusions on each side of the pump flange

unit after installation. This work requires the use of dedicated Vauxhall/Opel diagnostic equipment or a compatible alternative.

Removal

29 The yaw rate sensor is located under the centre console, behind the gearchange lever or selector lever.

30 Remove the centre console as described in Chapter 11, Section 26.

31 Lift up the sound deadening material, then undo the two nuts and withdraw the sensor from the floor.

32 Disconnect the sensor wiring connector and remove the sensor from the car.

Refitting

33 Refitting is a reversal of removal.

19 Vacuum pump – removal and refitting

Removal

Petrol engine models

1 Remove the battery and battery tray as described in Chapter 5A, Section 4.

2 Working at the left-hand rear of the engine compartment, disconnect the secondary vacuum hose quick-release fitting from the vacuum pump **(see illustration 12.1)**.

3 Disconnect the vacuum pump motor wiring connector and release the wiring harness from the vacuum pump mounting bracket.

4 Release the vacuum pump vent hose and grommet from the side member.

5 Undo the two retaining nuts and remove the vacuum pump and mounting bracket from the engine compartment.

6 Undo the two nuts, release the wiring connector and vent hose and separate the vacuum pump from the mounting bracket.

1.6 litre diesel engine models

7 The brake servo vacuum pump also incorporates the engine oil pump as an integral assembly. Refer to Chapter 2C, Section 13 for removal and refitting details.

1.7 litre diesel engine models

8 Remove the inlet manifold as described in Chapter 4B, Section 18.

9 Disconnect the quick-release fitting and detach the vacuum servo unit vacuum hose from the pump. Disconnect the smaller

vacuum hose from the outlet on the top of the pump **(see illustration)**.

10 Undo the three mounting bolts and remove the pump from the camshaft housing **(see illustration)**. Remove the sealing ring from the rear of the pump.

Refitting

11 Refitting is a reversal of removal, bearing in mind the following points:

a) *On diesel engine models, thoroughly clean the mating faces of the pump and camshaft housing.*

b) *Renew all disturbed gaskets and seals.*

c) *On 1.7 litre diesel engine models, apply a bead of silicone sealant over the sealing ring protrusions on each side of the pump flange* **(see illustration)**.

d) *On 1.7 litre diesel engine models, refit the inlet manifold as described in Chapter 4B, Section 18.*

e) *On petrol engine models, refit the battery tray and battery as described in Chapter 5A, Section 4.*

f) *Tighten all nuts and bolts to the specified torque (where given).*

Chapter 10
Suspension and steering

Contents

Degrees of difficulty

Easy, suitable for novice with little experience	Fairly easy, suitable for beginner with some experience	Fairly difficult, suitable for competent DIY mechanic	Difficult, suitable for experienced DIY mechanic	Very difficult, suitable for expert DIY or professional

Specifications

Front suspension
Type . . . . Independent, with MacPherson struts and anti-roll bar

Rear suspension
Type . . . . Semi-independent torsion beam, with coil springs and telescopic shock absorbers

Steering
Type . . . . Electric power-assisted rack and pinion

Torque wrench settings

	Nm	lbf ft
Front suspension		
Anti-roll bar link rod nut	65	48
Anti-roll bar-to-subframe	22	16
Driveshaft retaining nut*:		
Stage 1	50	37
Stage 2	Angle-tighten a further 60°	
Lower arm balljoint clamp bolt nut*:		
Stage 1	42	31
Stage 2	Slacken by 120°	
Stage 3	35	26
Stage 4	Angle-tighten a further 45°	
Lower arm balljoint-to-lower arm bolts/nuts	35	26
Lower arm front pivot bolt*:		
Stage 1	115	85
Stage 2	Slacken by 720°	
Stage 3	80	59
Stage 4	Angle-tighten a further 90°	
Lower arm rear pivot bolt*:		
Stage 1	80	59
Stage 2	Angle-tighten a further 75°	
Subframe:		
Bumper lower impact bar-to-subframe bolts:		
Inner bolts	22	16
Outer bolts	58	43
Centre mounting bolts	135	100
Front mounting bolts*:		
Stage 1	100	74
Stage 2	Angle-tighten a further 45°	
Rear mounting bolts*:		
Stage 1	160	118
Stage 2	Angle-tighten a further 45°	
Rear mounting/torque link-to-transmission bracket*:		
Stage 1	80	59
Stage 2	Angle-tighten a further 60°	
Suspension strut piston rod nut	65	48
Suspension strut-to-swivel hub	110	81
Suspension strut upper mounting nut	65	48
Rear suspension		
Hub bearing assembly-to-trailing arm and carrier plate	58	43
Rear axle front mounting bracket bolts*:		
Stage 1	160	118
Stage 2	Angle-tighten a further 38°	
Shock absorber lower mounting bolt*:		
Stage 1	150	111
Stage 2	Angle-tighten a further 65°	
Shock absorber upper mounting bolts	100	74
Shock absorber upper mounting nut	25	18
Steering		
Intermediate shaft clamp bolts	49	36
Steering column mounting bolts	22	16
Steering gear-to-subframe*:		
Stage 1	100	74
Stage 2	Angle-tighten a further 100°	
Steering wheel bolt	45	33
Track rod inner balljoint-to-steering rack	70	52
Track rod end-to-swivel hub*:		
Stage 1	30	22
Stage 2	Angle-tighten a further 120°	
Roadwheels		
Roadwheel nuts	140	103

*Use new fasteners

2.2 Knock back the staking securing the retaining nut to the driveshaft groove

2.3 Attach the holding tool to the wheel hub using two wheel nuts

1 General Information

1 The independent front suspension is of the MacPherson strut type, incorporating coil springs and integral telescopic shock absorbers. The MacPherson struts are located by transverse lower suspension arms, which utilise rubber inner mounting bushes, and incorporate a balljoint at the outer ends. The front swivel hubs, which carry the hub bearings, brake calipers and disc assemblies, are bolted to the MacPherson struts, and connected to the lower arms via the balljoints. A front anti-roll bar is fitted, which has link rods with balljoints at each end to connect it to the strut.

2 The rear suspension is of semi-independent type, consisting of a torsion beam and trailing arms, with coil springs and telescopic shock absorbers. The front ends of the trailing arms are attached to the vehicle underbody by horizontal bushes, and the rear ends are located by the shock absorbers, which are bolted to the underbody at their upper ends. The coil springs are mounted independently of the shock absorbers, and act directly between the trailing arms and the underbody. Each rear wheel bearing and hub assembly is manufactured as a sealed unit, and cannot be dismantled.

3 Electric power-assisted steering is fitted as standard, whereby an electric motor, drive gear assembly and torque sensor incorporated in the steering column provide a variable degree of power assistance according to roadspeed. The system is controlled by an electronic control unit with self-diagnostic capability, located on the steering column.

4 The steering column is linked to the steering gear by an intermediate shaft. The intermediate shaft has a universal joint fitted at each end, and is secured to the column and to the steering gear pinion shaft by means of clamp bolts.

2 Front swivel hub – removal and refitting

Removal

Note: *A new driveshaft retaining nut, lower arm balljoint clamp bolt and nut, swivel hub-to-suspension strut bolts and nuts, brake caliper mounting bracket bolts and a new track rod end retaining nut will be needed for refitting. The driveshaft outer joint splines may be a tight fit in the hub and it is possible that a puller/extractor will be required to draw the hub assembly off the driveshaft during removal.*

1 Firmly apply the handbrake, then jack up the front of the car and support it securely on axle stands (see *Jacking and vehicle support*). Remove the relevant front roadwheel.

2 Using a hammer and small chisel, knock back the staking securing the retaining nut to the driveshaft groove **(see illustration)**.

3 To prevent rotation of the wheel hub as the driveshaft retaining nut is slackened, make up a holding tool and attach the tool to the wheel hub using two wheel nuts **(see Tool Tip and illustration)**.

4 With the holding tool in place, slacken and remove the driveshaft retaining nut using a socket and long bar. This nut is very tight; make sure that there is no risk of pulling the car off the axle stands as the nut is slackened.

5 Slacken, but do not fully remove, the nut securing the track rod end to the steering arm on the swivel hub then use a two-legged puller to separate the track rod end balljoint taper **(see illustrations)**. Once the taper is released, remove the track rod end from the steering arm. Note that a new nut will be required for refitting.

6 Release the brake hydraulic hose from the suspension strut bracket **(see illustration)**.

2.5a Slacken the nut securing the track rod end to the steering arm...

2.5b ...separate the track rod end balljoint taper using a two-legged puller...

2.5c ...then unscrew the nut and remove the track rod end

2.6 Release the brake hydraulic hose from the suspension strut bracket

2.7a Unscrew the two bolts securing the brake caliper mounting bracket to the swivel hub...

2.7b ... and slide the caliper/bracket assembly from the swivel hub and brake disc

2.8a Undo the securing screw...

2.8b ...and withdraw the brake disc from the hub

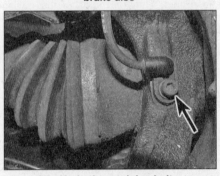

2.9a Undo the retaining bolt...

2.9b ...and remove the wheel speed sensor from the swivel hub

7 Unscrew the two bolts securing the brake caliper mounting bracket to the swivel hub, noting that new bolts will be required for

refitting. Slide the caliper/bracket assembly from the swivel hub and brake disc (there is no need to remove the brake pads). Suspend

the caliper/bracket assembly from the strut coil spring using wire or a cable tie – do not allow the caliper to hang on the brake hose **(see illustrations)**.

8 Undo the securing screw and remove the brake disc from the wheel hub **(see illustrations)**.

9 Undo the retaining bolt and remove the ABS wheel speed sensor from the swivel hub **(see illustrations)**.

10 Undo the three retaining bolts and remove the brake disc shield from the swivel hub.

11 Slacken and remove the two nuts and bolts securing the suspension strut to the swivel hub, noting that new nuts and bolts will be required for refitting. Disengage the swivel hub from the strut and pull the upper end outward **(see illustrations)**.

12 The hub must now be freed from the end of the driveshaft **(see illustration)**. It may be possible to pull the hub off the driveshaft, but if the end of the driveshaft is tight in the hub, temporarily refit the driveshaft retaining nut to protect the driveshaft threads, then tap the end of the driveshaft with a soft-faced hammer while pulling outwards on the swivel hub. Alternatively, use a suitable puller to press the driveshaft through the hub.

13 Collect the spacer from the end of the constant velocity joint **(see illustration)**.

14 Unscrew the nut and remove the clamp bolt securing the suspension lower arm balljoint to the swivel hub **(see illustrations)**. Note that a new clamp bolt and nut will be required for refitting.

2.11a Remove the two nuts and bolts securing the suspension strut to the swivel hub...

2.11b ...then disengage the swivel hub from the strut

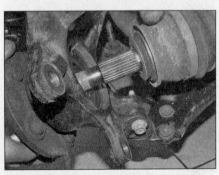

2.12 Pull the swivel hub outward and disengage the driveshaft

2.13 Collect the spacer from the end of the constant velocity joint

2.14a Unscrew the nut…

2.14b …and remove the clamp bolt securing the lower arm balljoint to the swivel hub

2.15 Lift the swivel hub up and disengage the lower arm balljoint

15 Support the swivel hub, then push down on the suspension lower arm to free the balljoint from the hub. Lift the swivel hub up and remove it from the car **(see illustration)**.

Refitting

16 Locate the lower arm balljoint in the swivel hub. Insert the new clamp bolt from the front of the swivel hub, so that its threads are facing to the rear. Fit the new nut to the clamp bolt, and tighten it to the specified torque and through the specified angle in the stages given in the Specifications, using a torque wrench and angle-tightening gauge.

17 Ensure that the driveshaft outer constant velocity joint and hub splines are clean, and make sure that the spacer washer is in place on the end of the constant velocity joint.

18 Engage the driveshaft splines with the hub and push the driveshaft through until sufficient threads are exposed to allow the nut to be fitted. Fit the new driveshaft retaining nut, tightening it by hand only at this stage.

19 Engage the swivel hub with the suspension strut, and insert the new bolts from the front of the strut so that their threads are facing to the rear. Fit the new nuts and tighten them to the specified torque and through the specified angle given in the Specifications, using a torque wrench and angle-tightening gauge.

20 Refit the ABS wheel speed sensor to the swivel hub, refit the bolt and tighten it securely.

21 Engage the track rod end in the swivel hub, then fit the new retaining nut and tighten it to the specified torque. Countrehold the balljoint shank using a second spanner to prevent rotation as the nut is tightened.

22 Refti the brake disc shield to the swivel hub and secure with the three retaining bolts tightened to the specified torque.

23 Refit the brake disc and tighten its retaining screw securely.

24 Slide the brake pads, caliper and mounting bracket over the disc and into position on the swivel hub. Fit the two new caliper mounting bracket retaining bolts and tighten them to the specified torque (see Chapter 9, Section).

25 Refit the brake hydraulic hose to the suspension strut.

26 Using the method employed on removal to prevent rotation, tighten the driveshaft retaining nut through the stages given in the Specifications.

27 Using a hammer and small chisel, securely stake the retaining nut into the driveshaft groove.

28 Refit the roadwheel, then lower the vehicle to the ground and tighten the roadwheel nuts to the specified torque.

---

3 Front hub bearings – checking and renewal

Checking

1 Firmly apply the handbrake, then jack up the front of the car and support it securely on axle stands (see *Jacking and vehicle support*). Remove the roadwheel.

2 A dial test indicator (DTI) will be required to measure the amount of play in the bearing. Locate the DTI on the suspension strut and zero the probe on the brake disc.

3 Lever the hub in and out and measure the amount of play in the bearing.

4 To measure the bearing lateral and radial run-out, unscrew the two bolts securing the brake caliper mounting bracket to the swivel hub, noting that new bolts will be required for refitting. Slide the caliper/bracket assembly from the swivel hub and brake disc (there is no need to remove the brake pads). Suspend the caliper/bracket assembly from the strut coil spring using wire or a cable tie – do not allow the caliper to hang on the brake hose.

3.13a Using a socket and puller arrangement…

5 Undo the securing screw and remove the brake disc from the wheel hub.

6 To check the lateral run-out, locate the DTI on the suspension strut and zero the probe on the front face of the hub flange. Rotate the hub and measure the run-out.

7 To check the radial run-out, zero the DTI probe on the upper face of the extended portion at the centre of the hub. Rotate the hub and measure the run-out.

8 If the play or run-out exceeds the specified amounts, renew the hub bearing as described below.

9 If the bearing is satisfactory, refit the brake disc and tighten its retaining screw securely.

10 Slide the brake pads, caliper and mounting bracket over the disc and into position on the swivel hub. Fit the two new caliper mounting bracket retaining bolts and tighten them to the specified torque (see Chapter 9, Section).

11 Refit the roadwheel, then lower the vehicle to the ground and tighten the roadwheel nuts to the specified torque.

Renewal

Note: *The bearing is sealed, pre-adjusted and pre-lubricated. Never overtighten the driveshaft nut beyond the specified torque setting in an attempt to 'adjust' the bearing.*

Note: *A press will be required to dismantle and rebuild the assembly; if such a tool is not available, a large bench vice and spacers (such as large sockets) will serve as an adequate substitute. The bearing's inner races are an interference fit on the hub; if the inner race remains on the hub flange when it is pressed out of the swivel hub, a knife-edged bearing puller will be required to remove it.*

12 Remove the swivel hub assembly as described in Section 2.

13 Using a tubular spacer/socket which bears only on the inner end of the hub flange, press the hub flange out of the bearing. If the bearing's inner race remains on the hub flange, remove it using a chisel or bearing puller – see note above **(see illustrations)**.

14 Extract the bearing retaining circlip from the swivel hub assembly **(see illustration)**.

15 Support the swivel hub on a suitable spacer and use a hammer and mandrel to drive the old bearing assembly out of the swivel hub **(see illustration)**.

3.13b ...to press the hub flange out from the bearing

3.13c Remove the bearing seal from the hub flange...

3.13d ...and the outer bearing from the hub carrier

3.13e Using a knife-edged bearing puller and socket arrangement...

3.13f ...to remove the bearing inner race from the hub flange

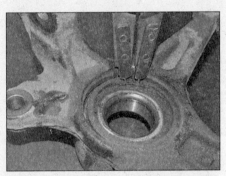

3.14 Use circlip pliers to remove the circlip

16 Thoroughly clean the hub flange and swivel hub, removing all traces of dirt and grease. Polish away any burrs or raised edges which might hinder reassembly. Check both assemblies for cracks or any other signs of wear or damage, and renew as necessary.

Renew the circlip regardless of its apparent condition.

17 Securely support the swivel hub, and locate the bearing in the hub. Ensure that the side of the bearing containing the brown plastic ABS wheel speed reluctor ring faces the transmission side of the hub. Press the bearing fully into position, ensuring that it enters the hub squarely, using a socket or mandrel **(see illustrations)**.

18 Once the bearing is correctly seated, secure the bearing in position with the new circlip. Make sure that the circlip is correctly located in its groove and with the open ends on either side of the ABS wheel speed sensor opening **(see illustrations)**.

19 Securely support the swivel hub in a vice, then locate the hub flange into the bearing inner race. With the underside of the inner race securely supported, press the

3.15 Using a drift to remove the bearing from the swivel hub

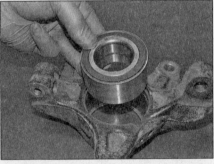

3.17a Locate the new bearing in the swivel hub with the brown reluctor ring facing the transmission side of the hub

3.17b Press the bearing fully into position using a socket or mandrel

3.18a Secure the bearing in position with a new circlip...

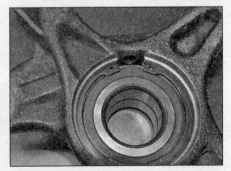

3.18b ... positioned with the open ends on either side of the ABS wheel speed sensor opening

3.19 Using a threaded rod and spacers/ sockets to press the hub flange into the bearing

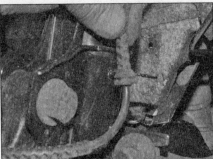

4.4 Release the wheel speed sensor wiring harness from the strut bracket

4.5a Unscrew the anti-roll bar link rod nut while using a Torx key to hold the link...

flange into the swivel hub bearing, using a tubular spacer/sockets and threaded rod, until it seats against the hub shoulder **(see illustration)**. Check that the hub flange rotates freely, and wipe off any excess oil or grease.

20 Refit the swivel hub assembly as described in Section 2.

4 Front suspension strut –
removal, overhaul and refitting

Removal

Note: *Ideally, both front suspension struts should be renewed at the same time in order to maintain good steering and suspension characteristics.*

1 Remove the windscreen cowl panel as described in Chapter 11, Section 21.
2 Apply the handbrake, then jack up the front of the vehicle and support it on axle stands (see *Jacking and vehicle support*). Remove the front roadwheel.
3 Release the brake hydraulic hose from the suspension strut bracket **(see illustration 2.6)**.
4 Release the ABS wheel speed sensor wiring harness from the suspension strut bracket **(see illustration)**.
5 Unscrew the nut and disconnect the anti-roll bar link rod from the strut. Use a suitable Torx key inserted into the link to hold the link while the nut is being loosened **(see illustrations)**.
6 Slacken and remove the two nuts and bolts securing the suspension strut to the swivel hub **(see illustrations)**. Disengage the top of the swivel hub from the strut.

7 Engage the help of an assistant to support the strut beneath the front wing. From within the engine compartment, lift off the plastic cap and unscrew the strut upper mounting nut **(see illustrations)**.
8 Lift off the upper mounting plate, then lower the strut and withdraw it from under the front wing **(see illustrations)**.

Overhaul

Note: *A spring compressor tool will be required for this operation. Before overhaul, mark the position of each component in relationship with each other for reassembly.*
9 With the suspension strut resting on a bench, or clamped in a vice, fit a spring compressor tool, and compress the coil spring to relieve the pressure on the spring seats **(see illustration)**. Ensure that the compressor tool is securely located on the spring, in

4.5b ...then disconnect the link rod from the strut

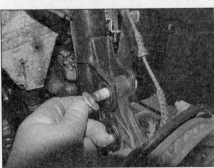

4.6a Undo the two nuts...

4.6b ...then remove the bolts securing the suspension strut to the swivel hub

4.7a Lift off the plastic cap...

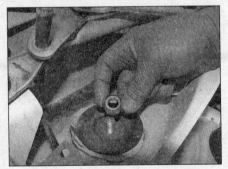

4.7b ...and unscrew the strut upper mounting nut

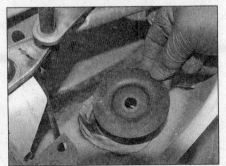

4.8a Lift off the upper mounting plate...

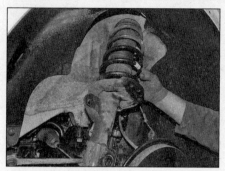

4.8b ...then withdraw the strut from under the front wing

4.9 Fit a spring compressor tool, and compress the coil spring

12 Remove the buffer and lower spring seat from the strut **(see illustrations)**.

13 With the strut assembly now completely dismantled, examine all the components for wear, damage or deformation. Renew any of the components as necessary.

14 Examine the strut for signs of fluid leakage. Check the strut piston for signs of pitting along its entire length, and check the strut body for signs of damage. While holding it in an upright position, test the operation of the strut by moving the piston through a full stroke, and then through short strokes of 50 to 100 mm. In both cases, the resistance felt should be smooth and continuous. If the resistance is jerky or uneven or if there is any visible sign of wear or damage to the strut, renewal is necessary.

15 If any doubt exists as to the condition of the coil spring, carefully remove the spring compressors and check the spring for distortion and signs of cracking. Renew the spring if it is damaged or distorted, or if there is any doubt as to its condition.

16 Inspect all other components for damage or deterioration, and renew any that are suspect.

4.10a Counter-hold the strut piston rod with a Torx key or suitable bit...

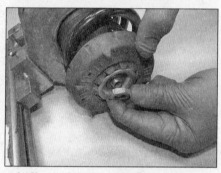

4.10b ...and unscrew the piston rod nut

17 Begin reassembly by placing the lower spring seat in position on the strut, ensuring that the end of the seat locates in the depression in the strut plate.

18 Refit the buffer, then insert the rubber gaiter into the spring.

19 With the spring compressed with the compressor tool, locate the spring on the strut making sure that it is correctly seated with its lower end against the raised stop on the lower spring seat **(see illustration)**.

accordance with the tool manufacturer's instructions.

10 Mark the position of the spring relevant to the top and bottom mountings, then counter-hold the strut piston rod with a Torx

key or suitable bit, and unscrew the piston rod nut **(see illustrations)**.

11 Remove the strut upper mounting and spring from the strut, then remove the rubber gaiter from the spring **(see illustrations)**.

4.11a Remove the strut upper mounting...

4.11b ...and spring from the strut...

4.11c ...then remove the rubber gaiter from the spring

4.12a Remove the buffer...

4.12b ...and lower spring seat from the strut

4.19 Make sure that the lower end of the spring is against the raised stop on the lower spring seat

20 Refit the upper mounting, then screw on the piston rod nut. Tighten the nut to the specified torque while counterholding the piston rod.

21 Slowly slacken the spring compressor tool to relieve the tension in the spring. Check that the end of the spring locates correctly against the stop on the lower spring seat. If necessary, turn the spring so that it locates correctly before the compressor tool is completely slackened. Remove the compressor tool when the spring is fully seated.

Refitting

22 Locate the strut in position under the front wing and place the upper mounting plate in position. Refit the upper mounting nut and tighten it to the specified torque. Refit the plastic cap to the strut piston rod.

23 Engage the swivel hub with the suspension strut, and insert the retaining bolts from the front of the strut so that their threads are facing to the rear. Refit the nuts and tighten them to the specified torque.

24 Connect the anti-roll bar link rod to the strut and use a Torx key to hold the link while tightening the nut to the specified torque.

25 Refit the brake hydraulic hose and ABS wheel speed sensor wiring to the suspension strut bracket.

26 Refit the roadwheel, then lower the vehicle to the ground and tighten the roadwheel nuts to the specified torque.

27 Refit the windscreen cowl panel as described in Chapter 11, Section 21.

5 Front lower arm – removal and refitting

Note: *New lower arm pivot bolts, and new lower arm balljoint nut/bolt, will be required.*

Removal

1 Using a tape measure, measure and record the distance between the upper edge of the wheel arch and the centre of the front hub. This dimension will be used during refitting to position the lower arm at the normal ride height level when tightening the mounting bolts.

2 Firmly apply the handbrake, then jack up the front of the car and support it securely on axle stands (see *Jacking and vehicle support*). Remove the appropriate front roadwheel.

3 Where fitted, remove the engine undertray as described in Chapter 11, Section 21.

4 Unscrew the nut and remove the clamp bolt securing the lower arm balljoint to the swivel hub **(see illustrations 2.14a and 2.14b)**. Note that a new bolt and nut will be required for refitting.

5 Using a lever, push down on the lower arm to free the balljoint from the swivel hub, then move the swivel hub to one side and release the arm, taking care not to damage the balljoint rubber boot **(see illustration 2.15)**. It

5.6 Unscrew the pivot bolt and remove the special nut securing the lower arm front mounting to the subframe

is advisable to place a protective cover over the rubber boot such as the plastic cap from an aerosol can, suitably cut to fit.

6 Unscrew the pivot bolt and remove the special nut securing the lower arm front mounting to the subframe **(see illustration)**. Note that a new pivot bolt will be required for refitting.

7 Unscrew the nut and remove the pivot bolt securing the lower arm rear mounting to the subframe **(see illustration)**. Note that a new pivot bolt will be required for refitting.

8 Disengage the mountings from the subframe and remove the lower arm from under the car.

9 Thoroughly clean the lower arm and the area around the arm mountings, removing all traces of dirt and underseal if necessary. Check carefully for cracks, distortion, or any other signs of wear or damage, paying particular attention to the mounting bushes and lower arm balljoint. If the mounting bushes are worn, the lower arm must be renewed as these components are not available separately. To renew the lower arm balljoint, see Section 6.

Refitting

10 Offer up the lower arm, aligning it with the subframe brackets. Insert the new pivot bolts, refit the retaining nuts, but only tighten the bolts/nuts hand tight at this stage.

11 Locate the lower arm balljoint in the swivel hub. Insert the new clamp bolt from the front of the swivel hub, so that its threads are facing to the rear. Fit the new nut to the clamp bolt,

6.2 Front suspension lower balljoint showing rivets securing it to the lower arm

5.7 Lower arm rear mounting pivot bolt and nut

and tighten it to the specified torque and through the specified angles given in the Specifications, using a torque wrench and angle-tightening gauge.

12 Position a trolley jack under the lower arm. Raise the arm until the distance between the upper edge of the wheel arch and the centre of the front hub is as recorded prior to removal.

13 With the lower arm now positioned at the normal ride height position, tighten the lower arm pivot bolts/nut to the specified torque and through the specified angle in the stages given in the Specifications, using a torque wrench and angle-tightening gauge.

14 Remove the trolley jack from under the lower arm.

15 Where applicable, refit the engine undertray as described in Chapter 11, Section 21.

16 Refit the roadwheel, lower the car to the ground, and tighten the roadwheel nuts to the specified torque.

6 Front lower arm balljoint – renewal

Note: *The original balljoint is riveted to the lower arm; service replacements are bolted in position.*

1 Remove the front lower arm as described in Section 5.

Note: *If the fitted balljoint is a service replacement, it is not necessary to completely remove the arm but only to disconnect the balljoint from the bottom of the swivel hub then unbolt the old balljoint.*

2 Mount the lower arm in a vice, then drill the heads from the three rivets that secure the balljoint to the lower arm, using a 10.0 mm diameter drill **(see illustration)**.

3 If necessary, tap the rivets from the lower arm, then remove the balljoint.

4 Clean any rust from the rivet holes, and apply rust inhibitor.

5 The new balljoint must be fitted using three special bolts, spring washers and nuts, available from a Vauxhall/Opel parts stockists.

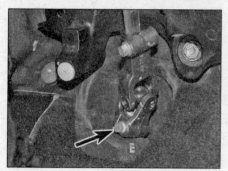

7.5 Unscrew the bolt securing the steering column intermediate shaft to the steering gear pinion

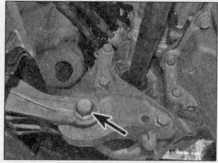

7.7 Undo and remove the through-bolt securing the rear engine mounting/torque link to the transmission bracket

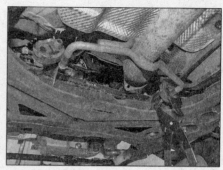

7.13 Release the exhaust system front pipe from the subframe rubber mountings

6 Ensure that the balljoint is fitted the correct way up, noting that the securing nuts are positioned on the underside of the lower arm. Tighten the nuts to the specified torque.

7 Refit the front lower arm as described in Section 5.

7 Front subframe – removal and refitting

Note: *Vauxhall/Opel special tool CH-51034 or a suitable alternative will be required to align the subframe when refitting.*

Note: *New subframe front and rear mounting bolts, new lower arm balljoint clamp bolts and nuts and new track rod end retaining nuts will be needed for refitting.*

Removal

1 Jack up the front and rear of the car and support it securely on axle stands (see *Jacking and vehicle support*). Preferably, position the car over an inspection pit, or on a lift. The help of an assistant will be needed for this procedure.

2 Set the front wheels in the straight-ahead position, then remove the ignition key and lock the column by turning the steering wheel as required. Remove both front roadwheels.

3 Where fitted, remove the engine undertray as described in Chapter 11, Section 21.

4 Remove the front bumper as described in Chapter 11, Section 6.

5 In the driver's footwell, unscrew the bolt securing the bottom of the steering column intermediate shaft to the steering gear pinion **(see illustration)**. Use paint or a suitable

marker pen to make alignment marks between the intermediate shaft and the steering gear pinion, then pull the shaft from the pinion and position to one side.

6 Use cable-ties or suitable straps to secure the radiator and condenser on each side to prevent them dropping when the subframe is removed.

7 Undo and remove the through-bolt securing the rear engine mounting/torque link to the transmission bracket **(see illustration)**.

8 Slacken, but do not fully remove, the nut securing the track rod end to the steering arm on the swivel hub then use a two-legged puller to separate the track rod end balljoint taper. Once the taper is released, remove the track rod end from the steering arm **(see illustrations 2.5a to 2.5c)**. Note that a new nut will be required for refitting.

9 Unscrew the nut and remove the clamp bolt each side securing the lower arm balljoint to the swivel hub **(see illustrations 2.14a and 2.14b)**. Note that new bolts and nuts will be required for refitting.

10 Working on one side at a time, use a lever to push down on the lower arm to free the balljoint from the swivel hub, then move the swivel hub to one side and release the arm, taking care not to damage the balljoint rubber boot. It is advisable to place a protective cover over the rubber boot such as the plastic cap from an aerosol can, suitably cut to fit.

11 Unscrew the nuts and disconnect the anti-roll bar link rods from the suspension struts on both sides. Use a suitable Torx key inserted into the link to hold the link while the nut is being loosened **(see illustrations 4.5a and 4.5b)**.

12 Release the wiring harness from the various clips around the subframe.

13 Release the exhaust system front pipe from the rubber mountings on the subframe **(see illustration)**.

14 Undo the three bolts each side securing the bumper lower impact bar to the subframe and withdraw the bar from under the car **(see illustrations)**.

15 On turbocharged engines, undo the retaining bolts and release the intercooler air duct support brackets from the subframe on each side **(see illustrations)**.

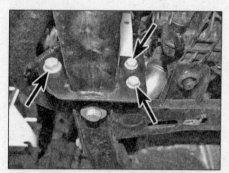

7.14a Undo the three bolts each side...

7.14b ...and remove the bumper lower impact bar

7.15a Undo the bolts securing the intercooler right-hand air duct...

7.15b ...and left-hand air duct to the subframe

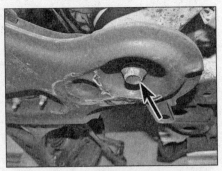

7.17 Undo the subframe front mounting bolt...

7.18 ...centre mounting bolt...

7.19 ...and rear mounting bolt, on each side

16 Support the subframe with a cradle across a trolley jack. Alternatively, two trolley jacks and the help of an assistant will be required.

17 Undo the subframe front mounting bolt on each side **(see illustration)**. Note that new bolts will be required for refitting

18 Undo the subframe centre mounting bolt on each side **(see illustration)**.

19 Undo the subframe rear mounting bolt on each side **(see illustration)**. Note that new bolts will be required for refitting

20 Slowly and carefully lower the subframe to the ground and remove it from under the car **(see illustration)**. As the subframe is lowered, make sure there are no cables or wiring still attached.

21 Remove the steering gear from the subframe with reference to Section 17, the lower suspension arms with reference to Section 5, the anti-roll bar with reference to Section 8, and the rear engine mounting/torque link with reference to Chapter 2A, Chapter 2B, Chapter 2C or Chapter 2D, as applicable.

Refitting

22 Refit any components removed from the subframe with reference to the Sections and Chapters listed in paragraph 21.

23 Manoeuvre the subframe back into position under the car and raise the trolley jack to bring the subframe into contact with the underbody.

24 Insert the new front mounting bolts, the centre mounting bolts and the new rear mounting bolts, but only tighten the bolts finger tight at this stage.

25 Insert the Vauxhall/Opel alignment tools (CH-51034) or suitable alternatives into the subframe alignment holes, then through and into the underbody, on each side **(see illustration)**.

26 With the subframe held in alignment with the tools, tighten the mounting bolts to the specified torque then, where applicable, through the specified angle, in the order front, then centre, then rear. When all the bolts have been tightened, remove the alignment tools.

27 The remainder of refitting is a reversal of removal, tightening all nuts and bolts to the specified torque and, where necessary, through the specified angle.

7.20 Lower the subframe to the ground and remove it from under the car

8 Front anti-roll bar – removal and refitting

Removal

1 Remove the front subframe assembly as described in Section 7.

2 Unscrew the two bolts from each anti-roll bar mounting clamp, and remove the anti-roll bar from the subframe **(see illustration)**.

3 If any damage or deterioration of the mounting clamp rubber bushes is detected, the complete anti-roll bar must be renewed; the rubber bushes are not available separately and must not be disturbed.

8.2 Unscrew the two bolts from each anti-roll bar mounting clamp

7.25 Vauxhall/Opel alignment tool inserted through the subframe and into the underbody

Refitting

4 Refit the anti-roll bar to the subframe, and fit the mounting clamp retaining bolts. Tighten the retaining bolts to the specified torque.

5 Refit the front subframe assembly as described in Section 7.

6 Refit the roadwheels if not already done, then lower the vehicle to the ground and tighten the roadwheel nuts to the specified torque.

9 Rear hub bearing – removal and refitting

Note: *The rear hub bearing is a sealed unit and no repairs are possible. If the bearing is worn, a new bearing assembly must be obtained.*

Removal

1 Chock the front wheels, then jack up the rear of the vehicle, and support it securely on axle stands (see *Jacking and vehicle support*). Remove the roadwheel, then ensure that the handbrake is released.

2 On 4WD models, disconnect the driveshaft from the rear hub as described in Chapter 8, Section 2.

3 Remove the rear brake disc as described in Chapter 9, Section 6.

4 Undo the retaining bolt and withdraw the ABS wheel speed sensor from the trailing arm carrier plate **(see illustrations)**.

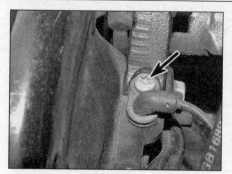

9.4a Undo the retaining bolt...

9.4b ...and withdraw the ABS wheel speed sensor from the trailing arm carrier plate

9.5 Hub bearing unit mounting bolts

5 Unscrew the four mounting bolts on the inside of the trailing arm, then withdraw the bearing unit, brake shield and carrier plate from the trailing arm **(see illustration)**.

Refitting

6 Thoroughly clean the trailing arm carrier plate, bearing unit and brake shield mating faces, then place the carrier plate and brake shield in position on the trailing arm.
7 Refit the bearing unit and screw on the four retaining bolts.
8 Tighten the retaining bolts progressively to the specified torque setting.
9 Refit the ABS wheel speed sensor and tighten the retaining bolt securely.
10 Refit the rear brake disc as described in Chapter 9, Section 6.
11 On 4WD models, reconnect the driveshaft to the rear hub as described in Chapter 8, Section 2.

10 Rear shock absorber – removal, inspection and refitting

Note: *Always renew shock absorbers in pairs to maintain good road handling.*
Note: *A new shock absorber lower mounting bolt will be needed for refitting.*

Removal

1 Chock the front wheels then jack up the rear of the car and securely support it on axle

stands (see *Jacking and vehicle support*). Remove the relevant rear roadwheel.
2 Using a trolley jack, slightly raise the trailing arm on the side being worked on. Take care not to damage the handbrake cable with the jack head.
3 Unscrew and remove the shock absorber lower mounting bolt from the trailing arm **(see illustration)**. Note that a new bolt will be required for refitting.
4 Unscrew the shock absorber upper mounting bracket bolts and remove the shock absorber from the car **(see illustration)**.
5 If required, unscrew the upper retaining nut and remove the upper mounting bracket from the shock absorber body.

Inspection

6 The shock absorber can be tested by clamping the lower mounting eye in a vice, then fully extending and compressing the shock absorber several times. Any evidence of jerky movement or lack of resistance indicates the need for renewal.
7 Examine the mounting rubbers in the shock absorber for excessive wear.
8 If the shock absorber or its mounting rubbers are worn excessively, renew the complete shock absorber.

Refitting

9 Place the shock absorber in position and refit the two upper mounting bracket bolts. Tighten the bolts to the specified torque.
10 Refit the new shock absorber lower

mounting bolt but only tighten it loosely at this stage.
11 Remove the trolley jack and refit the roadwheel.
12 Lower the car to the ground and tighten the roadwheel nuts to the specified torque.
13 Tighten the shock absorber lower mounting bolt to the specified torque, then through the specified angle.

11 Rear suspension coil spring – removal and refitting

Note: *Always renew coil springs in pairs to maintain good handling.*

Removal

1 Chock the front wheels then jack up the rear of the car and securely support it on axle stands (see *Jacking and vehicle support*). Remove the rear roadwheels.
2 Using trolley jacks positioned under the coil springs, slightly raise the trailing arms.
3 Unscrew and remove the shock absorber lower mounting bolts from the trailing arms on both sides **(see illustration 10.3)**. Note that new bolts will be required for refitting.
4 Carefully lower the trailing arms until all tension is removed from the coil springs.
5 Remove the coil spring and upper spring seat from the underbody and trailing arm, and withdraw from under the vehicle **(see illustration)**.

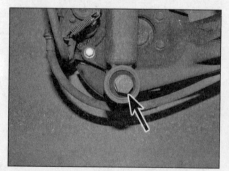

10.3 Unscrew and remove the shock absorber lower mounting bolt from the trailing arm

10.4 Unscrew the shock absorber upper mounting bracket bolts

11.5 Remove the coil spring and upper spring seat from the underbody and trailing arm

Refitting

6 Refitting is a reversal of removal, but note the following points.

a) Ensure that the spring locates correctly on the upper and lower seats, as well as on the trailing arm and underbody.

b) Tighten the new shock absorber lower mounting bolts to the specified torque and through the specified angle once the car has been lowered to the ground.

12 Rear axle –
removal and refitting

Removal

1 Chock the front wheels, then jack up the rear of the vehicle, and support it securely on axle stands (see *Jacking and vehicle support*). Remove the rear roadwheels.

2 On 4WD models, remove the final drive unit as described in Chapter 7C, Section 6.

3 Remove the exhaust system main section as described in Chapter 4A, Section 20 or Chapter 4B, Section 26 as applicable.

4 Undo the retaining bolt and withdraw the ABS wheel speed sensor from the trailing arm carrier plate on each side. Release the wiring harness retaining clips, then move the sensors to one side **(see illustrations 9.4a and 9.4b)**.

5 Remove the rear brake disc on each side as described in Chapter 9, Section 6.

6 Undo the retaining bolt and release each handbrake cable from the rear axle training arm **(see illustration)**.

7 Remove the rear suspension coil springs as described in Section 11.

8 Using paint or a suitable marker, mark the outline of the rear axle front mountings on the body on each side to ensure correct alignment when refitting.

9 Support the weight of the rear axle and trailing arms using two trolley jacks. Alternatively, one trolley jack and a length of wood may be used, but the help of an assistant will be required.

10 Make sure that the rear axle is securely supported, then unscrew and remove the two front mounting bracket bolts on each side **(see illustration)**. Note that new bolts will be required for refitting.

12.6 Undo the retaining bolts and release the handbrake cables from the training arms

11 Lower the rear axle to the ground and remove it from under the vehicle.

Refitting

12 Check the condition of the threads in the front mounting bracket captive nuts on the underbody. If necessary, use a tap to clean out the threads.

13 Support the rear axle on the trolley jacks, and position the assembly under the rear of the vehicle.

14 Raise the jacks, and fit the new mounting bracket bolts. Align the mounting brackets with the outline markings made during removal, then tighten the bolts to the specified torque and through the specified angle.

15 Place the coil springs and spring seats in position, then raise the trailing arms and insert the new shock absorber lower mounting bolts. Do not fully tighten the bolts at this stage.

16 Refit the rear brake disc on each side as described in Chapter 9, Section 6.

17 Refit and tighten the bolts securing the handbrake cable support bracket to the rear axle trailing arm on each side.

18 Refit the ABS wheel speed sensors to the trailing arm carrier plates on each side. Refit and tighten the retaining bolts, then refit the wiring harness retaining clips.

19 Refit the exhaust system main section as described in Chapter 4A, Section 20 or Chapter 4B, Section 26.

20 On 4WD models, refit the final drive unit as described in Chapter 7C, Section 6.

21 Remove the trolley jacks and refit the roadwheels.

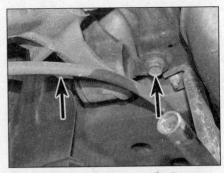

12.10 Unscrew and remove the two rear axle front mounting bracket bolts on each side

22 Lower the car to the ground and tighten the roadwheel nuts to the specified torque.

23 Tighten the shock absorber lower mounting bolts to the specified torque, then through the specified angle.

13 Steering wheel –
removal and refitting

⚠️ *Warning: Make sure that the airbag safety recommendations given in Chapter 12, Section 21 are followed, to prevent personal injury.*

Removal

1 Remove the airbag as described in Chapter 12, Section 22.

2 Remove the steering column shrouds as described in Chapter 11, Section 27.

3 Set the front wheels in the straight-ahead position, then lock the column in position after removing the ignition key.

4 Disconnect the wiring harness connector for the steering wheel switches **(see illustration)**.

5 Slacken the Torx retaining bolt securing the steering wheel to the column until two or three threads are left engaged **(see illustration)**.

6 Grip the steering wheel with both hands and carefully rock it from side-to-side to release it from the splines on the steering column. Once the wheel is free, remove the retaining bolt.

7 Check that there are alignment marks between the steering column shaft and steering wheel **(see illustration)**. If no marks

13.4 Disconnect the wiring harness connector for the steering wheel switches

13.5 Steering wheel retaining bolt

13.7 Alignment marks between the steering column shaft and steering wheel

13.8 Guide the wiring for the airbag through the aperture in the wheel, taking care not to damage the wiring connectors

13.9 Place a strip of adhesive tape across the airbag rotary connector to prevent the connector rotating

13.12 Apply thread locking compound to the steering wheel retaining bolt

are visible, centre punch the wheel and column shaft to ensure correct alignment when refitting.

8 Lift the steering wheel up and off the column. As the wheel is being removed, guide the wiring for the airbag through the aperture in the wheel, taking care not to damage the wiring connectors **(see illustration)**.

9 With the steering wheel removed, place a strip of adhesive tape across the side and front of the airbag rotary connector to prevent the connector rotating **(see illustration)**.

Refitting

10 Remove the adhesive tape used to prevent the airbag rotary connector rotating. If there is any doubt about the position of the rotary connector with regard to its centred position, centre the unit as described in Chapter 12, Section 22.

11 Refit the steering wheel, aligning the marks made prior to removal. Route the airbag wiring connectors through the steering wheel aperture.

12 Clean the threads on the retaining bolt and the threads in the steering column. Coat the retaining bolt with locking compound, then fit the retaining bolt and tighten it to the specified torque **(see illustration)**.

13 Reconnect the wiring connector for the steering wheel switches.

14 Release the steering lock, and refit the airbag as described in Chapter 12, Section 22.

15 Refit the steering column shrouds as described in Chapter 11, Section 27.

14 Ignition switch/ steering column lock – removal and refitting

Removal

1 Disconnect the battery negative lead as described in Chapter 5A Section 4.

2 Remove the steering column shrouds as described in Chapter 11, Section 27.

3 Undo the three screws and remove the security shield from the lock housing **(see illustrations)**.

4 Insert the ignition key into the ignition switch/lock, and turn it to position I.

5 Insert a thin rod into the hole in the top of the lock housing. Press the rod to release the detent spring, turn the key to the START position and pull out the lock cylinder using the key **(see illustration)**.

Refitting

6 Insert the ignition switch/lock into the lock housing, while the key is in the START position. Remove the rod from the lock housing.

7 Refit the security shield and tighten the retaining screws securely.

8 Refit steering column shrouds as described in Chapter 11, Section 27, then reconnect the battery.

15 Steering column – removal and refitting

⚠ *Warning: Make sure that the airbag safety recommendations given in Chapter 12, Section 21 are followed, to prevent personal injury.*

Removal

1 Set the front wheels in the straight-ahead position, then lock the column in position after removing the ignition key.

2 Remove the facia right-hand lower trim panel and the steering column shrouds as described in Chapter 11, Section 27.

3 Extract the centre pin and remove the plastic rivet securing the footwell air duct to the base of the facia. Disengage the duct from the air distribution housing and remove the duct.

4 Lock the steering column height adjuster in the fully up position.

5 Disconnect the wiring connectors at the steering column multi-function switches, ignition switch, airbag rotary connector, immobiliser transponder and any other electrical components according to equipment fitted. Release the wiring harness retaining clips and free the harness from the column.

6 Unscrew the bolt securing the bottom of

14.3a Undo the three retaining screws...

14.3b ...and remove the security shield from the lock housing

14.5 Insert a thin rod into the lock housing hole, press the rod to release the detent spring, and pull out the lock cylinder

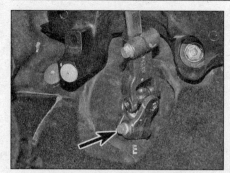

15.6 Unscrew the bolt securing the intermediate shaft to the steering gear pinion

15.7a Steering column upper retaining bolts...

15.7b ...and lower right-hand retaining bolt

the steering column intermediate shaft to the steering gear pinion **(see illustration)**. Use paint or a suitable marker pen to make alignment marks between the intermediate shaft and the steering gear pinion, then pull the shaft from the pinion and position to one side.

7 Suitably support the steering column and unscrew the four mounting bolts **(see illustrations)**. Withdraw the steering column from the facia crossmember, and remove the column from inside the vehicle.

8 With the column removed, take care to store it in a safe place where it is not likely to be knocked or damaged.

Refitting

9 Refitting is a reversal of removal, bearing in mind the following points:
a) *Tighten the steering column mounting bolts to the specified torque, starting with the two lower bolts, then the two upper bolts.*
b) *Ensure that the marks made on the intermediate shaft and steering gear pinion are correctly aligned, and tighten the intermediate shaft retaining bolt to the specified torque.*
c) *Refit the steering column shrouds and facia lower trim panel as described in Chapter 11, Section 27.*

16 Steering column intermediate shaft – removal and refitting

Removal

1 Remove the steering column as described in Section 15.

2 Using paint or similar, make alignment marks between the steering column and intermediate shaft upper universal joint, then slacken and remove the universal joint clamp bolt. Separate the intermediate shaft from the steering column shaft.

Refitting

3 Refitting is a reversal of removal, bearing in mind the following points:

a) *Tighten the intermediate shaft clamp bolt to the specified torque.*
b) *Ensure that the marks made on the intermediate shaft, steering column shaft.*
c) *Refit the steering column as described in Section 15.*

17 Steering gear assembly – removal and refitting

Note: *New steering gear mounting bolts and nuts will be required for refitting.*

Removal

1 Remove the front subframe as described in Section 7.

2 Undo the nuts from the two steering gear mounting bolts, while counter-holding the bolts from under the subframe **(see illustration)**. Lift off the two washers then remove the bolts. Note that new nuts and bolts will be required for refitting.

3 Twist the anti-roll bar upwards, then lift the steering gear off the subframe.

4 If a new steering gear assembly is to be fitted, then the track rod ends will need to be removed from each end of the steering track rods (see Section 19).

Refitting

5 Refitting is a reverse of the removal procedure, bearing in mind the following points:
a) *Tighten the steering gear mounting bolt*

17.2 Steering gear mounting bolt retaining nut

nuts to the specified torque, then through the specified angle. Note that new nuts/ bolts must be used.
b) *Set the steering gear in the straight-ahead position prior to refitting the front subframe.*
c) *Refit the front subframe as described in Section 7.*
d) *Where applicable, fit new track rod ends as described in Section 19.*
e) *Have the front wheel toe setting checked and if necessary adjusted at the earliest opportunity.*

18 Steering gear rubber gaiters – renewal

1 Remove the track rod end as described in Section 19.

2 Mark the correct fitted position of the gaiter on the track rod, then release the retaining clips, and slide the gaiter off the steering gear housing and track rod.

3 Thoroughly clean the track rod and the steering gear housing, clean off any corrosion, burrs or sharp edges which might damage the new gaiter's sealing lips on installation.

4 Carefully slide the new gaiter onto the track rod, and locate it on the steering gear housing. Align the outer edge of the gaiter with the mark made on the track rod prior to removal, then secure it in position with new retaining clips.

5 Refit the track rod end as described in Section 19.

19 Track rod end – removal and refitting

Note: *A new track rod end-to-swivel hub retaining nut will be required when refitting.*

Removal

1 Firmly apply the handbrake, then jack up the front of the car and support it securely on axle stands (see *Jacking and vehicle support*). Remove the appropriate front roadwheel.

19.4a Slacken the nut securing the track rod end to the steering arm...

19.4b ...separate the track rod end balljoint taper using a two-legged puller...

19.4c ...then unscrew the nut and remove the track rod end

2 If the track rod end is to be re-used, use a scriber, or similar, to mark its relationship to the track rod.

3 Hold the track rod arm, and unscrew the track rod end locknut by a quarter of a turn.

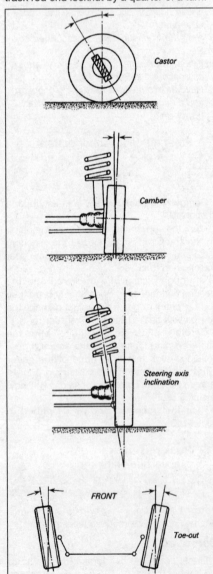

21.1 Front wheel geometry details

Castor

Camber

Steering axis inclination

FRONT

Toe-out

H23815

4 Slacken, but do not fully remove, the nut securing the track rod end to the steering arm on the swivel hub then use a two-legged puller to separate the track rod end balljoint taper (see illustrations). Once the taper is released, remove the track rod end from the steering arm. Note that a new nut will be required for refitting.

5 Counting the exact number of turns necessary to do so, unscrew the track rod end from the track rod.

6 Count the number of exposed threads between the end of the track rod and the locknut, and record this figure. If a new gaiter is to be fitted, unscrew the locknut from the track rod.

7 Carefully clean the track rod end and the track rod threads. Renew the track rod end if there is excessive free play of the balljoint shank, or if the shank is excessively stiff. If the balljoint gaiter is damaged, the complete track rod end assembly must be renewed; it is not possible to obtain the gaiter separately.

Refitting

8 If it was removed, screw the locknut onto the track rod threads, and position it so that the same number of exposed threads are visible as was noted prior to removal.

9 Screw the track rod end on to the track rod by the number of turns noted on removal. This should bring the track rod end to within approximately a quarter of a turn from the locknut, with the alignment marks that were noted on removal.

10 Refit the track rod end balljoint shank to the swivel hub, then fit a new retaining nut and tighten it to the specified torque and through the specified angle. If the balljoint shank turns as the nut is being tightened, counterhold the balljoint shank using a second spanner to prevent rotation.

11 Tighten the track rod end securing locknut on the track rod while holding the track rod stationary with a second spanner on the flats provided.

12 Refit the roadwheel, then lower the vehicle to the ground and tighten the roadwheel nuts to the specified torque.

13 Have the front wheel toe setting checked and if necessary adjusted at the earliest opportunity.

20 Track rod – renewal

Note: *When refitting, a new track rod end-to-swivel hub nut and new gaiter retaining clips will be required.*

1 Remove the steering gear assembly as described in Section 17.

2 Remove the track rod end as described in Section 19.

3 Release the retaining clips, and slide the steering gear gaiter off the end of the track rod as described in Section 18.

4 Turn the pinion to put the steering gear on full lock, so that the rack protrudes from the steering gear housing on the relevant side.

5 Using a suitable spanner to hold the rack, unscrew and remove the track rod inner balljoint from the end of the steering rack.

6 Remove the track rod assembly, and examine the track rod inner balljoint for signs of slackness or tight spots. Check that the track rod itself is straight and free from damage. If necessary, renew the track rod; it is also recommended that the steering gear gaiter is renewed.

7 Screw the balljoint into the end of the steering rack, then tighten the track rod inner balljoint to the specified torque.

8 Install the steering gaiter(s) as described in Section 18 and the track rod end as described in Section 19.

9 Refit the steering gear assembly as described in Section 17.

21 Wheel alignment and steering angles – general information

Definitions

1 A car's steering and suspension geometry is defined in four basic settings – all angles are expressed in degrees (toe settings are also sometimes expressed as a measurement); the steering axis is defined as an imaginary line drawn through the axis of the suspension strut, extended where necessary to contact the ground (see illustration).

2 *Camber* is the angle between each roadwheel and a vertical line drawn through its centre and tyre contact patch, when viewed from the front or rear of the car. Positive camber is when the roadwheels are tilted outwards from the vertical at the top; negative camber is when they are tilted inwards. The camber angle is not adjustable.

3 *Castor* is the angle between the steering axis and a vertical line drawn through each roadwheel's centre and tyre contact patch, when viewed from the side of the car. Positive castor is when the steering axis is tilted so that it contacts the ground ahead of the vertical; negative castor is when it contacts the ground behind the vertical. The castor angle is not adjustable.

4 *Toe* is the difference, viewed from above, between lines drawn through the roadwheel centres and the car's centre-line. 'Toe-in' is when the roadwheels point inwards, towards each other at the front, while 'toe-out' is when they splay outwards from each other at the front.

5 The front wheel toe setting is adjusted by screwing the track rods in or out of the track rod ends, to alter the effective length of the track rod assembly.

Checking and adjustment

6 Due to the special measuring equipment necessary to check the wheel alignment and steering angles, and the skill required to use it properly, the checking and adjustment of these settings is best left to a Vauxhall/ Opel dealer or similar expert. Note that most tyre-fitting shops now possess sophisticated checking equipment.

Chapter 11
Bodywork and fittings

Contents

Degrees of difficulty

Easy, suitable for novice with little experience	**Fairly easy,** suitable for beginner with some experience	**Fairly difficult,** suitable for competent DIY mechanic	**Difficult,** suitable for experienced DIY mechanic	**Very difficult,** suitable for expert DIY or professional

Specifications

Torque wrench settings

	Nm	lbf ft
Facia crossmember centre brace retaining bolts	22	16
Facia crossmember mounting bolts/nuts	22	16
Front seat mounting bolts	40	30
Seat belt component mountings	45	33

1 General Information

1 The bodyshell is of five-door Hatchback configuration, and is made of pressed-steel sections. The bodyshell and underframe feature variable thickness steel, achieved by laser-welded technology, used to join steel panels of different gauges. This gives a stiffer structure, with mounting points being more rigid, which gives an improved crash performance.

2 An additional safety crossmember is incorporated between the A-pillars in the upper area of the bulkhead, and the facia and steering column are secured to it. The lower bulkhead area is reinforced by additional systems of members connected to the front of the vehicle. The body side rocker panels (sills) have been divided along the length of the vehicle by internal reinforcement, this functions like a double tube which increases its strength. All doors are reinforced and incorporate side impact protection, which is secured in the door structure. There are additional impact

absorbers to the front and rear of the vehicle, behind the bumper assemblies.

3 All sheet metal surfaces which are prone to corrosion are galvanised. The painting process includes a base colour which closely matches the final topcoat, so that any stone damage is not as noticeable. The front wings are of a bolt-on type to ease their renewal if required.

4 Extensive use is made of plastic materials, mainly in the interior, but also in exterior components. The front and rear bumpers and the front grille are injection-moulded from a synthetic material which is very strong, and yet light. Plastic components such as wheel arch liners are fitted to the underside of the car, to improve the body's resistance to corrosion.

2 Maintenance – bodywork and underframe

1 The general condition of a vehicle's body-work is the one thing that significantly affects its value. Maintenance is easy, but needs to be regular. Neglect, particularly

after minor damage, can lead quickly to further deterioration and costly repair bills. It is important also to keep watch on those parts of the vehicle not immediately visible, for instance the underside, inside all the wheel arches, and the lower part of the engine compartment.

2 The basic maintenance routine for the bodywork is washing – preferably with a lot of water, from a hose. This will remove all the loose solids which may have stuck to the vehicle. It is important to flush these off in such a way as to prevent grit from scratching the finish. The wheel arches and underframe need washing in the same way, to remove any accumulated mud, which will retain moisture and tend to encourage rust. Paradoxically enough, the best time to clean the underframe and wheel arches is in wet weather, when the mud is thoroughly wet and soft. In very wet weather, the underframe is usually cleaned of large accumulations automatically, and this is a good time for inspection.

3 Periodically, except on vehicles with a wax-based underbody protective coating, it is a good idea to have the whole of the

underframe of the vehicle steam-cleaned, engine compartment included, so that a thorough inspection can be carried out to see what minor repairs and renovations are necessary. Steam-cleaning is available at many garages, and is necessary for the removal of the accumulation of oily grime, which sometimes is allowed to become thick in certain areas. If steam-cleaning facilities are not available, there are some excellent grease solvents available which can be brush-applied; the dirt can then be simply hosed off. Note that these methods should not be used on vehicles with wax-based underbody protective coating, or the coating will be removed. Such vehicles should be inspected annually, preferably just prior to Winter, when the underbody should be washed down, and any damage to the wax coating repaired. Ideally, a completely fresh coat should be applied. It would also be worth considering the use of such wax-based protection for injection into door panels, sills, box sections, etc, as an additional safeguard against rust damage, where such protection is not provided by the vehicle manufacturer.

4 After washing paintwork, wipe off with a chamois leather to give an unspotted clear finish. A coat of clear protective wax polish will give added protection against chemical pollutants in the air. If the paintwork sheen has dulled or oxidised, use a cleaner/polisher combination to restore the brilliance of the shine. This requires a little effort, but such dulling is usually caused because regular washing has been neglected. Care needs to be taken with metallic paintwork, as special non-abrasive cleaner/polisher is required to avoid damage to the finish. Always check that the door and ventilator opening drain holes and pipes are completely clear, so that water can be drained out. Brightwork should be treated in the same way as paintwork. Windscreens and windows can be kept clear of the smeary film which often appears, by the use of proprietary glass cleaner. Never use any form of wax or other body or chromium polish on glass.

3 Maintenance of upholstery and carpets – general

1 Mats and carpets should be brushed or vacuum-cleaned regularly, to keep them free of grit. If they are badly stained, remove them from the vehicle for scrubbing or sponging, and make quite sure they are dry before refitting. Seats and interior trim panels can be kept clean by wiping with a damp cloth. If they do become stained (which can be more apparent on light-coloured upholstery), use a little liquid detergent and a soft nail brush to scour the grime out of the grain of the material. Do not forget to keep the headlining clean in the same way as the upholstery. When using liquid cleaners inside the vehicle,

do not over-wet the surfaces being cleaned. Excessive damp could get into the seams and padded interior, causing stains, offensive odours or even rot.

2 If the inside of the vehicle gets wet accidentally, it is worthwhile taking some trouble to dry it out properly, particularly where carpets are involved. Do not leave oil or electric heaters inside the vehicle for this purpose.

4 Minor body damage – repair

Minor scratches

1 If the scratch is very superficial, and does not penetrate to the metal of the bodywork, repair is very simple. Lightly rub the area of the scratch with a paintwork renovator, or a very fine cutting paste, to remove loose paint from the scratch, and to clear the surrounding bodywork of wax polish. Rinse the area with clean water.

2 Apply touch-up paint to the scratch using a fine paint brush; continue to apply fine layers of paint until the surface of the paint in the scratch is level with the surrounding paintwork. Allow the new paint at least two weeks to harden, then blend it into the surrounding paintwork by rubbing the scratch area with a paintwork renovator or a very fine cutting paste. Finally, apply wax polish.

3 Where the scratch has penetrated right through to the metal of the bodywork, causing the metal to rust, a different repair technique is required. Remove any loose rust from the bottom of the scratch with a penknife, then apply rust-inhibiting paint to prevent the formation of rust in the future. Using a rubber or nylon applicator, fill the scratch with bodystopper paste. If required, this paste can be mixed with cellulose thinners to provide a very thin paste which is ideal for filling narrow scratches. Before the stopper-paste in the scratch hardens, wrap a piece of smooth cotton rag around the top of a finger. Dip the finger in cellulose thinners, and quickly sweep it across the surface of the stopper-paste in the scratch; this will ensure that the surface of the stopper-paste is slightly hollowed. The scratch can now be painted over as described earlier in this Section.

Dents

4 When deep denting of the vehicle's bodywork has taken place, the first task is to pull the dent out, until the affected bodywork almost attains its original shape. There is little point in trying to restore the original shape completely, as the metal in the damaged area will have stretched on impact, and cannot be reshaped fully to its original contour. It is better to bring the level of the dent up to a point which is about 3 mm below the level of the surrounding bodywork. In cases where

the dent is very shallow anyway, it is not worth trying to pull it out at all. If the underside of the dent is accessible, it can be hammered out gently from behind, using a mallet with a wooden or plastic head. Whilst doing this, hold a suitable block of wood firmly against the outside of the panel, to absorb the impact from the hammer blows and thus prevent a large area of the bodywork from being 'belled-out'.

5 Should the dent be in a section of the bodywork which has a double skin, or some other factor making it inaccessible from behind, a different technique is called for. Drill several small holes through the metal inside the area – particularly in the deeper section. Then screw long self-tapping screws into the holes, just sufficiently for them to gain a good purchase in the metal. Now the dent can be pulled out by pulling on the protruding heads of the screws with a pair of pliers.

6 The next stage of the repair is the removal of the paint from the damaged area, and from an inch or so of the surrounding 'sound' bodywork. This is accomplished most easily by using a wire brush or abrasive pad on a power drill, although it can be done just as effectively by hand, using sheets of abrasive paper. To complete the preparation for filling, score the surface of the bare metal with a screwdriver or the tang of a file, or alternatively, drill small holes in the affected area. This will provide a really good 'key' for the filler paste.

7 To complete the repair, see the Section on filling and respraying.

Rust holes or gashes

8 Remove all paint from the affected area, and from an inch or so of the surrounding 'sound' bodywork, using an abrasive pad or a wire brush on a power drill. If these are not available, a few sheets of abrasive paper will do the job most effectively. With the paint removed, you will be able to judge the severity of the corrosion, and therefore decide whether to renew the whole panel (if this is possible) or to repair the affected area. New body panels are not as expensive as most people think, and it is often quicker and more satisfactory to fit a new panel than to attempt to repair large areas of corrosion.

9 Remove all fittings from the affected area, except those which will act as a guide to the original shape of the damaged bodywork (eg headlight shells etc). Then, using tin snips or a hacksaw blade, remove all loose metal and any other metal badly affected by corrosion. Hammer the edges of the hole inwards, in order to create a slight depression for the filler paste.

10 Wire-brush the affected area to remove the powdery rust from the surface of the remaining metal. Paint the affected area with rust-inhibiting paint, if the back of the rusted area is accessible, treat this also.

11 Before filling can take place, it will be necessary to block the hole in some way. This

can be achieved by the use of aluminium or plastic mesh, or aluminium tape.

12 Aluminium or plastic mesh, or glass-fibre matting, is probably the best material to use for a large hole. Cut a piece to the approximate size and shape of the hole to be filled, then position it in the hole so that its edges are below the level of the surrounding bodywork. It can be retained in position by several blobs of filler paste around its periphery.

13 Aluminium tape should be used for small or very narrow holes. Pull a piece off the roll, trim it to the approximate size and shape required, then pull off the backing paper (if used) and stick the tape over the hole; it can be overlapped if the thickness of one piece is insufficient. Burnish down the edges of the tape with the handle of a screwdriver or similar, to ensure that the tape is securely attached to the metal underneath.

Filling and respraying

14 Before using this Section, see the Sections on dent, deep scratch, rust holes and gash repairs.

15 Many types of bodyfiller are available, but generally speaking, those proprietary kits which contain a tin of filler paste and a tube of resin hardener are best for this type of repair. A wide, flexible plastic or nylon applicator will be found invaluable for imparting a smooth and well-contoured finish to the surface of the filler.

16 Mix up a little filler on a clean piece of card or board – measure the hardener carefully (follow the maker's instructions on the pack), otherwise the filler will set too rapidly or too slowly. Using the applicator, apply the filler paste to the prepared area; draw the applicator across the surface of the filler to achieve the correct contour and to level the surface. As soon as a contour that approximates to the correct one is achieved, stop working the paste – if you carry on too long, the paste will become sticky and begin to 'pick-up' on the applicator. Continue to add thin layers of filler paste at 20-minute intervals, until the level of the filler is just proud of the surrounding bodywork.

17 Once the filler has hardened, the excess can be removed using a metal plane or file. From then on, progressively-finer grades of abrasive paper should be used, starting with a 40-grade production paper, and finishing with a 400-grade wet-and-dry paper. Always wrap the abrasive paper around a flat rubber, cork, or wooden block – otherwise the surface of the filler will not be completely flat. During the smoothing of the filler surface, the wet-and-dry paper should be periodically rinsed in water. This will ensure that a very smooth finish is imparted to the filler at the final stage.

18 At this stage, the 'dent' should be surrounded by a ring of bare metal, which in turn should be encircled by the finely 'feathered' edge of the good paintwork. Rinse the repair area with clean water, until all of the

dust produced by the rubbing-down operation has gone.

19 Spray the whole area with a light coat of primer – this will show up any imperfections in the surface of the filler. Repair these imperfections with fresh filler paste or bodystopper, and once more smooth the surface with abrasive paper. Repeat this spray-and-repair procedure until you are satisfied that the surface of the filler, and the feathered edge of the paintwork, are perfect. Clean the repair area with clean water, and allow to dry fully.

20 The repair area is now ready for final spraying. Paint spraying must be carried out in a warm, dry, windless and dust-free atmosphere. This condition can be created artificially if you have access to a large indoor working area, but if you are forced to work in the open, you will have to pick your day very carefully. If you are working indoors, dousing the floor in the work area with water will help to settle the dust which would otherwise be in the atmosphere. If the repair area is confined to one body panel, mask off the surrounding panels; this will help to minimise the effects of a slight mis-match in paint colours. Bodywork fittings (eg chrome strips, door handles etc) will also need to be masked off. Use genuine masking tape, and several thicknesses of newspaper, for the masking operations.

21 Before commencing to spray, agitate the aerosol can thoroughly, then spray a test area (an old tin, or similar) until the technique is mastered. Cover the repair area with a thick coat of primer; the thickness should be built up using several thin layers of paint, rather than one thick one. Using 400-grade wet-and-dry paper, rub down the surface of the primer until it is really smooth. While doing this, the work area should be thoroughly doused with water, and the wet-and-dry paper periodically rinsed in water. Allow to dry before spraying on more paint.

22 Spray on the top coat, again building up the thickness by using several thin layers of paint. Start spraying at one edge of the repair area, and then, using a side-to-side motion, work until the whole repair area and about 2 inches of the surrounding original paintwork is covered. Remove all masking material 10 to 15 minutes after spraying on the final coat of paint.

23 Allow the new paint at least two weeks to harden, then, using a paintwork renovator, or a very fine cutting paste, blend the edges of the paint into the existing paintwork. Finally, apply wax polish.

Plastic components

24 With the use of more and more plastic body components by the vehicle manufacturers (eg bumpers. spoilers, and in some cases major body panels), rectification of more serious damage to such items has become a matter of either entrusting repair work to a specialist in this field, or renewing complete components. Repair of such damage by the

DIY owner is not really feasible, owing to the cost of the equipment and materials required for effecting such repairs. The basic technique involves making a groove along the line of the crack in the plastic, using a rotary burr in a power drill. The damaged part is then welded back together, using a hot-air gun to heat up and fuse a plastic filler rod into the groove. Any excess plastic is then removed, and the area rubbed down to a smooth finish. It is important that a filler rod of the correct plastic is used, as body components can be made of a variety of different types (eg polycarbonate, ABS, polypropylene).

25 Damage of a less serious nature (abrasions, minor cracks etc) can be repaired by the DIY owner using a two-part epoxy filler repair material. Once mixed in equal proportions, this is used in similar fashion to the bodywork filler used on metal panels. The filler is usually cured in twenty to thirty minutes, ready for sanding and painting.

26 If the owner is renewing a complete component himself, or if he has repaired it with epoxy filler, he will be left with the problem of finding a suitable paint for finishing which is compatible with the type of plastic used. At one time, the use of a universal paint was not possible, owing to the complex range of plastics encountered in body component applications. Standard paints, generally speaking, will not bond to plastic or rubber satisfactorily. However, it is now possible to obtain a plastic body parts finishing kit which consists of a pre-primer treatment, a primer and coloured top coat. Full instructions are normally supplied with a kit, but basically, the method of use is to first apply the pre-primer to the component concerned, and allow it to dry for up to 30 minutes. Then the primer is applied, and left to dry for about an hour before finally applying the special-coloured top coat. The result is a correctly-coloured component, where the paint will flex with the plastic or rubber, a property that standard paint does not normally possess.

5 Major body damage repair
– general

1 Where serious damage has occurred, or large areas need renewal due to neglect, it means that complete new panels will need welding-in, and this is best left to professionals. If the damage is due to impact, it will also be necessary to check completely the alignment of the bodyshell, and this can only be carried out accurately by a Vauxhall/Opel dealer, or accident repair specialist, using special jigs. If the body is left misaligned, it is primarily dangerous as the car will not handle properly, and secondly, uneven stresses will be imposed on the steering, suspension and possibly transmission, causing abnormal wear, or complete failure, particularly to such items as the tyres.

6.3a Remove the front wheel arch outer moulding...

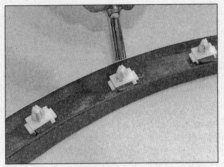

6.3b ...by sliding an 8mm socket over the ends of the retainers to collapse the retaining tabs...

6.3c ...then lift the moulding up to disengage the locating tab

6 Front bumper – removal and refitting

Removal

1 Firmly apply the handbrake, and then jack up the front of the vehicle and support it securely on axle stands (see *Jacking and vehicle support*). Remove the front roadwheels.
2 Remove the front wheel arch liners on both sides as described in Section 21.
3 Remove the front wheel arch outer moulding on each side by sliding an 8mm socket, or equivalent, over the ends of the retainers to collapse the retaining tabs. Disengage the six retainers from the front wing and then the three retainers from the front bumper. Lift the moulding up to disengage the locating tab from the bumper, then remove the moulding (see illustrations).
4 Undo the retaining screw on each side securing the upper corner of the bumper to the front wing (see illustration).
5 Extract the centre pins, then remove the eight plastic expanding rivets securing the bumper upper panel to the bumper and bonnet lock platform. Lift up the upper panel and remove it from the engine compartment (see illustrations).
6 Undo the four bolts securing the upper edge of the radiator grille to the bumper upper support, and the screw each side securing the upper edge of the bumper to the headlight unit (see illustration).

7 Undo the six bolts securing the engine undertray front extension panel to the underside of the bumper.
8 With the aid of an assistant, pull the bumper out at the sides to disengage the bumper from the guide rails then pull the bumper forward (see illustrations).
9 As applicable, disconnect the foglight, exterior temperature sensor and parking distance sensor wiring connectors, and the headlight washer hoses. Carefully remove the bumper from the car.

Refitting

10 Refitting is a reversal of the removal procedure, ensuring that all bumper fasteners are securely tightened.

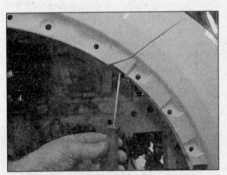

6.4 Undo the screw each side securing the upper corner of the bumper to the wing

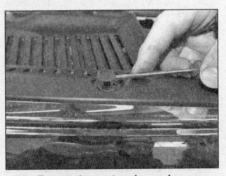

6.5a Extract the centre pins and remove the eight expanding rivets...

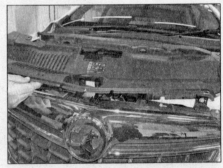

6.5b ...then lift away the bumper upper panel

6.6 Undo the radiator grille retaining bolts and the bumper upper edge retaining screws

6.8a Pull the bumper out at the sides to disengage it from the guide rails...

6.8b ...then pull the bumper forward

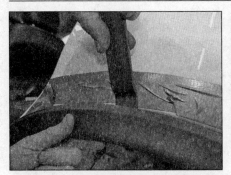

7.3a Using a forked tool, prise off the wheel arch outer moulding to release the retaining clips...

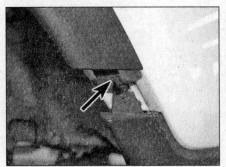

7.3b ...then disengage the retaining tab and remove the moulding

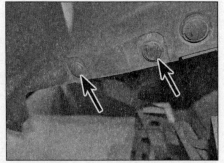

7.5 Remove the two plastic expanding rivets on each side from the lower edge of the bumper

7 Rear bumper – removal and refitting

Removal

1 Chock the front wheels, then jack up the rear of the vehicle and support on axle stands (see *Jacking and vehicle support*). Remove the roadwheels.

2 Remove the rear wheel arch liners on both sides as described in Section 21.

3 Apply two layers of masking tape or similar around the edge of the rear wheel arch outer moulding on each side to prevent any paint damage. Using a suitable forked tool, prise the moulding off the rear wing and bumper on each side to release the eight retaining clips. Lift the moulding up to disengage the locating tab from the bumper, then remove the moulding **(see illustrations)**.

Note: *The rear wheel arch outer moulding is likely to be damaged, or the retaining clips are likely to be broken during removal. Be prepared to obtain a new moulding (which is supplied complete with new retaining clips) for refitting.*

4 Remove the rear light cluster each side as described in Chapter 12, Section 7.

5 Extract the centre pins, then remove the two plastic expanding rivets on each side from the lower edge of the bumper **(see illustration)**.

6 Working under the bumper, carefully push a screwdriver between the bumper and the rear guide. Insert a second small screwdriver into the bumper slot and depress the snaps one at a time, while pulling on the bumper to gradually remove the bumper from the rear guide **(see illustration)**.

7 With the aid of an assistant, pull the bumper out at the sides to disengage the support bracket. Disengage the bumper rear locating pegs from the guides on each side and carefully pull the bumper rearwards **(see illustration)**.

8 Where applicable, disconnect the parking distance sensor and light unit wiring connectors, then remove the bumper from the car.

Refitting

9 Refitting is a reversal of the removal procedure, ensuring that all bumper fasteners are secure.

8 Bonnet – removal, refitting and adjustment

Removal

1 Open the bonnet, and have an assistant support it.

2 Using a marker pen or paint, mark around the hinge positions on the bonnet.

3 With the aid of the assistant, unscrew the two nuts securing the bonnet to the hinges on both sides.

4 Lift off the bonnet taking care not to damage the vehicle paintwork.

Refitting

5 Align the marks made on the bonnet before removal with the hinges, then refit and tighten the bonnet securing nuts.

6 Check the bonnet adjustment as follows.

Adjustment

7 Close the bonnet, and check that there is an equal gap (approximately 3.0 mm) at each side, between the bonnet and the wing panels. Check also that the bonnet sits flush in relation to the surrounding body panels.

8 The bonnet should close smoothly and

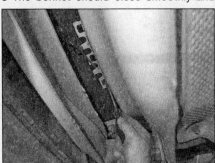

7.6 Using two screwdrivers to release the bumper from the rear guide

positively without excessive pressure. If this is not the case, adjustment will be required.

9 To adjust the bonnet alignment, loosen the bonnet-to-hinge mounting nuts, and move the bonnet on the studs as required (the holes in the hinges are enlarged). If necessary, the scissor-type hinge mounting bolts may be loosened as well. To adjust the bonnet front height in relation to the front wings, adjustable rubber bump stops are fitted to the front corners of the bonnet. These may be screwed in or out as necessary.

9 Bonnet lock – removal and refitting

Removal

1 Extract the centre pins, then remove the eight plastic expanding rivets securing the bumper upper panel to the bumper and bonnet lock platform. Lift up the upper panel and remove it from the engine compartment **(see illustrations 6.5a and 6.5b)**.

2 Using a marker pen or paint, mark around the bonnet lock position on the bonnet lock platform.

3 Undo the two bonnet lock retaining bolts and withdraw the lock from its location.

4 Disengage the bonnet release cable end fitting from the lock lever, then depress the tabs and withdraw the outer cable from the lock. Remove the lock from the car.

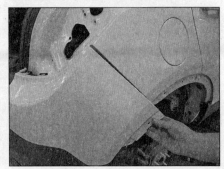

7.7 Pull the bumper out at the sides to disengage the support bracket

11.2 Carefully prise free the wiring harness rubber boot and the boot frame from the door pillar

11.5 Undo the door check strap retaining bolt

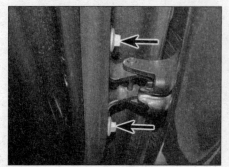

11.6 Undo the two bolts securing the lower hinge to the door

11.7 Undo the two bolts securing the upper hinge to the door and lift away the door

Refitting

5 Refitting is a reversal of removal, aligning the lock with the marks made before removal.

10 Bonnet release cable – removal and refitting

Removal

1 Remove the side sill inner trim panel on the right-hand side as described in Section 25.
2 Remove the windscreen cowl panel as described in Section 21.
3 Remove the bonnet lock as described in Section 9.
4 Insert a small screwdriver between the body of the bonnet release handle and the facia crossmember to free the detent lugs, then pull the handle downward to disengage it from the crossmember.
5 Disengage the outer cable from the release handle, then slide the inner cable end fitting out of the handle.
6 Release the cable from the support clips and brackets in the engine compartment. Release the grommet from the bulkhead then withdraw the cable into the engine compartment. As an aid to refitting, tie a length of string to the cable before removing it and leave the string in position ready for refitting.

Refitting

7 Refitting is a reversal of removal, but tie the string to the end of the cable, and use the string to pull the cable into position. Ensure that the cable is routed as noted before removal, and make sure that the grommet is correctly seated. On completion, refit the bonnet lock as described in Section 9, the windscreen cowl panel as described in Section 21 and the side sill inner trim panel as described in Section 18.

11 Door – removal, refitting and adjustment

Removal

1 Open the door and support it under its lower edge on a trolley jack covered with pads of rags.
2 Carefully prise free the wiring harness rubber boot and the boot frame from the door pillar. Refit the boot to the frame allowing the frame (and boot) to be simply pushed back into position when refitting **(see illustration)**.
3 Disconnect the wiring harness connector from the door pillar socket.
4 Using a marker pen or paint, mark around the lower door hinge position on the door.
5 Undo the bolt securing the door check strap to the vehicle body **(see illustration)**.
6 Ensure that the door is well supported, then undo the two bolts securing the lower hinge to the door **(see illustration)**.
7 Undo the two bolts securing the upper hinge to the door and with the aid of an assistant, lift away the door **(see illustration)**.

Refitting and adjustment

8 Refit the door using the reverse of the removal procedure, then close the door and check that it fits correctly in its aperture, with equal gaps at all points between it and the surrounding bodywork.
9 If adjustment is required, slacken the bolts securing the hinges to the door and reposition the door as required. Securely tighten the bolts after completing the adjustment.
10 The striker alignment should be checked after either the door or the lock has been disturbed. To adjust a striker, slacken its screws, reposition it and securely tighten the screws.

12 Door inner trim panel – removal and refitting

Front door

Removal

1 Stick a piece of folded-over masking tape onto the interior handle trim plate and use the tape to pull off the plate **(see illustration)**.
2 Undo the screw now exposed after removal of the trim plate **(see illustration)**.

12.1 Use a piece of masking tape to pull off the interior handle trim plate

12.2 Undo the screw now exposed after removal of the trim plate

12.3a Lift up the trim cap...

12.3b ...then undo the screw in the door grab handle aperture

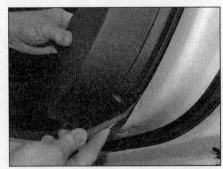

12.4a Carefully prise the bottom and sides of the panel away from door...

3 Using a small screwdriver, lift up the trim cap, then undo the retaining screw in the door grab handle aperture **(see illustrations)**.

4 Using a wide-bladed screwdriver or trim removal tool, carefully prise the bottom and sides of the panel away from door to release the internal clips. Lift the panel upward to release it from the window aperture **(see illustrations)**.

5 Once the panel is free, reach behind, and using a small screwdriver, lift the front edge of the door lock operating cable retainer. Slide the outer cable to the rear and disengage the inner cable end from the interior handle **(see illustration)**.

6 Disconnect the wiring connectors, then remove the panel from the door.

7 Check the condition of the panel retaining clips and renew any that are broken or distorted.

Refitting

8 Refitting is a reversal of removal.

Rear door

Removal

9 Where a manual window regulator is fitted, use a proprietary regulator spring clip removal tool to remove the spring clip. Remove the handle from the splined shaft then remove the circular spacer. Refit the spring clip to the handle **(see illustrations)**.

10 Stick a piece of folded-over masking tape onto the interior handle trim plate and use the tape to pull off the plate **(see illustration)**.

12.4b ...then lift the panel upward to release it from the window aperture

11 Undo the screw now exposed after removal of the trim plate **(see illustration)**.

12 Using a small screwdriver, lift up the

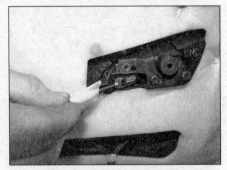

12.5 Disengage the door lock operating cable from the interior handle

trim cap, then undo the retaining screw in the door grab handle aperture **(see illustrations)**.

12.9a Where a manual window regulator is fitted, use a spring clip removal tool to remove the spring clip

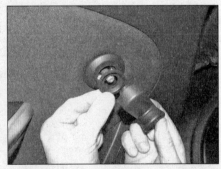

12.9b Remove the handle from the splined shaft then remove the circular spacer

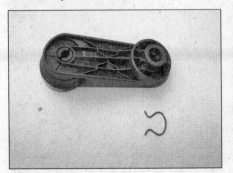

12.9c Refit the spring clip to the handle

12.10 Use a piece of masking tape to pull off the interior handle trim plate

12.11 Undo the screw now exposed after removal of the trim plate

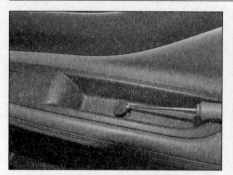

12.12a Lift up the trim cap...

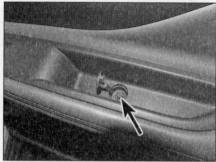

12.12b ...then undo the screw in the door grab handle aperture

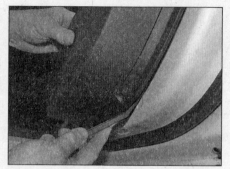

12.13a Carefully prise the bottom and sides of the panel away from door...

12.13b ...then lift the panel upward to release it from the window aperture

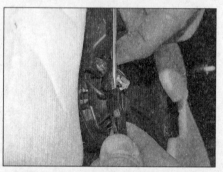

12.14a Lift the front edge of the door lock operating cable retainer and slide the outer cable to the rear...

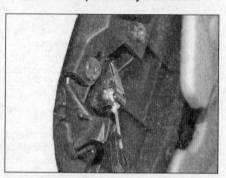

12.14b ...then disengage the inner cable end from the interior handle

13 Using a wide-bladed screwdriver or trim removal tool, carefully prise the bottom and sides of the panel away from door to release the internal clips. Lift the panel upward to release it from the window aperture (see illustrations).

14 Once the panel is free, reach behind, and using a small screwdriver, lift the front edge of the door lock operating cable retainer. Slide the outer cable to the rear and disengage the inner cable end from the interior handle (see illustrations).

15 Disconnect the wiring connectors, then remove the panel from the door and check the condition of the panel retaining clips. Renew any that are broken or distorted.

Refitting

16 Refitting is a reversal of removal.

13 Door handles and lock components – removal and refitting

Door interior handle

1 The door interior handle is an integral part of the door inner trim panel and cannot be individually removed. If there are any problems with the interior handle, a new inner trim panel will be required.

Front door exterior handle

Removal

2 Open the door and carefully prise out the blanking cap from the rear edge of the door to gain access to the handle locking screw (see illustration).

3 Pull the exterior door handle outwards and hold it in that position. With the exterior door handle in the open position, turn the handle locking screw anti-clockwise until it reaches its stop. Withdraw the fixed part of the handle, (containing the lock cylinder on the driver's side), from the door (see illustration).

4 Slide the exterior door handle to the rear, disengage the front pivot from the handle frame and remove the handle from the door (see illustration).

Refitting

5 Engage the handle front pivot with the frame and move the handle back into position.

6 Refit the fixed part of the handle to the door.

7 Hold the exterior handle and turn the handle locking screw clockwise to retain the handle.

8 Check the operation of the handle then refit the blanking cap to the edge of the door.

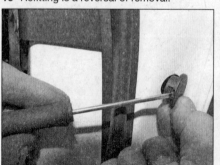

13.2 Prise out the blanking cap from the rear edge of the door to gain access to the handle locking screw

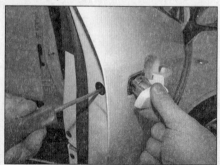

13.3 Turn the handle locking screw anti-clockwise, then withdraw the fixed part of the handle from the door

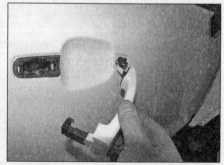

13.4 Slide the handle to the rear, disengage the front pivot from the frame and remove the handle

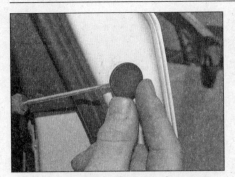

13.9 Prise out the blanking cap from the rear edge of the door to gain access to the handle locking screw

13.10 Turn the handle locking screw anti-clockwise, then withdraw the fixed part of the handle from the door

13.11 Slide the handle to the rear, disengage the front pivot from the frame and remove the handle

Rear door exterior handle

Removal

9 Open the door and carefully prise out the blanking cap from the rear edge of the door to gain access to the handle locking screw **(see illustration)**.

10 Pull the exterior door handle outwards and hold it in that position. With the exterior door handle in the open position, turn the handle locking screw anti-clockwise until it reaches its stop. Withdraw the fixed part of the handle from the door **(see illustration)**.

11 Slide the exterior door handle to the rear, disengage the front pivot from the handle frame and remove the handle from the door **(see illustration)**.

Refitting

12 Engage the handle front pivot with

the frame and move the handle back into position.

13 Refit the fixed part of the handle to the door.

14 Hold the exterior handle and turn the handle locking screw clockwise to retain the handle.

15 Check the operation of the handle then refit the blanking cap to the edge of the door.

Front door lock cylinder

Removal

16 Remove the exterior handle as described previously in this Section.

17 Unclip the trim cap and remove the cap from the lock cylinder housing **(see illustration)**.

18 The lock cylinder body is an integral part of the housing and no further dismantling is possible.

Refitting

19 Refitting is a reversal of removal.

Front door lock

Removal

20 Remove the exterior handle as described previously in this Section.

21 Remove the door inner trim panel as described in Section 12.

22 Undo the two bolts and remove the inner trim panel support bracket from the door panel **(see illustration)**.

23 Carefully peel back the protective plastic sheet from the door, starting at the top. Feed the door lock operating cable through the sheet as the sheet is removed **(see illustration)**.

24 Ensure that the door window glass is fully closed, then undo the upper bolt and lower nut securing the window glass rear guide to the door. Remove the guide out through the door aperture **(see illustrations)**.

25 Where fitted, undo the three bolts

13.17 Unclip the trim cap and remove the cap from the lock cylinder housing

13.22 Undo the two bolts and remove the inner trim panel support bracket

13.23 Carefully peel back the plastic sheet from the door, starting at the top

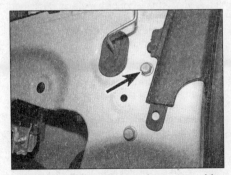

13.24a Undo the window glass rear guide upper retaining bolt...

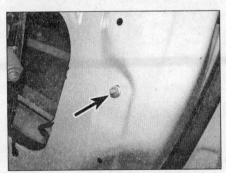

13.24b ...and lower retaining nut...

13.24c ...then remove the guide out through the door aperture

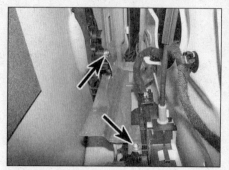

13.25a Undo the two inner retaining bolts…

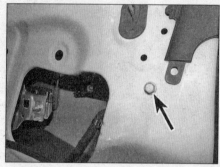

13.25b …and outer retaining bolt…

13.25c …then remove the security cover out through the door aperture

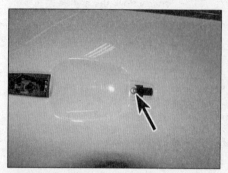

13.26 Undo the exterior handle frame front retaining screw

13.29 Undo the three screws securing the lock to the door

13.30 Remove the lock assembly and exterior handle frame through the door aperture

13.35 Undo the two bolts and remove the inner trim panel support bracket

13.36 Carefully peel back the protective plastic sheet from the door

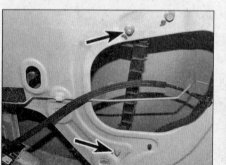

13.37a Undo the two window glass rear guide retaining nuts…

13.37b …and remove the guide out through the door aperture

securing the security cover to the exterior handle frame, door lock and door, and remove the cover through the door aperture **(see illustrations).**

26 Undo the exterior handle frame front retaining screw **(see illustration).** Slide the handle frame forward to release the rear locating lugs and free the frame from the door.

27 Release the interior lock button operating rod grommet from the door panel.

28 Lift the locking bar and disconnect the wiring connector from the door lock.

29 Undo the three screws securing the lock to the door **(see illustration).**

30 Lower the lock assembly together with the exterior handle frame down into the door, bringing the interior lock button and rod in through the hole provided. Remove the lock assembly and exterior handle frame through the door aperture **(see illustration).**

31 If required, disconnect the operating cable and rods from the lock linkage and bracket.

Refitting

32 Refitting is a reversal of removal. Ensure that the protective plastic sheet is firmly stuck with no air bubbles. If the sheet was damaged during removal it should be renewed.

Rear door lock

Removal

33 Remove the exterior handle as described previously in this Section.

34 Remove the door inner trim panel as described in Section 12.

35 Undo the two bolts and remove the inner trim panel support bracket from the door panel **(see illustration).**

36 Carefully peel back the protective plastic sheet from the door, starting at the top **(see illustration).** Feed the door lock operating cable through the sheet as the sheet is removed.

37 Ensure that the door window glass is fully closed, then undo the two nuts securing the window glass rear guide to the door. Remove the guide out through the door aperture **(see illustrations).**

13.38 Open the retaining clip and release the exterior handle operating rod from the door lock

13.39 Release the interior lock button operating rod bellcrank from the door panel and release the rod from the guides

13.40 Undo the three screws securing the lock to the door

38 Open the orange exterior handle operating rod retaining clip and release the operating rod from the door lock **(see illustration)**.

39 Release the interior lock button operating rod bellcrank from the door panel and release the operating rod from the guides **(see illustration)**.

40 Undo the three screws securing the lock to the door **(see illustration)**.

41 Lower the lock assembly down into the door until the wiring connector becomes accessible. Lift the locking bar and disconnect the wiring connector, then remove the lock from the door **(see illustrations)**.

42 If required, disconnect the operating cable and rods from the lock linkage.

Refitting

43 Refitting is a reversal of removal. Ensure that the protective plastic sheet is firmly stuck with no air bubbles. If the sheet was damaged during removal it should be renewed.

13.41a Lift the locking bar…

13.41b …then disconnect the wiring connector and remove the lock from the door

---

> **14 Door window glass and regulator –** removal and refitting

Front door sliding window glass

Removal

1 Remove the door inner trim panel as described in Section 12.

2 Undo the two bolts and remove the inner

trim panel support bracket from the door panel **(see illustration 13.22)**.

3 Carefully peel back the protective plastic sheet from the door, starting at the top. Feed the door lock operating cable through the sheet as the sheet is removed **(see illustration 13.23)**.

4 Ensure that the door window glass is fully closed, then undo the upper bolt and lower nut securing the window glass rear guide to the door. Remove the guide out through the door aperture **(see illustrations 13.24a to 13.24c)**.

5 Starting at the front and moving slowly rearward, pull outward on the upper door frame inner trim to release the integral tabs from the door inner panel **(see illustration)**.

6 Slacken the two retaining nuts on the window regulator to enable the glass to be removed from the regulator guides **(see**

illustration). If necessary, reconnect the wiring connector to the window control switch and reposition the window glass so the two nuts are accessible.

7 Lower the regulator, then tilt the window glass down at the front, lift it upwards and remove it from the outside of the door **(see illustration)**.

Refitting

8 Refitting is a reversal of removal.

Rear door sliding window glass

Removal

9 Remove the door inner trim panel as described in Section 12.

10 Undo the two bolts and remove the inner trim panel support bracket from the door panel **(see illustration 13.35)**.

14.5 Remove the upper door frame inner trim

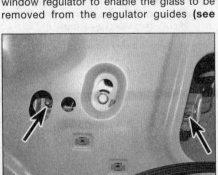

14.6 Slacken the two nuts on the regulator to enable the glass to be removed

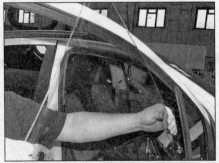

14.7 Tilt the window glass down at the front, lift it upwards and remove it from the outside of the door

14.13 Extract the plastic retaining stud securing the upper door frame inner trim

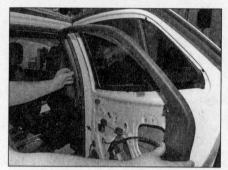

14.14 Remove the upper door frame inner trim

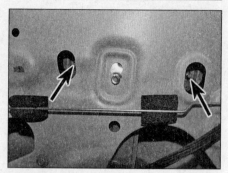

14.15 Slacken the two nuts on the regulator to enable the glass to be removed

11 Carefully peel back the protective plastic sheet from the door, starting at the top **(see illustration 13.36)**. Feed the door lock operating cable through the sheet as the sheet is removed.

12 Ensure that the door window glass is fully closed, then undo the two nuts securing the window glass rear guide to the door. Remove the guide out through the door aperture **(see illustrations 13.37a and 13.37b)**.

13 Using a forked tool, extract the plastic retaining stud securing the upper door frame inner trim to the rear of the door **(see illustration)**.

14 Starting at the front and moving slowly rearward, pull outward on the upper door frame inner trim to release the integral tabs from the door inner panel **(see illustration)**.

15 Slacken the two retaining nuts on the window regulator to enable the glass to be

removed from the regulator guides **(see illustration)**. If necessary, temporarily refit the regulator handle or reconnect the wiring connector to the window control switch and reposition the window glass so the two nuts are accessible.

16 Disengage the window glass from the regulator guides, then fully close the window and secure it in the closed position using adhesive tape **(see illustration)**.

17 If electric windows are fitted, disconnect the window regulator wiring connector **(see illustration)**.

18 Undo the three retaining bolts (electric windows) or five retaining bolts (manual windows) securing the window regulator to the door, then manipulate the regulator out through the door aperture **(see illustrations)**.

19 Support the window glass and remove the

masking tape securing it to the door. Rotate the glass through ninety degrees so the bottom of the glass is now facing forward, then lift it upwards and remove it from the outside of the door **(see illustration)**.

Refitting

20 Refitting is a reversal of removal.

Front door window regulator

Removal

21 Remove the door inner trim panel as described in Section 12.

22 Carefully peel back the protective plastic sheet from the door, starting at the top. Feed the door lock operating cable through the sheet as the sheet is removed **(see illustrations 13.23)**.

23 Slacken the two retaining nuts on the window regulator to enable the glass to be removed from the regulator guides **(see illustration 14.06)**. If necessary, reconnect the wiring connector to the window control switch and reposition the window glass so the two nuts are accessible.

24 Release the window glass from the regulator and slide the glass to the fully closed position. Retain the window glass in the closed position using adhesive tape over the top of the door frame.

25 Disconnect the wiring connector from the window regulator motor **(see illustration)**.

26 Undo the four bolts securing the regulator to the door, then manipulate the regulator out through the door aperture **(see illustrations)**.

14.16 Secure the glass in the closed position using adhesive tape

14.17 Disconnect the wiring connector from the regulator

14.18a Undo the three retaining bolts (electric windows)...

14.18b ...then manipulate the regulator out through the door aperture

14.19 Remove the window glass from the outside of the door

14.25 Disconnect the wiring connector from the window regulator motor

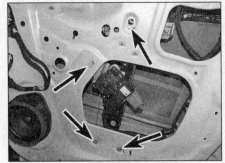

14.26a Undo the four bolts securing the regulator to the door...

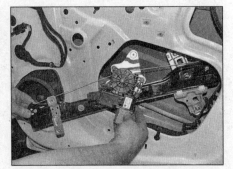

14.26b ...then manipulate the regulator out through the door aperture

Refitting

27 Refitting is a reversal of removal.

Rear door window regulator

Removal

28 Remove the door inner trim panel as described in Section 12.

29 Undo the two bolts and remove the inner trim panel support bracket from the door panel (see illustration 13.35).

30 Carefully peel back the protective plastic sheet from the door, starting at the top (see illustration 13.36). Feed the door lock operating cable through the sheet as the sheet is removed.

31 Slacken the two retaining nuts on the window regulator to enable the glass to be removed from the regulator guides (see illustration 14.15). If necessary, temporarily

refit the regulator handle or reconnect the wiring connector to the window control switch and reposition the window glass so the two nuts are accessible.

32 Disengage the window glass from the regulator guides, then fully close the window and secure it in the closed position using adhesive tape (see illustration 14.16).

33 If electric windows are fitted, disconnect the window regulator wiring connector (see illustration 14.17).

34 Undo the three retaining bolts (electric windows) or five retaining bolts (manual windows) securing the window regulator to the door, then manipulate the regulator out through the door aperture (see illustrations 14.18a and 14.18b).

Refitting

35 Refitting is a reversal of removal.

Front door fixed window glass

Removal

36 Remove the front door sliding window glass as described previously in this Section.

37 Move aside the door weatherstrip above the fixed window and undo the screw securing the window guide channel to the door frame (see illustration).

38 Working through the aperture in the door, undo the bolt securing the window guide channel to the door panel (see illustration).

39 Pull the window guide channels out of the door frame, then pull the fixed window rearward from its location (see illustrations). Remove the window and guide channels from the door.

Refitting

40 Refitting is a reversal of removal.

Rear door fixed window glass

Removal

41 Remove the rear door sliding window glass as described previously in this Section.

42 Using a plastic spatula or similar tool lift up the front of the door outer waist seal to disengage the retaining tab. Continue lifting the waist seal, pull it forward clear of the fixed window, and remove it from the door (see illustrations).

43 Move aside the door weatherstrip above the fixed window and undo the screw securing

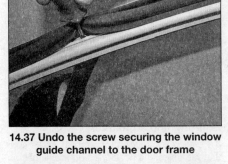

14.37 Undo the screw securing the window guide channel to the door frame

14.38 Undo the bolt securing the window guide channel to the door panel

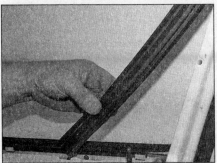

14.39a Pull the window guide channels out of the door frame...

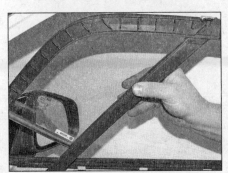

14.39b ...then pull the fixed window rearward from its location

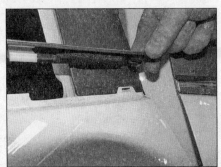

14.42a Lift up the front of the door outer waist seal...

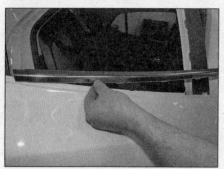

14.42b ...then pull it forward clear of the fixed window, and remove it from the door

14.43 Undo the screw securing the window guide channel to the door frame

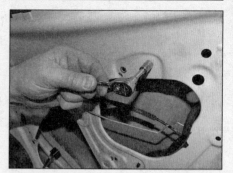

14.44 Undo the bolt securing the window guide channel to the door panel

14.45a Pull the window guide channels out of the door frame...

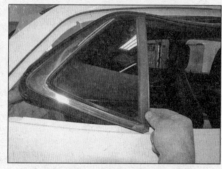

14.45b ...then pull the fixed window forward from its location

Refitting

46 Refitting is a reversal of removal.

15 Tailgate and support struts – removal, refitting and adjustment

Tailgate

Removal

1 Remove the tailgate trim panels as described in Section 25.
2 Disconnect the tailgate washer jet hose at the in-line connector and disconnect the adjacent wiring harness connectors (see illustration).
3 Using a hooked tool, carefully release the rubber gaiter at the top of the tailgate from the plastic frame. Depress the retaining tabs and release the plastic frame from the tailgate. Refit the gaiter to the frame allowing the frame (and gaiter) to be simply pushed back into position when refitting (see illustration).
4 Pull the wiring harness and washer hose out of the tailgate (see illustration).
5 Using a pencil or marker pen, mark the position of the hinges on the tailgate to aid refitting.
6 Engage the help of an assistant to support the tailgate, then disconnect the support struts as described later in this Section.
7 Unscrew the four bolts securing the hinges to the tailgate, and lift the tailgate from the car (see illustration).

Refitting and adjustment

8 Refitting is a reversal of removal. Make sure that the tailgate is positioned on the hinges as noted on removal and tighten the mounting bolts securely. With the tailgate closed, check that it is positioned centrally within the body aperture. If adjustment is necessary, loosen the tailgate mounting bolts, reposition the tailgate, then retighten the bolts. Check that the striker enters the lock centrally, and if necessary adjust the striker position by loosening the mounting bolts. Tighten the bolts on completion.

Support struts

Removal

9 Open the tailgate and note which way round the struts are fitted. Have an assistant support the tailgate in its open position.

the window guide channel to the door frame (see illustration).
44 Undo the bolt securing the window guide channel to the door panel (see illustration).

45 Pull the window guide channels out of the door frame, then pull the fixed window forward from its location (see illustrations). Remove the window and guide channels from the door.

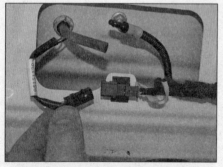

15.2 Disconnect the tailgage washer jet hose and adjacent wiring connectors

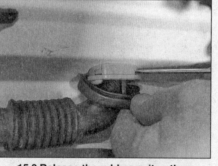

15.3 Release the rubber gaiter, then depress the tabs and release the plastic frame from the tailgate

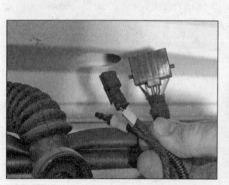

15.4 Pull the wiring harness and washer hose out of the tailgate

15.7 Hinge-to-tailgate retaining bolts

10 Using a small screwdriver, prise the spring clip from the bottom of the strut and disconnect it from the ball on the body **(see illustration)**.
11 Similarly prise the spring clip from the top of the strut and disconnect it from the ball on the tailgate. Withdraw the strut.

Refitting

12 Refitting is a reversal of removal.

16 Tailgate lock components – removal and refitting

Lock assembly

Removal

1 Remove the tailgate trim panels as described in Section 25.
2 Undo the four bolts and remove the metal shield over the exterior handle **(see illustrations)**.

3 Disconnect the lock wiring connector **(see illustration)**.
4 Undo the two lock assembly retaining screws and remove the lock from the tailgate **(see illustrations)**.

Refitting

5 Refitting is a reversal of removal. Check that when closed the tailgate lock engages the lock striker centrally. If necessary loosen the bolts and adjust the position of the striker, then tighten the bolts.

Release switch

Removal

6 Remove the tailgate trim panels as described in Section 25.
7 Undo the four bolts and remove the metal shield over the exterior handle **(see illustrations 16.2a and 16.2b)**.
8 Disconnect the release switch wiring connector **(see illustration)**.
9 Using a small screwdriver, depress the

15.10 Prise off the spring clips from the top and bottom of the support strut

internal retaining tab on each side and remove the exterior handle from the tailgate **(see illustration)**.
10 Undo the two screws and remove the release switch from the handle **(see illustrations)**.

Refitting

11 Refitting is a reversal of removal.

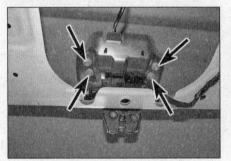

16.2a Undo the four bolts…

16.2b …and remove the metal shield over the exterior handle

16.3 Disconnect the tailgate lock wiring connector

16.4a Undo the retaining screws…

16.4b …and remove the lock from the tailgate

16.8 Disconnect the release switch wiring connector

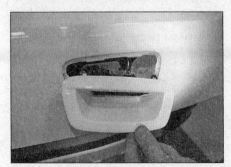

16.9 Remove the exterior handle from the tailgate

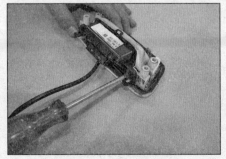

16.10a Undo the two screws…

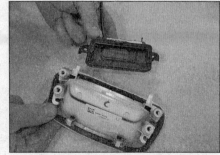

16.10b …and remove the release switch from the exterior handle

17.4 Disconnect the mirror wiring harness connector

17.5 Undo the three nuts securing the mirror to the door

17.6 Withdraw the mirror, wiring harness and grommet from the door

17 Exterior mirrors and associated components – removal and refitting

Exterior mirror

Removal

1 Remove the door inner trim panel as described in Section 12.

2 Undo the two bolts and remove the inner trim panel support bracket from the door panel **(see illustration 13.22)**.

3 Carefully peel back the protective plastic sheet from the door, starting at the top. Feed the door lock operating cable through the sheet as the sheet is removed **(see illustration 13.23)**.

4 Disconnect the mirror wiring harness connector, then release the wiring harness grommet from the door **(see illustration)**.

5 Support the mirror and undo the three retaining nuts **(see illustration)**.

6 Withdraw the mirror from the door, then pull out the wiring harness and grommet **(see illustration)**.

Refitting

7 Refitting is a reversal of removal.

Mirror glass

Removal

Caution: It is advisable to cover the mirror glass with adhesive tape and wear gloves during this operation.

17.9 Prise the lower edge of the glass outwards to release the internal retaining clips

8 Press the top of the glass inwards so that the glass is forced out from the bottom.

9 Using a plastic wedge, prise the lower edge of the glass outwards to release the internal retaining clips **(see illustration)**.

10 Withdraw the mirror glass and disconnect the wiring connectors **(see illustration)**.

Refitting

11 Refitting is a reversal of removal. Carefully press the mirror glass into the housing until the centre retainer clips are engaged.

Mirror motor

Removal

12 Remove the mirror glass as described previously.

13 Undo the three retaining screws and withdraw the motor assembly from the mirror body.

14 Disconnect the wiring connector and remove the mirror motor from the mirror body.

Refitting

15 Refitting is a reversal of removal.

Outer cover

Removal

16 Move the mirror body to the fully forward position and undo the two mirror bracket cover retaining screws.

17 Carefully prise apart the two halves of the mirror bracket cover and remove the cover.

18 Undo the four retaining screws and remove the mirror housing bezel.

17.10 Withdraw the mirror glass and disconnect the wiring connectors

19 Unscrew the outer cover retaining bolt and lift away the outer cover. Where applicable, disconnect the wiring connector.

Refitting

20 Refitting is a reversal of removal.

18 Interior mirror – removal and refitting

Removal

1 If fitted, carefully ease the right-hand side mirror base trim sideways and lift off the trim. Remove the trim on the left-hand side in the same way.

2 Undo the screw securing the mirror mount to the mirror base.

3 Where applicable, disconnect the mirror wiring connector.

4 Grasp the mirror in both hands and rock it from side to side while at the same time pushing upwards to release the internal mounting tab.

Refitting

5 Refitting is the reverse of removal.

19 Windscreen and fixed window glass – general information

1 The windscreen and tailgate glass is cemented in position with a special adhesive and require the use of specialist equipment for their removal and refitting. Renewal of such fixed glass is considered beyond the scope of the home mechanic. Owners are strongly advised to have the work carried out by one of the many specialist windscreen fitting specialists.

20 Sunroof – general information

1 An electric sunroof is offered as an optional extra on most models, and is fitted as standard equipment on some models.

21.5 Removing a front wheel arch liner

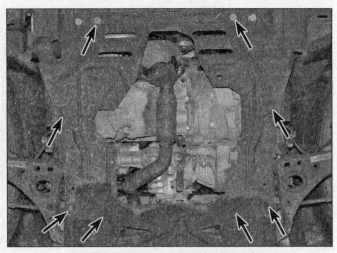

21.7 Engine undertray attachment points

2 Due to the complexity of the sunroof mechanism, considerable expertise is needed to repair, renew or adjust the sunroof components successfully. Removal of the roof first requires the headlining to be removed, which is a complex and tedious operation in itself, and not a task to be undertaken lightly. Therefore, any problems with the sunroof should be referred to a Vauxhall/Opel dealer.

21 Body exterior fittings – removal and refitting

Radiator grille

1 Remove the front bumper as described in Section 6.
2 The radiator grille panel and the other detachable panels on the bumper are retained by a series of screws and plastic clips, the location of which will become obvious on visual inspection.
3 Carefully disengage all the relevant clips using a small screwdriver, while at the same time pulling the panel from its location.
4 Refit the grille to the bumper ensuring it is securely held by the retaining tabs, then refit the front bumper as described in Section 6.

Wheel arch liners

5 The plastic wheel arch liners fitted to the underside of the front and rear wings are secured in position by a mixture of nuts, screws and plastic rivets, and removal will be fairly obvious on inspection. Work methodically around the liner, removing its retaining nuts, screws and plastic rivets until it is free to be removed from under the relevant wheel arch (see illustration). The wheel arch liner plastic rivets are removed by either prising out the centre portion using a small screwdriver, or tapping in the centre pin slightly, then removing the rivet body.
6 Refitting is the reverse of removal.

Engine undertray

7 On certain models an undertray is fitted beneath the engine/transmission assembly. The undertray is secured to the front subframe by a combination of bolts, screws and plastic rivets (see illustration). The plastic rivets are removed by prising out the centre pin using a small screwdriver, then removing the rivet body.
8 Refitting is the reverse of removal.

Windscreen cowl panel

9 Remove the windscreen wiper arms as described in Chapter 12, Section 12.
10 Working on one side at a time, grasp the cowl panel side extension and disengage the retainers from the top edge of the wing and end edge of the cowl panel (see illustration).
11 Using a forked tool, remove the six plastic rivets securing the windscreen cowl panel to the plenum chamber (see illustration).
12 Using a plastic spatula or similar tool, ease the upper edge of the cowl panel away from the windscreen to release the flange from the windscreen channel (see illustration).
13 Turn the panel over, disconnect the washer hose from the washer jet and remove the panel.
14 Refitting is the reverse of removal.

Body trim strips and badges

15 The various body trim strips and badges are held in position with a special adhesive tape. Removal requires the trim/badge to be heated, to soften the adhesive, and then cut away from the surface. Due to the high risk of damage to the vehicle's paintwork during this operation, it is recommended that this task should be entrusted to a Vauxhall/Opel dealer.

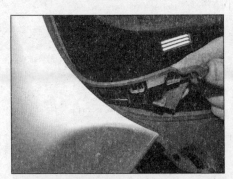

21.10 Remove the cowl panel side extension on each side

21.11 Remove the six plastic rivets securing the windscreen cowl panel to the plenum chamber

21.12 Ease the upper edge of the cowl panel away from the windscreen to release the flange from the windscreen channel

22.3a Prise off the trim cap on the seat belt tensioner connector...

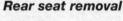

22.3b ...undo the retaining bolt...

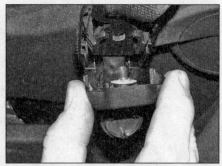

22.3c ...lift off the connector cover...

22 Seats – removal and refitting

⚠️ **Warning:** *The front seats are equipped with seat belt tensioners, and side airbags are incorporated into the outer sides of the seats. The seat belt tensioners and side airbags may cause injury if triggered accidentally. If the tensioner has been triggered due to a sudden impact or accident, the unit must be renewed, as it cannot be reset. If a seat is to be disposed of, the tensioner must be triggered before the seat is removed from the vehicle. Due to safety considerations, tensioner or seat disposal must be entrusted to a Vauxhall/Opel dealer. Where side airbags* are fitted, refer to Chapter 12, Section 21 for the precautions which should be observed when dealing with an airbag system.

Front seat removal

1 Set the seat height adjustment to its middle position.
2 Disconnect the battery negative lead as described in Chapter 5A Section 4. Wait 2 minutes for the capacitors to discharge, before working on the seat electrics.
3 Using a small screwdriver, carefully prise off the trim cap on the seat belt tensioner connector. Undo the retaining bolt, lift off the connector cover and separate the seat belt from the tensioner **(see illustrations)**.
4 Slide the seat adjustment fully forward.
5 Remove the trim cover over the outer retaining bolt by sliding it away from the seat and lifting off. Slacken and remove the seat retaining bolts from the rear of the guide rails **(see illustrations)**.
6 Pull the seat backwards to disengage the guide rail front mounting lugs from the floor.
7 Pull out the locking bar and disconnect the wiring connector from the socket on the floor. The seat can now be lifted out of the vehicle. Note that the seat is very heavy and the help of an assistant may be required.

Rear seat removal

Cushion

8 Pull on the lower release strap to raise the front of the cushion and fold the cushion forward to gain access to the mounting nuts.
9 Undo the cushion frame mounting nuts, lift the frame off the studs and remove it from the car **(see illustration)**.

Backrest – right-hand side

10 Remove the rear seat cushion as described previously in this Section.
11 If the backrest side padding incorporates an airbag, disconnect the battery negative terminal (see Chapter 5A, Section 4). Wait 2 minutes for the capacitors to discharge before continuing.
12 Pull the upper pad area of the backrest side padding forward then use a suitable flat-bladed tool to release the retainer from the side padding's retaining bar. Pull the side padding up to release it from the seat, disconnect the wiring connector and remove the side padding from the car.
13 Fold the backrest forward.

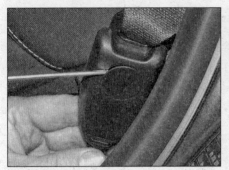

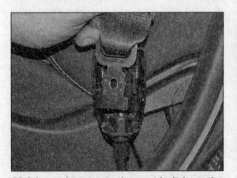

22.3d ...and separate the seat belt from the tensioner

22.5a Remove the trim cover over the outer guide rail retaining bolt...

22.5b ...then undo the guide rail outer retaining bolt...

22.5c ...and inner retaining bolt

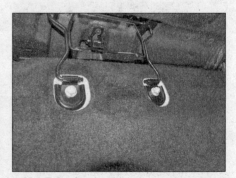

22.9 Rear seat cushion frame mounting nuts

22.14 Undo the retaining bolt and remove the backrest hinge stop

22.15 Lift up the outer side of the backrest to disengage the outer hinge, then pull the backrest off the hinge pin

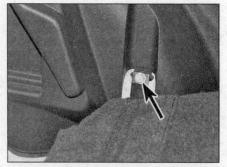

22.21 Undo the bolt and remove the backrest hinge stop

14 Undo the retaining bolt and remove the backrest hinge stop **(see illustration)**.
15 Lift up the outer side of the backrest to disengage the outer hinge, then pull the backrest off the hinge pin **(see illustration)**. Remove the backrest from the vehicle.

Backrest – left-hand side

16 Remove the backrest on the right-hand side as described previously.
17 If the backrest side padding incorporates an airbag, disconnect the battery negative terminal (see Chapter 5A, Section 4). Wait 2 minutes for the capacitors to discharge before continuing.
18 Pull the upper pad area of the backrest side padding forward then use a suitable flat-bladed tool to release the retainer from the side padding's retaining bar. Pull the side padding up to release it from the seat, disconnect the wiring connector and remove the side padding from the car.
19 Unscrew the retaining nut and remove the centre seat belt lower anchorage.
20 Fold the backrest forward.
21 Undo the retaining bolt and remove the backrest hinge stop **(see illustration)**.
22 Remove the spacer washer from the backrest hinge pin **(see illustration)**.
23 Lift up the outer side of the backrest to disengage the outer hinge, then disengage the hinge pin from the mounting bracket **(see illustration)**. Remove the backrest from the vehicle.

Front and rear seat refitting

24 Refitting is a reverse of the removal procedure, tightening the retaining bolts and nuts (where applicable) to the specified torque.

23 Seat belt tensioning mechanism – general information

1 All models covered in this manual are fitted with a front seat belt pyrotechnic tensioner system. The system is designed to instantaneously take up any slack in the seat belt in the case of a sudden frontal impact, therefore reducing the possibility of injury to the front seat occupants. Each front seat is

22.22 Remove the spacer washer from the backrest hinge pin

fitted with its own system, the components of which are mounted on the seat frame.
2 The seat belt tensioner is triggered by a frontal impact causing a deceleration of six times the force of gravity or greater. Lesser impacts, including impacts from behind, will not trigger the system.
3 When the system is triggered, the material within the tensioner canister is ignited, producing a rapid generation of gas. The gas produced from this reaction deploys the seat belt pretensioners which removes all of the slack in the seat belts.
4 There is a risk of injury if the system is triggered inadvertently when working on the vehicle, and it is therefore strongly recommended that any work involving the seat belt tensioner system is entrusted to a Vauxhall/Opel dealer. Refer to the warning given at the beginning of Section 22 before contemplating any work on the front seats.

24.2 Undo the front seat belt upper mounting bolt

22.23 Disengage the backrest hinge pin from the mounting bracket

24 Seat belt components – removal and refitting

⚠️ *Warning: The front seats are fitted with pyrotechnic seat belt tensioners which are triggered by the airbag control system. Before removing a seat belt, disconnect the battery and wait at least 2 minutes to allow the system capacitors to discharge.*

Front belt and reel

Removal

1 Remove the B-pillar inner trim panel as described in Section 25.
2 Undo the seat belt upper mounting bolt **(see illustration)**.
3 Disconnect the wiring connector from the seat belt reel **(see illustration)**.

24.3 Seat belt reel wiring connector and retaining bolt

24.8 Disconnect the wiring connector, then undo the retaining bolt and remove the stalk from the seat

4 Release the seat belt guide from the B-pillar.
5 Undo the bolt securing the seat belt reel to the B-pillar, then remove the reel from the inside of the pillar.

Refitting

6 Refitting is a reversal of removal, but tighten the mounting bolts to the specified torque.

Front belt stalk

Removal

7 Remove the front seat as described in Section 22.
8 Disconnect the wiring connector, then undo the retaining bolt and remove the stalk from the seat **(see illustration)**.

Refitting

9 Refitting is a reversal of removal, but tighten the retaining bolt to the specified torque.

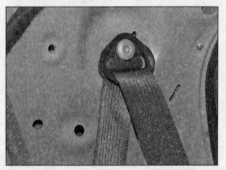

24.18a Rear outer seat belt upper mounting bolt...

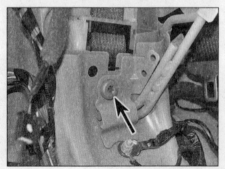

24.19 Rear outer seat belt reel mounting bolt

24.14 Undo the retaining bolt and remove the tensioner from the seat frame

Front belt tensioner

Removal

10 Disconnect the battery negative lead as described in Chapter 5A Section 4. Wait 2 minutes for the capacitors to discharge, before working on the seat electrics.
11 Using a small screwdriver, carefully prise off the trim cap on the seat belt tensioner connector. Undo the retaining bolt, lift off the connector cover and separate the seat belt from the tensioner **(see illustrations 22.3a to 22.3d)**.
12 Undo the tensioner trim cover retaining screw.
13 Remove the tensioner cover by applying light pressure to the front of the cover, then push the cover rearward to remove it from the seat frame.
14 Disconnect the tensioner wiring

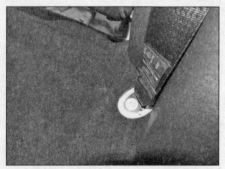

24.18b ...and lower mounting bolt

24.23 Undo the retaining nut and remove the relevant seat belt buckle from the floor

connector, then undo the retaining bolt and remove the tensioner from the seat frame **(see illustration)**.

Refitting

15 Refitting is a reversal of removal, but tighten the tensioner retaining bolt to the specified torque.

Rear outer belt and reel

Removal

16 Remove the rear seat cushion as described in Section 22.
17 Remove the luggage compartment lower side trim panel as described in Section 25.
18 Undo the seat belt upper and lower mounting bolts **(see illustrations)**.
19 Undo the seat belt reel mounting bolt and remove the reel and belt from the car **(see illustration)**.

Refitting

20 Refitting is a reversal of removal, but tighten the mounting bolts to the specified torque.

Rear centre belt and reel

21 The inertia reel for the centre rear seat belt is located internally within the rear seat backrest. To gain access, the backrest must be removed and completely dismantled. This is a complex operation and considerable expertise is needed to remove and refit the seat upholstery and internal components without damage. Therefore, any problems with the centre seat belt and reel should be referred to a Vauxhall/Opel dealer.

Rear belt buckle

Removal

22 Fold the rear seat cushion forward for access to the buckle retaining nut.
23 Undo the retaining nut and remove the relevant buckle from the floor **(see illustration)**.

Refitting

24 Refitting is a reversal of removal, but tighten the mounting nuts to the specified torque.

25 Interior trim – removal and refitting

1 The interior trim panels are secured by a combination of clips and screws. Removal and refitting is generally self-explanatory, noting that it may be necessary to remove or loosen surrounding panels to allow a particular panel to be removed. The following paragraphs describe the general removal and refitting details of the major panels.

A-pillar trim panel

2 Pull the trim panel away from the A-pillar at the top and centre to release the two retaining

clips, then lift it up to disengage the bottom locating tabs.

3 Reach behind the panel and slide the tether clip from its location **(see illustration)**.

4 Disconnect the speaker wiring connector and remove the panel from the car **(see illustration)**.

5 Refitting is a reversal of removal. Note that the manufacturers state that the tether clip must be renewed every time the panel is removed.

Side sill inner trim panel

6 Using a plastic spatula or similar tool, pull the relevant panel upward from the sill to release the retaining clips, then remove the panel from the car **(see illustrations)**.

7 Refitting is a reversal of removal.

B-pillar trim panel

Lower trim panel

8 Remove the front and rear side sill inner trim panels as described previously in this Section.

9 Move the front seat fully forward.

10 Starting at the bottom, pull the panel away from the B-pillar to release the lower, middle and upper retaining clips, then remove the panel from the car **(see illustration)**.

11 Refitting is a reversal of removal.

Upper trim panel

12 Remove the B-pillar lower trim panel as described previously.

13 Using a small screwdriver, carefully prise off the trim cap on the seat belt tensioner connector. Undo the retaining bolt, lift off the connector cover and separate the seat belt from the tensioner **(see illustrations 22.3a to 22.3d)**.

14 Extract the trim cap and undo the trim panel upper retaining screw **(see illustration)**.

15 Pull the panel away from the B-pillar to release the retaining clips **(see illustration)**. Feed the seat belt through the aperture in the panel, then remove the panel from the car.

16 Refitting is a reversal of removal.

Luggage compartment upper side trim panel

17 Undo the rear seat belt lower mounting bolt.

18 Remove the parcel shelf, then extract

the trim cap and undo the trim panel upper retaining screw **(see illustrations)**.

19 Pull the panel away to release the internal

clips **(see illustration)**. Feed the seat belt through the aperture in the panel, then remove the panel from the car.

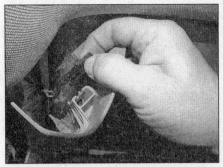

25.3 Reach behind the panel and slide the tether clip from its location

25.4 Disconnect the speaker wiring connector

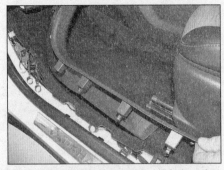

25.6a Removing a front side sill inner trim panel

25.6b Removing a rear side sill inner trim panel

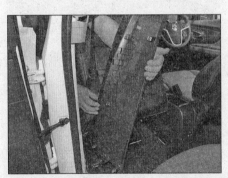

25.10 Pull the panel away from the B-pillar to release the retaining clips

25.14 Extract the trim cap and undo the trim panel upper retaining screw

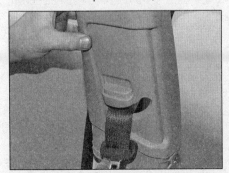

25.15 Pull the panel away from the B-pillar to release the retaining clips

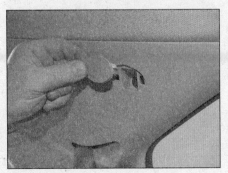

25.18a Extract the trim cap...

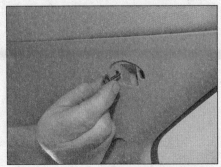

25.18b ...and undo the trim panel upper retaining screw

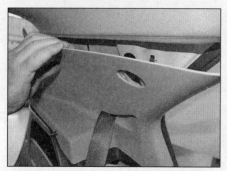

25.19 Pull the panel away to release the internal clips, then feed the seat belt through the aperture in the panel

25.23a Unscrew the retainer…

25.23b …then remove the load floor stowage compartment on the side concerned

25.25 Undo the retaining bolt and remove the cargo tie-down hook

25.26 Withdraw the panel from its location and remove it from the car

20 Refitting is a reversal of removal, tightening the seat belt lower mounting bolt to the specified torque.

Luggage compartment lower side trim panel

21 Remove the rear seat backrest as described in Section 22.
22 Remove the rear side sill inner trim panel as described previously in this Section.
23 Unscrew the retainer, then remove the load floor stowage compartment on the side concerned **(see illustrations)**.
24 Remove the tailgate aperture lower centre trim panel as described later in this Section.
25 Undo the retaining bolt and remove the cargo tie-down hook **(see illustration)**.
26 Withdraw the panel from its location and, if working on the left-hand side, disconnect the luggage compartment light wiring connector **(see illustration)**.
27 Refitting is a reversal of removal.

Tailgate trim panel

28 Open the tailgate and remove the parcel shelf.
29 Using a plastic spatula or similar tool, pull the upper centre trim panel away from the tailgate to release the internal clips and remove the panel from the tailgate **(see illustration)**.
30 Using pliers, pull out the parcel shelf lifting pegs on each side **(see illustration)**. Note that these pegs are a one-time use component and will break on removal. Therefore new pegs will be required for refitting.
31 Pull the side trim panels away from the tailgate to release the internal clips and remove them from the tailgate **(see illustration)**. Note that the lower end of each panel engages with the lower main trim panel.
32 Squeeze and pull the inside handle to release it from the tailgate **(see illustration)**.
33 Starting at the top clip on the right-hand side, pull the main panel away from the tailgate to release the internal clips and remove the panel from the tailgate **(see illustrations)**.
34 Refitting is a reversal of removal.

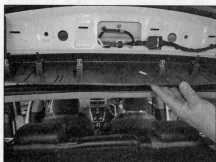

25.29 Pull the upper centre trim panel away from the tailgate to release the internal clips

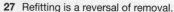

25.30 Pull out the parcel shelf lifting pegs on each side

25.31 Pull the side trim panels away from the tailgate to release the internal clips

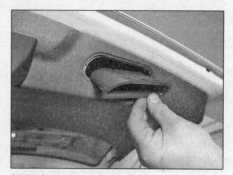

25.32 Squeeze and pull the inside handle to release it from the tailgate

Tailgate aperture lower centre trim panel

35 Open the tailgate.
36 Undo the retaining bolt and remove the cargo tie-down hook on each side (**see illustration**).
37 Pull the panel away to release the internal clips, and remove the panel from the car (**see illustration**).
38 Refitting is a reversal of removal.

26 Centre console – removal and refitting

Removal

1 Disconnect the battery negative lead as described in Chapter 5A Section 4.
2 Remove the facia left-hand and right-hand lower trim panels by carefully pulling them away from the facia and centre console to release the internal retaining clips, then sliding them rearward to disengage the upper locating pegs (**see illustrations**).
3 On manual transmission models, carefully prise up the gear lever trim surround using a plastic spatula or similar tool, then disengage the sides of the trim from the centre console and facia (**see illustrations**).
4 On automatic transmission models, carefully prise up and remove the selector lever trim surround, using a plastic spatula or similar tool.
5 Pull the handbrake lever fully up to provide

clearance for removal of the trim cover. Use a plastic spatula or similar tool on the outside edge of the cover and lift the cover up and out of the centre console (**see illustration**).

6 Release the cup holder upper flange from the centre console, then lift out the cup holder and remove the support base (**see illustrations**).

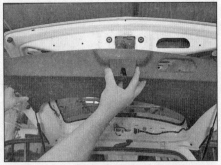

25.33a Pull the main panel away from the tailgate…

25.33b …to release the internal retaining clips

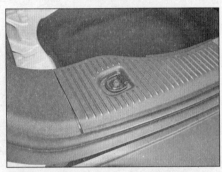

25.36 Undo the retaining bolts and remove the cargo tie-down hooks

25.37 Pull the panel away to release the internal clips, and remove the panel from the car

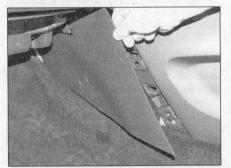

26.2a Carefully pull the lower trim panels away from the facia and centre console…

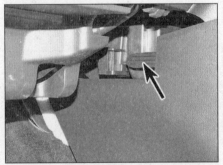

26.2b …then slide them rearward to disengage the upper locating pegs

26.3a On manual transmission models, carefully prise up the gear lever trim surround…

26.3b …then disengage the sides of the trim from the centre console and facia

26.5 Carefully prise free and remove the handbrake lever trim cover

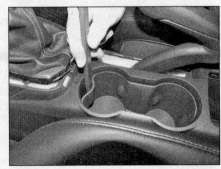

26.6a Release the cup holder upper flange from the console…

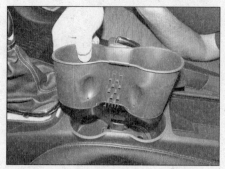

26.6b ...then lift out the cup holder...

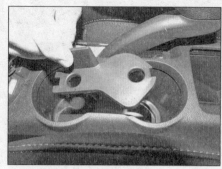

26.6c ...and remove the support base

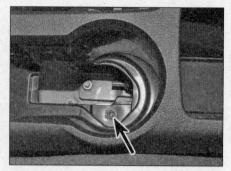

26.7 Undo the centre console retaining bolt in the cup holder aperture

26.8 Undo the bolts securing the front of the centre console to the facia

26.9 Undo the bolt each side securing the centre console sides to the facia crossmember

access to the centre console rear mounting bolts.

11 Prise off the trim caps and undo the rear retaining bolt on each side (see illustrations).

12 Lift the console up and slide it to the rear, then remove the console from the car (see illustration).

Refitting

13 Refitting is a reversal of removal.

27 Facia panel components – removal and refitting

Glovebox

Removal

1 Open the glovebox fully, then detach the damper from the retaining peg by depressing the tabs and sliding the damper off.

2 Press together the sides of the glovebox door and lower the door past the stops (see illustration).

3 Undo the glovebox hinge retaining bolt each side and remove the glovebox from the facia (see illustrations).

Refitting

4 Refitting is a reversal of removal.

Facia right-hand lower trim panel

Removal

5 Undo the two screws on the right-hand side

7 Undo the console retaining bolt now exposed in the cup holder aperture (see illustration).

8 Undo the four bolts securing the front of the centre console to the facia (see illustration).

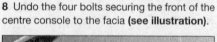

26.11a Prise off the trim caps...

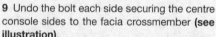

26.11b ...and undo the console rear retaining bolt on each side

9 Undo the bolt each side securing the centre console sides to the facia crossmember (see illustration).

10 Move the front seats fully forward for

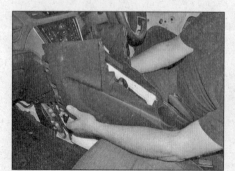

26.12 Lift the console up and slide it to the rear, then remove the console from the car

27.2 Press the sides of the glovebox door and lower it past the stops

27.3a Undo the glovebox hinge retaining bolt each side...

27.3b ...and remove the glovebox from the facia

27.5 Undo the two screws on the right-hand side securing the lower trim panel to the facia

27.7 Remove the right-hand lower trim panel from under the facia

securing the lower trim panel to the facia (**see illustration**).

6 Undo the screw on the left-hand side securing the lower trim panel to the facia.

7 Using a plastic spatula or similar tool, carefully prise free the facia right-hand lower trim panel. Lower the panel, disconnect the diagnostic socket, then remove the panel from the facia (**see illustration**).

Refitting

8 Refitting is a reversal of removal.

Facia centre trim surround

Removal

9 Using a plastic spatula or similar tool, carefully prise free the centre trim surround, starting at the bottom (**see illustration**).

10 Release the retaining clips and remove the trim surround from the facia (**see illustration**).

Refitting

11 Refitting is a reversal of removal.

Facia lower centre panel

Removal

12 Remove the facia centre trim surround as described previously.

13 Undo the four screws securing the lower centre panel to the facia (**see illustration**).

14 Withdraw the lower centre panel from the facia, disconnect the wiring connectors and remove the panel (**see illustrations**).

Refitting

15 Refitting is a reversal of removal.

Facia centre accessory switch panel

Removal

16 On manual transmission models, carefully prise up the gear lever trim surround using a plastic spatula or similar tool, then disengage the sides of the trim from the centre console and facia (**see illustrations 26.3a and 26.3b**).

17 On automatic transmission models, carefully prise up and remove the selector lever trim surround, using a plastic spatula or similar tool.

18 Remove the facia centre trim surround as described previously.

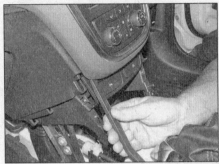

27.9 Carefully prise free the centre trim surround...

27.10 ...then release the retaining clips and remove the trim surround from the facia

19 Undo the two screws securing the accessory switch panel to the facia (**see illustration**).

20 Withdraw the panel from the facia and disconnect the wiring connectors (**see illustrations**).

27.13 Undo the four screws securing the lower centre panel to the facia

27.14a Withdraw the lower centre panel from the facia...

27.14b ...and disconnect the wiring connectors

27.19 Undo the two screws securing the accessory switch panel to the facia

27.20a Withdraw the accessory switch panel from the facia...

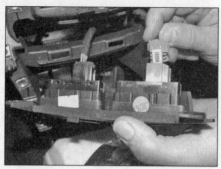

27.20b ...and disconnect the wiring connectors

27.22a Carefully prise up the front of the upper centre panel cover...

27.22b ...to release the front and rear retaining clips

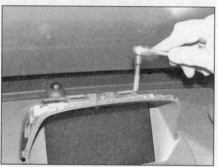

27.23 Undo the two screws securing the front of the upper centre panel to the facia

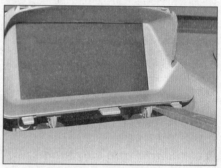

27.24a Carefully prise up the rear of the upper centre panel...

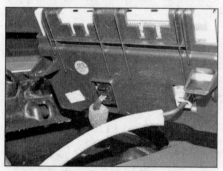

27.24b ...then disconnect the wiring connectors from the information display unit and remove the panel

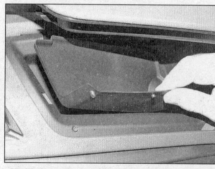

27.26 Open the stowage compartment lid and take out the rubber mat

Refitting

21 Refitting is a reversal of removal.

Facia upper centre panel

Removal

22 Using a plastic spatula or similar tool, carefully prise up the front of the upper centre panel cover to release the retaining clips, then remove the cover **(see illustrations)**.

23 Undo the two screws securing the front of the upper centre panel to the facia **(see illustration)**.

24 Using a plastic spatula or similar tool, carefully prise up the rear of the upper centre panel to release the retaining clips. Disconnect the wiring connectors from the information display unit and remove the panel **(see illustrations)**.

Refitting

25 Refitting is a reversal of removal.

Facia upper stowage compartment

Removal

26 Open the stowage compartment lid and take out the rubber mat **(see illustration)**.

27 Undo the three screws securing the stowage compartment to the facia **(see illustration)**.

28 Using a plastic spatula or similar tool, carefully prise the stowage compartment from its location and remove it from the facia **(see illustration)**.

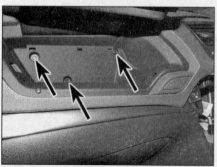

27.27 Undo the three screws securing the stowage compartment to the facia

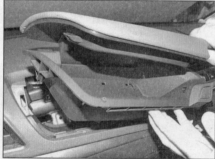

27.28 Carefully prise the stowage compartment free and remove it from the facia

27.31a Unclip the upper shroud from the lower shroud...

27.31b ...disengage the upper shroud from the front flexible section and remove the upper shroud

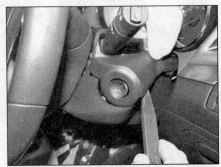

27.32 Unclip the ignition switch lock cylinder bezel and remove it from the lower shroud

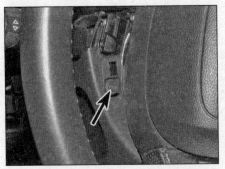

27.33a Unscrew the lower shroud retaining screw on the left-hand side...

27.33b ...and right-hand side

27.34 Undo the retaining screw securing the lower shroud to the underside of the steering column

Refitting

29 Refitting is a reversal of removal.

Steering column shrouds

Removal

30 Fully extend the steering column using the telescopic function.

31 Unclip the upper shroud from the lower shroud, disengage the upper shroud from the front flexible section and remove the upper shroud **(see illustrations)**.

32 Unclip the ignition switch lock cylinder bezel and remove it from the lower shroud **(see illustration)**.

33 Turn the steering wheel as necessary for access and unscrew the lower shroud retaining screw on each side **(see illustrations)**.

34 Undo the retaining screw securing the lower shroud to the underside of the steering column **(see illustration)**.

35 Disengage the lower shroud from the peg on the steering column lock assembly, then remove the lower shroud from the steering column **(see illustration)**.

Refitting

36 Refitting is a reversal of removal.

Complete facia assembly

Note: *This is an involved operation entailing the removal of numerous components and assemblies, and the disconnection of a multitude of wiring connectors. Make notes on the location of all disconnected wiring, or attach labels to the connectors, to avoid confusion when refitting. Taking a series of photographs throughout the removal procedure will prove invaluable when refitting, particularly as an aid to the routing and location of the various wiring looms.*

Removal

37 Disconnect the battery negative lead as described in Chapter 5A Section 4.

38 Remove the A-pillar trim panels on both sides as described in Section 25.

39 Remove the centre console as described in Section 26.

40 Remove the following facia panel components as described previously in this Section:

a) Glovebox.
b) Facia right-hand lower trim panel.
c) Facia centre trim surround.
d) Facia lower centre panel.
e) Facia centre accessory switch panel.
f) Facia upper centre panel.
g) Facia upper stowage compartment.
h) Steering column shrouds.

41 Undo the two bolts and withdraw the lower stowage compartment from the facia. Disconnect the wiring connectors and remove the stowage compartment **(see illustrations)**.

42 Using a plastic spatula or similar tool carefully prise free the facia outer trim panels on each side **(see illustration)**. Disconnect the wiring connector when removing the left-hand panel.

27.35 Disengage the shroud from the peg on the steering column then remove the lower shroud

27.41a Undo the two bolts...

27.41b ...withdraw the lower stowage compartment from the facia...

27.41c ...then disconnect the wiring connectors and remove the compartment

27.42 Carefully prise free the facia outer trim panels on each side

27.48 Lift out the control unit located in the centre of the facia and disconnect the wiring connectors

27.50 Undo the passenger's airbag module bracket retaining bolts and disconnect the airbag wiring connector

47 Remove the exterior light switch as described in Chapter 12, Section 4.

48 Lift out the control unit located in the centre of the facia and disconnect the wiring connectors **(see illustration)**.

49 Similarly, remove the unit located in the central air vent aperture.

50 Undo the passenger's airbag module bracket retaining bolts and disconnect the airbag wiring connector **(see illustration)**.

51 Undo the glovebox lid lock striker retaining bolt, then remove the glovebox light and disconnect the wiring connector **(see illustrations)**.

52 Remove the sunlight sensor from the top of the facia and disconnect the wiring connector **(see illustration)**.

53 Undo the following fasteners securing the facia to the crossmember **(see illustrations)** :
a) One bolt at the left-hand and right-hand end.
b) Two bolts in the driver's footwell area.

43 Remove the radio/CD player as described in Chapter 12, Section 16.

44 Remove the facia centre air vent panel as described in Chapter 3, Section 11.

45 Remove the steering column as described in Chapter 10, Section 15.

46 Remove the instrument panel as described in Chapter 12, Section 9.

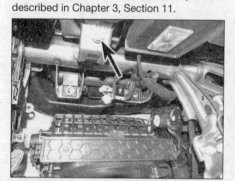

27.51a Undo the glovebox lid lock striker retaining bolt...

27.51b ...then remove the glovebox light and disconnect the wiring connector

27.52 Remove the sunlight sensor and disconnect the wiring connector

27.53a Undo the facia retaining bolt at each end...

27.53b ...the two retaining bolts in the driver's footwell area...

27.53c ...the retaining bolt in the instrument panel aperture...

c) *One bolt in the instrument panel aperture.*
d) *Two bolts in the upper centre panel aperture.*
e) *Two bolts in the lower centre panel aperture.*
f) *Three bolts in the glovebox aperture.*

54 With the help of an assistant, carefully lift the facia from its location **(see illustration)**. Check that all wiring has been disconnected, then remove the facia from the car.

55 Check that all wiring has been disconnected, then remove the facia from the car.

Refitting

56 Refitting is a reversal of removal ensuring that all wiring is correctly reconnected and all mountings securely tightened.

28 Facia crossmember – removal and refitting

Removal

1 Remove the facia assembly as described in Section 27.
2 Undo the three retaining screws (where fitted) and remove the facia lower panel on the passenger's side **(see illustration)**.
3 Remove the windscreen cowl panel as described in Section 21.
4 Unclip the instrument panel fuse block from the facia crossmember and position it to one side.
5 Working in the engine compartment, undo the facia crossmember retaining nut located beneath the windscreen wiper linkage location **(see illustration)**.
6 Undo the two upper bolts securing the brake pedal mounting bracket to the facia crossmember **(see illustration)**.
7 Undo the lower bolt securing the brake pedal mounting bracket to the facia crossmember.
8 Working around the facia crossmember, centre console area, air distribution housing and body, disconnect all the relevant wiring connectors and earth leads to enable the wiring harnesses to be removed with the facia crossmember.
9 Undo the three bolts each side and remove

27.53d ...the retaining bolts at each side of the upper centre panel aperture...

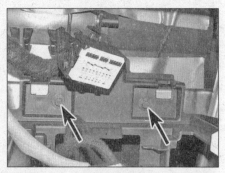

27.53e ...the two bolts in the lower centre panel aperture...

27.53f ...the three retaining bolts in the glovebox aperture

27.54 With the help of an assistant, carefully lift the facia from its location

the facia crossmember centre braces **(see illustrations)**.
10 Undo the bolt securing the heater blower

28.2 Undo the three retaining screws and remove the facia lower panel on the passenger's side

motor housing to the facia crossmember bracket **(see illustration)**.
11 Undo the two upper bolts and the lower

28.5 Undo the facia crossmember retaining nut located beneath the windscreen wiper linkage location

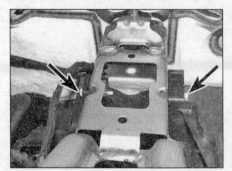

28.6 Undo the two upper bolts securing the brake pedal mounting bracket to the facia crossmember

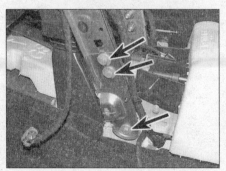

28.9a Undo the three bolts each side...

28.9b ...and remove the facia crossmember centre braces

28.10 Undo the bolt securing the heater blower motor housing to the facia crossmember bracket

28.11a Undo the two upper bolts…

28.11b …and the lower bolt securing the air distribution housing to the facia crossmember

28.12 Undo the two bolts each side securing the facia crossmember to the A-pillars

28.13 With the help of an assistant, carefully withdraw the facia crossmember and remove it from the car

bolt securing the air distribution housing to the facia crossmember **(see illustrations)**.

12 Undo the two bolts each side securing the facia crossmember to the A-pillars **(see illustration)**.

13 With the help of an assistant, carefully withdraw the facia crossmember from the bulkhead and air distribution housing. Check that all wiring has been disconnected, then remove the facia crossmember from the car **(see illustration)**.

Refitting

14 Refitting is a reversal of removal ensuring that all wiring is correctly reconnected and all mountings securely tightened.

Chapter 12
Body electrical systems

Contents

Degrees of difficulty

Easy, suitable for novice with little experience	**Fairly easy,** suitable for beginner with some experience	**Fairly difficult,** suitable for competent DIY mechanic	**Difficult,** suitable for experienced DIY mechanic	**Very difficult,** suitable for expert DIY or professional

Specifications

General
System type .... 12 volt, negative earth

Fuses
Refer to the labels on the fuse/relay box covers and to the wiring diagrams at the end of this Chapter

Bulbs

Note: *The following is a list of the principle bulbs used on Astra models, based on information available at the time of writing. For bulbs not included, consult a Vauxhall/Opel dealer for latest specifications.*

Bulbs	Type	Wattage
Courtesy light...................................	W5WLL	5
Front direction indicator light	WY21W	21
Front direction indicator side repeater light	WY5W	5
Front foglight	H11	42
Front sidelight/daytime running light	-	21/5
Glovebox light	W5WLL	5
Headlight cornering light (xenon type headlight)	H11	55
Headlight dipped beam (halogen type headlight)	H7	55
Headlight dipped/main beam (xenon type headlight)	D3S	35
Headlight main beam (halogen type headlight)	HB3	55
Number plate light	W5W	5
Rear light cluster:		
Rear direction indicator light	W16W	16
Rear foglight..	H21W	21
Reversing light	H21W	21
Stop light	W16W	16
Tail light...	W5W	5

1 General information and precautions

Warning: Before carrying out any work on the electrical system, read through the precautions given in 'Safety first!', and in Chapter 5A, Section 1.

1 The electrical system is of the 12 volt negative earth type. Power for the lights and all electrical accessories is supplied by a lead-acid type battery, which is charged by the engine-driven alternator.

2 This Chapter covers repair and service procedures for the various electrical components not associated with the engine. Information on the battery, alternator and starter motor can be found in Chapter 5A.

3 It should be noted that, prior to working on any component in the electrical system, the battery negative terminal should first be disconnected, to prevent the possibility of electrical short-circuits and/or fires, as described in Chapter 5A Section 4.

2 Electrical fault finding – general information

Note: *Refer to the precautions given in 'Safety first!' and in Section 1 before starting work. The following tests relate to testing of the main electrical circuits, and should not be used to test delicate electronic circuits (such as the anti-lock braking system or fuel injection system), particularly where an electronic control unit is used.*

General

1 A typical electrical circuit consists of an electrical component, any switches, relays, motors, fuses, fusible links or circuit breakers related to that component, and the wiring and connectors which link the component to both the battery and the vehicle body. To help to pinpoint a problem in an electrical circuit, wiring diagrams are shown at the end of this Chapter.

2 Before attempting to diagnose an electrical fault, first study the appropriate wiring diagram to obtain a complete understanding of the components included in the particular circuit concerned. The possible sources of a fault can be narrowed down by noting if other components related to the circuit are operating properly. If several components or circuits fail at one time, the problem is likely to be related to a shared fuse or earth connection.

3 Electrical problems usually stem from simple causes, such as loose or corroded connections, a faulty earth connection, a blown fuse, a melted fusible link, or a faulty relay (refer to Section 3 for details of testing relays). Inspect the condition of all fuses, wires and connections in a problem circuit before testing the components. Use the wiring diagrams to determine which terminal connections will need to be checked in order to pinpoint the trouble-spot.

4 The basic tools required for electrical fault finding include a circuit tester or voltmeter (a 12 volt bulb with a set of test leads can also be used for certain tests); a self-powered test light (sometimes known as a continuity tester); an ohmmeter (to measure resistance); a battery and set of test leads; and a jumper wire, preferably with a circuit breaker or fuse incorporated, which can be used to bypass suspect wires or electrical components. Before attempting to locate a problem with test instruments, use the wiring diagram to determine where to make the connections.

5 To find the source of an intermittent wiring fault (usually due to a poor or dirty connection, or damaged wiring insulation), a 'wiggle' test can be performed on the wiring. This involves wiggling the wiring by hand to see if the fault occurs as the wiring is moved. It should be possible to narrow down the source of the fault to a particular section of wiring. This method of testing can be used in conjunction with any of the tests described in the following sub-Sections.

6 Apart from problems due to poor connections, two basic types of fault can occur in an electrical circuit – open-circuit, or short-circuit.

7 Open-circuit faults are caused by a break somewhere in the circuit, which prevents current from flowing. An open-circuit fault will prevent a component from working, but will not cause the relevant circuit fuse to blow.

8 Short-circuit faults are caused by a 'short' somewhere in the circuit, which allows the current flowing in the circuit to 'escape' along an alternative route, usually to earth. Short-circuit faults are normally caused by a breakdown in wiring insulation, which allows a feed wire to touch either another wire, or an earthed component such as the bodyshell. A short-circuit fault will normally cause the relevant circuit fuse to blow.

Finding an open-circuit

9 To check for an open-circuit, connect one lead of a circuit tester or voltmeter to either the negative battery terminal or a known good earth.

10 Connect the other lead to a connector in the circuit being tested, preferably nearest to the battery or fuse.

11 Switch on the circuit, bearing in mind that some circuits are live only when the ignition switch is turned to a particular position.

12 If voltage is present (indicated either by the tester bulb lighting or a voltmeter reading, as applicable), this means that the section of the circuit between the relevant connector and the battery is problem-free.

13 Continue to check the remainder of the circuit in the same fashion.

14 When a point is reached at which no voltage is present, the problem must lie between that point and the previous test point with voltage. Most problems can be traced to a broken, corroded or loose connection.

Finding a short-circuit

15 To check for a short-circuit, first disconnect the load(s) from the circuit (loads are the components which draw current from a circuit, such as bulbs, motors, heating elements, etc).

16 Remove the relevant fuse from the circuit, and connect a circuit tester or voltmeter to the fuse connections.

17 Switch on the circuit, bearing in mind that some circuits are live only when the ignition switch is turned to a particular position.

18 If voltage is present (indicated either by the tester bulb lighting or a voltmeter reading, as applicable), this means that there is a short-circuit.

19 If no voltage is present, but the fuse still blows with the load(s) connected, this indicates an internal fault in the load(s).

Finding an earth fault

20 The battery negative terminal is connected to 'earth' – the metal of the engine/transmission unit and the car body – and most systems are wired so that they only receive a positive feed, the current returning via the metal of the car body. This means that the component mounting and the body form part of that circuit. Loose or corroded mountings can therefore cause a range of electrical faults, ranging from total failure of a circuit, to a puzzling partial fault. In particular, lights may shine dimly (especially when another circuit sharing the same earth point is in operation), motors (eg, wiper motors or the radiator cooling fan motor) may run slowly, and the operation of one circuit may have an apparently-unrelated effect on another. Note that on many vehicles, earth straps are used between certain components, such as the engine/transmission and the body, usually where there is no metal-to-metal contact between components, due to flexible rubber mountings, etc.

21 To check whether a component is properly earthed, disconnect the battery, and connect one lead of an ohmmeter to a known good earth point. Connect the other lead to the wire or earth connection being tested. The resistance reading should be zero; if not, check the connection as follows.

22 If an earth connection is thought to be faulty, dismantle the connection, and clean back to bare metal both the bodyshell and the wire terminal or the component earth connection mating surface. Be careful to remove all traces of dirt and corrosion, then use a knife to trim away any paint, so that a clean metal-to-metal joint is made. On reassembly, tighten the joint fasteners securely; if a wire terminal is being refitted, use serrated washers between the terminal and the bodyshell, to ensure a clean and

3.2 Lift off the fuse/relay box cover to access the engine compartment fuses

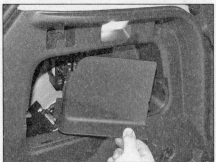

3.3a Press the sides of the glovebox door and lower it past the stops...

3.3b ...to gain access to the fuses behind

secure connection. When the connection is remade, prevent the onset of corrosion in the future by applying a coat of petroleum jelly or silicone-based grease. Alternatively, at regular intervals, spray on a proprietary ignition sealer or a water-dispersant lubricant.

3 Fuses and relays – general information

Fuses

1 The main fuses are located in the fuse/relay box on the left-hand side of the engine compartment, with additional fuses located behind the glovebox, and in the left-hand side of the luggage compartment.

2 To gain access to the engine compartment fuses, depress the two tabs and lift off the fuse/relay box cover **(see illustration)**.

3 To gain access to the fuses behind the glovebox, open the glovebox fully, then detach the damper from the retaining peg by depressing the tabs and sliding the damper off. Press together the sides of the glovebox door and lower the door past the stops **(see illustrations)**.

4 Access to the fuses in the luggage compartment can be gained by opening the stowage compartment cover **(see illustration)**.

5 To remove a fuse, first switch off the circuit concerned (or the ignition), then pull the fuse out of its terminals using the plastic removal tool provided **(see illustration)**. The wire within the fuse is clearly visible; if the fuse is blown, it will be broken or melted.

6 Always renew a fuse with one of an identical rating; never use a fuse with a different rating from the original, nor substitute anything else. Never renew a fuse more than once without tracing the source of the trouble. The fuse rating is stamped on top of the fuse; note that the fuses are also colour-coded for easy recognition.

7 If a new fuse blows immediately, find the cause before renewing it again; a short to earth as a result of faulty insulation is most likely. Where a fuse protects more than one circuit, try to isolate the defect by switching on

3.4 Access to the fuses in the luggage compartment can be gained by opening the stowage compartment cover

each circuit in turn (if possible) until the fuse blows again. Always carry a supply of spare fuses of each relevant rating on the vehicle, a spare of each rating should be clipped into the base of the fuse/relay box.

Relays

8 Most of the relays are located in the fuse/relay box in the engine compartment **(see illustration 3.2)**.

9 If a circuit or system controlled by a relay develops a fault and the relay is suspect, operate the system; if the relay is functioning, it should be possible to hear it click as it is energised. If this is the case, the fault lies with the components or wiring of the system. If the relay is not being energised, then either the relay is not receiving a main supply or a switching voltage, or the relay itself is faulty. Testing is by the substitution of a known

4.5a Depress the tabs on the top and bottom of the switch...

good unit, but be careful; while some relays are identical in appearance and in operation, others look similar but perform different functions.

10 To renew a relay, first ensure that the ignition switch is off. The relay can then simply be pulled out from the socket and the new relay pressed in.

4 Switches – removal and refitting

Note: *Disconnect the battery negative terminal (refer to battery disconnection and reconnection in Chapter 5A, Section 4) before removing any switch, and reconnect the terminal after refitting.*

Ignition switch/ steering column lock

1 Refer to Chapter 10, Section 14.

Steering column switches

2 Fully extend the steering column using the telescopic function.

3 Remove the steering column shrouds as described in Chapter 11, Section 27.

4 Disconnect the wiring connector from the rear of the relevant switch.

5 Depress the tabs on the top and bottom of the switch, then slide the switch from the housing **(see illustrations)**.

6 Refitting is a reversal of removal.

4.5b ...then slide the switch from the housing

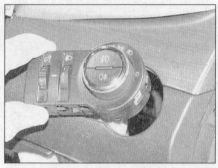

4.8a Reach up behind the light switch and push it out of the facia...

4.8b ...then disconnect the wiring connector and remova the switch

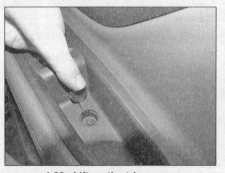

4.11 Carefully lift the upper and lower tabs securing the switch block to the vent panel, while pushing the switch block out

4.12 Once the tabs are released, remove the switch block from the vent panel

Exterior light switch

7 Remove the facia right-hand lower trim panel as described in Chapter 11, Section 27.
8 Reach up behind the switch and push it out of the facia. Disconnect the wiring connector and remove the switch **(see illustrations)**.
9 Refitting is a reversal of removal.

Facia centre vent panel switches

10 Remove the facia centre air vent panel as described in Chapter 3, Section 11.
11 Using a screwdriver, carefully lift the upper and lower tabs securing the switch block to the vent panel, while at the same time, pushing the switch block out **(see illustration)**.
12 Once the tabs are released, remove the switch block from the vent panel **(see illustration)**.
13 Refitting is a reversal of removal.

Facia lower centre panel switches

14 Remove the facia lower centre panel as described in Chapter 11, Section 27.
15 The accessory switches and audio control switches are part of a sealed unit and cannot be separately removed.
16 The heater and air conditioning system controls and switches are an integral part of the heater/air conditioning control unit. Removal and refitting procedures are contained in Chapter 3, Section 11.

Heater/air conditioning switches

17 The heater and air conditioning system controls and switches are an integral part of the heater/air conditioning control unit. Removal and refitting procedures are contained in Chapter 3, Section 11.

Handbrake warning light switch

18 Refer to Chapter 9, Section 16.

Stop-light switch

19 Refer to Chapter 9, Section 15.

Electric mirror switch

20 Using a small screwdriver, lift up the trim cap, then undo the retaining screw in the door grab handle aperture **(see illustrations)**.
21 Using a plastic spatula or similar tool, carefully prise up the switch panel/grab handle assembly and withdraw it from the door panel.
22 Disconnect the wiring connectors and remove the assembly.
23 Using a small screwdriver, carefully lift the three retaining tabs and remove the mirror switch from the switch panel.
24 Refitting is a reversal of removal.

Electric window switches

25 Remove the switch panel/grab handle assembly from the door triom panel as described previously in paragraphs 20 to 22.
26 Release the tabs on the switch block and remove the switch block from the switch panel/grab handle assembly.
27 Refitting is a reversal of removal.

Steering wheel switches

28 Remove the driver's airbag as described in Section 22.
29 Disconnect the two wiring connectors for the steering wheel switches **(see illustration)**.

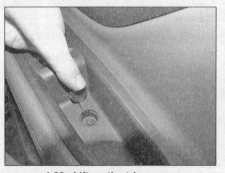

4.20a Lift up the trim cap...

4.20b ...then undo the screw in the door grab handle aperture

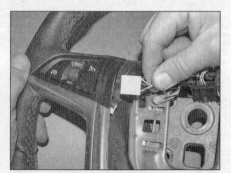

4.29 Disconnect the wiring connectors for the steering wheel switches

30 Using a plastic spatula or similar tool, carefully prise free and remove the plastic trim surround from the steering wheel **(see illustration)**.

31 Undo the two retaining screws and remove the relevant switch from the trim surround **(see illustration)**.

32 Refitting is a reversal of removal.

5 Bulbs (exterior lights) – renewal

General

1 Whenever a bulb is renewed, note the following points:

a) *Make sure the switch is in the OFF position, for the respective bulb you are working on.*

b) *Remember that if the light has just been in use, the bulb may be extremely hot.*

c) *Always check the bulb contacts and holder, ensuring that there is clean metal-to-metal contact between the bulb and its live(s) and earth. Clean off any corrosion or dirt before fitting a new bulb.*

d) *Wherever bayonet-type bulbs are fitted, ensure that the live contact(s) bear firmly against the bulb contact.*

e) *Always ensure that the new bulb is of the correct rating, and that it is completely clean before fitting it; this applies particularly to headlight/foglight bulbs.*

f) *Access to the rear of the headlight unit for bulb renewal is limited. Depending on the bulb being renewed, access can be improved by removing the windscreen washer reservoir filler tube (left-hand light unit), or removing the air cleaner assembly as described in Chapter 4A, Section 2 or Chapter 4B, Section 3 (right-hand light unit). Alternatively (and preferably) remove the relevant headlight unit from the car as described in Section 7).*

Halogen type headlight unit

Dipped beam

2 Rotate the plastic cover anti-clockwise and

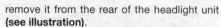

4.30 Carefully prise free and remove the plastic trim surround from the steering wheel

remove it from the rear of the headlight unit **(see illustration)**.

3 Disconnect the wiring connector from the rear of the bulb **(see illustration)**.

4 Push in on the right side of the headlamp bulb spring retainer and push the retainer up and away from the bulb **(see illustration)**.

5 Remove the bulb from the headlight unit **(see illustration)**.

6 When handling the new bulb, use a tissue or clean cloth to avoid touching the glass with the fingers; moisture and grease from the skin can cause blackening and rapid failure of this type of bulb. If the glass is accidentally touched, wipe it clean using methylated spirit.

7 Place the new bulb in the light unit,

4.31 Undo the two retaining screws and remove the relevant switch from the trim surround

engaging the tab on the bulb base with the cutout in the bulbholder.

8 Bring the spring retainer down and engage the retainer leg with the hook on the bulbholder.

9 Reconnect the wiring connector, then refit the plastic cover to the rear of the headlight unit.

Main beam

10 Rotate the plastic cover anti-clockwise and remove it from the rear of the headlight unit **(see illustration)**.

11 Rotate the bulb anti-clockwise and withdraw it from the light unit **(see illustration)**.

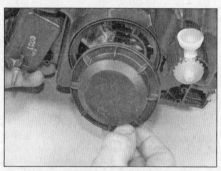

5.2 Rotate the plastic cover anti-clockwise and remove it from the rear of the headlight unit

5.3 Disconnect the wiring connector from the rear of the bulb

5.4 Push the spring retainer up and away from the bulb

5.5 Remove the bulb from the headlight unit

5.10 Rotate the plastic cover anti-clockwise and remove it from the rear of the headlight unit

5.11 Rotate the bulb anti-clockwise and withdraw it from the light unit

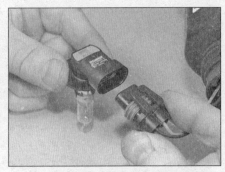

5.12 Hold the bulb by its base and disconnect the wiring connector

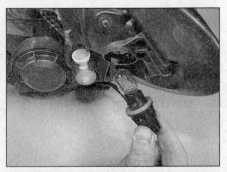

5.15 Rotate the bulbholder anti-clockwise and remove it from the rear of the headlight

5.16 Remove the capless (push fit) bulb from the bulbholder

5.18 Rotate the bulbholder anti-clockwise and remove it from the rear of the headlight unit

5.19 Remove the capless (push fit) bulb from the bulbholder

12 Hold the bulb by its base and disconnect the wiring connector **(see illustration)**. When handling the new bulb, use a tissue or clean cloth to avoid touching the glass with the fingers; moisture and grease from the skin can cause blackening and rapid failure of this type of bulb. If the glass is accidentally touched, wipe it clean using methylated spirit.

13 Connect the wiring connector to the new bulb, then fit the bulb to the light unit. Turn the bulb clockwise to secure.

14 Refit the plastic cover to the rear of the headlight unit.

Front sidelight/daytime running light

15 Rotate the bulbholder anti-clockwise and remove it from the rear of the headlight unit **(see illustration)**.

16 Remove the capless (push fit) bulb from the bulbholder **(see illustration)**.

17 Fit the new bulb to the bulbholder, then refit the bulbholder to the light unit.

Front direction indicator

18 Rotate the bulbholder anti-clockwise and remove it from the rear of the headlight unit **(see illustration)**.

19 Remove the capless (push fit) bulb from the bulbholder **(see illustration)**.

20 Fit the new bulb to the bulbholder, then refit the bulbholder to the light unit.

Xenon type headlight unit

⚠️ *Warning: Xenon headlights operate at very high voltage. Do not touch the associated wiring when the headlights are switched on.*

Dipped/main beam

21 Remove the headlight unit (see Section 7).

22 Undo the three screws and remove the access cover at the rear of the headlight unit.

23 Undo the two screws and remove the headlight bulb cover.

24 Disconnect the wiring connector and lift out the xenon starter unit and bulb assembly from the headlight unit. Note that the starter unit and bulb are a single unit and cannot be separated.

25 Refitting is the reverse of the removal procedure.

Cornering light

26 Renewal of the cornering light bulb is the same as renewal of the halogen type headlight main beam bulb described previously in this Section.

Front direction indicator

27 Renewal of the direction indicator bulb is the same as renewal of the halogen type headlight direction indiactor bulb described previously in this Section.

Front foglight

Note: *The front foglight bulbs are accessible from under the front bumper.*

28 Remove the wheel arch liner on the side concerned as described in Chapter 11, Section 21.

29 Twist the foglight bulbholder anti-clockwise, and remove it from the rear of the light unit **(see illustration)**.

30 Disconnect the wiring connector from the bulbholder **(see illustration)**.

31 When handling the new bulb, use a tissue or clean cloth to avoid touching the glass with

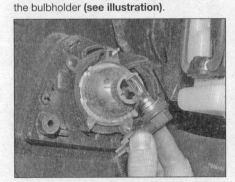

5.29 Twist the foglight bulbholder anti-clockwise, and remove it from the rear of the light unit

5.30 Disconnect the wiring connector from the bulbholder

5.34 Pull the capless bulb out of the bulbholder

5.36a If working on the left-hand side, lift up the trim caps...

5.36b ...undo the two screws...

5.36c ...and remove the stowage pocket door from the trim panel

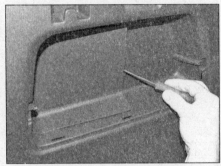

5.36d Release the edge of the access cover...

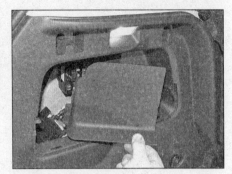

5.36e ...then remove the cover from the trim panel

the fingers; moisture and grease from the skin can cause blackening and rapid failure of this type of bulb. If the glass is accidentally touched, wipe it clean using methylated spirit.

32 Refitting is a reverse of the removal procedure.

Indicator side repeater

33 Push the light unit forward, then using a plastic spatula or similar tool, carefully release the rear edge of the light unit and withdraw the unit from the front wing. Take great care not damage the painted finish of the wing.

34 Turn the bulbholder anti-clockwise and remove it from the light unit. The bulb is of the capless (push-fit) type, and can be removed by simply pulling it out of the bulbholder **(see illustration)**.

35 Refitting is a reverse of the removal procedure.

Rear light cluster

36 If working on the left-hand side, lift up the trim caps, undo the two screws and remove the stowage pocket door from the trim panel. Using a small screwdriver, release the edge of the access cover, then remove the cover from the trim panel **(see illustrations)**.

37 If working on the right-hand side, remove the access cover from the side trim panel and, where applicable, take out the puncture repair kit **(see illustration)**.

38 Reach in through the opening in the trim panel and remove the relevant bulbholder from the light cluster. All the bulbholders have a bayonet fitting and are removed by turning clockwise **(see illustration)**.

39 Remove the capless (push-fit) bulb from the bulbholder **(see illustration)**.

40 Refitting is a reverse of the removal procedure.

Rear foglight

41 Remove the wheel arch liner on the side concerned as described in Chapter 11, Section 21.

42 Twist the foglight bulbholder anti-clockwise, and remove it from the the light unit.

43 The bulb is of the capless (push-fit) type, and can be removed by simply pulling it out of the bulbholder **(see illustration)**.

44 Refitting is a reverse of the removal procedure.

Number plate light

45 Using a plastic spatula or similar tool, carefully prise the light unit out from its location **(see illustration)**.

5.37 If working on the right-hand side, remove the access cover from the side trim panel

5.38 Turn the relevant bulbholder anti-clockwise and remove it from the light cluster

5.39 Remove the capless (push-fit) bulb from the bulbholder

5.43 Remove the capless bulb from the bulbholder

5.45 Carefully prise the light unit out from its location

6 Bulbs (interior lights) – renewal

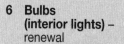

5.46a Twist the bulbholder to remove it from the light unit...

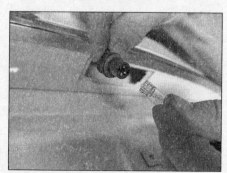

5.46b ...and remove the push-fit bulb

46 Twist the bulbholder to remove it from the light unit and remove the capless (push fit) bulb from the bulbholder **(see illustrations)**.
47 Refitting is a reverse of the removal procedure.

High-level stop-light

48 The high-level stop-light bulbs are of the LED (light emitting diode) type and cannot be individually renewed. Remove the complete light unit as described in Section 7.

General

1 Refer to Section 5, paragraph 1.

Front courtesy lights

2 Using a plastic spatula or similar tool, carefully prise the light unit lens from the overhead console **(see illustration)**.
3 Pull the relevant bulb from its socket **(see illustration)**.
4 Install the new bulb, ensuring that it is securely held in position by the contacts, and clip the light unit lens back into position.

Rear courtesy lights

5 Using a small screwdriver, carefully prise the light unit from its location **(see illustration)**.
6 Withdraw the festoon bulb from the light unit contacts **(see illustration)**.

Vanity light

7 Fold down the sun visor and carefully prise out the light unit lens **(see illustration)**.
8 Using a small screwdriver, release the bulb from its contacts **(see illustration)**.
9 Install the new bulb, ensuring that it is securely held in position by the contacts, and clip the light unit lens back into position.

Luggage compartment light

10 Using a suitable screwdriver, carefully prise the light unit out of position, and release

6.2 Carefully prise free the light unit lens

6.3 Pull the relevant bulb from its socket

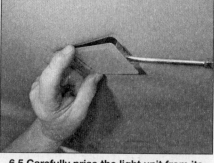

6.5 Carefully prise the light unit from its location

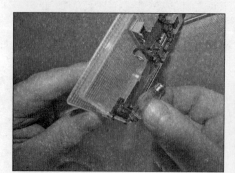

6.6 Withdraw the festoon bulb from the light unit contacts

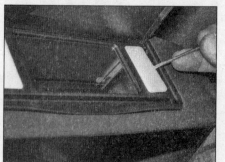

6.7 Fold down the sun visor and carefully prise out the light unit lens

6.8 Using a small screwdriver, release the bulb from its contacts

the bulb from the light unit contacts **(see illustrations)**.

11 Install the new bulb, ensuring that it is securely held in position by the contacts, and clip the light unit back into position.

Glovebox light

12 Open the glovebox lid and carefully prise the light unit out of position **(see illustration)**.
13 Release the bulb from its contacts, install the new bulb, ensuring it is securely held in position by the contacts **(see illustration)**. Push the light unit back into position.

Switch illumination

14 All the switches are fitted with illumination bulbs; some are also fitted with a bulb to show when the circuit concerned is operating. These bulbs are an integral part of the switch assembly, and cannot be obtained separately.

---

7 Exterior light units – removal and refitting

Note: *Disconnect the battery negative terminal as described in Chapter 5A Section 4 before removing any light unit, and reconnect the terminal after refitting.*

Headlight

⚠️ *Warning: Xenon dipped beam headlights operate at very high voltage. Do not touch the associated wiring when the headlights are switched on.*

6.10a Using a suitable screwdriver, carefully prise the light unit out of position...

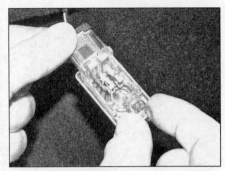
6.10b ...and release the bulb from the light unit contacts

6.12 Open the glovebox lid and carefully prise the light unit out of position

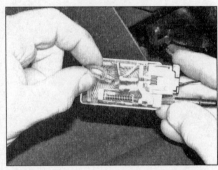
6.13 Release the bulb from its contacts

1 Remove the front bumper as described in Chapter 11, Section 6.
2 Undo the four retaining bolts and withdraw the headlight unit **(see illustrations)**.

3 Disconnect the wiring connector from the rear of the headlight unit, then remove the unit from the car **(see illustrations)**.
4 Refitting is a reverse of the removal

7.2a Undo the two upper bolts...

7.2b ...the inner bolt...

7.2c ...and the outer bolt...

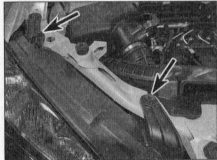
7.2d ...then withdraw the headlight unit

7.3a Disconnect the main wiring connector...

7.3b ...and release the support clip, then remove the headlight unit

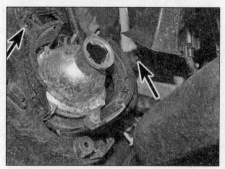

7.7 Undo the two retaining screws and remove the light unit from the front bumper

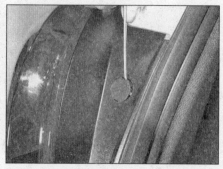

7.10a Remove the trim cap over the light unit outer retaining screw...

7.10b ...then undo the screw

7.13a Undo the nuts from the four light unit studs (shown with light unit removed)...

7.13b ...then remove the light unit from the rear wing

procedure. On completion, check the headlight beam alignment using the information given in Section 8.

Front foglight

5 Remove the wheel arch liner on the side concerned as described in Chapter 11, Section 21.
6 Twist the foglight bulbholder anti-clockwise, and remove it from the rear of the light unit.
7 Undo the two retaining screws and remove the light unit from the front bumper **(see illustration).**
8 Refitting is a reverse of the removal procedure. On completion, check the foglight beam alignment with reference to the information given in Section 8.

Indicator side repeater light

9 The procedure is described as part of the bulb renewal procedure in Section 5.

Rear light cluster

10 Open the tailgate and remove the trim cap over the light unit outer retaining screw, then undo the screw **(see illustrations).**
11 Working in the luggage compartment remove the light unit access covers on the side concerned as described in Section 5, paragraphs 36 and 37
12 Reach in behind the trim panel and disconnect the wiring connector from the light unit.
13 Undo the four light unit retaining nuts and remove the light unit from the rear wing **(see illustrations).**

14 Refitting is a reverse of the removal procedure.

Number plate light

15 The procedure is described as part of the bulb renewal procedure in Section 5.

High-level stop-light

16 Remove the tailgate upper centre trim panel as described in Chapter 11, Section 25.
17 Disconnect the tailgate washer jet hose at the in-line connector and disconnect the adjacent wiring harness connectors.
18 Undo the bolt at each end securing the spoiler to the tailgate **(see illustration).**
19 Working through the apertures in the tailgage, undo the three spoiler retaining nuts **(see illustration).**
20 Using a plastic spatula or similar tool, work around the spoiler and carefully prise it free from the tailgate **(see illustration).** Take great care when doing this so as not to scratch the paintwork. Preferably, protect the paint using masking tape.
21 Undo the three screws and remove the high-level stop-light from the spoiler.
22 Refitting is a reverse of the removal procedure.

Rear foglight

23 Remove the wheel arch liner on the side concerned as described in Chapter 11, Section 21.

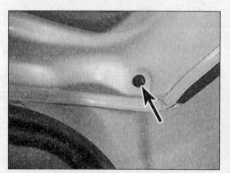

7.18 Undo the bolt at each end securing the spoiler to the tailgate

7.19 Undo the three nuts securing the spoiler to the tailgate

7.20 Use a plastic spatula to release the spoiler retaining clips

24 Disconnect the foglight wiring connector and release the wiring harness clip from the foglight body **(see illustration)**.

25 Undo the three screws and remove the foglight from the bumper **(see illustration)**.

26 Refitting is a reverse of the removal procedure.

8 Headlight beam alignment – general information

1 Accurate adjustment of the headlight beam is only possible using optical beam-setting equipment, and this work should therefore be carried out by a Vauxhall/Opel dealer or suitably-equipped workshop.

2 For reference, the headlights can be adjusted using the adjuster assemblies fitted to the top of each light unit. The outer adjuster, alters the vertical position of the beam. The inner adjuster alters the horizontal aim of the beam.

3 Most models have an electrically-operated headlight beam adjustment system, controlled via a switch in the facia. The recommended settings are as follows.

a) 0 Front seat(s) occupied
b) 1 All seats occupied
c) 2 All seats occupied, and load in luggage compartment
d) 3 Driver's seat occupied and load in the luggage compartment

Note: When adjusting the headlight aim, ensure that the switch is set to position 0.

9 Instrument panel – removal and refitting

Note: The instrument panel is a complete sealed assembly, and no dismantling of the instrument panel is possible.

Removal

1 Disconnect the battery negative lead as described in Chapter 5A Section 4.

2 Remove the steering column upper shroud as described in Chapter 11, Section 27.

3 Using a plastic spatula or similar tool,

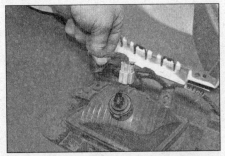

7.24 Disconnect the foglight wiring connector and release the wiring harness clip from the foglight body

7.25 Undo the three foglight retaining screws

9.3a Carefully prise up the instrument panel hood...

9.3b ...to release the internal retaining clips

carefully prise up the instrument panel hood to release the internal retaining clips **(see illustrations)**.

4 Carefully prise free the steering column upper shroud flexible section and remove it from below the instrument panel **(see illustrations)**.

5 Undo the instrument panel upper retaining bolt and the two lower bolts **(see illustrations)**.

9.4a Carefully prise free the steering column upper shroud flexible section...

9.4b ...and remove it from below the instrument panel

9.5a Undo the instrument panel upper retaining bolt...

9.5b ...and the lower retaining bolt on each side

9.6a Withdraw the instrument panel from the facia...

9.6b ...then disconnect the wiring connector and remove the panel

10.3a Undo the two lower retaining bolts...

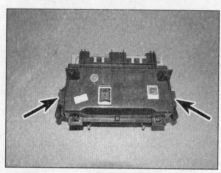

10.3b ...and two rear retaining bolts

6 Withdraw the instrument panel from the facia, disconnect the wiring connector and remove the panel (see illustrations).

11.2 Upper horn retaining nut

12.2 Stick a piece of masking tape to the glass, then mark it to use as an alignment aid on refitting

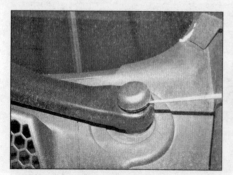

12.3a Carefully prise off the wiper arm spindle nut cover (windscreen wiper arm)...

12.3b ...or pull the blade cover of the spindle end of the arm (tailgate wiper arm)

10 Information display unit – removal and refitting

Removal

1 Disconnect the battery negative lead as described in Chapter 5A Section 4.
2 Remove the facia upper centre panel as described in Chapter 11, Section 27.
3 Undo the two lower retaining bolts and the two rear retaining bolts (see illustrations).
4 Spread the legs of the retaining catches and withdraw the upper centre panel components from the information display unit.

Refitting

5 Refitting is a reversal of removal.

11 Horn – removal and refitting

Removal

1 Remove the front bumper as described in Chapter 11, Section 6.
2 Disconnect the wiring connector, undo the retaining nuts and remove both horns from the mounting bracket (see illustration).

Refitting

3 Refitting is the reverse of removal.

12 Wiper arm – removal and refitting

Removal

1 Operate the wiper motor, then switch it off so that the wiper arm returns to the at-rest (parked) position.
2 Stick a piece of masking tape to the windscreen or tailgate glass, then make a mark on the tape, in line with the wiper blade, to use as an alignment aid on refitting (see illustration).
3 If removing a windscreen wiper arm, carefully prise off the wiper arm spindle nut cover. If removing a tailgate wiper arm, pull the blade cover off the spindle end of the wiper arm then disengage the retaining tab and remove the cover (see illustrations).
4 Slacken and remove the spindle nut and washer.
5 If removing a windscreen wiper arm, rock the arm from side-to-side to release it from the spindle. If removing a tailgate wiper arm, use a small puller to release the arm (see illustration).

Refitting

6 Ensure that the wiper arm and spindle splines are clean and dry, then refit the arm to

the spindle, aligning the wiper blade with the tape fitted on removal. Refit the spindle nut, tightening it securely, and clip the nut cover back in position.

13 Windscreen wiper motor and linkage – removal and refitting

Removal

1 Remove the wiper arms as described in the Section 12.
2 Remove the windscreen cowl panel as described in Chapter 11, Section 21.
3 Undo the two retaining bolts, and withdraw the wiper motor and linkage assembly from its location **(see illustration)**.
4 Disconnect the wiring connector from the wiper motor and remove the wiper motor and linkage from the car.
5 To remove the motor, release the crank arm from the motor linkage arm balljoint using a large screwdriver, then undo the two bolts and separate the motor from the linkage frame **(see illustrations)**.

Refitting

6 Refitting is the reverse of removal.

14 Tailgate wiper motor – removal and refitting

Removal

1 Remove the wiper arm as described in Section 12.
2 Remove the tailgate trim panel as described in Chapter 11, Section 25.
3 Disconnect the wiring connector, then slacken and remove the wiper motor mounting bolts and remove the wiper motor **(see illustrations)**.

Refitting

4 Refitting is the reverse of removal, ensuring the wiper motor retaining bolts are securely tightened.

12.5 If removing a tailgate wiper arm, use a small puller to release the arm

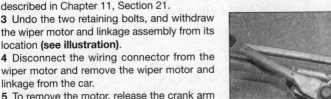

13.5a Release the crank arm from the motor linkage arm balljoint using a large screwdriver…

15 Windscreen/tailgate washer system components – removal and refitting

Washer system reservoir

1 Remove the front wheel arch liner on the left-hand side as described in Chapter 11, Section 21.
2 From within the engine compartment, extract the plastic expanding rivet, then pull up and remove the filler neck for the washer reservoir.
3 Disconnect the wiring connectors at the washer pump(s) and release the wiring from the reservoir.
4 Being prepared for fluid spillage, disconnect

13.3 Windscreen wiper linkage retaining bolts

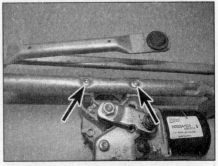

13.5b …then undo the two bolts and separate the motor from the linkage frame

the washer hoses from the washer pump(s) and free the hoses from the guides on the reservoir.
5 Undo the three nuts and remove the reservoir from under the car **(see illustration)**.
6 Refitting is the reverse of removal, ensuring that the wiring and washer hoses are securely connected.

Washer pump

7 Remove the front wheel arch liner on the left-hand side as described in Chapter 11, Section 21.
8 Being prepared for fluid spillage, disconnect the washer hoses from the washer pump **(see illustration)**.
9 Disconnect the washer pump wiring connector, then carefully ease the pump out

14.3a Disconnect the wiring connector…

14.3b …then slacken and remove the wiper motor mounting bolts and remove the wiper motor

15.5 Undo the three nuts and remove the reservoir from under the car

15.8 Disconnect the washer hoses from the washer pump

15.12 Disconnect the washer hose(s) from the washer jet

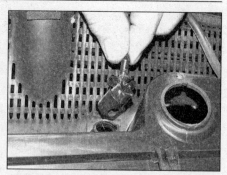

15.13 Compress the tab on the cowl panel, while pulling up the jet

from the reservoir and recover its sealing grommet.

10 Refitting is the reverse of removal, using a new sealing grommet if the original one shows signs of damage or deterioration.

Windscreen washer jets

11 Remove the windscreen cowl panel as described in Chapter 11, Section 21.
12 Disconnect the washer hose(s) from the washer jet **(see illustration)**.
13 Using a small screwdriver, compress the tab on the cowl panel, while pulling up the jet. Remove the jet from the panel **(see illustration)**.
14 Refitting is the reverse of removal.

Tailgate washer jet

15 Remove the spoiler from the tailgate as described in Section 7 paragraphs 16 to 20.
16 Disconnect the washer hose from the washer jet.

17 Compress the two tabs on the washer jet and push the jet out of the spoiler.
18 Refitting is the reverse of removal.

16 Radio/CD player – removal and refitting

Note: *The following removal and refitting procedure is for the range of radio/CD units which Vauxhall/Opel fit as standard equipment. Removal and refitting procedures of non-standard units may differ slightly.*
Note: *If a new radio/CD player is to be fitted, the unit must be programmed and configured to the car using Vauxhall/Opel diagnostic equipment. Sound quality may be impaired until this work is carried out.*

Removal

1 Disconnect the battery negative lead as described in Chapter 5A, Section 4.
2 Remove the facia lower centre panel as described in Chapter 11, Section 27.
3 Undo the two retaining bolts and withdraw the radio/CD player from the facia. Disconnect the wiring connector and aerial lead and remove the radio/CD player **(see illustrations)**.

Refitting

4 Refitting is the reverse of removal.

17 Speakers – removal and refitting

A-pillar trim side speaker

1 Remove the A-pillar trim on the side concerned as described in Chapter 11, Section 25.
2 Depress the retaining tabs and remove the speaker from the A-pillar trim.

Front/rear door speaker

3 Remove the door inner trim panel as described in Chapter 11, Section 12.
4 Undo the retaining screw, then free the speaker from the door. Disconnect the wiring connector and remove the speaker **(see illustrations)**.
5 Refitting is the reverse of removal.

18 Radio aerial – general information

1 Removal of the radio aerial entails removal of the headlining, which is a complicated operation, considered to be outside the scope of this manual. Therefore, any problems relating to the aerial or wiring should be entrusted to a Vauxhall/Opel dealer.

19 Anti-theft alarm system – general information

Note: *This information is applicable only to the anti-theft alarm system fitted by Vauxhall/Opel as standard equipment.*

16.3a Undo the two retaining bolts and withdraw the radio/CD player from the facia

16.3b Disconnect the wiring connector and aerial lead and remove the radio/CD player

17.4a Undo the retaining screw, then free the speaker from the door

17.4b Disconnect the wiring connector and remove the speaker

1 All models in the range are fitted with an anti-theft alarm system as standard equipment. The alarm is automatically armed and disarmed when the deadlocks are operated using the driver's door lock or remote control key. The alarm has switches on all the doors (including the tailgate), the bonnet, the radio/CD player and the ignition and starter circuits. If the tailgate, bonnet or any of the doors are opened whilst the alarm is set, the alarm horn will sound and the hazard warning lights will flash. The alarm also has an immobiliser function which makes the ignition and starter circuits inoperable whilst the alarm is triggered.

2 The alarm system performs a self-test every time it is switched on; this test takes approximately 30 seconds. During the self-test, the LED (light emitting diode) on the top of the facia will come on. If the LED flashes quickly, then either the tailgate, bonnet or one of the doors is open, or there is a fault in the circuit. After the initial 30-second period, the LED will flash slowly to indicate that the alarm is switched on. On unlocking the driver's door lock, the LED will illuminate for approximately 1 second, then go out, indicating that the alarm has been switched off.

3 Should the alarm system develop a fault, the vehicle should be taken to a Vauxhall/Opel dealer for examination.

20 Heated seat components – general information

1 On models with heated seats, a heater mat is fitted to both the seat back and seat cushion. Renewal of either heater mat involves peeling back the upholstery, removing the old mat, sticking the new mat in position and then refitting the upholstery. Note that upholstery removal and refitting requires considerable skill and experience if it is to be carried out successfully, and is therefore best entrusted to your Vauxhall/Opel dealer. In practice, it will be very difficult for the home mechanic to carry out the job without ruining the upholstery.

21 Airbag system – general information and precautions

General information

1 A driver's airbag is fitted as standard equipment on all models. The airbag is fitted in the steering wheel centre pad. Additionally, a passenger's airbag located in the facia, side airbags located in the front seats, and curtain airbags located in the headlining are standard or optionally available.

2 The system is armed only when the ignition is switched on, however, a reserve power source maintains a power supply to the system in the event of a break in the main electrical supply. The steering wheel and facia airbags

are activated by a 'g' sensor (deceleration sensor), and controlled by an electronic control unit located under the centre console. The side airbags and curtain airbags are activated by severe side impact and operate independently of the main system. A separate electrical supply, control unit and sensor is provided for the side/curtain airbags on each side of the car.

3 The airbags are inflated by a gas generator, which forces the bag out from its location in the steering wheel, facia, seat back frame, or headlining.

Precautions

⚠ Warning: The following precautions must be observed when working on vehicles equipped with an airbag system, to prevent the possibility of personal injury.

General precautions

4 The following precautions must be observed when carrying out work on a vehicle equipped with an airbag:

a) Do not disconnect the battery with the engine running.
b) Before carrying out any work in the vicinity of the airbag, removal of any of the airbag components, or any welding work on the vehicle, de-activate the system as described in the following sub-Section.
c) Do not attempt to test any of the airbag system circuits using test meters or any other test equipment.
d) If the airbag warning light comes on, or any fault in the system is suspected, consult a Vauxhall/Opel dealer without delay.
e) Do not attempt to carry out fault diagnosis, or any dismantling of the components.

Precautions when handling an airbag

a) Transport the airbag by itself, bag upward.
b) Do not put your arms around the airbag.
c) Carry the airbag close to the body, bag outward.
d) Do not drop the airbag or expose it to impacts.
e) Do not attempt to dismantle the airbag unit.
f) Do not connect any form of electrical equipment to any part of the airbag circuit.

Precautions when storing an airbag

a) Store the unit in a cupboard with the airbag upward.

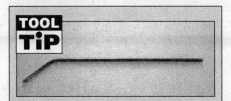

TOOL TIP

To make an airbag removal tool, use a coat hanger or metal rod 3 to 4 mm in diameter and cut to approximately 180 mm in length. Bend the rod by approximately 30° at a point approximately 30 mm from one end of the rod.

b) Do not expose the airbag to temperatures above 80°C.
c) Do not expose the airbag to flames.
d) Do not attempt to dispose of the airbag – consult a Vauxhall/Opel dealer.
e) Never refit an airbag which is known to be faulty or damaged.

De-activation of airbag system

5 The system must be de-activated before carrying out any work on the airbag components or surrounding area:

a) Switch on the ignition and check the operation of the airbag warning light on the instrument panel. The light should illuminate when the ignition is switched on, then extinguish.
b) Switch off the ignition.
c) Remove the ignition key.
d) Switch off all electrical equipment.
e) Disconnect the battery negative lead as described in Chapter 5A Section 4.
f) Insulate the battery negative terminal and the end of the battery negative lead to prevent any possibility of contact.
g) Wait for at least two minutes before carrying out any further work. Wait at least ten minutes if the airbag warning light did not operate correctly.

Activation of airbag system

6 To activate the system on completion of any work, proceed as follows:

a) Ensure that there are no occupants in the vehicle, and that there are no loose objects around the vicinity of the steering wheel. Close the vehicle doors and windows.
b) Ensure that the ignition is switched off then reconnect the battery negative terminal.
c) Open the driver's door and switch on the ignition, without reaching in front of the steering wheel. Check that the airbag warning light illuminates briefly then extinguishes.
d) Switch off the ignition.
e) If the airbag warning light does not operate as described in paragraph c), consult a Vauxhall/Opel dealer before driving the vehicle.

22 Airbag system components – removal and refitting

⚠ Warning: Refer to the precautions given in Section 21 before attempting to carry out work on any of the airbag components.

1 De-activate the airbag system as described in the previous Section, then proceed as described under the relevant heading.

Driver's airbag

2 To release the airbag retaining wire spring, make up a removal tool as shown (see Tool Tip).

3 Lower the steering column and move it to the maximum extended position.

22.4 Turn the steering wheel so that the two holes in the back of the lower spoke are at the top

22.5a Insert the removal tool into one of the airbag holes and bring the tool into contact with the wire spring – shown with airbag removed

22.5b Pull the tool away from the steering column, and pull the airbag up to release it on that side

22.6a Release the locking clip...

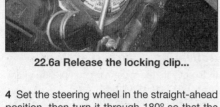

22.6b ...and disconnect the airbag wiring connector

4 Set the steering wheel in the straight-ahead position, then turn it through 180° so that the two holes in the back of the lower spoke are at the top **(see illustration)**.

5 Insert the bent end of the removal tool into one of the holes on the rear of the steering wheel and bring the tool into contact with the wire spring. Pull the other end of the tool away from the steering column, and at the same time pull the airbag away from the steering wheel to release it on that side **(see illustrations)**. Repeat the procedure on the other side of the steering wheel.

6 Release the locking clip, then disconnect the wiring connector at the rear of the airbag unit **(see illustrations)**. Remove the airbag unit. Note that the airbag must not be knocked or dropped, and should be stored the correct way up, with its padded surface uppermost.

7 Refitting is a reversal of the removal procedure.

Passenger's airbag

8 Remove the complete facia assembly as described in Chapter 11, Section 27.

9 Undo the four retaining nuts and remove the airbag from the underside of the facia **(see illustration)**.

10 Note that the airbag must not be knocked or dropped, and should be stored the correct way up (as mounted in the vehicle).

11 Refitting is a reversal of the removal procedure, tightening the retaining nuts securely.

Side airbags

12 The side airbags are located internally within the front seat back and no attempt should be made to remove them. Any

suspected problems with the side airbag system should be referred to a Vauxhall/Opel dealer.

Curtain airbags

13 The curtain airbags are located behind the headlining above the doors on each side and no attempt should be made to remove them. Any suspected problems with the curtain airbag system should be referred to a Vauxhall/Opel dealer.

Airbag control unit

14 The airbag control unit is located beneath the centre console and no attempt should be made to remove it. Any suspected problems with the control unit should be referred to a Vauxhall/Opel dealer.

Airbag rotary connector

15 Disconnect the battery negative lead as described in Chapter 5A Section 4.

16 Remove the steering column shrouds as described in Chapter 11, Section 27.

17 Set the roadwheels in the straight-ahead position and ensure they remain in that position during the removal and refitting procedures.

18 Remove the steering wheel as described in Chapter 10, Section 13.

19 Disconnect the two upper wiring connectors and, where fitted, the lower wiring connector from the airbag rotary connector **(see illustrations)**.

22.9 Undo the four retaining nuts and remove the airbag from the underside of the facia

22.19a Disconnect the two upper wiring connectors...

22.19b ...and, where fitted, the lower wiring connector from the airbag rotary connector

20 Undo the four retaining screws and remove the rotary connector from the steering column **(see illustrations)**.

21 Refitting is a reversal of the removal procedure.

22 Before refitting the steering wheel, the rotary connector should be centralised (unless it is known absolutely that it was centralised before removal, and that the rotary connector has not been turned during or since its removal).

23 Ensure that the roadwheels are in the straight-ahead position and the centering mark on the end of the steering column shaft is in the 6 o'clock position

24 Turn the rotary connector clockwise gently, until resistance is felt. Now turn the rotary connector about three turns anti-clockwise, until the yellow indicator is visible in the connector window.

22.20a Undo the four retaining screws…

22.20b …and remove the rotary connector from the steering column

Wiring diagram colour codes

BE – Beige
BK – Black
BN – Brown
BU – Blue
DG – Dark Green
DB – Dark Blue
GN – Green
GY – Grey
LB – Light Blue
LG – Light Green
OG – Orange
PK – Pink
RD – Red
VT – Violet
WH – White
YE – Yellow

FUSE BLOCK – UNDERHOOD

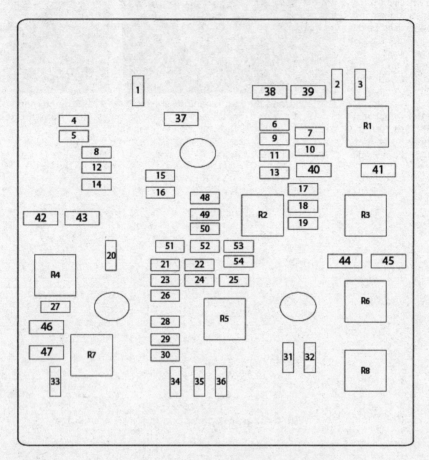

FUSE / RELAY	VALUE	DESCRIPTION	OEM NAME
1	25 A	Sunroof control module	F1UA
2	7.5 A	Outside rearview mirror switch	F2UA
3	-	Not used	-
4	-	Not used	-
5	30 A	Electronic brake control module	F5UA
6	5 A	Battery sensor module	F6UA
7	-	Not used	-
8	15 A	Automatic transmission assembly	F8UA
9	-	Not used	-
10	10 A	Headlamp assembly left and right, Fuse block-Underhood	F10UA
11	20 A	Rear wiper relay, Rear wiper motor	F11UA
12	25 A	Rear defogger grid	F12UA
13	10 A	Headlamp assembly left	F13UA
14	7.5 A	Outside rearview mirror driver and passenger	F14UA
15	-	Not used	-
16	25 A	Seat memory control module	F16UA
17	5 A	Automatic transmission assembly	F17UA

18	5 A	Engine control module	F18UA
19	20 A	Fuel pump relay, Fuel pump and level sensor assembly	F19UA
20	-	Not used	-
21	15 A	Cooling fan low speed relay, Cooling fan medium speed relay, Cooling fan high speed relay, Cooling fan speed control relay, Cooling fan relay	F21UA
22	-	Not used	-
23	15 A	Engine control module, Ignition coil module, Fuel injector 1, Fuel injector 2, Fuel injector 3, Fuel injector 4 for Petrol	F23UA
	7.5 A	Engine control module for Diesel	
24	10 A	Rear window washer pump relay, Windshield washer pump	F24UA
25	10 A	Headlamp assembly right	F25UA
26	10 A	Heated oxygen sensor 1, Heated oxygen sensor 2, Canister purge solenoid and Turbo bypass solenoid, Wastegate control solenoid for 1.4 Petrol or Inlet manifold tuning valve, Thermostat body heating element for 1.6 Petrol	F26UA
		Oil pressure control solenoid, Piston cooling jet solenoid, Inlet manifold tuning valve for 1.6 Diesel or Swirl control motor for 1.7 Diesel	
27	-	Not used	-
28	10 A	Engine control module for Diesel	F28UA
29	20 A	Engine control module	F29UA
30	10 A	Multifunction intake air sensor for Petrol	F30UA
		Heated oxygen sensor 1, Heated oxygen sensor 2 for Diesel	
31	15 A	Headlamp left high beam for manual headlamps or High beam solenoid actuator left, headlamp ballast left for automatic headlamps	F31UA
32	15 A	Headlamp right high beam for manual headlamps or High beam solenoid actuator right, headlamp ballast right for automatic headlamps	F32UA
33	15 A	Engine control module 5 A also used	F33UA
34	15 A	Horn high note, Horn low note	F34UA
35	10 A	A/C compressor clutch relay, A/C compressor	F35UA
36	15 A	Fog lamp left front, fog lamp right front	F36UA
37	40 A	Electronic brake control module	F1UB
38	30 A	Windshield wiper relay, Windshield wiper speed control relay	F2UB
39	25 A	Blower motor control module	F3UB
40	30 A	Fuse block - Instrument panel	F4UB
41	-	Not used	-
42	30 A	Fuel heater relay, Fuel heater for Diesel	F6UB
43	-	Not used	-
44	40 A	Cooling fan low speed relay, Cooling fan medium speed relay, 30 A also used	F8UB
45	60 A	Cooling fan high speed relay 40 A also used	F9UB
46	60 A	Brake booster pump motor relay, brake booster pump motor, Engine control module 30 A also used	F10UB
47	30 A	Starter relay, Starter motor	F11UB
R1	-	Ignition main relay	KR73
R2	-	Fuel pump relay	KR23A
R3	-	Cooling fan medium speed relay	KR20P
R4	-	Not used	-
R5	-	Engine controls ignition relay	KR75
R6	-	Cooling fan high speed relay	KR20D
R7	-	Starter relay	KR27
R8	-	Cooling fan low speed relay	KR20C

RELAY BLOCK – UNDERHOOD

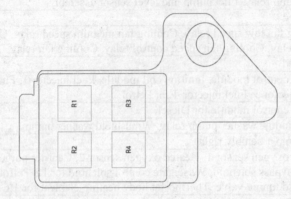

RELAY	VALUE	DESCRIPTION	OEM NAME
R1	-	Brake booster pump motor relay	R1
R2	-	Cooling fan speed control relay	R2
R3	-	Cooling fan relay	R3
R4	-	Headlamp washer pump relay	R4

FUSE BLOCK – BATTERY

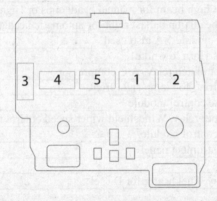

FUSE	VALUE	DESCRIPTION	OEM NAME
1	80 A	Power steering control module	F1BA
2	430 A	Generator, Starter motor, 400 A also used	F2BA
3	100 A	Fuse block - Instrument panel, 80 A also used	F3BA
4	150 A	Fuse block – Underhood	F4BA
5	80 A	Fuse block - Rear body	F5BA

FUSE BLOCK - INSTRUMENT PANEL

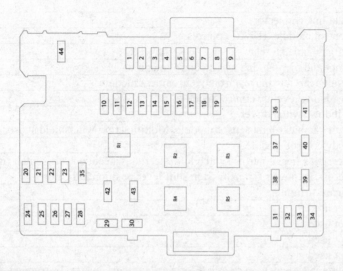

FUSE / RELAY	VALUE	DESCRIPTION	OEM NAME
1	15 A	Body control module: Center high mounted stop lamp, License plate lamp Left, License plate lamp right, Trailer interface control module, Multifunction switch instrument panel	F1DA
2	15 A	Body control module: Daytime running lamp relay left	F2DA
3	15 A	Body control module: Side marker lamp right rear, Park/ Daytime running lamp left, Turn signal repeater lamp left, Tail/Stop lamp left, Tail lamp right, Fog lamp left rear, Fog lamp right rear, Daytime running lamp relay right, Trailer interface control module, Fuse Block - Underhood	F3DA
4	15 A	Body control module: Side marker lamp left rear, Park/ Daytime running lamp right, Tail lamp left, Tail/Stop lamp right, Rear compartment courtesy lamp, Fuse block - Underhood	F4DA
5	15 A	Body control module: Rear wiper relay, Multifunction switch 1 - Instrument Panel, Fuel type select switch, Fuel economy mode switch, Fuse Block – Underhood, Relay Block Underhood	F5DA
6	15 A	Body control module: Inside rearview mirror, Rearview camera, Turn signal lamp right rear, Backup lamp left, backup lamp right, Windshield washer pump relay, Windshield wiper relay, Ignition main relay, Door dead lock relay, Door lock switch driver, Headlamp switch, Multifunction switch instrument panel, Outside rearview mirror switch, Sunroof switch, Window switch driver, Window switch left rear, Window switch passenger, Window switch right rear, Sunroof tilt switch, Fuel type select switch, Fuel economy mode switch, Fuse block - Instrument panel	F6DA
7	15 A	Sunshade left, Sunshade right, Turn signal lamp left rear, Instrument panel compartment lamp, dome lamp, Dome/Reading lamp, Body control module, Fuse block - Underhood	F7DA
8	30 A	Body control module	F8DA
9	2 A	Headlamp switch, Ignition switch, Steering wheel control switch left, Steering wheel control switch right, Steering wheel air bag coil	F9DA
10	10 A	Inflatable restraint sensing and Diagnostic module	F10DA

11	7.5 A	Data link connector	F11DA
12	10 A	Control module, HVAC controls	F12DA
13	10 A	Liftgate unlatch relay	F13DA
14	10 A	Front and rear parking assist control module	F14DA
15	10 A	Inside rearview mirror, front view camera module	F15DA
16	10 A	Headlamp leveling control module	F16DA
17	7.5 A	Window switch driver	F17DA
18	5 A	Rain / ambient light sensor module, Multifunction windshield sensor	F18DA
19	5 A	Body control module	F19DA
20	2 A	Steering wheel controls switch left, Steering wheel controls switch right	F20DA
21	10 A	Accessory power, Transmission shift lever position indicator	F21DA
22	20 A	Cigarette lighter	F22DA
23	-	Not used	-
24	-	Not used	-
25	-	Not used	-
26	-	Not used	-
27	10 A	Clutch pedal position switch, Electrical auxiliary heater, Instrument cluster	F27DA
28	10 A	Headlamp switch, Power supply, Headlamp leveling control module	F28DA
29	-	Not used	-
30	-	Not used	-
31	5 A	Instrument panel	F31DA
32	15 A	Radio, Alarm control module	F32DA
33	10 A	Radio controls, info display module	F33DA
34	10 A	Digital radio receiver control module	F34DA
35	-	Not used	-
36	-	Not used	-
37	-	Not used	-
38	30 A	Window motor driver, Window motor passenger	F3DB
39	30 A	Window motor left rear, Window motor right rear	F4DA
40	30 A	Battery saver relay 1	-
41	-	Not used	-
42	-	Not used	-
43	-	Not used	-
44	80 A	Electrical auxiliary heater	F1DC
R1	-	Retained accessory power relay	KR76
R2	-	Liftgate unlatch relay	KR95A
R3	-	Not used	-
R4	-	Not used	-
R5	-	Battery saver relay 1	KR104A

FUSE BLOCK - READ BODY

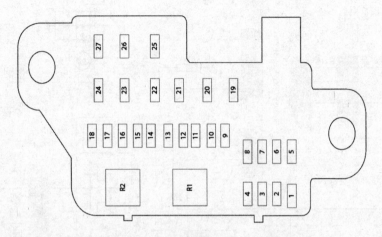

FUSE / RELAY	VALUE	DESCRIPTION	OEM NAME
1	15 A	Seat lumbar support switch driver	F1RA
2	15 A	Seat lumbar support switch passenger	F2RA
3	25 A	Audio amplifier	F3RA
4	20 A	Trailer connector	F4RA
5	10 A	Rear differential clutch module	F5RA
6	-	Not used	-
7	20 A	Fuel injector control module	F7RA
8	-	Not used	-
9	-	Not used	-
10	-	Not used	-
11	10 A	Trailer interface control module	F11RA
12	-	Not used	-
13	7.5 A	Steering wheel heating control module	F13RA
14	20 A	Trailer connector	F14RA
15	-	Not used	-
16	10 A	Fuel heater / water in fuel sensor	F16RA
17	10 A	Inside rearview mirror, Rearview camera	F17RA
18	10 A	Fuel injector control module	F18RA
19	30 A	Seat adjuster switch driver	F1RB
20	30 A	Seat adjuster switch passenger	F2RB
21	40 A	Trailer interface control module	F3RB
22	30 A	Power inverter module	F4RB
23	40 A	Ignition main relay, Ignition run relay	F5RB
24	30 A	Headlamp washer pump relay	F6RB
25	-	Not used	-
26	-	Not used	-
27	-	Not used	-
R1	30 A	Ignition main relay	KR73
R2	25 A	Ignition run relay	KR74

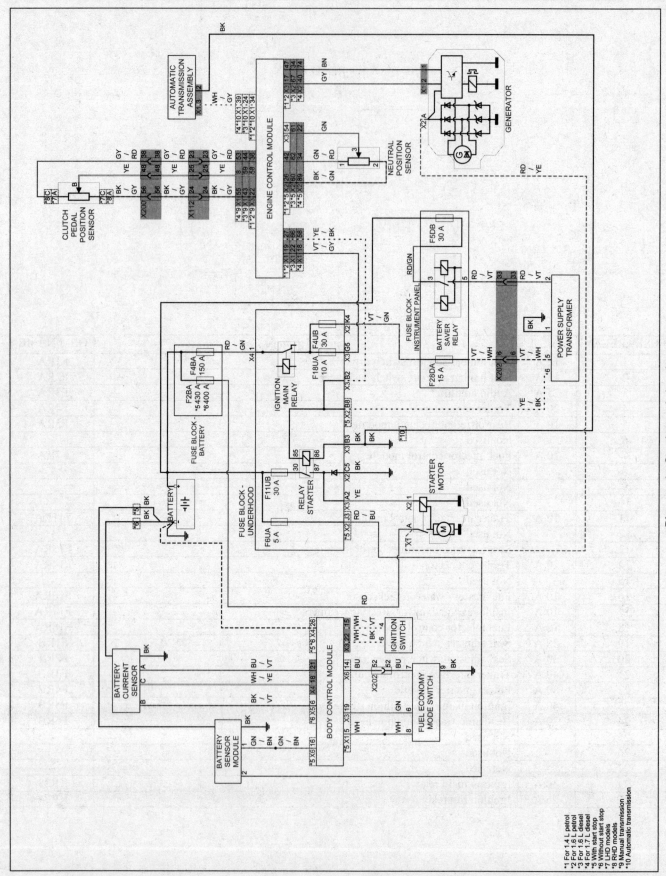

Diagram 1 – Starting and charging

*1 For 1.4 L petrol
*2 For 1.6 L petrol
*3 For 1.6 L diesel
*4 For 1.7 L diesel
*5 With start stop
*6 Without start stop
*7 LHD models
*8 RHD models
*9 Manual transmission
*10 Automatic transmission

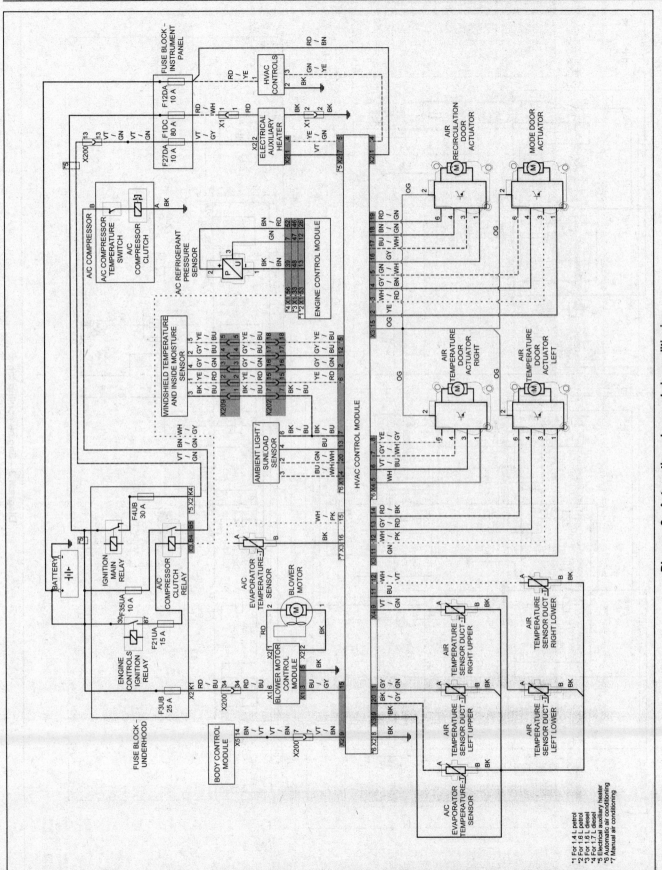

Diagram 2 – Automatic and manual air conditioning

*1 For 1.4 L petrol
*2 For 1.6 L petrol
*3 For 1.6 L diesel
*4 For 1.7 L diesel
*5 Electrical auxiliary heater
*6 Automatic air conditioning
*7 Manual air conditioning

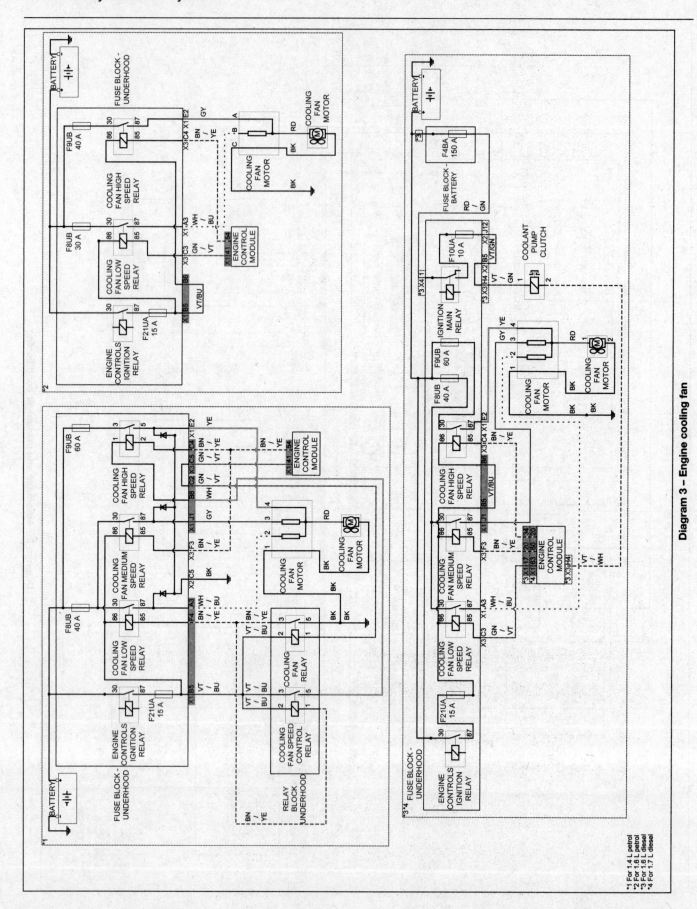

Diagram 3 – Engine cooling fan

*1 For 1.4 L petrol
*2 For 1.6 L petrol
*3 For 1.6 L diesel
*4 For 1.7 L diesel

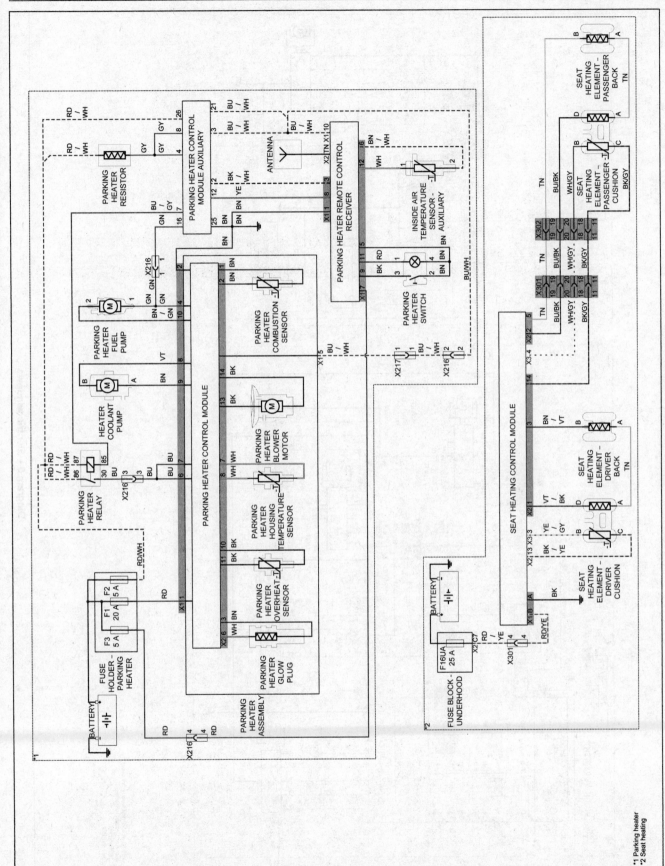

Diagram 4 – Seat heating and parking heater

*1 Parking heater
*2 Seat heating

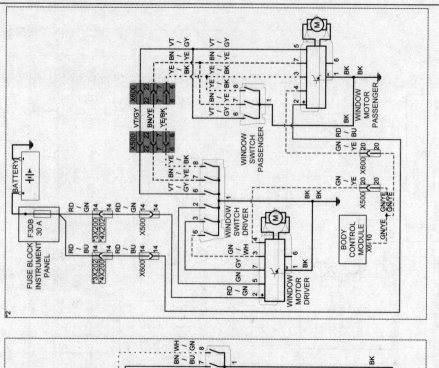

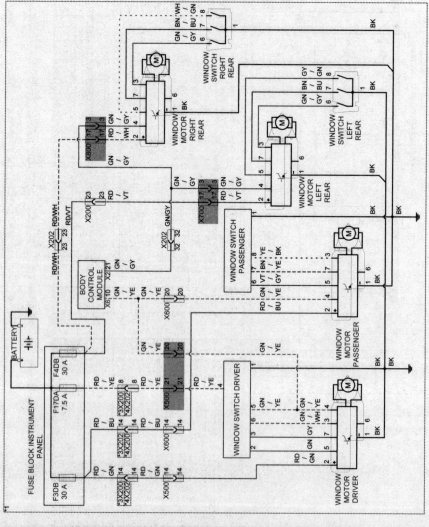

Diagram 5 – Electric windows

*1 With front and rear power windows
*2 With front power windows
*3 LHD models
*4 RHD models

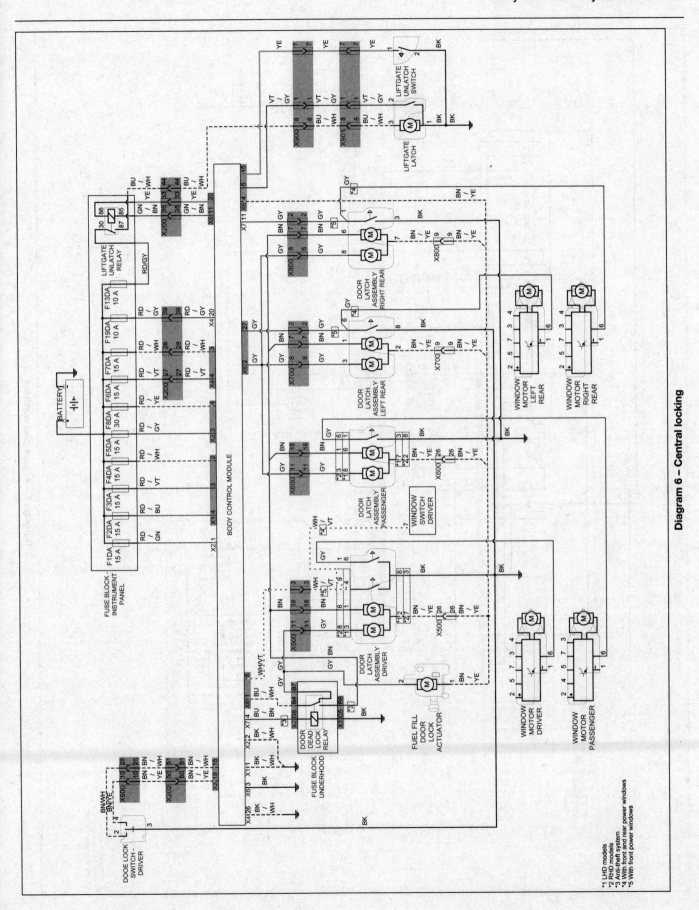

Diagram 6 – Central locking

*1 LHD models
*2 RHD models
*3 Anti-theft system
*4 With front and rear power windows
*5 With front power windows

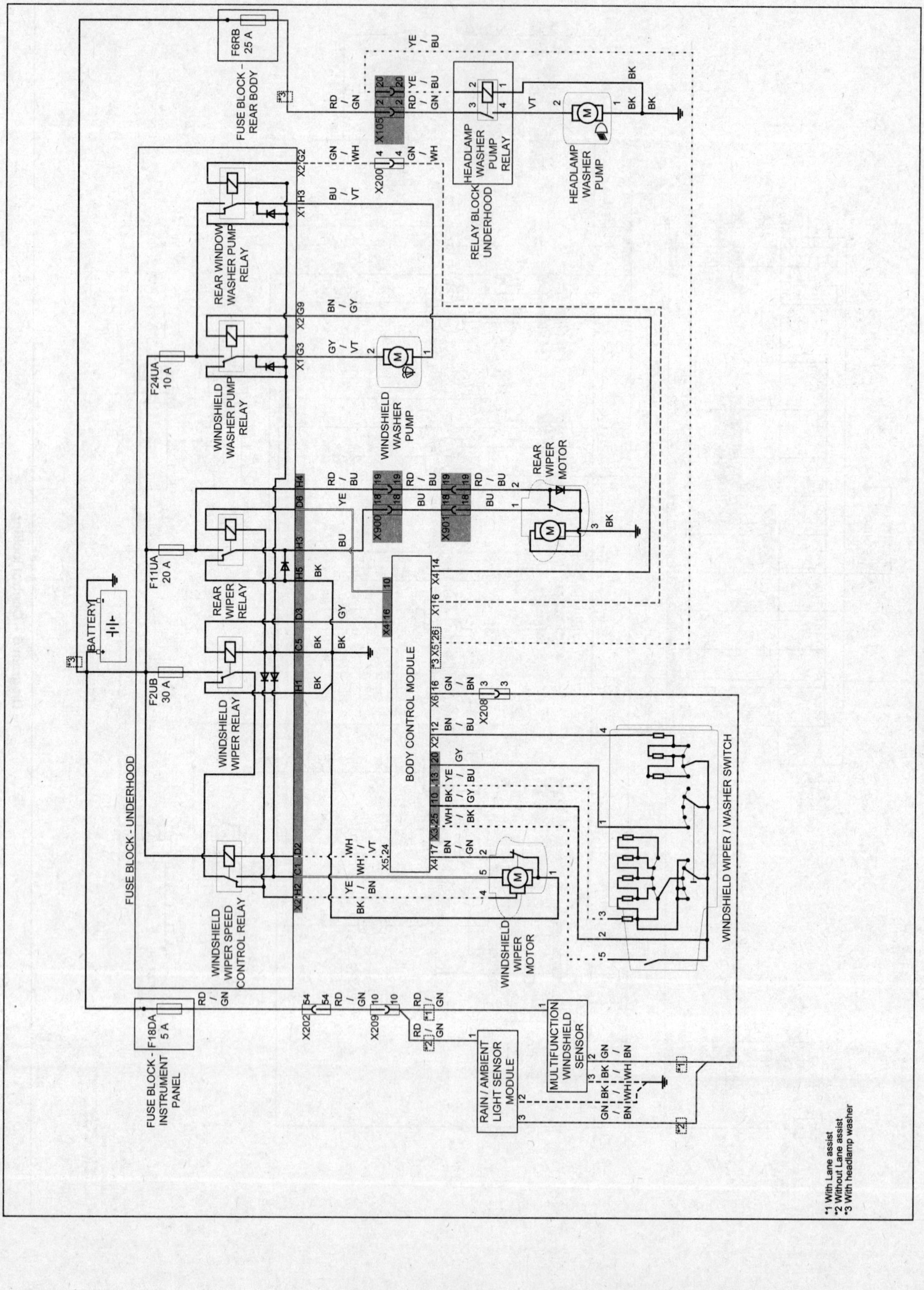

Diagram 7 – Wipers and washers

*1 With Lane assist
*2 Without Lane assist
*3 With headlamp washer

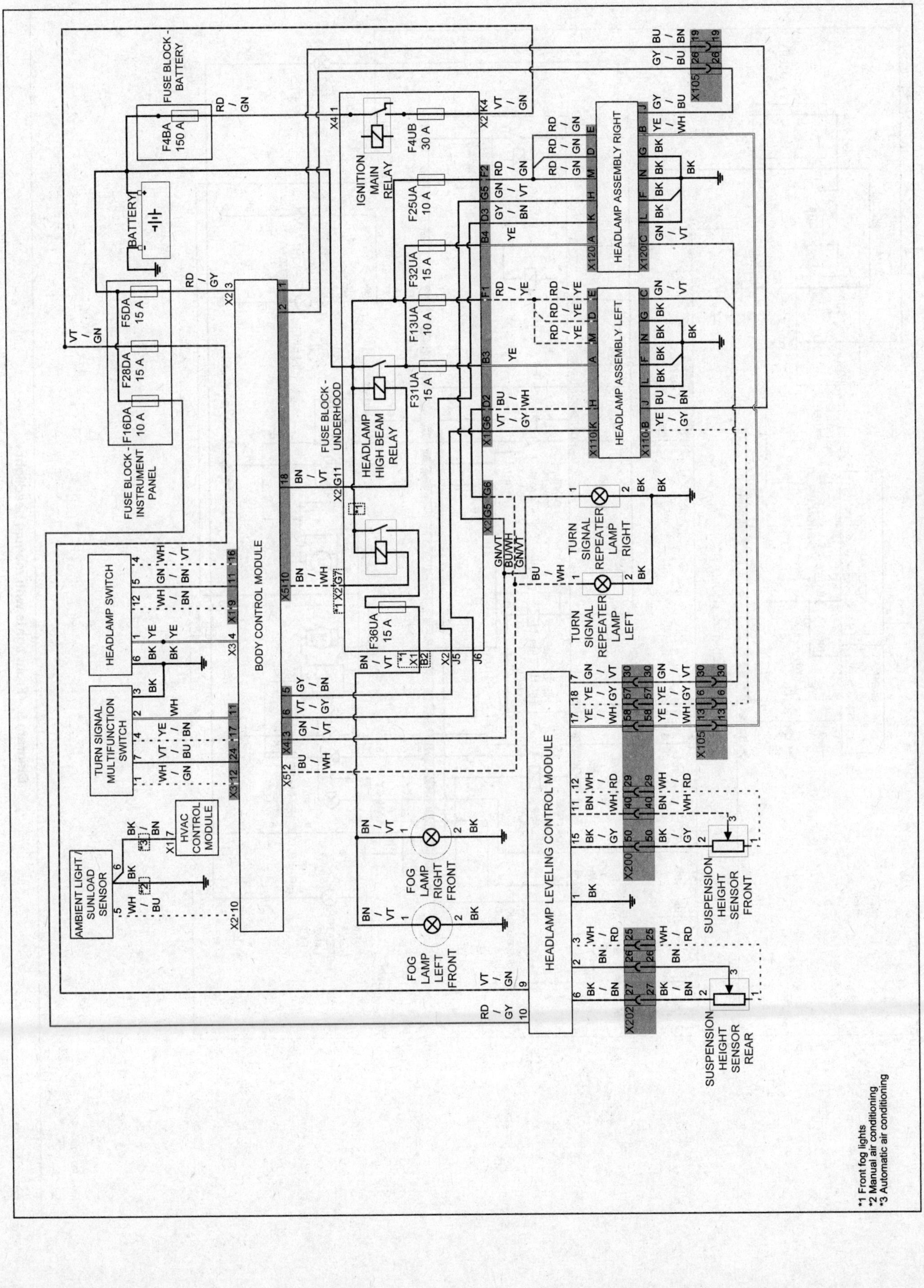

Diagram 8 – Front lights with automatic headlights

*1 Front fog lights
*2 Manual air conditioning
*3 Automatic air conditioning

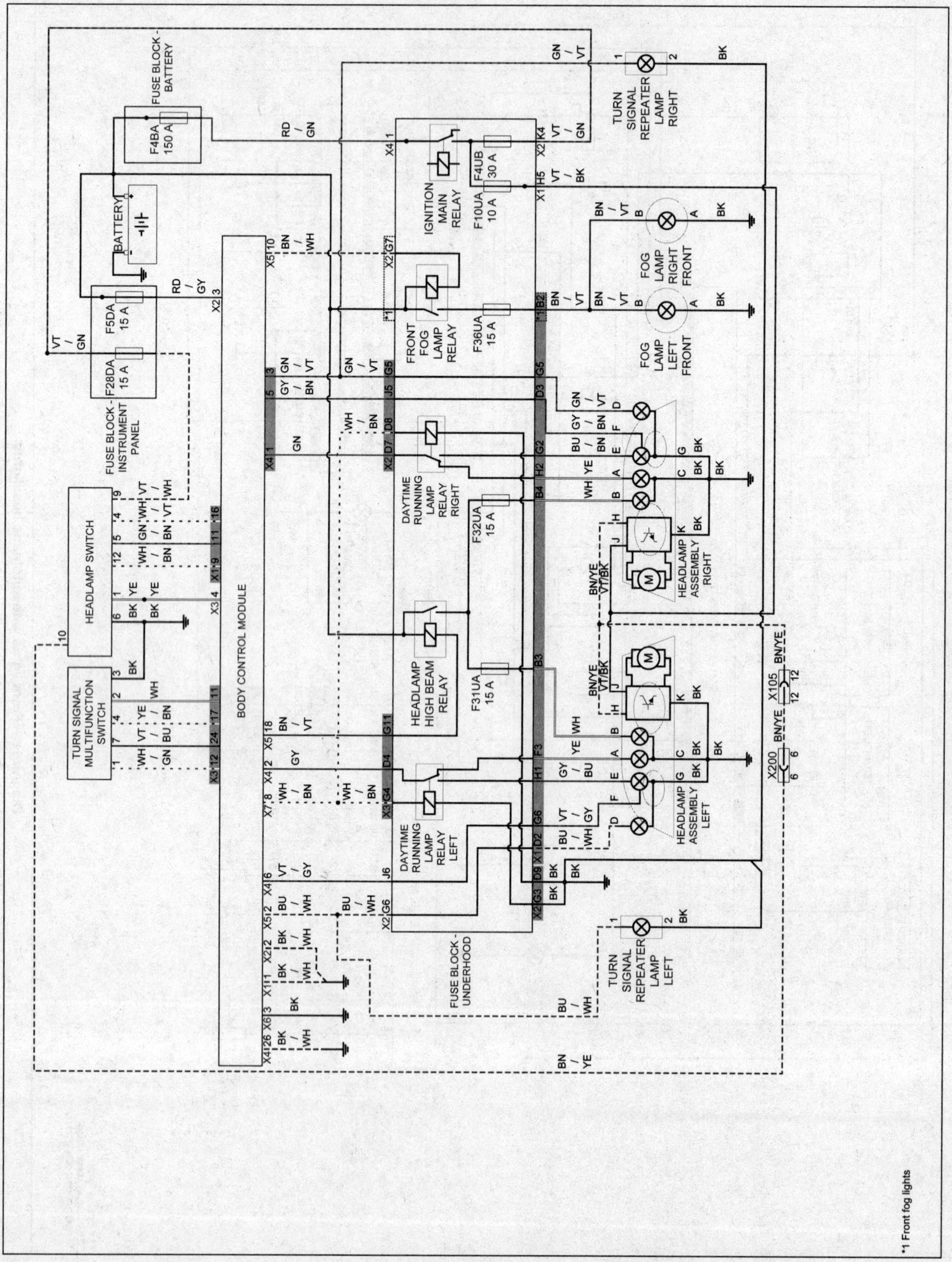

Diagram 9 – Front lights with manual headlights

*1 Front fog lights

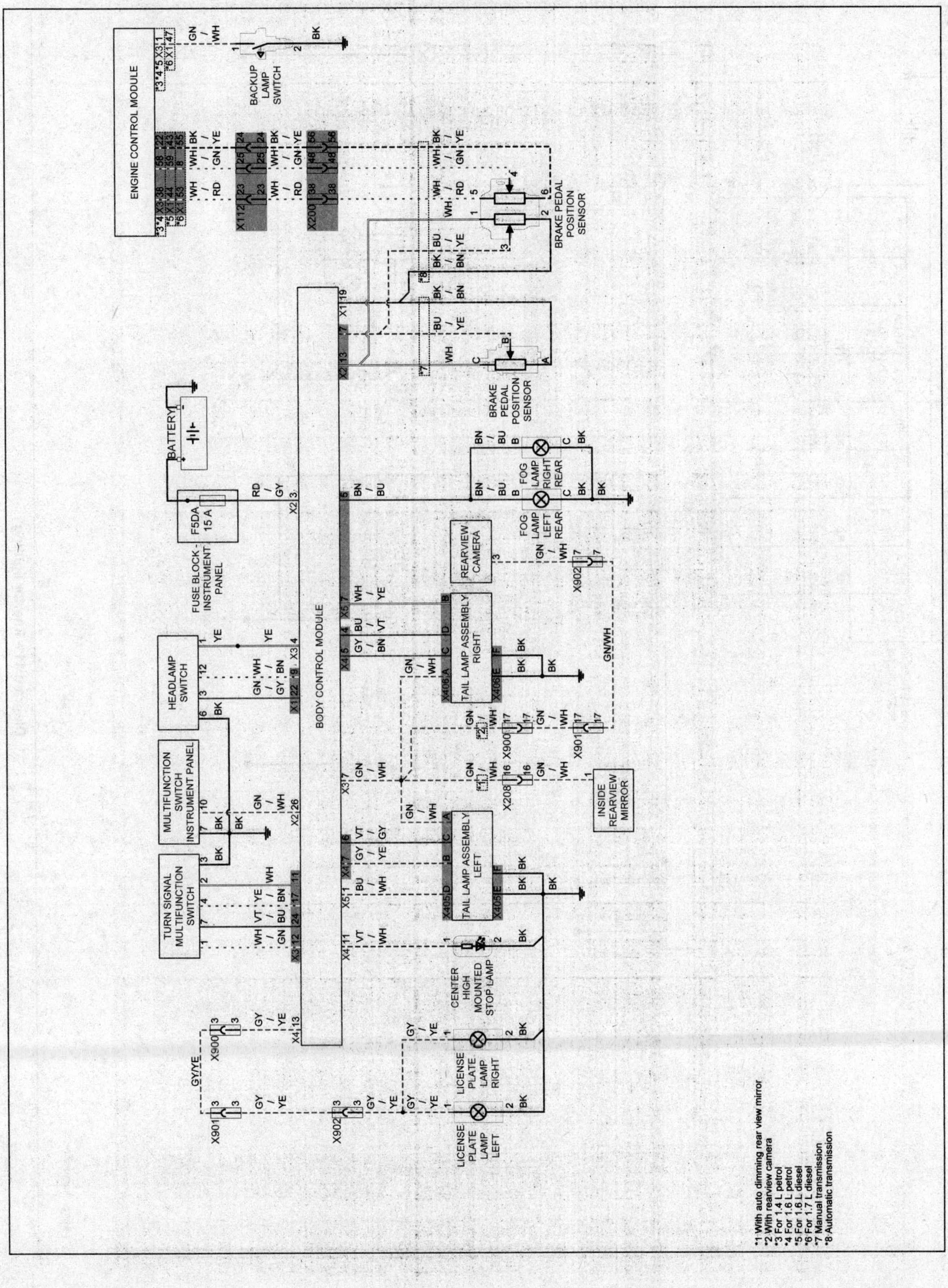

Diagram 10 – Rear lights

*1 With auto dimming rear view mirror
*2 With rearview camera
*3 For 1.4 L petrol
*4 For 1.6 L petrol
*5 For 1.6 L diesel
*6 For 1.7 L diesel
*7 Manual transmission
*8 Automatic transmission

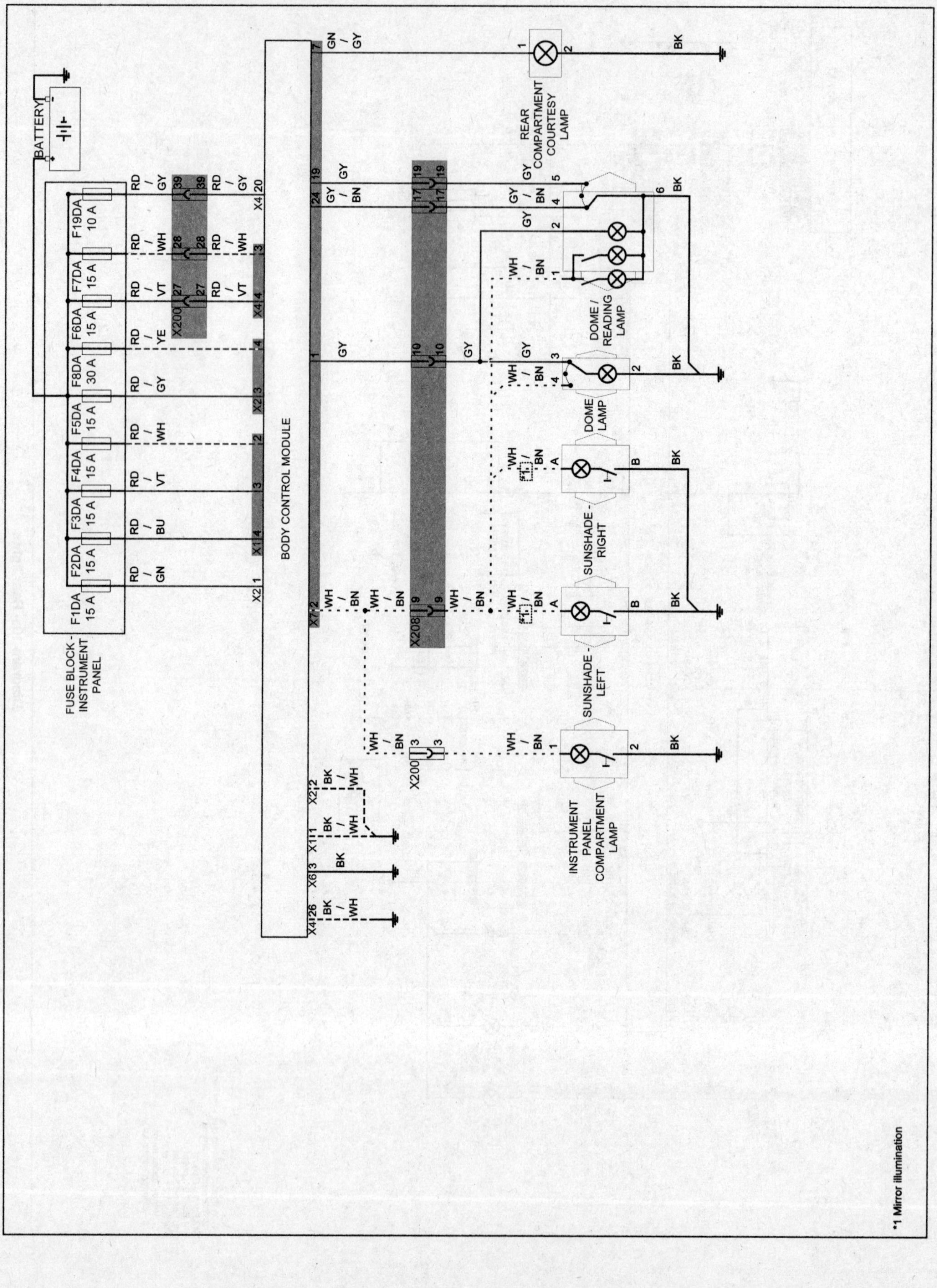

Diagram 11 – Interior lighting

*1 Mirror illumination

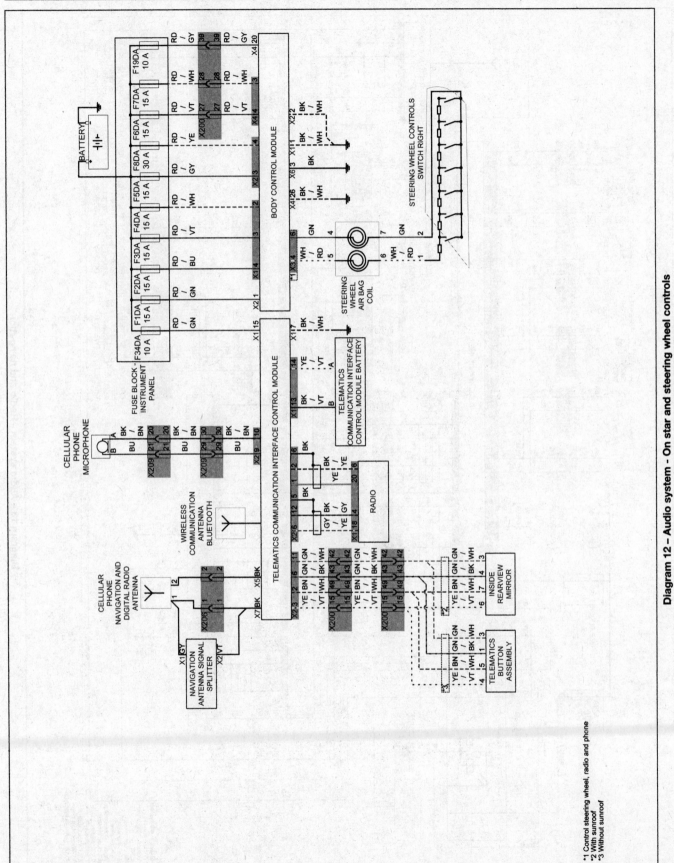

Diagram 12 – Audio system - On star and steering wheel controls

*1 Control steering wheel, radio and phone
*2 With sunroof
*3 Without sunroof

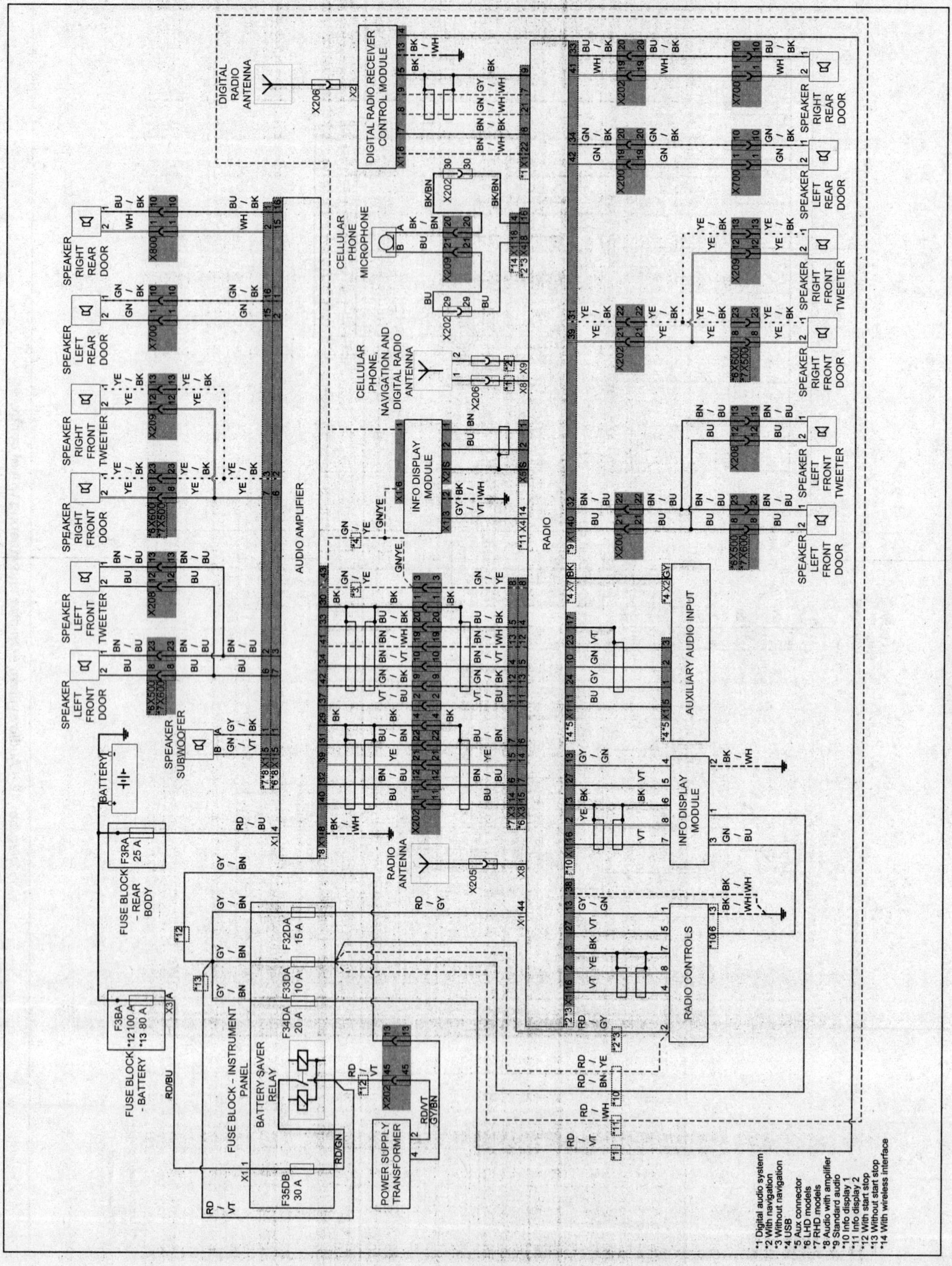

Diagram 13 – Audio system - Radio and navigation

*1 Digital audio system
*2 With navigation
*3 Without navigation
*4 USB
*5 Aux connector
*6 LHD models
*7 RHD models
*8 Audio with amplifier
*9 Standard audio
*10 Info display 1
*11 Info display 2
*12 With start stop
*13 Without start stop
*14 With wireless interface

FUEL LEVEL SENSOR SECONDARY

ENGINE CONTROL MODULE

*1*2 X3 X3 44
*3*5 X3 14
*4*5 X1 49

WH
/
BU

15
15
BU
/
WH

E
E
BU
/
WH
4

27
3
42
BK
/
GN

22
22
BK
/
GN

M
M
BK
/
GN

BU
/
VT
64
18
10

8
8
BU
/
VT

D
D
BU
/
VT

X3
X112
X350

*1*2 X1 28
*3 X1 50
*4 X1 59

GN/GY

X3 F2

85

30

86

87

FUEL PUMP RELAY

X2 H7 C5

GY   BK

IGNITION MAIN RELAY

F19UA
20 A

X4 1

RD
/
GN

F4BA
150 A

FUSE BLOCK - BATTERY

BATTERY

FUSE BLOCK - UNDERHOOD

X350 A
A
A
GY

FUEL PUMP AND LEVEL SENSOR ASSEMBLY

4
3
1
M
2

BK

BK

GN

*5 BK/GN

*5 BK/GN

*5 BK

*6 /
GN

*5

*5

BK/GN

FUEL LEVEL SENSOR SECONDARY

2

Diagram 14 – Fuel pump

*1 For 1.4 L petrol
*2 For 1.6 L petrol
*3 For 1.6 L diesel
*4 For 1.7 L diesel
*5 For AWD models
*6 Except AWD models

Notes

Dimensions and Weights

Note: *All figures are approximate and may vary according to model. Refer to manufacturer's data for exact figures.*

Dimensions

Overall length	.4278 mm
Overall width (including mirrors)	.2038 mm
Overall height (unladen)	.1658 mm
Wheelbase	.2555 mm

Weights

Vehicle weights are listed on the VIN label on the side door pillar **(see illustration)**

First (kg) figure is	Gross vehicle weight
Second (kg) figure is	Gross vehicle weight + Gross trailer weight
Third (kg) figure is	Gross axle weight (front)
Fourth (kg) figure is	Gross axle weight (rear)

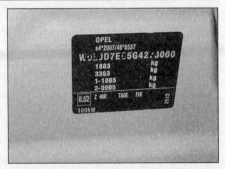

Vehicle weights are listed on the VIN label on the side door pillar

Fuel economy

Although depreciation is still the biggest part of the cost of motoring for most car owners, the cost of fuel is more immediately noticeable. These pages give some tips on how to get the best fuel economy.

Working it out

Manufacturer's figures

Car manufacturers are required by law to provide fuel consumption information on all new vehicles sold. These 'official' figures are obtained by simulating various driving conditions on a rolling road or a test track. Real life conditions are different, so the fuel consumption actually achieved may not bear much resemblance to the quoted figures.

How to calculate it

Many cars now have trip computers which will

display fuel consumption, both instantaneous and average. Refer to the owner's handbook for details of how to use these.

To calculate consumption yourself (and maybe to check that the trip computer is accurate), proceed as follows.

1. Fill up with fuel and note the mileage, or zero the trip recorder.
2. Drive as usual until you need to fill up again.
3. Note the amount of fuel required to refill the tank, and the mileage covered since the previous fill-up.
4. Divide the mileage by the amount of fuel used to obtain the consumption figure.

For example:

Mileage at first fill-up (a) = 27,903
Mileage at second fill-up (b) = 28,346
Mileage covered (b - a) = 443
Fuel required at second fill-up = 48.6 litres

The half-completed changeover to metric units in the UK means that we buy our fuel in litres, measure distances in miles and talk about fuel consumption in miles per gallon. There are two ways round this: the first is to convert the litres to gallons before doing the calculation (by dividing by 4.546, or see Table 1). So in the example:

48.6 litres ÷ 4.546 = 10.69 gallons
443 miles ÷ 10.69 gallons = 41.4 mpg

The second way is to calculate the consumption in miles per litre, then multiply that figure by 4.546 (or see Table 2).

So in the example, fuel consumption is:

443 miles ÷ 48.6 litres = 9.1 mpl
9.1 mpl x 4.546 = 41.4 mpg

The rest of Europe expresses fuel consumption in litres of fuel required to travel 100 km (l/100 km). For interest, the conversions are given in Table 3. In practice it doesn't matter what units you use, provided you know what your normal consumption is and can spot if it's getting better or worse.

Table 1: conversion of litres to Imperial gallons

litres	1	2	3	4	5	10	20	30	40	50	60	70
gallons	0.22	0.44	0.66	0.88	1.10	2.24	4.49	6.73	8.98	11.22	13.47	15.71

Table 2: conversion of miles per litre to miles per gallon

miles per litre	5	6	7	8	9	10	11	12	13	14
miles per gallon	23	27	32	36	41	46	50	55	59	64

Table 3: conversion of litres per 100 km to miles per gallon

litres per 100 km	4	4.5	5	5.5	6	6.5	7	8	9	10
miles per gallon	71	63	56	51	47	43	40	35	31	28

Maintenance

A well-maintained car uses less fuel and creates less pollution. In particular:

Filters

Change air and fuel filters at the specified intervals.

Oil

Use a good quality oil of the lowest viscosity specified by the vehicle manufacturer (see *Lubricants and fluids*). Check the level often and be careful not to overfill.

Spark plugs

When applicable, renew at the specified intervals.

Tyres

Check tyre pressures regularly. Under-inflated tyres have an increased rolling resistance. It is generally safe to use the higher pressures specified for full load conditions even when not fully laden, but keep an eye on the centre band of tread for signs of wear due to over-inflation.

When buying new tyres, consider the 'fuel saving' models which most manufacturers include in their ranges.

Driving style

Acceleration

Acceleration uses more fuel than driving at a steady speed. The best technique with modern cars is to accelerate reasonably briskly to the desired speed, changing up through the gears as soon as possible without making the engine labour.

Air conditioning

Air conditioning absorbs quite a bit of energy from the engine – typically 3 kW (4 hp) or so. The effect on fuel consumption is at its worst in slow traffic. Switch it off when not required.

Anticipation

Drive smoothly and try to read the traffic flow so as to avoid unnecessary acceleration and braking.

Automatic transmission

When accelerating in an automatic, avoid depressing the throttle so far as to make the transmission hold onto lower gears at higher speeds. Don't use the 'Sport' setting, if applicable.

When stationary with the engine running, select 'N' or 'P'. When moving, keep your left foot away from the brake.

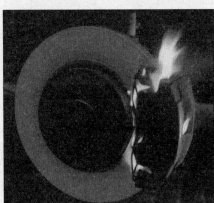

Braking

Braking converts the car's energy of motion into heat – essentially, it is wasted. Obviously some braking is always going to be necessary, but with good anticipation it is surprising how much can be avoided, especially on routes that you know well.

Carshare

Consider sharing lifts to work or to the shops. Even once a week will make a difference.

Electrical loads

Electricity is 'fuel' too; the alternator which charges the battery does so by converting some of the engine's energy of motion into electrical energy. The more electrical accessories are in use, the greater the load on the alternator. Switch off big consumers like the heated rear window when not required.

Freewheeling

Freewheeling (coasting) in neutral with the engine switched off is dangerous. The effort required to operate power-assisted brakes and steering increases when the engine is not running, with a potential lack of control in emergency situations.

In any case, modern fuel injection systems automatically cut off the engine's fuel supply on the overrun (moving and in gear, but with the accelerator pedal released).

Gadgets

Bolt-on devices claiming to save fuel have been around for nearly as long as the motor car itself. Those which worked were rapidly adopted as standard equipment by the vehicle manufacturers. Others worked only in certain situations, or saved fuel only at the expense of unacceptable effects on performance, driveability or the life of engine components.

The most effective fuel saving gadget is the driver's right foot.

Journey planning

Combine (eg) a trip to the supermarket with a visit to the recycling centre and the DIY store, rather than making separate journeys.

When possible choose a travelling time outside rush hours.

Load

The more heavily a car is laden, the greater the energy required to accelerate it to a given speed. Remove heavy items which you don't need to carry.

One load which is often overlooked is the contents of the fuel tank. A tankful of fuel (55 litres / 12 gallons) weighs 45 kg (100 lb) or so. Just half filling it may be worthwhile.

Lost?

At the risk of stating the obvious, if you're going somewhere new, have details of the route to hand. There's not much point in achieving record mpg if you also go miles out of your way.

Parking

If possible, carry out any reversing or turning manoeuvres when you arrive at a parking space so that you can drive straight out when you leave. Manoeuvering when the engine is cold uses a lot more fuel.

Driving around looking for free on-street parking may cost more in fuel than buying a car park ticket.

Premium fuel

Most major oil companies (and some supermarkets) have premium grades of fuel which are several pence a litre dearer than the standard grades. Reports vary, but the consensus seems to be that if these fuels improve economy at all, they do not do so by enough to justify their extra cost.

Roof rack

When loading a roof rack, try to produce a wedge shape with the narrow end at the front. Any cover should be securely fastened – if it flaps it's creating turbulence and absorbing energy.

Remove roof racks and boxes when not in use – they increase air resistance and can create a surprising amount of noise.

Short journeys

The engine is at its least efficient, and wear is highest, during the first few miles after a cold start. Consider walking, cycling or using public transport.

Speed

The engine is at its most efficient when running at a steady speed and load at the rpm where it develops maximum torque. (You can find this figure in the car's handbook.) For most cars this corresponds to between 55 and 65 mph in top gear.

Above the optimum cruising speed, fuel consumption starts to rise quite sharply. A car travelling at 80 mph will typically be using 30% more fuel than at 60 mph.

Supermarket fuel

It may be cheap but is it any good? In the UK all supermarket fuel must meet the relevant British Standard. The major oil companies will say that their branded fuels have better additive packages which may stop carbon and other deposits building up. A reasonable compromise might be to use one tank of branded fuel to three or four from the supermarket.

Switch off when stationary

Switch off the engine if you look like being stationary for more than 30 seconds or so. This is good for the environment as well as for your pocket. Be aware though that frequent restarts are hard on the battery and the starter motor.

Windows

Driving with the windows open increases air turbulence around the vehicle. Closing the windows promotes smooth airflow and

reduced resistance. The faster you go, the more significant this is.

And finally . . .

Driving techniques associated with good fuel economy tend to involve moderate acceleration and low top speeds. Be considerate to the needs of other road users who may need to make brisker progress; even if you do not agree with them this is not an excuse to be obstructive.

Safety must always take precedence over economy, whether it is a question of accelerating hard to complete an overtaking manoeuvre, killing your speed when confronted with a potential hazard or switching the lights on when it starts to get dark.

Conversion factors

Length (distance)

Inches (in)	x 25.4	= Millimetres (mm)	x 0.0394	=	Inches (in)
Feet (ft)	x 0.305	= Metres (m)	x 3.281	=	Feet (ft)
Miles	x 1.609	= Kilometres (km)	x 0.621	=	Miles

Volume (capacity)

Cubic inches (cu in; in³)	x 16.387	= Cubic centimetres (cc; cm³)	x 0.061	=	Cubic inches (cu in; in³)
Imperial pints (Imp pt)	x 0.568	= Litres (l)	x 1.76	=	Imperial pints (Imp pt)
Imperial quarts (Imp qt)	x 1.137	= Litres (l)	x 0.88	=	Imperial quarts (Imp qt)
Imperial quarts (Imp qt)	x 1.201	= US quarts (US qt)	x 0.833	=	Imperial quarts (Imp qt)
US quarts (US qt)	x 0.946	= Litres (l)	x 1.057	=	US quarts (US qt)
Imperial gallons (Imp gal)	x 4.546	= Litres (l)	x 0.22	=	Imperial gallons (Imp gal)
Imperial gallons (Imp gal)	x 1.201	= US gallons (US gal)	x 0.833	=	Imperial gallons (Imp gal)
US gallons (US gal)	x 3.785	= Litres (l)	x 0.264	=	US gallons (US gal)

Mass (weight)

Ounces (oz)	x 28.35	= Grams (g)	x 0.035	=	Ounces (oz)
Pounds (lb)	x 0.454	= Kilograms (kg)	x 2.205	=	Pounds (lb)

Force

Ounces-force (ozf; oz)	x 0.278	= Newtons (N)	x 3.6	=	Ounces-force (ozf; oz)
Pounds-force (lbf; lb)	x 4.448	= Newtons (N)	x 0.225	=	Pounds-force (lbf; lb)
Newtons (N)	x 0.1	= Kilograms-force (kgf; kg)	x 9.81	=	Newtons (N)

Pressure

Pounds-force per square inch (psi; lbf/in²; lb/in²)	x 0.070	= Kilograms-force per square centimetre (kgf/cm²; kg/cm²)	x 14.223	=	Pounds-force per square inch (psi; lbf/in²; lb/in²)
Pounds-force per square inch (psi; lbf/in²; lb/in²)	x 0.068	= Atmospheres (atm)	x 14.696	=	Pounds-force per square inch (psi; lbf/in²; lb/in²)
Pounds-force per square inch (psi; lbf/in²; lb/in²)	x 0.069	= Bars	x 14.5	=	Pounds-force per square inch (psi; lbf/in²; lb/in²)
Pounds-force per square inch (psi; lbf/in²; lb/in²)	x 6.895	= Kilopascals (kPa)	x 0.145	=	Pounds-force per square inch (psi; lbf/in²; lb/in²)
Kilopascals (kPa)	x 0.01	= Kilograms-force per square centimetre (kgf/cm²; kg/cm²)	x 98.1	=	Kilopascals (kPa)
Millibar (mbar)	x 100	= Pascals (Pa)	x 0.01	=	Millibar (mbar)
Millibar (mbar)	x 0.0145	= Pounds-force per square inch (psi; lbf/in²; lb/in²)	x 68.947	=	Millibar (mbar)
Millibar (mbar)	x 0.75	= Millimetres of mercury (mmHg)	x 1.333	=	Millibar (mbar)
Millibar (mbar)	x 0.401	= Inches of water (inH₂O)	x 2.491	=	Millibar (mbar)
Millimetres of mercury (mmHg)	x 0.535	= Inches of water (inH₂O)	x 1.868	=	Millimetres of mercury (mmHg)
Inches of water (inH₂O)	x 0.036	= Pounds-force per square inch (psi; lbf/in²; lb/in²)	x 27.68	=	Inches of water (inH₂O)

Torque (moment of force)

Pounds-force inches (lbf in; lb in)	x 1.152	= Kilograms-force centimetre (kgf cm; kg cm)	x 0.868	=	Pounds-force inches (lbf in; lb in)
Pounds-force inches (lbf in; lb in)	x 0.113	= Newton metres (Nm)	x 8.85	=	Pounds-force inches (lbf in; lb in)
Pounds-force inches (lbf in; lb in)	x 0.083	= Pounds-force feet (lbf ft; lb ft)	x 12	=	Pounds-force inches (lbf in; lb in)
Pounds-force feet (lbf ft; lb ft)	x 0.138	= Kilograms-force metres (kgf m; kg m)	x 7.233	=	Pounds-force feet (lbf ft; lb ft)
Pounds-force feet (lbf ft; lb ft)	x 1.356	= Newton metres (Nm)	x 0.738	=	Pounds-force feet (lbf ft; lb ft)
Newton metres (Nm)	x 0.102	= Kilograms-force metres (kgf m; kg m)	x 9.804	=	Newton metres (Nm)

Power

Horsepower (hp)	x 745.7	= Watts (W)	x 0.0013	=	Horsepower (hp)

Velocity (speed)

Miles per hour (miles/hr; mph)	x 1.609	= Kilometres per hour (km/hr; kph)	x 0.621	=	Miles per hour (miles/hr; mph)

Fuel consumption*

Miles per gallon, Imperial (mpg)	x 0.354	= Kilometres per litre (km/l)	x 2.825	=	Miles per gallon, Imperial (mpg)
Miles per gallon, US (mpg)	x 0.425	= Kilometres per litre (km/l)	x 2.352	=	Miles per gallon, US (mpg)

Temperature

Degrees Fahrenheit = ($°C \times 1.8$) + 32   Degrees Celsius (Degrees Centigrade; °C) = ($°F - 32$) x 0.56

It is common practice to convert from miles per gallon (mpg) to litres/100 kilometres (l/100km), where mpg x l/100 km = 282

Spare parts are available from many sources, including maker's appointed garages, accessory shops, and motor factors. To be sure of obtaining the correct parts, it will sometimes be necessary to quote the vehicle identification number. If possible, it can also be useful to take the old parts along for positive identification. Items such as starter motors and alternators may be available under a service exchange scheme – any parts returned should be clean.

Our advice regarding spare parts is as follows.

Officially appointed garages

This is the best source of parts which are peculiar to your car, and which are not otherwise generally available (eg, badges, interior trim, certain body panels, etc). It is also the only place at which you should buy parts if the vehicle is still under warranty.

Accessory shops

These are very good places to buy materials and components needed for the maintenance of your car (oil, air and fuel filters, light bulbs, drivebelts, greases, brake pads, touch-up paint, etc). Components of this nature sold by a reputable shop are of the same standard as those used by the car manufacturer.

Besides components, these shops also sell tools and general accessories, usually have convenient opening hours, charge lower prices, and can often be found close to home. Some accessory shops have parts counters where components needed for almost any repair job can be purchased or ordered.

Motor factors

Good factors will stock all the more important components which wear out comparatively quickly, and can sometimes supply individual components needed for the overhaul of a larger assembly (eg, brake seals and hydraulic parts, bearing shells, pistons, valves). They may also handle work such as cylinder block reboring, crankshaft regrinding, etc.

Tyre and exhaust specialists

These outlets may be independent, or members of a local or national chain. They frequently offer competitive prices when compared with a main dealer or local garage, but it will pay to obtain several quotes before making a decision. When researching prices, also ask what 'extras' may be added – for instance fitting a new valve and balancing the wheel are both commonly charged on top of the price of a new tyre.

Other sources

Beware of parts or materials obtained from market stalls, car boot sales or similar outlets. Such items are not invariably sub-standard, but there is little chance of compensation if they do prove unsatisfactory. In the case of safety-critical components such as brake pads, there is the risk not only of financial loss, but also of an accident causing injury or death.

Second-hand components or assemblies obtained from a car breaker can be a good buy in some circumstances, but his sort of purchase is best made by the experienced DIY mechanic.

Whenever servicing, repair or overhaul work is carried out on the car or its components, observe the following procedures and instructions. This will assist in carrying out the operation efficiently and to a professional standard of workmanship.

Joint mating faces and gaskets

When separating components at their mating faces, never insert screwdrivers or similar implements into the joint between the faces in order to prise them apart. This can cause severe damage which results in oil leaks, coolant leaks, etc upon reassembly. Separation is usually achieved by tapping along the joint with a soft-faced hammer in order to break the seal. However, note that this method may not be suitable where dowels are used for component location.

Where a gasket is used between the mating faces of two components, a new one must be fitted on reassembly; fit it dry unless otherwise stated in the repair procedure. Make sure that the mating faces are clean and dry, with all traces of old gasket removed. When cleaning a joint face, use a tool which is unlikely to score or damage the face, and remove any burrs or nicks with an oilstone or fine file.

Make sure that tapped holes are cleaned with a pipe cleaner, and keep them free of jointing compound, if this is being used, unless specifically instructed otherwise.

Ensure that all orifices, channels or pipes are clear, and blow through them, preferably using compressed air.

Oil seals

Oil seals can be removed by levering them out with a wide flat-bladed screwdriver or similar implement. Alternatively, a number of self-tapping screws may be screwed into the seal, and these used as a purchase for pliers or some similar device in order to pull the seal free.

Whenever an oil seal is removed from its working location, either individually or as part of an assembly, it should be renewed.

The very fine sealing lip of the seal is easily damaged, and will not seal if the surface it contacts is not completely clean and free from scratches, nicks or grooves. If the original sealing surface of the component cannot be restored, and the manufacturer has not made provision for slight relocation of the seal relative to the sealing surface, the component should be renewed.

Protect the lips of the seal from any surface which may damage them in the course of fitting. Use tape or a conical sleeve where possible. Where indicated, lubricate the seal lips with oil before fitting and, on dual-lipped seals, fill the space between the lips with grease.

Unless otherwise stated, oil seals must be fitted with their sealing lips toward the lubricant to be sealed.

Use a tubular drift or block of wood of the appropriate size to install the seal and, if the seal housing is shouldered, drive the seal down to the shoulder. If the seal housing is unshouldered, the seal should be fitted with its face flush with the housing top face (unless otherwise instructed).

Screw threads and fastenings

Seized nuts, bolts and screws are quite a common occurrence where corrosion has set in, and the use of penetrating oil or releasing fluid will often overcome this problem if the offending item is soaked for a while before attempting to release it. The use of an impact driver may also provide a means of releasing such stubborn fastening devices, when used in conjunction with the appropriate screwdriver bit or socket. If none of these methods works, it may be necessary to resort to the careful application of heat, or the use of a hacksaw or nut splitter device. Before resorting to extreme methods, check that you are not dealing with a left-hand thread!

Studs are usually removed by locking two nuts together on the threaded part, and then using a spanner on the lower nut to unscrew the stud. Studs or bolts which have broken off below the surface of the component in which they are mounted can sometimes be removed using a stud extractor.

Always ensure that a blind tapped hole is completely free from oil, grease, water or other fluid before installing the bolt or stud. Failure to do this could cause the housing to crack due to the hydraulic action of the bolt or stud as it is screwed in.

For some screw fastenings, notably cylinder head bolts or nuts, torque wrench settings are no longer specified for the latter stages of tightening, "angle-tightening" being called up instead. Typically, a fairly low torque wrench setting will be applied to the bolts/nuts in the correct sequence, followed by one or more stages of tightening through specified angles.

When checking or retightening a nut or bolt to a specified torque setting, slacken the nut or bolt by a quarter of a turn, and then retighten to the specified setting. However, this should not be attempted where angular tightening has been used.

Locknuts, locktabs and washers

Any fastening which will rotate against a component or housing during tightening should always have a washer between it and the relevant component or housing.

Spring or split washers should always be renewed when they are used to lock a critical component such as a big-end bearing retaining bolt or nut. Locktabs which are folded over to retain a nut or bolt should always be renewed.

Self-locking nuts can be re-used in non-critical areas, providing resistance can be felt when the locking portion passes over the bolt or stud thread. However, it should be noted that self-locking stiffnuts tend to lose their effectiveness after long periods of use, and should then be renewed as a matter of course.

Split pins must always be replaced with new ones of the correct size for the hole.

When thread-locking compound is found on the threads of a fastener which is to be re-used, it should be cleaned off with a wire brush and solvent, and fresh compound applied on reassembly.

Special tools

Some repair procedures in this manual entail the use of special tools such as a press, two or three-legged pullers, spring compressors, etc. Wherever possible, suitable readily-available alternatives to the manufacturer's special tools are described, and are shown in use. In some instances, where no alternative is possible, it has been necessary to resort to the use of a manufacturer's tool, and this has been done for reasons of safety as well as the efficient completion of the repair operation. Unless you are highly-skilled and have a thorough understanding of the procedures described, never attempt to bypass the use of any special tool when the procedure described specifies its use. Not only is there a very great risk of personal injury, but expensive damage could be caused to the components involved.

Environmental considerations

When disposing of used engine oil, brake fluid, antifreeze, etc, give due consideration to any detrimental environmental effects. Do not, for instance, pour any of the above liquids down drains into the general sewage system, or onto the ground to soak away, as this is likely to pollute your local environment. Many local council refuse tips provide a facility for waste oil disposal, as do some garages. You can find your nearest disposal point by calling the Environment Agency on 03708 506 506 or by visiting www.oilbankline.org.uk.

Note: It is illegal and anti-social to dump oil down the drain. To find the location of your local oil recycling bank, call 03708 506 506 or visit www.oilbankline.org.uk.

The jack supplied with the vehicle tool kit should only be used for changing roadwheels – see *Wheel changing*. Ensure the jack head is correctly engaged before attempting to raise the vehicle. When carrying out any other kind of work, raise the vehicle using a hydraulic jack, and always supplement the jack with axle stands positioned under the vehicle jacking points.

When jacking up the vehicle with a trolley jack, position the jack head under one of the relevant jacking points. Use a block of wood between the jack or axle stand and the sill – the block of wood should have a groove cut into it, in which the welded flange of the sill will locate. Do not jack the vehicle under the sump or any of the steering or suspension components.

Supplement the jack using axle stands **(see illustrations)**.

 ⚠️ *Warning: Never work under, around, or near a raised vehicle, unless it is adequately supported in at least two places.*

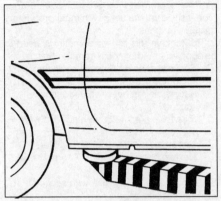

Front jacking point for hydraulic jack or axle stands

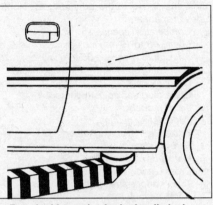

Rear jacking point for hydraulic jack or axle stands

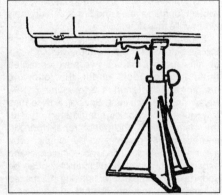

Axle stands should be placed under, or adjacent to the jacking point (arrowed)

Vehicle identification

Modifications are a continuing and unpublished process in vehicle manufacture, quite apart from major model changes. Spare parts manuals and lists are compiled upon a numerical basis, the individual vehicle numbers being essential to correct identification of the component required.

When ordering spare parts, always give as much information as possible. Quote the car model, year of manufacture and vehicle identification and/or engine numbers as appropriate.

The *vehicle identification label* is attached to the side door pillar and includes the Vehicle Identification Number (VIN) and vehicle weight information.

The *Vehicle Identification Number* (VIN) is given on the vehicle identification label and may also be viewed through the base of the windscreen on the passenger's side.

The *engine number* is stamped on a horizontal flat located on the front of the cylinder block, at the transmission end. The first part of the engine number gives the engine code – eg A14NET (LUJ).

Vauxhall/Opel use a 'Car pass' scheme for vehicle identification. This is a card which is issued to the customer when the car is first purchased. It contains important information, eg, VIN, key number and radio code. It also includes a special code for diagnostic equipment, therefore it must be kept in a secure place and not in the vehicle.

Introduction

A selection of good tools is a fundamental requirement for anyone contemplating the maintenance and repair of a motor vehicle. For the owner who does not possess any, their purchase will prove a considerable expense, offsetting some of the savings made by doing-it-yourself. However, provided that the tools purchased meet the relevant national safety standards and are of good quality, they will last for many years and prove an extremely worthwhile investment.

To help the average owner to decide which tools are needed to carry out the various tasks detailed in this manual, we have compiled three lists of tools under the following headings: *Maintenance and minor repair, Repair and overhaul*, and *Special*. Newcomers to practical mechanics should start off with the *Maintenance and minor repair* tool kit, and confine themselves to the simpler jobs around the vehicle. Then, as confidence and experience grow, more difficult tasks can be undertaken, with extra tools being purchased as, and when, they are needed. In this way, a *Maintenance and minor repair* tool kit can be built up into a *Repair and overhaul* tool kit over a considerable period of time, without any major cash outlays. The experienced do-it-yourselfer will have a tool kit good enough for most repair and overhaul procedures, and will add tools from the *Special* category when it is felt that the expense is justified by the amount of use to which these tools will be put.

Maintenance and minor repair tool kit

The tools given in this list should be considered as a minimum requirement if routine maintenance, servicing and minor repair operations are to be undertaken. We recommend the purchase of combination spanners (ring one end, open-ended the other); although more expensive than open-ended ones, they do give the advantages of both types of spanner.

☐ *Combination spanners:*
  Metric - 8 to 19 mm inclusive
☐ *Adjustable spanner - 35 mm jaw (approx.)*
☐ *Spark plug spanner (with rubber insert) - petrol models*
☐ *Spark plug gap adjustment tool - petrol models*
☐ *Set of feeler gauges*
☐ *Brake bleed nipple spanner*
☐ *Screwdrivers:*
  Flat blade - 100 mm long x 6 mm dia
  Cross blade - 100 mm long x 6 mm dia
  Torx - various sizes (not all vehicles)
☐ *Combination pliers*
☐ *Hacksaw (junior)*
☐ *Tyre pump*
☐ *Tyre pressure gauge*
☐ *Oil can*
☐ *Oil filter removal tool (if applicable)*
☐ *Fine emery cloth*
☐ *Wire brush (small)*
☐ *Funnel (medium size)*
☐ *Sump drain plug key (not all vehicles)*

Repair and overhaul tool kit

These tools are virtually essential for anyone undertaking any major repairs to a motor vehicle, and are additional to those given in the *Maintenance and minor repair* list. Included in this list is a comprehensive set of sockets. Although these are expensive, they will be found invaluable as they are so versatile - particularly if various drives are included in the set. We recommend the half-inch square-drive type, as this can be used with most proprietary torque wrenches.

The tools in this list will sometimes need to be supplemented by tools from the *Special* list:

☐ *Sockets to cover range in previous list (including Torx sockets)*
☐ *Reversible ratchet drive (for use with sockets)*
☐ *Extension piece, 250 mm (for use with sockets)*
☐ *Universal joint (for use with sockets)*
☐ *Flexible handle or sliding T "breaker bar" (for use with sockets)*
☐ *Torque wrench (for use with sockets)*
☐ *Self-locking grips*
☐ *Ball pein hammer*
☐ *Soft-faced mallet (plastic or rubber)*
☐ *Screwdrivers:*
  Flat blade - long & sturdy, short (chubby), and narrow (electrician's) types
  Cross blade – long & sturdy, and short (chubby) types
☐ *Pliers:*
  Long-nosed
  Side cutters (electrician's)
  Circlip (internal and external)
☐ *Cold chisel - 25 mm*
☐ *Scriber*
☐ *Scraper*
☐ *Centre-punch*
☐ *Pin punch*
☐ *Hacksaw*
☐ *Brake hose clamp*
☐ *Brake/clutch bleeding kit*
☐ *Selection of twist drills*
☐ *Steel rule/straight-edge*
☐ *Allen keys (inc. splined/Torx type)*
☐ *Selection of files*
☐ *Wire brush*
☐ *Axle stands*
☐ *Jack (strong trolley or hydraulic type)*
☐ *Light with extension lead*
☐ *Universal electrical multi-meter*

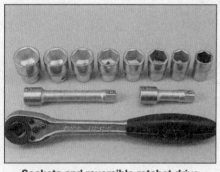

Sockets and reversible ratchet drive

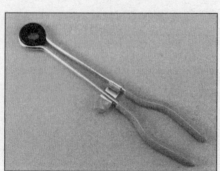

Brake bleeding kit

Torx key, socket and bit

Hose clamp

Angular-tightening gauge

Special tools

The tools in this list are those which are not used regularly, are expensive to buy, or which need to be used in accordance with their manufacturers' instructions. Unless relatively difficult mechanical jobs are undertaken frequently, it will not be economic to buy many of these tools. Where this is the case, you could consider clubbing together with friends (or joining a motorists' club) to make a joint purchase, or borrowing the tools against a deposit from a local garage or tool hire specialist.

The following list contains only those tools and instruments freely available to the public, and not those special tools produced by the vehicle manufacturer specifically for its dealer network. You will find occasional references to these manufacturers' special tools in the text of this manual. Generally, an alternative method of doing the job without the vehicle manufacturers' special tool is given. However, sometimes there is no alternative to using them. Where this is the case and the relevant tool cannot be bought or borrowed, you will have to entrust the work to a dealer.

☐ Angular-tightening gauge
☐ Valve spring compressor
☐ Valve grinding tool
☐ Piston ring compressor
☐ Piston ring removal/installation tool
☐ Cylinder bore hone
☐ Balljoint separator
☐ Coil spring compressors (where applicable)
☐ Two/three-legged hub and bearing puller
☐ Impact screwdriver
☐ Micrometer and/or vernier calipers
☐ Dial gauge
☐ Tachometer
☐ Fault code reader
☐ Cylinder compression gauge
☐ Hand-operated vacuum pump and gauge
☐ Clutch plate alignment set
☐ Brake shoe steady spring cup removal tool
☐ Bush and bearing removal/installation set
☐ Stud extractors
☐ Tap and die set
☐ Lifting tackle

Buying tools

Reputable motor accessory shops and superstores often offer excellent quality tools at discount prices, so it pays to shop around.

Remember, you don't have to buy the most expensive items on the shelf, but it is always advisable to steer clear of the very cheap tools. Beware of 'bargains' offered on market stalls, on line or at car boot sales. There are plenty of good tools around at reasonable prices, but always aim to purchase items which meet the relevant national safety standards. If in doubt, ask the proprietor or manager of the shop for advice before making a purchase.

Care and maintenance of tools

Having purchased a reasonable tool kit, it is necessary to keep the tools in a clean and serviceable condition. After use, always wipe off any dirt, grease and metal particles using a clean, dry cloth, before putting the tools away. Never leave them lying around after they have been used. A simple tool rack on the garage or workshop wall for items such as screwdrivers and pliers is a good idea. Store all normal spanners and sockets in a metal box. Any measuring instruments, gauges, meters, etc, must be carefully stored where they cannot be damaged or become rusty.

Take a little care when tools are used. Hammer heads inevitably become marked, and screwdrivers lose the keen edge on their blades from time to time. A little timely attention with emery cloth or a file will soon restore items like this to a good finish.

Working facilities

Not to be forgotten when discussing tools is the workshop itself. If anything more than routine maintenance is to be carried out, a suitable working area becomes essential.

It is appreciated that many an owner-mechanic is forced by circumstances to remove an engine or similar item without the benefit of a garage or workshop. Having done this, any repairs should always be done under the cover of a roof.

Wherever possible, any dismantling should be done on a clean, flat workbench or table at a suitable working height.

Any workbench needs a vice; one with a jaw opening of 100 mm is suitable for most jobs. As mentioned previously, some clean dry storage space is also required for tools, as well as for any lubricants, cleaning fluids, touch-up paints etc, which become necessary.

Another item which may be required, and which has a much more general usage, is an electric drill with a chuck capacity of at least 8 mm. This, together with a good range of twist drills, is virtually essential for fitting accessories.

Last, but not least, always keep a supply of old newspapers and clean, lint-free rags available, and try to keep any working area as clean as possible.

Micrometers

Dial test indicator ("dial gauge")

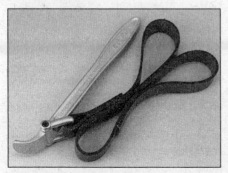
Oil filter removal tool (strap wrench type)

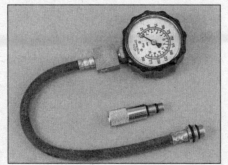
Compression tester

Bearing puller

This is a guide to getting your vehicle through the MOT test. Obviously it will not be possible to examine the vehicle to the same standard as the professional MOT tester. However, working through the following checks will enable you to identify any problem areas before submitting the vehicle for the test.

It has only been possible to summarise the test requirements here, based on the regulations in force at the time of printing. Test standards are becoming increasingly stringent, although there are some exemptions for older vehicles.

An assistant will be needed to help carry out some of these checks.

The checks have been sub-divided into four categories, as follows:

1 Checks carried out **FROM THE VEHICLE INTERIOR**

2 Checks carried out **WITH THE VEHICLE ON THE GROUND**

3 Checks carried out **WITH THE VEHICLE RAISED AND THE WHEELS FREE TO TURN**

4 Checks carried out on **YOUR VEHICLE'S EXHAUST EMISSION SYSTEM**

1 Checks carried out **FROM THE VEHICLE INTERIOR**

Handbrake (parking brake)

☐ Test the operation of the handbrake. Excessive travel (too many clicks) indicates incorrect brake or cable adjustment.

☐ Check that the handbrake cannot be released by tapping the lever sideways. Check the security of the lever mountings.

☐ If the parking brake is foot-operated, check that the pedal is secure and without excessive travel, and that the release mechanism operates correctly.

☐ Where applicable, test the operation of the electronic handbrake. The brake should engage and disengage without excessive delay. If the warning light does not extinguish, or a warning message is displayed when the brake is disengaged, this could indicate a fault which will need further investigation.

Footbrake

☐ Depress the brake pedal and check that it does not creep down to the floor, indicating a master cylinder fault. Release the pedal, wait a few seconds, then depress it again. If the pedal travels nearly to the floor before firm resistance is felt, brake adjustment or repair is necessary. If the pedal feels spongy, there is air in the hydraulic system which must be removed by bleeding.

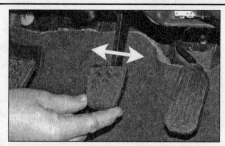

☐ Check that the brake pedal is secure and in good condition. Check also for signs of fluid leaks on the pedal, floor or carpets, which would indicate failed seals in the brake master cylinder.

☐ Check the servo unit (when applicable) by operating the brake pedal several times, then keeping the pedal depressed and starting the engine. As the engine starts, the pedal will move down. If not, the vacuum hose or the servo itself may be faulty.

Steering wheel and column

☐ Examine the steering wheel for fractures or looseness of the hub, spokes or rim.

☐ Move the steering wheel from side to side and then up and down. Check that the steering wheel is not loose on the column, indicating wear or a loose retaining nut. Continue moving the steering wheel as before, but also turn it slightly from left to right.

☐ Check that the steering wheel is not loose on the column, and that there is no abnormal movement of the steering wheel, indicating wear in the column support bearings or couplings.

☐ Check that the ignition lock (where fitted) engages and disengages correctly.

☐ Steering column adjustment mechanisms (where fitted) must be able to lock the column securely in place with no play evident.

Windscreen, mirrors and sunvisor

☐ The windscreen must be free of cracks or other significant damage within the 'swept area' of the windscreen. This is the area swept by the windscreen wipers. A second test area, known as 'Zone A', is the part of the swept area 290 mm wide, centred on the steering wheel centre line. Any damage in Zone A that cannot be contained in a 10 mm diameter circle, or any damage in the remainder of the swept area that cannot be contained in a 40 mm diameter circle, may cause the vehicle to fail the test.

☐ Any items that may obscure the drivers view, such as stickers, sat-navs, anything hanging from the interior mirror, should be removed prior to the test.

☐ Vehicles registered after 1st August 1978 must have a drivers side mirror, and either an interior mirror, or a passenger's side mirror. Cameras (or indirect vision devices) may replace the mirrors, but they must function correctly.

☐ The driver's sunvisor must be capable of being stored in the "up" position.

Seat belts, seats and supplementary restraint systems (SRS)

Note: *The following checks are applicable to all seat belts, front and rear.*

☐ Examine the webbing of all the belts (including rear belts if fitted) for cuts, serious fraying or deterioration. Fasten and unfasten each belt to check the buckles. If applicable, check the retracting mechanism. Check the security of all seat belt mountings accessible from inside the vehicle, ensuring any height adjustable mountings lock securely in place.

☐ Where the seat belt is attached to a seat, the frame and mountings of the seat form part of the belt mountings, and are to be inspected as such.

☐ Any airbag, or SRS warning light must extinguish a few seconds after the ignition is switched on. Failure to do so indicates a fault which must be investigated.

☐ Seat belts with pre-tensioners, once activated, have a "flag" or similar showing on the seat belt stalk. This, in itself, is a reason for test failure.

☐ Check that the original airbag(s) is/are present, and not obviously defective.

☐ The seats themselves must be securely attached and the backrests must lock in the upright position. The driver's seat must also be able to slide forwards/rearwards, and lock in several positions.

Doors

☐ Both front doors must be able to be opened and closed from outside and inside, and must latch securely when closed.

☐ The rear doors must open from the outside.

☐ Examine all door hinges, catches and striker plates for missing, deteriorated, or insecure parts that could effect the opening and closing of the doors.

Speedometer

☐ The vehicle speedometer must be present, and appear operative. The figures on the speedometer must be legible, and illuminated when the lights are switched on.

2 Checks carried out WITH THE VEHICLE ON THE GROUND

Vehicle identification

☐ Number plates must be in good condition, secure and legible, with letters and numbers correctly spaced – spacing at (A) should be 33 mm and at (B) 11 mm. At the front, digits must be black on a white background and at the rear

black on a yellow background. Other background designs (such as honeycomb) are not permitted.

☐ The VIN plate and/or homologation plate must be permanently displayed and legible.

Electrical equipment

☐ Switch on the ignition and check the operation of the horn.

☐ Check the windscreen washers and wipers, examining the wiper blades; renew damaged or perished blades. The wiper blades must clear a large enough area of the windscreen to provide an 'adequate' view of the road, and be able to be parked in a position where they will not affect the drivers' view.

☐ On vehicles first used from 1st September 2009, the headlight washers (where fitted) must operate correctly.

☐ Check the operation of the stop-lights. This includes any lights that appear to be connected – Eg. high-level lights.

☐ Check the operation of the sidelights and number plate lights. The lenses and reflectors must be secure, clean and undamaged.

☐ Check the operation and alignment of the headlights. The headlight reflectors must not be tarnished and the lenses must be undamaged. Where plastic lenses are fitted, check they haven't deteriorated to the extent where they affect the light ouput or beam image. It's often possible to restore the plastic lens using a suitable polish or aftermarket treatment.

☐ Where HID or LED headlights are fitted, check the operation of the cleaning and self-levelling functions.

☐ The headlight main beam warning lamp must be functional.

☐ On vehicles first used from 1st March 2018, the daytime running lights (where fitted) must operate correctly.

☐ Switch on the ignition and check the operation of the direction indicators (including the instrument panel tell-tale) and the hazard warning lights. Operation of the sidelights and stop-lights must not affect the indicators – if it does, the cause is usually a bad earth at the rear light cluster. Indicators should flash at a rate of between 60 and 120 times per minute – faster or slower than this could indicate a fault with the flasher unit or a bad earth at one of the light units.

☐ The hazard warning lights must operate with the ignition on and off.

☐ Check the operation of the rear foglight(s), including the warning light on the instrument panel or in the switch. Note that the foglight

must be positioned in the centre or driver's side of the vehicle. If only the passenger's side illuminates, the test will fail.

☐ The warning lights must illuminate in accordance with the manufacturers' design (this includes any warning messages). For most vehicles, the ABS and other warning lights should illuminate when the ignition is switched on, and (if the system is operating properly) extinguish after a few seconds. Refer to the owner's handbook.

☐ On vehicles first used from 1st September 2009, the reversing lights must operate correctly when reverse gear is selected.

☐ Check the vehicle battery for security and leakage.

☐ Check the visible/accessible vehicle wiring is adequately supported, with no evidence of damage or deterioration that could result in a short-circuit.

Footbrake

☐ Examine the master cylinder, brake pipes and servo unit for leaks, loose mountings, corrosion or other damage. If ABS is fitted, this unit should also be examined for signs of leaks or corrosion.

☐ The fluid reservoir must be secure and the fluid level must be between the upper (A) and lower (B) markings.

☐ Check the fluid in the reservoir for signs of contamination.

☐ Inspect both front brake flexible hoses for cracks or deterioration of the rubber. Turn the steering from lock to lock, and ensure that the hoses do not contact the wheel, tyre, or any part of the steering or suspension mechanism. With the brake pedal firmly depressed, check the hoses for bulges or leaks under pressure.

Steering and suspension

☐ Have your assistant turn the steering wheel from side to side slightly, up to the point where the steering gear just begins to transmit this movement to the roadwheels. Check for excessive free play between the steering wheel and the steering gear, indicating wear or insecurity of the steering column joints, the column-to-steering gear coupling, or the steering gear itself. With a standard (380 mm diameter) steering wheel, there should be no more than 13 mm of free play for rack-and-pinion systems, and no more than 75 mm for non-rack-and-pinion designs.

☐ Have your assistant turn the steering

wheel more vigorously in each direction, so that the roadwheels just begin to turn. As this is done, examine all the steering joints, linkages, fittings and attachments. Renew any component that shows signs of wear or damage. On vehicles with hydraulic power steering, check the security and condition of the steering pump, drivebelt and hoses.

☐ Note that all movement checks on power steering systems are carried out with the engine running.

☐ Check that the vehicle is standing level, and at approximately the correct ride height.

Exhaust system

☐ Start the engine. With your assistant holding a rag over the tailpipe, check the entire system for leaks. Repair or renew leaking sections.

3 Checks carried out WITH THE VEHICLE RAISED AND THE WHEELS FREE TO TURN

Jack up the front and rear of the vehicle, and securely support it on axle stands. Position the stands clear of the suspension assemblies. Ensure that the wheels are clear of the ground and that the steering can be turned from lock to lock.

Steering mechanism

☐ Have your assistant turn the steering from lock to lock. Check that the steering turns smoothly, and that no part of the steering mechanism, including a wheel or tyre, fouls any brake hose or pipe or any part of the body structure.

☐ Examine the steering rack rubber gaiters for damage or insecurity of the retaining clips. If power steering is fitted, check for signs of damage or leakage of the fluid hoses, pipes or connections. Also check for excessive stiffness or binding of the steering, a missing split pin or locking device, or severe corrosion of the body structure within 30 cm of any steering component attachment point.

☐ Check the track rod end ball joint dust covers. Any covers that are missing, seriously damaged, deteriorated or insecure, may fail inspection.

Front and rear suspension and wheel bearings

☐ Starting at the front right-hand side, grasp the roadwheel at the 3 o'clock and 9 o'clock positions and rock gently but firmly. Check for free play or insecurity at the wheel bearings, suspension balljoints, or suspension mountings, pivots and attachments.

☐ Now grasp the wheel at the 12 o'clock and 6 o'clock positions and repeat the previous inspection. Spin the wheel, and check for roughness or tightness of the front wheel bearing.

☐ If excess free play is suspected at a component pivot point, this can be confirmed by using a large screwdriver or similar tool and levering between the mounting and the component attachment. This will confirm whether the wear is in the pivot bush, its retaining bolt, or in the mounting itself (the bolt holes can often become elongated).

☐ Carry out all the above checks at the other front wheel, and then at both rear wheels.

Springs and shock absorbers

☐ Examine the suspension struts (when applicable) for serious fluid leakage, corrosion, or damage to the casing. Also check the security of the mounting points.

☐ If coil springs are fitted, check that the spring ends locate in their seats, and that the spring is not corroded, cracked or broken.

☐ If leaf springs are fitted, check that all leaves are intact, that the axle is securely attached to each spring, and that there is no deterioration of the spring eye mountings, bushes, and shackles.

☐ The same general checks apply to vehicles fitted with other suspension types, such as torsion bars, hydraulic displacer units, etc. Ensure that all mountings and attachments are secure, that there are no signs of excessive wear, corrosion or damage, and (on hydraulic types) that there are no fluid leaks or damaged pipes.

☐ Check any suspension and anti-roll bar link ball joint dust covers. Any covers that are missing, seriously damaged, deteriorated or insecure, may fail inspection.

☐ Examine each shock absorber for signs of leakage, corrosion of the casing, missing, detached or worn pivots and/or rubber bushes.

Driveshafts (fwd vehicles only)

☐ Rotate each front wheel in turn and inspect the inner and outer joint gaiters for splits or damage. Also check that each driveshaft is straight and undamaged.

Braking system

☐ If possible without dismantling, check brake pad wear and disc condition. Ensure that the friction lining material has not worn excessively, (A) and that the discs are not fractured, pitted, scored or badly worn (B). As a general rule, if the friction material is less than 1.5 mm thick, the inspection will fail.

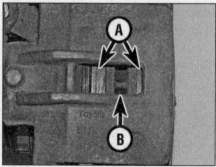

☐ Examine all the rigid brake pipes underneath the vehicle, and the flexible hose(s) at the rear. Look for corrosion, chafing or insecurity of the pipes, and for signs of bulging under pressure, chafing, splits or deterioration of the flexible hoses.

☐ Look for signs of fluid leaks at the brake calipers or on the brake backplates. Repair or renew leaking components.

☐ Slowly spin each wheel, while your assistant depresses and releases the footbrake. Ensure that each brake is operating and does not bind when the pedal is released.

☐ Examine the handbrake mechanism, checking for frayed or broken cables, excessive corrosion, or wear or insecurity of the linkage. Check that the mechanism works on each relevant wheel, and releases fully, without binding.

☐ Check the ABS sensors' wiring for signs of damage, deterioration or insecurity.

☐ It is not possible to test brake efficiency without special equipment, but a road test can be carried out later to check that the vehicle pulls up in a straight line.

Fuel and exhaust systems

☐ Inspect the fuel tank (including the filler cap), fuel pipes, hoses and unions. All components must be secure and free from leaks. Locking fuel caps must lock securely and the key must be provided for the MOT test.

☐ Examine the exhaust system over its entire length, checking for any damaged, broken or missing mountings, security of the retaining clamps and rust or corrosion.

☐ If the vehicle was originally equipped with a catalytic converter or particulate filter, one must be fitted.

Wheels and tyres

☐ Examine the sidewalls and tread area of each tyre in turn. Check for cuts, tears, lumps, bulges, separation of the tread, and exposure of the ply or cord due to wear or damage. Check that the tyre bead is correctly seated on the wheel rim, that the valve is sound and properly seated, and that the wheel is not distorted or damaged.

☐ Check that the tyres are of the correct size for the vehicle, that they are of the same size and type on each axle, and that the pressures are correct. The vehicle will fail the test if the tyres are obviously under-inflated.

☐ Check the tyre tread depth. The legal minimum at the time of writing is 1.6 mm over the central three-quarters of the tread width. Abnormal tread wear may indicate incorrect front wheel alignment or wear in steering or suspension components.

☐ Check that all wheel bolts/nuts are present.

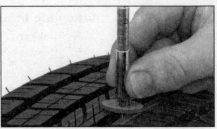

☐ If the spare wheel is fitted externally or in a separate carrier beneath the vehicle, check that mountings are secure and free of excessive corrosion.

Body corrosion

☐ Check the condition of the entire vehicle structure for signs of corrosion in load-bearing areas. (These include chassis box sections, side sills, cross-members, pillars, and all suspension, steering, braking system and seat belt mountings and anchorages.) Any corrosion which has seriously reduced the thickness of a load-bearing area (or is within 30 cm of safety-related components such as steering or suspension) is likely to cause the vehicle to fail. In this case professional repairs are likely to be needed.

☐ Damage or corrosion which causes sharp or otherwise dangerous edges to be exposed will also cause the vehicle to fail.

Towbars

☐ Check the condition of mounting points (both beneath the vehicle and within boot/hatchback areas) for signs of corrosion, ensuring that all fixings are secure and not worn or damaged. There must be no excessive play in detachable tow ball arms or quick-release mechanisms.

☐ Examine the security and condition of the towbar electrics socket. If the later 13-pin socket is fitted, the MOT tester will check its' wiring functions/connections are correct.

General leaks

☐ The vehicle will fail the test if there is a fluid leak of any kind that poses an environmental risk.

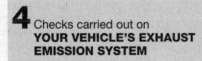

4 Checks carried out on **YOUR VEHICLE'S EXHAUST EMISSION SYSTEM**

Petrol models

☐ The engine should be warmed up, and running well (ignition system in good order, air filter element clean, etc).

☐ Before testing, run the engine at around 2500 rpm for 20 seconds. Let the engine drop to idle, and watch for smoke from the exhaust. If the idle speed is too high, or if dense blue or black smoke emerges for more than 5 seconds, the vehicle will fail. Typically, blue smoke signifies oil burning (engine wear); black smoke means unburnt fuel (dirty air cleaner element, or other fuel system fault).

☐ An exhaust gas analyser for measuring carbon monoxide (CO) and hydrocarbons (HC) is now needed. If one cannot be hired or borrowed, have a local garage perform the check.

CO emissions (mixture)

☐ The MOT tester has access to the CO limits for all vehicles from 1st August 1992. The CO level is measured at idle speed, and at 'fast idle' (2500 to 3000 rpm). The following limits are given as a general guide:

At idle speed – Less than 0.3% CO
At 'fast idle' – Less than 0.2% CO
Lambda reading – 0.97 to 1.03

☐ If the CO level is too high, this may point to poor maintenance, a fuel injection system problem, faulty lambda (oxygen) sensor or catalytic converter. Try an injector cleaning treatment, and check the vehicle's ECU for fault codes.

HC emissions

☐ The MOT tester has access to HC limits for all vehicles. The HC level is measured at 'fast idle' (2500 to 3000 rpm). The following limits are given as a general guide:

At 'fast idle' – Less than 200 ppm

☐ Excessive HC emissions are typically caused by oil being burnt (worn engine), or by a blocked crankcase ventilation system ('breather'). If the engine oil is old and thin, an oil change may help. If the engine is running badly, check the vehicle's ECU for fault codes.

Diesel models

☐ If the vehicle was fitted with a DPF (Diesel Particulate Filter) when it left the factory, it will fail the test if the MOT tester can see smoke of any colour emitting from the exhaust, or finds evidence that the filter has been tampered with.

☐ The only emission test for diesel engines is measuring exhaust smoke density, using a calibrated smoke meter.

☐ This test involves accelerating the engine to its maximum unloaded speed a minimum of once, and a maximum of 6 times. With the smoke meter connected, the engine is accelerated quickly to its maximum speed. If the smoke level is at or below the limit specified, the vehicle will pass. If the level is more than the specified limit then two further accelerations are carried out, and an average of the readings calculated. If the vehicle is still over the limit, a further three accelerations are carried out, with the average of the last three calculated after each check.

Note: *On engines with a timing belt, it is VITAL that the belt is in good condition before the test is carried out.*

Vehicles registered after 1st July 2008
Smoke level must not exceed 1.5m-1 – Turbo-charged and non-Turbocharged engines

Vehicles registered before 1st July 2008
Smoke level must not exceed 2.5m-1 – Non-turbo vehicles
Smoke level must not exceed 3.0m-1 – Turbocharged vehicles:

☐ If excess smoke is produced, try fitting a new air cleaner element, or using an injector cleaning treatment. If the engine is running badly, where applicable, check the vehicle's ECU for fault codes. Also check the vehicle's EGR system, where applicable. At high mileages, the injectors may require professional attention.

Engine

- ☐ Engine fails to rotate when attempting to start
- ☐ Engine rotates, but will not start
- ☐ Engine difficult to start when cold
- ☐ Engine difficult to start when hot
- ☐ Starter motor noisy or excessively-rough in engagement
- ☐ Engine starts, but stops immediately
- ☐ Engine idles erratically
- ☐ Engine misfires at idle speed
- ☐ Engine misfires throughout the driving speed range
- ☐ Engine hesitates on acceleration
- ☐ Engine stalls
- ☐ Engine lacks power
- ☐ Engine backfires
- ☐ Oil pressure warning light illuminated with engine running
- ☐ Engine runs-on after switching off
- ☐ Engine noises

Cooling system

- ☐ Overheating
- ☐ Overcooling
- ☐ External coolant leakage
- ☐ Internal coolant leakage
- ☐ Corrosion

Fuel and exhaust systems

- ☐ Excessive fuel consumption
- ☐ Fuel leakage and/or fuel odour
- ☐ Excessive noise or fumes from exhaust system

Clutch

- ☐ Pedal travels to floor – no pressure or very little resistance
- ☐ Clutch fails to disengage (unable to select gears)
- ☐ Clutch slips (engine speed increases, with no increase in vehicle speed)
- ☐ Judder as clutch is engaged
- ☐ Noise when depressing or releasing clutch pedal

Manual transmission

- ☐ Noisy in neutral with engine running
- ☐ Noisy in one particular gear
- ☐ Difficulty engaging gears
- ☐ Jumps out of gear
- ☐ Vibration
- ☐ Lubricant leaks

Automatic transmission

- ☐ Fluid leakage
- ☐ Transmission fluid brown, or has burned smell
- ☐ General gear selection problems
- ☐ Transmission will not downshift (kickdown) with accelerator pedal fully depressed
- ☐ Engine will not start in any gear, or starts in gears other than Park or Neutral
- ☐ Transmission slips, shifts roughly, is noisy, or has no drive in forward or reverse gears

Driveshafts

- ☐ Vibration when accelerating or decelerating
- ☐ Clicking or knocking noise on turns (at slow speed on full-lock)

Braking system

- ☐ Vehicle pulls to one side under braking
- ☐ Noise (grinding or high-pitched squeal) when brakes applied
- ☐ Excessive brake pedal travel
- ☐ Brake pedal feels spongy when depressed
- ☐ Excessive brake pedal effort required to stop vehicle
- ☐ Judder felt through brake pedal or steering wheel when braking
- ☐ Brakes binding
- ☐ Rear wheels locking under normal braking

Suspension and steering

- ☐ Vehicle pulls to one side
- ☐ Wheel wobble and vibration
- ☐ Excessive pitching and/or rolling around corners, or during braking
- ☐ Wandering or general instability
- ☐ Excessively-stiff steering
- ☐ Excessive play in steering
- ☐ Tyre wear excessive

Electrical system

- ☐ Battery will not hold a charge for more than a few days
- ☐ Ignition/no-charge warning light remains illuminated with engine running
- ☐ Lights inoperative
- ☐ Instrument readings inaccurate or erratic
- ☐ Horn inoperative, or unsatisfactory in operation
- ☐ Windscreen wipers inoperative, or unsatisfactory in operation
- ☐ Windscreen washers inoperative, or unsatisfactory in operation
- ☐ Electric windows inoperative, or unsatisfactory in operation
- ☐ Central locking system inoperative, or unsatisfactory in operation

Introduction

The vehicle owner who does his or her own maintenance according to the recommended service schedules should not have to use this section of the manual very often. Modern component reliability is such that, provided those items subject to wear or deterioration are inspected or renewed at the specified intervals, sudden failure is comparatively rare. Faults do not usually just happen as a result of sudden failure, but develop over a period of time. Major mechanical failures in particular are usually preceded by characteristic symptoms over hundreds or even thousands of miles. Those components which do occasionally fail without warning are often small and easily carried in the vehicle.

With any fault-finding, the first step is to decide where to begin investigations. Sometimes this is obvious, but on other occasions, a little detective work will be necessary. The owner who makes half a dozen haphazard adjustments or replacements may be successful in curing a fault (or its symptoms), but will be none the wiser if the fault recurs, and ultimately may have spent more time and money than was necessary. A calm and logical approach will be found to be more satisfactory in the long run. Always take into account any warning signs or abnormalities that may have been noticed in the period preceding the fault – power loss, high or low gauge readings, unusual smells, etc – and remember that failure of components such as fuses or spark plugs may only be pointers to some underlying fault.

The pages which follow provide an easy-reference guide to the more common problems which may occur during the operation of the vehicle. These problems and their possible causes are grouped under headings denoting various components or systems, such as Engine, Cooling system, etc. The general Chapter which deals with the problem is also shown in brackets; refer to the relevant part of that Chapter for system-specific information. Whatever the fault, certain basic principles apply. These are as follows:

Verify the fault. This is simply a matter of

being sure that you know what the symptoms are before starting work. This is particularly important if you are investigating a fault for someone else, who may not have described it very accurately.

Don't overlook the obvious. For example, if the vehicle won't start, is there fuel in the tank? (Don't take anyone else's word on this particular point, and don't trust the fuel gauge either!) If an electrical fault is indicated, look for loose or broken wires before digging out the test gear.

Cure the disease, not the symptom. Substituting a flat battery with a fully-charged one will get you off the hard shoulder, but if the underlying cause is not attended to, the new battery will go the same way. Similarly, changing oil-fouled spark plugs for a new set will get you moving again, but remember that the reason for the fouling (if it wasn't simply an incorrect grade of plug) will have to be established and corrected.

Don't take anything for granted. Particularly, don't forget that a 'new' component may itself be defective (especially if it's been rattling around in the boot for months), and don't leave components out of a fault diagnosis sequence just because they are new or recently-fitted. When you do finally diagnose a difficult fault, you'll probably realise that all the evidence was there from the start.

Consider what work, if any, has recently been carried out. Many faults arise through careless or hurried work. For instance, if any work has been performed under the bonnet, could some of the wiring have been dislodged or incorrectly routed, or a hose trapped? Have all the fasteners been properly tightened? Were new, genuine parts and new gaskets used? There is often a certain amount of detective work to be done in this case, as an apparently-unrelated task can have far-reaching consequences.

Engine

Engine fails to rotate when attempting to start

☐ Battery terminal connections loose or corroded (see *Weekly checks*)
☐ Battery discharged or faulty (Chapter 5A)
☐ Broken, loose or disconnected wiring in the starting circuit (Chapter 5A)
☐ Defective starter solenoid or ignition switch (Chapter 5A)
☐ Defective starter motor (Chapter 5A)
☐ Starter pinion or flywheel ring gear teeth loose or broken (Chapter 2A, 2B, 2C or 2D)
☐ Engine earth strap broken or disconnected (Chapter 5A)

Engine rotates, but will not start

☐ Fuel tank empty
☐ Battery discharged (engine rotates slowly) (Chapter 5A)
☐ Battery terminal connections loose or corroded (see *Weekly checks*)
☐ Ignition components damp or damaged – petrol models (Chapter 5B)
☐ Broken, loose or disconnected wiring in the ignition circuit – petrol models (Chapter 12)
☐ Worn, faulty or incorrectly-gapped spark plugs – petrol models (Chapter 1A)
☐ Fuel injection/engine management system fault (Chapter 4A or 4B)
☐ Major mechanical failure (eg camshaft drive) (Chapter 2A, 2B, 2C or 2D)

Engine difficult to start when cold

☐ Battery discharged (Chapter 5A)
☐ Battery terminal connections loose or corroded (see *Weekly checks*)
☐ Worn, faulty or incorrectly-gapped spark plugs – petrol models (Chapter 1A)
☐ Fuel injection/engine management system fault (Chapter 4A or 4B)
☐ Other ignition system fault (Chapter 1A or 5B)
☐ Low cylinder compression (Chapter 2A, 2B, 2C or 2D)

Engine difficult to start when hot

☐ Air filter element dirty or clogged (Chapter 1A or 1B)
☐ Fuel injection/engine management system fault (Chapter 4A or 4B)
☐ Other ignition system fault (Chapter 1A or 5B)
☐ Low cylinder compression (Chapter 2A, 2B, 2C or 2D)

Starter motor noisy or excessively-rough in engagement

☐ Starter pinion or flywheel ring gear teeth loose or broken (Chapter 2A, 2B, 2C or 2D)
☐ Starter motor mounting bolts loose or missing (Chapter 5A)
☐ Starter motor internal components worn or damaged (Chapter 5A)

Engine starts, but stops immediately

☐ Loose or faulty electrical connections in the ignition circuit – petrol models (Chapter 12)
☐ Vacuum leak at the throttle housing or inlet manifold – petrol models (Chapter 4A or 4B)
☐ Fuel injection/engine management system fault (Chapter 4A or 4B)

Engine idles erratically

☐ Air filter element clogged (Chapter 1A or 1B)
☐ Vacuum leak at the throttle housing, inlet manifold or associated hoses – petrol models (Chapter 4A or 4B)
☐ Worn, faulty or incorrectly-gapped spark plugs – petrol models (Chapter 1A)
☐ Uneven or low cylinder compression (Chapter 2A, 2B, 2C or 2D)
☐ Camshaft lobes worn (Chapter 2A, 2B, 2C or 2D)
☐ Timing chain incorrectly fitted (Chapter 2A or 2C)
☐ Timing belt incorrectly fitted (Chapter 2B or 2D)
☐ Fuel injection/engine management system fault (Chapter 4A or 4B)

Engine misfires at idle speed

☐ Worn, faulty or incorrectly-gapped spark plugs – petrol models (Chapter 1A)
☐ Vacuum leak at the throttle housing, inlet manifold or associated hoses (Chapter 4A or 4B)
☐ Fuel injection/engine management system fault (Chapter 4A or 4B)
☐ Uneven or low cylinder compression (Chapter 2A, 2B, 2C or 2D)
☐ Disconnected, leaking, or perished crankcase ventilation hoses (Chapter 4C)

Engine (continued)

Engine misfires throughout the driving speed range

- ☐ Fuel filter choked – diesel models (Chapter 1B)
- ☐ Fuel pump faulty, or delivery pressure low (Chapter 4A or 4B)
- ☐ Fuel tank vent blocked, or fuel pipes restricted (Chapter 4A or 4B)
- ☐ Vacuum leak at the throttle housing, inlet manifold or associated hoses (Chapter 4A or 4B)
- ☐ Worn, faulty or incorrectly-gapped spark plugs – petrol models (Chapter 1A)
- ☐ Faulty injector(s) – diesel models (Chapter 4B)
- ☐ Faulty ignition coil – petrol models (Chapter 5B)
- ☐ Uneven or low cylinder compression (Chapter 2A, 2B, 2C or 2D)
- ☐ Fuel injection/engine management system fault (Chapter 4A or 4B)

Engine hesitates on acceleration

- ☐ Worn, faulty or incorrectly-gapped spark plugs – petrol models (Chapter 1A)
- ☐ Vacuum leak at the throttle housing, inlet manifold or associated hoses (Chapter 4A or 4B)
- ☐ Fuel injection/engine management system fault (Chapter 4A or 4B)

Engine stalls

- ☐ Vacuum leak at the throttle housing, inlet manifold or associated hoses (Chapter 4A or 4B)
- ☐ Fuel filter choked – diesel models (Chapter 1B)
- ☐ Fuel pump faulty, or delivery pressure low – petrol models (Chapter 4A or 4B)
- ☐ Fuel tank vent blocked, or fuel pipes restricted (Chapter 4A or 4B)
- ☐ Fuel injection/engine management system fault (Chapter 4A or 4B)

Engine lacks power

- ☐ Timing chain incorrectly fitted (Chapter 2A or 2C)
- ☐ Timing belt incorrectly fitted (Chapter 2B or 2D)
- ☐ Air filter element blocked (Chapter 1A or 1B)
- ☐ Fuel filter choked – diesel models (Chapter 1B)
- ☐ Fuel pump faulty, or delivery pressure low (Chapter 4A or 4B)
- ☐ Uneven or low cylinder compression (Chapter 2A, 2B, 2C or 2D)
- ☐ Worn, faulty or incorrectly-gapped spark plugs – petrol models (Chapter 1A)
- ☐ Vacuum leak at the throttle housing, inlet manifold or associated hoses (Chapter 4A or 4B)
- ☐ Fuel injection/engine management system fault (Chapter 4A or 4B)
- ☐ Brakes binding (Chapter 1A, 1B or 9)
- ☐ Clutch slipping (Chapter 1A or 1B)

Engine backfires

- ☐ Timing chain incorrectly fitted (Chapter 2A or 2C)
- ☐ Timing belt incorrectly fitted (Chapter 2B or 2D)
- ☐ Vacuum leak at the throttle housing, inlet manifold or associated hoses (Chapter 4A or 4B)
- ☐ Fuel injection/engine management system fault (Chapter 4A or 4B)

Oil pressure warning light illuminated with engine running

- ☐ Low oil level, or incorrect oil grade (see *Weekly checks*)
- ☐ Faulty oil pressure sensor, or wiring damaged (Chapter 5A)
- ☐ Worn engine bearings and/or oil pump (Chapter 2A, 2B, 2C or 2D)
- ☐ High engine operating temperature (Chapter 3)

Engine runs-on after switching off

- ☐ Excessive carbon build-up in engine (Chapter 2E)
- ☐ High engine operating temperature (Chapter 3)
- ☐ Fuel injection/engine management system fault (Chapter 4A or 4B)

Engine noises

Pre-ignition (pinking) or knocking during acceleration or under load

- ☐ Ignition timing incorrect/ignition system fault (Chapter 5B)
- ☐ Incorrect grade of spark plug – petrol models (Chapter 1A)
- ☐ Incorrect grade of fuel (Chapter 4A)
- ☐ Vacuum leak at the throttle housing, inlet manifold or associated hoses (Chapter 4A or 4B)
- ☐ Excessive carbon build-up in engine (Chapter 2E)
- ☐ Fuel injection/engine management system fault (Chapter 4A or 4B)

Whistling or wheezing noises

- ☐ Leaking inlet manifold or throttle housing gasket (Chapter 4A or 4B)
- ☐ Leaking exhaust manifold gasket or pipe-to-manifold joint (Chapter 4A or 4B)
- ☐ Leaking vacuum hose (Chapter 4A, 4B, 4C or 9)
- ☐ Blowing cylinder head gasket (Chapter 2A, 2B, 2C or 2D)

Tapping or rattling noises

- ☐ Worn valve gear or camshaft (Chapter 2A, 2B, 2C or 2D)
- ☐ Ancillary component fault (coolant pump, alternator, etc) (Chapter 3, 5A, etc)

Knocking or thumping noises

- ☐ Worn big-end bearings (regular heavy knocking, perhaps less under load) (Chapter 2E)
- ☐ Worn main bearings (rumbling and knocking, perhaps worsening under load) (Chapter 2E)
- ☐ Piston slap – most noticeable when cold, caused by piston/bore wear (Chapter 2E)
- ☐ Ancillary component fault (coolant pump, alternator, etc) (Chapter 3, 5A, etc)

Cooling system

Overheating

☐ Insufficient coolant in system (see *Weekly checks*)
☐ Thermostat faulty (Chapter 3)
☐ Radiator core blocked, or grille restricted (Chapter 1A or 1B)
☐ Cooling fan faulty (Chapter 3)
☐ Pressure cap faulty (Chapter 1A or 1B)
☐ Inaccurate coolant temperature sensor (Chapter 3)
☐ Airlock in cooling system (Chapter 1A or 1B)

Overcooling

☐ Thermostat faulty (Chapter 3)
☐ Inaccurate coolant temperature sensor (Chapter 3)

External coolant leakage

☐ Deteriorated or damaged hoses or hose clips (Chapter 3)
☐ Radiator core or heater matrix leaking (Chapter 3)
☐ Pressure cap faulty (Chapter 1A or 1B)
☐ Coolant pump seal leaking (Chapter 3)
☐ Boiling due to overheating (Chapter 3)
☐ Cylinder block core plug leaking (Chapter 2E)

Internal coolant leakage

☐ Leaking cylinder head gasket (Chapter 2A, 2B, 2C or 2D)
☐ Cracked cylinder head or cylinder block (Chapter 2E)

Corrosion

☐ Infrequent draining and flushing (Chapter 1A or 1B)
☐ Incorrect coolant mixture or inappropriate coolant type (Chapter 1A or 1B)

Fuel and exhaust systems

Excessive fuel consumption

☐ Air filter element dirty or clogged (Chapter 1A or 1B)
☐ Fuel injection/engine management system fault (Chapter 4A or 4B)
☐ Tyres under-inflated (see *Weekly checks*)

Fuel leakage and/or fuel odour

☐ Damaged or corroded fuel tank, pipes or connections (Chapter 1A or 1B)

Excessive noise or fumes from exhaust system

☐ Leaking exhaust system or manifold joints (Chapter 4A or 4B)
☐ Leaking, corroded or damaged silencers or pipe (Chapter 4A or 4B)
☐ Broken mountings causing body or suspension contact (Chapter 4A or 4B)

Clutch

Pedal travels to floor – no pressure or very little resistance

☐ Leaking hydraulic fluid (Chapter 6)
☐ Faulty slave or master cylinder (Chapter 6)
☐ Broken diaphragm spring in clutch pressure plate (Chapter 6)

Clutch fails to disengage (unable to select gears)

☐ Leaking hydraulic fluid (Chapter 6)
☐ Clutch plate sticking on gearbox input shaft splines (Chapter 6)
☐ Clutch plate sticking to flywheel or pressure plate (Chapter 6)
☐ Faulty pressure plate assembly (Chapter 6)
☐ Clutch release mechanism worn or incorrectly assembled (Chapter 6)

Clutch slips (engine speed increases, with no increase in vehicle speed)

☐ Leaking hydraulic fluid (Chapter 6)
☐ Clutch plate linings excessively worn (Chapter 6)
☐ Clutch plate linings contaminated with oil or grease (Chapter 6)
☐ Faulty pressure plate or weak diaphragm spring (Chapter 6)

Judder as clutch is engaged

☐ Clutch plate linings contaminated with oil or grease (Chapter 6)
☐ Clutch plate linings excessively worn (Chapter 6)
☐ Leaking hydraulic fluid (Chapter 6)
☐ Faulty or distorted pressure plate or diaphragm spring (Chapter 6).
☐ Worn or loose engine or transmission mountings (Chapter 2A, 2B, 2C or 2D)
☐ Clutch plate hub or gearbox input shaft splines worn (Chapter 6)

Noise when depressing or releasing clutch pedal

☐ Faulty clutch release cylinder (Chapter 6)
☐ Worn or dry clutch pedal bushes (Chapter 6)
☐ Faulty pressure plate assembly (Chapter 6)
☐ Pressure plate diaphragm spring broken (Chapter 6)
☐ Broken clutch disc cushioning springs (Chapter 6)

Manual transmission

Noisy in neutral with engine running

☐ Input shaft bearings worn (noise apparent with clutch pedal released, but not when depressed) (Chapter 7A)*
☐ Clutch release bearing worn (noise apparent with clutch pedal depressed, possibly less when released) (Chapter 6)

Noisy in one particular gear

☐ Worn, damaged or chipped gear teeth (Chapter 7A)*

Difficulty engaging gears

☐ Clutch fault (Chapter 6)
☐ Oil level low (Chapter 7A)
☐ Worn or damaged gearchange linkage (Chapter 7A)
☐ Worn synchroniser units (Chapter 7A)*

Jumps out of gear

☐ Worn or damaged gearchange linkage (Chapter 7A)
☐ Worn synchroniser units (Chapter 7A)*
☐ Worn selector forks (Chapter 7A)*

Vibration

☐ Lack of oil (Chapter 7A)
☐ Worn bearings (Chapter 7A)*

Lubricant leaks

☐ Leaking driveshaft oil seal (Chapter 7A)
☐ Leaking housing joint (Chapter 7A)*
☐ Leaking input shaft oil seal (Chapter 7A)*

Although the corrective action necessary to remedy the symptoms described is beyond the scope of the home mechanic, the above information should be helpful in isolating the cause of the condition, so that the owner can communicate clearly with a professional mechanic.

Driveshafts

Vibration when accelerating or decelerating

☐ Worn inner constant velocity joint (Chapter 8)
☐ Bent or distorted driveshaft (Chapter 8)

Clicking or knocking noise on turns (at slow speed on full-lock)

☐ Lack of constant velocity joint lubricant, possibly due to damaged gaiter (Chapter 8)
☐ Worn outer constant velocity joint (Chapter 8)

Braking system

Note: *Before assuming that a brake problem exists, make sure that the tyres are in good condition and correctly inflated, that the front wheel alignment is correct, and that the vehicle is not loaded with weight in an unequal manner. Apart from checking the condition of all pipe and hose connections, any faults occurring on the anti-lock braking system should be referred to a Vauxhall/Opel dealer for diagnosis.*

Vehicle pulls to one side under braking

☐ Worn, defective, damaged or contaminated brake pads on one side (Chapter 9)
☐ Seized or partially-seized front brake caliper piston (Chapter 9)
☐ A mixture of brake pad lining materials fitted between sides (Chapter 9)
☐ Brake caliper or backplate mounting bolts loose (Chapter 9)
☐ Worn or damaged steering or suspension components (Chapter 1A, 1B or 10)

Noise (grinding or high-pitched squeal) when brakes applied

☐ Brake pad friction lining material worn down to metal backing (Chapter 9)
☐ Excessive corrosion of brake disc (may be apparent after the vehicle has been standing for some time (Chapter 9)
☐ Foreign object (stone chipping, etc) trapped between brake disc and shield (Chapter 1A, 1B or 9)

Excessive brake pedal travel

☐ Faulty master cylinder (Chapter 9)
☐ Air in hydraulic system (Chapter 9)
☐ Faulty vacuum servo unit (Chapter 1A or 1B)

Brake pedal feels spongy when depressed

☐ Air in hydraulic system (Chapter 9)
☐ Deteriorated flexible rubber brake hoses (Chapter 9)
☐ Master cylinder mounting nuts loose (Chapter 9)
☐ Faulty master cylinder (Chapter 9)

Excessive brake pedal effort required to stop vehicle

☐ Faulty vacuum servo unit (Chapter 9)
☐ Disconnected, damaged or insecure brake servo vacuum hose (Chapter 9)
☐ Primary or secondary hydraulic circuit failure (Chapter 9)
☐ Seized brake caliper piston(s) (Chapter 9)
☐ Brake pads incorrectly fitted (Chapter 9)
☐ Incorrect grade of brake pads fitted (Chapter 9)
☐ Brake pad linings contaminated (Chapter 9)

Judder felt through brake pedal or steering wheel when braking

Note: *Under heavy braking on vehicles equipped with ABS, vibration may be felt through the brake pedal. This is a normal feature of ABS operation, and does not constitute a fault.*

☐ Excessive run-out or distortion of discs (Chapter 9)
☐ Brake pad linings worn (Chapter 9)
☐ Brake caliper or backplate mounting bolts loose (Chapter 9)
☐ Wear in suspension or steering components or mountings (Chapter 1A, 1B or 10)
☐ Front wheels out of balance (see *Weekly checks*)

Brakes binding

☐ Seized brake caliper piston(s) (Chapter 9)
☐ Incorrectly-adjusted handbrake mechanism (Chapter 1A, 1B or 9)
☐ Faulty master cylinder (Chapter 9)

Rear wheels locking under normal braking

☐ Rear brake pad linings contaminated or damaged (Chapter 9)

Suspension and steering

Note: *Before diagnosing suspension or steering faults, be sure that the trouble is not due to incorrect tyre pressures, mixtures of tyre types, or binding brakes.*

Vehicle pulls to one side

- ☐ Defective tyre (see *Weekly checks*)
- ☐ Excessive wear in suspension or steering components (Chapter 1A, 1B or 10)
- ☐ Incorrect front wheel alignment (Chapter 10)
- ☐ Accident damage to steering or suspension components

Wheel wobble and vibration

- ☐ Front wheels out of balance (vibration felt mainly through the steering wheel) (see *Weekly checks*)
- ☐ Rear wheels out of balance (vibration felt throughout the vehicle) (see *Weekly checks*)
- ☐ Roadwheels damaged or distorted (see *Weekly checks*)
- ☐ Faulty or damaged tyre (see *Weekly checks*)
- ☐ Worn steering or suspension joints, bushes or components (Chapter 1A, 1B or 10)
- ☐ Wheel nuts loose (Chapter 1A or 1B)

Excessive pitching and/or rolling around corners, or during braking

- ☐ Defective shock absorbers (Chapter 1A, 1B or 10)
- ☐ Broken or weak spring and/or suspension component (Chapter 1A, 1B or 10)
- ☐ Worn or damaged anti-roll bar or mountings (Chapter 1A, 1B or 10)

Wandering or general instability

- ☐ Incorrect front wheel alignment (Chapter 10)
- ☐ Worn steering or suspension joints, bushes or components (Chapter 1A, 1B or 10)
- ☐ Roadwheels out of balance (see *Weekly checks*)
- ☐ Faulty or damaged tyre (see *Weekly checks*)
- ☐ Wheel nuts loose (Chapter 1A or 1B)
- ☐ Defective shock absorbers (Chapter 1A, 1B or 10)

Excessively-stiff steering

- ☐ Lack of steering gear lubricant (Chapter 10)
- ☐ Seized track rod end balljoint or suspension balljoint (Chapter 1A, 1B or 10)
- ☐ Incorrect front wheel alignment (Chapter 10)
- ☐ Steering rack or column bent or damaged (Chapter 10)

Excessive play in steering

- ☐ Worn steering track rod end balljoints (Chapter 1A, 1B or 10)
- ☐ Worn rack and pinion steering gear (Chapter 1A, 1B or 10)
- ☐ Worn steering or suspension joints, bushes or components (Chapter 1A, 1B or 10)

Tyre wear excessive

Tyres worn on inside or outside edges

- ☐ Tyres under-inflated (wear on both edges) (see *Weekly checks*)
- ☐ Incorrect camber or castor angles (wear on one edge only) (Chapter 10)
- ☐ Worn steering or suspension joints, bushes or components (Chapter 1A, 1B or 10)
- ☐ Excessively-hard cornering or braking
- ☐ Accident damage

Tyre treads exhibit feathered edges

- ☐ Incorrect toe-setting (Chapter 10)

Tyres worn in centre of tread

- ☐ Tyres over-inflated (see *Weekly checks*)

Tyres worn on inside and outside edges

- ☐ Tyres under-inflated (see *Weekly checks*)

Tyres worn unevenly

- ☐ Tyres/wheels out of balance (see *Weekly checks*)
- ☐ Excessive wheel or tyre run-out
- ☐ Worn shock absorbers (Chapter 1A, 1B or 10)
- ☐ Faulty tyre (see *Weekly checks*)

Electrical system

Note: *For problems associated with the starting system, refer to the faults listed under 'Engine' earlier in this Section.*

Battery will not hold a charge for more than a few days

☐ Battery defective internally (Chapter 5A)
☐ Battery terminal connections loose or corroded (see *Weekly checks*)
☐ Auxiliary drivebelt worn or faulty automatic adjuster (Chapter 1A or 1B)
☐ Alternator not charging at correct output (Chapter 5A)
☐ Alternator or voltage regulator faulty (Chapter 5A)
☐ Short-circuit causing continual battery drain (Chapter 12)

Ignition/no-charge warning light remains illuminated with engine running

☐ Auxiliary drivebelt broken, worn, or incorrectly adjusted (Chapter 1A or 1B)
☐ Alternator brushes worn, sticking or dirty (Chapter 5A)
☐ Alternator brush springs weak or broken (Chapter 5A)
☐ Internal fault in alternator or voltage regulator (Chapter 5A)
☐ Broken, disconnected, or loose wiring in charging circuit (Chapter 5A)

Lights inoperative

☐ Bulb blown (Chapter 12)
☐ Corrosion of bulb or bulbholder contacts (Chapter 12)
☐ Blown fuse (Chapter 12)
☐ Faulty relay (Chapter 12)
☐ Broken, loose, or disconnected wiring (Chapter 12)
☐ Faulty switch (Chapter 12)

Instrument readings inaccurate or erratic

Fuel or temperature gauges give no reading

☐ Faulty gauge sender unit or temperature sensor (Chapter 4A or 4B)
☐ Wiring open-circuit (Chapter 12)
☐ Faulty instrument panel (Chapter 12)

Fuel or temperature gauges give continuous maximum reading

☐ Faulty gauge sender unit or temperature sensor (Chapter 4A, Section 6, 4B, Section 6 or 3, Section 7).
☐ Wiring short-circuit (Chapter 12, Section 2).
☐ Faulty instrument panel (Chapter 12, Section 9).

Horn inoperative, or unsatisfactory in operation

Horn operates all the time

☐ Horn switch contacts faulty (Chapter 12)

Horn fails to operate

☐ Blown fuse (Chapter 12)
☐ Cable or connections loose, broken or disconnected (Chapter 12)
☐ Faulty horn (Chapter 12)

Horn emits intermittent or unsatisfactory sound

☐ Cable connections loose (Chapter 12)
☐ Horn mountings loose (Chapter 12)
☐ Faulty horn (Chapter 12)

Windscreen/tailgate wipers inoperative, or unsatisfactory in operation

Wipers fail to operate, or operate very slowly

☐ Wiper blades stuck to screen, or linkage seized or binding (Chapter 12)
☐ Blown fuse (Chapter 12)
☐ Cable or connections loose, broken or disconnected (Chapter 12)
☐ Faulty relay (Chapter 12)
☐ Faulty wiper motor (Chapter 12)

Wiper blades sweep over too large or too small an area of the glass

☐ Wiper arms incorrectly positioned on spindles (Chapter 12)
☐ Excessive wear of wiper linkage (Chapter 12)
☐ Wiper motor or linkage mountings loose or insecure (Chapter 12)

Wiper blades fail to clean the glass effectively

☐ Wiper blade rubbers dirty, worn or perished (see *Weekly checks*)
☐ Wiper arm tension springs broken, or arm pivots seized (Chapter 12)
☐ Insufficient windscreen washer additive to adequately remove road film (see *Weekly checks*)

Windscreen washers inoperative, or unsatisfactory in operation

One or more washer jets inoperative

☐ Blocked washer jet
☐ Disconnected, kinked or restricted fluid hose (Chapter 12)
☐ Insufficient fluid in washer reservoir (see *Weekly checks*)

Washer pump fails to operate

☐ Broken or disconnected wiring or connections (Chapter 12)
☐ Blown fuse (Chapter 12)
☐ Faulty washer switch (Chapter 12)
☐ Faulty washer pump (Chapter 12)

Washer pump runs for some time before fluid is emitted from jets

☐ Faulty one-way valve in fluid supply hose (Chapter 12)

Electric windows inoperative, or unsatisfactory in operation

Window glass will only move in one direction

☐ Faulty switch (Chapter 12)

Window glass slow to move

☐ Incorrectly-adjusted door glass guide channels (Chapter 11)
☐ Regulator seized or damaged, or in need of lubrication (Chapter 11)
☐ Door internal components or trim fouling regulator (Chapter 11)
☐ Faulty motor (Chapter 11)

Window glass fails to move

☐ Incorrectly-adjusted door glass guide channels (Chapter 11)
☐ Blown fuse (Chapter 12)
☐ Faulty relay (Chapter 12)
☐ Broken or disconnected wiring or connections (Chapter 12)
☐ Faulty motor (Chapter 11)

Central locking system inoperative, or unsatisfactory in operation

Complete system failure

☐ Blown fuse (Chapter 12)
☐ Faulty relay (Chapter 12)
☐ Broken or disconnected wiring or connections (Chapter 12)

Latch locks but will not unlock, or unlocks but will not lock

☐ Faulty master switch (Chapter 12)
☐ Broken or disconnected latch operating rods or levers (Chapter 11)
☐ Faulty relay (Chapter 12)

One solenoid/motor fails to operate

☐ Broken or disconnected wiring or connections (Chapter 12)
☐ Faulty solenoid/motor (Chapter 11)
☐ Broken, binding or disconnected latch operating rods or levers (Chapter 11)
☐ Fault in door latch (Chapter 11)

A

ABS (Anti-lock brake system) A system, usually electronically controlled, that senses incipient wheel lockup during braking and relieves hydraulic pressure at wheels that are about to skid.

Air bag An inflatable bag hidden in the steering wheel (driver's side) or the dash or glovebox (passenger side). In a head-on collision, the bags inflate, preventing the driver and front passenger from being thrown forward into the steering wheel or windscreen.

Air cleaner A metal or plastic housing, containing a filter element, which removes dust and dirt from the air being drawn into the engine.

Air filter element The actual filter in an air cleaner system, usually manufactured from pleated paper and requiring renewal at regular intervals.

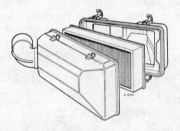

Air filter

Allen key A hexagonal wrench which fits into a recessed hexagonal hole.

Alligator clip A long-nosed spring-loaded metal clip with meshing teeth. Used to make temporary electrical connections.

Alternator A component in the electrical system which converts mechanical energy from a drivebelt into electrical energy to charge the battery and to operate the starting system, ignition system and electrical accessories.

Ampere (amp) A unit of measurement for the flow of electric current. One amp is the amount of current produced by one volt acting through a resistance of one ohm.

Anaerobic sealer A substance used to prevent bolts and screws from loosening. Anaerobic means that it does not require oxygen for activation. The Loctite brand is widely used.

Antifreeze A substance (usually ethylene glycol) mixed with water, and added to a vehicle's cooling system, to prevent freezing of the coolant in winter. Antifreeze also contains chemicals to inhibit corrosion and the formation of rust and other deposits that would tend to clog the radiator and coolant passages and reduce cooling efficiency.

Anti-seize compound A coating that reduces the risk of seizing on fasteners that are subjected to high temperatures, such as exhaust manifold bolts and nuts.

Asbestos A natural fibrous mineral with great heat resistance, commonly used in the composition of brake friction materials.

Asbestos is a health hazard and the dust created by brake systems should never be inhaled or ingested.

Axle A shaft on which a wheel revolves, or which revolves with a wheel. Also, a solid beam that connects the two wheels at one end of the vehicle. An axle which also transmits power to the wheels is known as a live axle.

Axleshaft A single rotating shaft, on either side of the differential, which delivers power from the final drive assembly to the drive wheels. Also called a driveshaft or a halfshaft.

B

Ball bearing An anti-friction bearing consisting of a hardened inner and outer race with hardened steel balls between two races.

Bearing The curved surface on a shaft or in a bore, or the part assembled into either, that permits relative motion between them with minimum wear and friction.

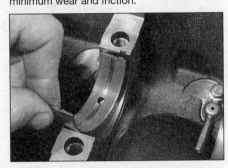

Bearing

Big-end bearing The bearing in the end of the connecting rod that's attached to the crankshaft.

Bleed nipple A valve on a brake wheel cylinder, caliper or other hydraulic component that is opened to purge the hydraulic system of air. Also called a bleed screw.

Brake bleeding Procedure for removing air from lines of a hydraulic brake system.

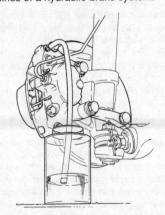

Brake bleeding

Brake disc The component of a disc brake that rotates with the wheels.

Brake drum The component of a drum brake that rotates with the wheels.

Brake linings The friction material which contacts the brake disc or drum to retard the vehicle's speed. The linings are bonded or riveted to the brake pads or shoes.

Brake pads The replaceable friction pads that pinch the brake disc when the brakes are applied. Brake pads consist of a friction material bonded or riveted to a rigid backing plate.

Brake shoe The crescent-shaped carrier to which the brake linings are mounted and which forces the lining against the rotating drum during braking.

Braking systems For more information on braking systems, consult the *Haynes Automotive Brake Manual*.

Breaker bar A long socket wrench handle providing greater leverage.

Bulkhead The insulated partition between the engine and the passenger compartment.

C

Caliper The non-rotating part of a disc-brake assembly that straddles the disc and carries the brake pads. The caliper also contains the hydraulic components that cause the pads to pinch the disc when the brakes are applied. A caliper is also a measuring tool that can be set to measure inside or outside dimensions of an object.

Camshaft A rotating shaft on which a series of cam lobes operate the valve mechanisms. The camshaft may be driven by gears, by sprockets and chain or by sprockets and a belt.

Canister A container in an evaporative emission control system; contains activated charcoal granules to trap vapours from the fuel system.

Canister

Carburettor A device which mixes fuel with air in the proper proportions to provide a desired power output from a spark ignition internal combustion engine.

Castellated Resembling the parapets along the top of a castle wall. For example, a castellated balljoint stud nut.

Castor In wheel alignment, the backward or forward tilt of the steering axis. Castor is positive when the steering axis is inclined rearward at the top.

Catalytic converter A silencer-like device in the exhaust system which converts certain pollutants in the exhaust gases into less harmful substances.

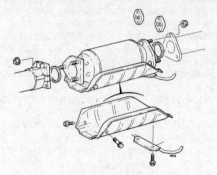

Catalytic converter

Circlip A ring-shaped clip used to prevent endwise movement of cylindrical parts and shafts. An internal circlip is installed in a groove in a housing; an external circlip fits into a groove on the outside of a cylindrical piece such as a shaft.

Clearance The amount of space between two parts. For example, between a piston and a cylinder, between a bearing and a journal, etc.

Coil spring A spiral of elastic steel found in various sizes throughout a vehicle, for example as a springing medium in the suspension and in the valve train.

Compression Reduction in volume, and increase in pressure and temperature, of a gas, caused by squeezing it into a smaller space.

Compression ratio The relationship between cylinder volume when the piston is at top dead centre and cylinder volume when the piston is at bottom dead centre.

Constant velocity (CV) joint A type of universal joint that cancels out vibrations caused by driving power being transmitted through an angle.

Core plug A disc or cup-shaped metal device inserted in a hole in a casting through which core was removed when the casting was formed. Also known as a freeze plug or expansion plug.

Crankcase The lower part of the engine block in which the crankshaft rotates.

Crankshaft The main rotating member, or shaft, running the length of the crankcase, with offset "throws" to which the connecting rods are attached.

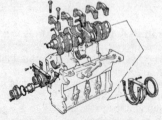

Crankshaft assembly

Crocodile clip See Alligator clip

D

Diagnostic code Code numbers obtained by accessing the diagnostic mode of an engine management computer. This code can be used to determine the area in the system where a malfunction may be located.

Disc brake A brake design incorporating a rotating disc onto which brake pads are squeezed. The resulting friction converts the energy of a moving vehicle into heat.

Double-overhead cam (DOHC) An engine that uses two overhead camshafts, usually one for the intake valves and one for the exhaust valves.

Drivebelt(s) The belt(s) used to drive accessories such as the alternator, water pump, power steering pump, air conditioning compressor, etc. off the crankshaft pulley.

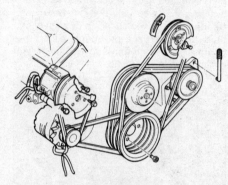

Accessory drivebelts

Driveshaft Any shaft used to transmit motion. Commonly used when referring to the axleshafts on a front wheel drive vehicle.

Drum brake A type of brake using a drum-shaped metal cylinder attached to the inner surface of the wheel. When the brake pedal is pressed, curved brake shoes with friction linings press against the inside of the drum to slow or stop the vehicle.

E

EGR valve A valve used to introduce exhaust gases into the intake air stream.

Electronic control unit (ECU) A computer which controls (for instance) ignition and fuel injection systems, or an anti-lock braking system. For more information refer to the *Haynes Automotive Electrical and Electronic Systems Manual*.

Electronic Fuel Injection (EFI) A computer controlled fuel system that distributes fuel through an injector located in each intake port of the engine.

Emergency brake A braking system, independent of the main hydraulic system, that can be used to slow or stop the vehicle if the primary brakes fail, or to hold the vehicle stationary even though the brake pedal isn't depressed. It usually consists of a hand lever that actuates either front or rear brakes mechanically through a series of cables and linkages. Also known as a handbrake or parking brake.

Endfloat The amount of lengthwise movement between two parts. As applied to a crankshaft, the distance that the crankshaft can move forward and back in the cylinder block.

Engine management system (EMS) A computer controlled system which manages the fuel injection and the ignition systems in an integrated fashion.

Exhaust manifold A part with several passages through which exhaust gases leave the engine combustion chambers and enter the exhaust pipe.

F

Fan clutch A viscous (fluid) drive coupling device which permits variable engine fan speeds in relation to engine speeds.

Feeler blade A thin strip or blade of hardened steel, ground to an exact thickness, used to check or measure clearances between parts.

Feeler blade

Firing order The order in which the engine cylinders fire, or deliver their power strokes, beginning with the number one cylinder.

Flywheel A heavy spinning wheel in which energy is absorbed and stored by means of momentum. On cars, the flywheel is attached to the crankshaft to smooth out firing impulses.

Free play The amount of travel before any action takes place. The "looseness" in a linkage, or an assembly of parts, between the initial application of force and actual movement. For example, the distance the brake pedal moves before the pistons in the master cylinder are actuated.

Fuse An electrical device which protects a circuit against accidental overload. The typical fuse contains a soft piece of metal which is calibrated to melt at a predetermined current flow (expressed as amps) and break the circuit.

Fusible link A circuit protection device consisting of a conductor surrounded by heat-resistant insulation. The conductor is smaller than the wire it protects, so it acts as the weakest link in the circuit. Unlike a blown fuse, a failed fusible link must frequently be cut from the wire for replacement.

G

Gap The distance the spark must travel in jumping from the centre electrode to the side electrode in a spark plug. Also refers to the spacing between the points in a contact breaker assembly in a conventional points-type ignition, or to the distance between the reluctor or rotor and the pickup coil in an electronic ignition.

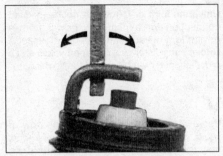

Adjusting spark plug gap

Gasket Any thin, soft material - usually cork, cardboard, asbestos or soft metal - installed between two metal surfaces to ensure a good seal. For instance, the cylinder head gasket seals the joint between the block and the cylinder head.

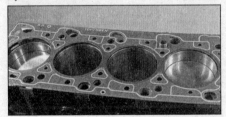

Gasket

Gauge An instrument panel display used to monitor engine conditions. A gauge with a movable pointer on a dial or a fixed scale is an analogue gauge. A gauge with a numerical readout is called a digital gauge.

H

Halfshaft A rotating shaft that transmits power from the final drive unit to a drive wheel, usually when referring to a live rear axle.

Harmonic balancer A device designed to reduce torsion or twisting vibration in the crankshaft. May be incorporated in the crankshaft pulley. Also known as a vibration damper.

Hone An abrasive tool for correcting small irregularities or differences in diameter in an engine cylinder, brake cylinder, etc.

Hydraulic tappet A tappet that utilises hydraulic pressure from the engine's lubrication system to maintain zero clearance (constant contact with both camshaft and valve stem). Automatically adjusts to variation in valve stem length. Hydraulic tappets also reduce valve noise.

I

Ignition timing The moment at which the spark plug fires, usually expressed in the number of crankshaft degrees before the piston reaches the top of its stroke.

Inlet manifold A tube or housing with passages through which flows the air-fuel mixture (carburettor vehicles and vehicles with throttle body injection) or air only (port fuel-injected vehicles) to the port openings in the cylinder head.

J

Jump start Starting the engine of a vehicle with a discharged or weak battery by attaching jump leads from the weak battery to a charged or helper battery.

L

Load Sensing Proportioning Valve (LSPV) A brake hydraulic system control valve that works like a proportioning valve, but also takes into consideration the amount of weight carried by the rear axle.

Locknut A nut used to lock an adjustment nut, or other threaded component, in place. For example, a locknut is employed to keep the adjusting nut on the rocker arm in position.

Lockwasher A form of washer designed to prevent an attaching nut from working loose.

M

MacPherson strut A type of front suspension system devised by Earle MacPherson at Ford of England. In its original form, a simple lateral link with the anti-roll bar creates the lower control arm. A long strut - an integral coil spring and shock absorber - is mounted between the body and the steering knuckle. Many modern so-called MacPherson strut systems use a conventional lower A-arm and don't rely on the anti-roll bar for location.

Multimeter An electrical test instrument with the capability to measure voltage, current and resistance.

N

NOx Oxides of Nitrogen. A common toxic pollutant emitted by petrol and diesel engines at higher temperatures.

O

Ohm The unit of electrical resistance. One volt applied to a resistance of one ohm will produce a current of one amp.

Ohmmeter An instrument for measuring electrical resistance.

O-ring A type of sealing ring made of a special rubber-like material; in use, the O-ring is compressed into a groove to provide the sealing action.

Overhead cam (ohc) engine An engine with the camshaft(s) located on top of the cylinder head(s).

Overhead valve (ohv) engine An engine with the valves located in the cylinder head, but with the camshaft located in the engine block.

Oxygen sensor A device installed in the engine exhaust manifold, which senses the oxygen content in the exhaust and converts this information into an electric current. Also called a Lambda sensor.

P

Phillips screw A type of screw head having a cross instead of a slot for a corresponding type of screwdriver.

Plastigage A thin strip of plastic thread, available in different sizes, used for measuring clearances. For example, a strip of Plastigage is laid across a bearing journal. The parts are assembled and dismantled; the width of the crushed strip indicates the clearance between journal and bearing.

Plastigage

Propeller shaft The long hollow tube with universal joints at both ends that carries power from the transmission to the differential on front-engined rear wheel drive vehicles.

Proportioning valve A hydraulic control valve which limits the amount of pressure to the rear brakes during panic stops to prevent wheel lock-up.

R

Rack-and-pinion steering A steering system with a pinion gear on the end of the steering shaft that mates with a rack (think of a geared wheel opened up and laid flat). When the steering wheel is turned, the pinion turns, moving the rack to the left or right. This movement is transmitted through the track rods to the steering arms at the wheels.

Radiator A liquid-to-air heat transfer device designed to reduce the temperature of the coolant in an internal combustion engine cooling system.

Refrigerant Any substance used as a heat transfer agent in an air-conditioning system. R-12 has been the principle refrigerant for many years; recently, however, manufacturers have begun using R-134a, a non-CFC substance that is considered less harmful to the ozone in the upper atmosphere.

Rocker arm A lever arm that rocks on a shaft or pivots on a stud. In an overhead valve engine, the rocker arm converts the upward movement of the pushrod into a downward movement to open a valve.

Rotor In a distributor, the rotating device inside the cap that connects the centre electrode and the outer terminals as it turns, distributing the high voltage from the coil secondary winding to the proper spark plug. Also, that part of an alternator which rotates inside the stator. Also, the rotating assembly of a turbocharger, including the compressor wheel, shaft and turbine wheel.

Runout The amount of wobble (in-and-out movement) of a gear or wheel as it's rotated. The amount a shaft rotates "out-of-true." The out-of-round condition of a rotating part.

S

Sealant A liquid or paste used to prevent leakage at a joint. Sometimes used in conjunction with a gasket.

Sealed beam lamp An older headlight design which integrates the reflector, lens and filaments into a hermetically-sealed one-piece unit. When a filament burns out or the lens cracks, the entire unit is simply replaced.

Serpentine drivebelt A single, long, wide accessory drivebelt that's used on some newer vehicles to drive all the accessories, instead of a series of smaller, shorter belts. Serpentine drivebelts are usually tensioned by an automatic tensioner.

Serpentine drivebelt

Shim Thin spacer, commonly used to adjust the clearance or relative positions between two parts. For example, shims inserted into or under bucket tappets control valve clearances. Clearance is adjusted by changing the thickness of the shim.

Slide hammer A special puller that screws into or hooks onto a component such as a shaft or bearing; a heavy sliding handle on the shaft bottoms against the end of the shaft to knock the component free.

Sprocket A tooth or projection on the periphery of a wheel, shaped to engage with a chain or drivebelt. Commonly used to refer to the sprocket wheel itself.

Starter inhibitor switch On vehicles with an automatic transmission, a switch that prevents starting if the vehicle is not in Neutral or Park.

Strut See MacPherson strut.

T

Tappet A cylindrical component which transmits motion from the cam to the valve stem, either directly or via a pushrod and rocker arm. Also called a cam follower.

Thermostat A heat-controlled valve that regulates the flow of coolant between the cylinder block and the radiator, so maintaining optimum engine operating temperature. A thermostat is also used in some air cleaners in which the temperature is regulated.

Thrust bearing The bearing in the clutch assembly that is moved in to the release levers by clutch pedal action to disengage the clutch. Also referred to as a release bearing.

Timing belt A toothed belt which drives the camshaft. Serious engine damage may result if it breaks in service.

Timing chain A chain which drives the camshaft.

Toe-in The amount the front wheels are closer together at the front than at the rear. On rear wheel drive vehicles, a slight amount of toe-in is usually specified to keep the front wheels running parallel on the road by offsetting other forces that tend to spread the wheels apart.

Toe-out The amount the front wheels are closer together at the rear than at the front. On front wheel drive vehicles, a slight amount of toe-out is usually specified.

Tools For full information on choosing and using tools, refer to the *Haynes Automotive Tools Manual*.

Tracer A stripe of a second colour applied to a wire insulator to distinguish that wire from another one with the same colour insulator.

Tune-up A process of accurate and careful adjustments and parts replacement to obtain the best possible engine performance.

Turbocharger A centrifugal device, driven by exhaust gases, that pressurises the intake air. Normally used to increase the power output from a given engine displacement, but can also be used primarily to reduce exhaust emissions (as on VW's "Umwelt" Diesel engine).

U

Universal joint or U-joint A double-pivoted connection for transmitting power from a driving to a driven shaft through an angle. A U-joint consists of two Y-shaped yokes and a cross-shaped member called the spider.

V

Valve A device through which the flow of liquid, gas, vacuum, or loose material in bulk may be started, stopped, or regulated by a movable part that opens, shuts, or partially obstructs one or more ports or passageways. A valve is also the movable part of such a device.

Valve clearance The clearance between the valve tip (the end of the valve stem) and the rocker arm or tappet. The valve clearance is measured when the valve is closed.

Vernier caliper A precision measuring instrument that measures inside and outside dimensions. Not quite as accurate as a micrometer, but more convenient.

Viscosity The thickness of a liquid or its resistance to flow.

Volt A unit for expressing electrical "pressure" in a circuit. One volt that will produce a current of one ampere through a resistance of one ohm.

W

Welding Various processes used to join metal items by heating the areas to be joined to a molten state and fusing them together. For more information refer to the *Haynes Automotive Welding Manual*.

Wiring diagram A drawing portraying the components and wires in a vehicle's electrical system, using standardised symbols. For more information refer to the *Haynes Automotive Electrical and Electronic Systems Manual*.

Note: *References throughout this index are in the form "**Chapter number**" • "Page number". So, for example, 2C•15 refers to page 15 of Chapter 2C.*

Note: *References throughout this index are in the form* **"Chapter number"** • **"Page number".** *So, for example, 2C•15 refers to page 15 of Chapter 2C.*

Note: *References throughout this index are in the form* **"Chapter number"** • **"Page number"**. *So, for example, 2C•15 refers to page 15 of Chapter 2C.*

Note: *References throughout this index are in the form "**Chapter number**" • "**Page number**". So, for example, 2C•15 refers to page 15 of Chapter 2C.*

*Note: References throughout this index are in the form "**Chapter number**" • "**Page number**". So, for example, 2C•15 refers to page 15 of Chapter 2C.*